HEAT TRANSFER

HEAT TRANSFER

THIRD EDITION

Alan J. Chapman

Rice University

Macmillan Publishing Co., Inc.
New York
Collier Macmillan Publishers
London

Macmillan Publishing Co., Inc.
866 Third Avenue, New York, New York 10022

Collier-Macmillan Canada, Ltd., Toronto, Ontario

Library of Congress Cataloging in Publication Data

Chapman, Alan Jesse
  Heat Transfer

  Includes bibliographies.
  1. Heat—Transmission. I. Title.
QC320.C5   1974   536′.2   72–92001
ISBN 0-02-321450-3

Printing: 1 2 3 4 5 6 7 8   Year: 4 5 6 7 8 9 0

Dedicated to
Isabel and Wallace Chapman

Preface

This third edition follows the basic philosophy set forth in the earlier editions of this book. Heat transfer, a field comprising a synthesis of a number of engineering–science disciplines, is significantly influenced by the rapidly occurring changes in modern technology. To provide sufficient flexibility to permit the engineer to work effectively in a great variety of new and emerging fields, as well as in the mature, established areas of engineering, it is important to teach the fundamental concepts of a subject as well as the details of application. This textbook presents these fundamental concepts in a suitably rigorous manner, while showing how the analytically obtained facts can be applied with meaningful results to a real physical problem. Sound physical insight and the way in which it supplements analysis are stressed as being of equal importance.

This book is, then, meant to be a teaching text and not a treatise on heat transfer. As such, many worked-out examples are included throughout, and numerous problems for the student to work as exercises are supplied with most chapters. Also, in keeping with the basic idea that heat transfer is a subject of meaning only when real problems are solved, the appendixes provide extensive collections of the thermal properties of a wide variety of gases, liquids, and solids frequently encountered in engineering practice. A separate appendix on SI units and conversions has been added to this edition. The approach of the presentation presumes that the subject is being studied at the introductory level by advanced undergraduates. Mathematical preparation through advanced calculus and differential equations is presumed, as well as basic thermodynamics and fluid dynamics. However, some appendix material is provided on certain mathematical techniques of heat conduction, and Chapter 6 is devoted to a brief treatment of those topics of viscous fluid dynamics pertinent to convective heat transfer.

The work on conduction includes one-dimensional cases having engineering significance in addition to multidimensional and transient problems.

Extended surfaces are thoroughly treated; this edition adds some consideration of problems involving internal heat generation. Numerical methods
of analyzing heat conduction problems are the subject of an entire chapter.
Emphasis has been placed on techniques frequently used in industrial
calculations using digital computers, rather than on numerical methods as
such.

Convective heat transfer offers an excellent opportunity to show how
powerful analytic techniques can be combined with sound engineering
judgment to solve relatively complex physical problems. The importance
of boundary layer theory is emphasized. This edition treats the subject
of turbulent flow and heat transfer more extensively than did the previous
editions, and the universal velocity distribution is introduced. The previous
concentration on dimensional analysis has been dropped in favor of the
presently more often used methods of similitude resulting from an examination of the governing differential equations.

The exposition of the subject of free convection has been completely
reworked in this edition, with the inclusion of a more extensive presentation
of the analytical methods as well as empirical correlations.

Radiative heat transfer is explained in terms of the behavior of real
surfaces and also the idealized analysis of black or gray surfaces. Included
are treatments of enclosure theory, solar radiation, and applications to
industrial and aerospace problems of significance. The often-tedious
problem of shape factor algebra is presented in an appendix.

The final two chapters are devoted to the ultimate aim of a heat transfer
course: the solution of problems involving the combined modes of conduction, convection, and radiation. Included here are the iterative techniques
so often required in such problems. A significant amount of space is devoted
to the analysis and design of heat exchangers, by both the log-mean and
effectiveness methods. The analysis of thermometric errors incurred by
conductive and radiative interactions is also presented. Modern applications such as space radiators and heat pipes are explained. The latter subject,
one of the most actively pursued developments in recent years, is new in
this edition.

Throughout the text, acknowledgment has been made of the many
authors of books and journal articles that formed the sources from which
many data were taken. The author would also like to acknowledge the
many helpful suggestions of students and other users of the earlier editions.
Their comments were invaluable in the preparation of this volume. Particular acknowledgment is due Mrs. Jackie Bourne for her excellent typing
of the manuscript.

Houston, Texas   A. J. C.

Contents

Appendixes

CHAPTER 1

Introduction

1.1 Introductory Remarks

In the science of thermodynamics, which deals with energy in its various forms and with its transformation from one form to another, two particularly important *transient* forms are defined, work and heat. These energies are termed transient since, by definition, they exist only when there is an exchange of energy between two systems or between a system and its surroundings, i.e., when an energy form of one system (such as kinetic energy, potential energy, internal energy, flow energy, chemical energy, etc.) is transformed into an energy form of another system or of the surroundings. When such an exchange takes place without the transfer of mass from the system and not by means of a temperature difference, the energy is said to have been transferred through the performance of *work*. If, on the other hand, the exchange is due to a temperature difference, the energy is said to have been transferred by a flow of *heat*. It is this form, heat, and the basic physical laws governing its exchange with which this book is concerned. It should be noted that the existence of a temperature difference is a distinguishing feature of the energy form known as heat.

In most instances the problems of engineering importance involving an exchange of energy by the flow of heat are those in which there is a transfer of internal energy (or enthalpy in the case of flow processes) between two systems. The two systems may be different parts of the same body. In general this internal energy transfer is called *heat transfer*, although, thermodynamically speaking, this is incorrect. The flow of heat is the mechanism of the transfer of the internal energy, not the quantity transferred. However, it is convenient to use this expression and to speak of heat as "flowing," as has been done here in spite of the implied contradiction with thermodynamics. Indeed, the old and discredited caloric theory of heat, in which heat is defined as a weightless, colorless, odorless fluid flowing from one body to another, would form an adequate basis for the science of heat transfer.

When such exchanges of internal energy or heat take place, the first law of thermodynamics requires that the heat given up by one body must equal that taken up by the other. The second law of thermodynamics demands that the transfer of heat take place from the hotter system to the colder system.

1.2 The Importance of Heat Transfer

The importance of a thorough knowledge of the science of heat transfer and the necessity of being able to analyze, quantitatively, problems involving a transfer of heat have become increasingly important as modern technology has become more and more complex. In almost every phase of scientific and engineering work, processes involving the exchange of energy through a flow of heat are encountered.

Mechanical and chemical engineers are particularly concerned with problems of heat transfer. Modern power generation involves the production of work from either a combustible fuel or a nuclear reaction. This energy is converted into useful work by means of boilers, turbines, condensers, air heaters, water preheaters, pumps, etc. All these pieces of apparatus involve a transfer of heat by one means or another, as does almost every piece of apparatus found in a chemical process industry or a petroleum refinery. Certainly, designing the familiar internal combustion engine, gas turbine, and jet engine requires a complete understanding of heat transfer for a thorough analysis of the combustion and cooling processes.

The so-called "thermal barrier" involves finding means of transferring away from the aircraft the enormous amounts of heat produced by the dissipative effect of the viscosity of the air. Indeed, since all processes in nature have been observed to be irreversible, it follows that all natural processes involve a dissipation of the various forms of mechanical energy into thermal energy with consequent heat transfer processes taking place.

It is this dissipative aspect of the second law of thermodynamics which leads to the much-discussed problem of "thermal pollution," created by the inevitable discharge of waste heat into the environment (air and water). The development of effective solutions to the problems of thermal pollution is perhaps one of the most challenging applications of the knowledge of heat transfer at the present time.

The importance of heat transfer in the production of comfort cooling or comfort heating is readily apparent. This influences the design of building structures of all kinds.

1.3   The Fundamental Concepts and the Basic Modes of Heat Transfer

It is customary to categorize the various heat transfer processes into three basic types or modes, although, as will become apparent as one studies the subject, it is certainly a rare instance when one encounters a problem of practical importance which does not involve at least two, and sometimes all three, of these modes occurring simultaneously. The three modes are conduction, convection, and radiation.

Heat *conduction* is the term applied to the mechanism of internal energy exchange from one body to another, or from one part of a body to another part, by the exchange of the kinetic energy of motion of the molecules by direct communication or by the drift of free electrons in the case of heat conduction in metals. This flow of energy or heat passes from the higher energy molecules to the lower energy ones (i.e., from a high temperature region to a low temperature region). The distinguishing feature of conduction is that it takes place within the boundaries of a body, or across the boundary of a body into another body placed in contact with the first, without an appreciable displacement of the matter comprising the body.

A metal bar heated on one end will, in time, become hot at its other end. This is the simplest illustration of conduction. The laws governing conduction can be expressed in concise mathematical terms, and the analysis of the heat flow can be treated analytically in many instances.

Convection is the term applied to the heat transfer mechanism which occurs in a fluid by the mixing of one portion of the fluid with another portion due to gross movements of the mass of fluid. The actual process of energy transfer from one fluid particle or molecule to another is still one of conduction, but the energy may be transported from one point in space to another by the displacement of the fluid itself.

The fluid motion may be caused by external mechanical means (e.g., by a fan, pump, etc.), in which case the process is called *forced convection*. If the fluid motion is caused by density differences which are created by the temperature differences existing in the fluid mass, the process is termed *free convection* or *natural convection*. The circulation of the water in a pan heated on a stove is an example of free convection. The important heat transfer problems of condensing and boiling are also examples of convection—involving the additional complication of a latent heat exchange.

It is virtually impossible to observe pure heat conduction in a fluid because as soon as a temperature difference is imposed on a fluid, natural convection currents will occur as a result of density differences.

The basic laws of heat conduction must be coupled with those of fluid motion in order to describe, mathematically, the process of heat convection. The mathematical analysis of the resulting system of differential equations is perhaps one of the most complex fields of applied mathematics. Thus, for engineering applications, convection analysis will be seen to be a subtle combination of powerful mathematical techniques and the intelligent use of empiricism and experience.

Thermal *radiation* is the term used to describe the electromagnetic radiation which has been observed to be emitted at the surface of a body which has been thermally excited. This electromagnetic radiation is emitted in all directions; and when it strikes another body, part may be reflected, part may be transmitted, and part may be absorbed. If the incident radiation is thermal radiation (i.e., if it is of the proper wavelength), the absorbed radiation will appear as heat within the absorbing body.

Thus, in a manner completely different from the two modes discussed above, heat may pass from one body to another without the need of a medium of transport between them. In some instances there may be a separating medium, such as air, which is unaffected by this passage of energy. The heat of the sun is the most obvious example of thermal radiation.

There will be a continuous interchange of energy between two radiating bodies, with a net exchange of energy from the hotter to the colder. Even in the case of thermal equilibrium, an energy exchange occurs, although the net exchange will be zero.

1.4 The Fundamental Laws of Conduction

Thermal Conductivity and Thermal Conductance

The basic law governing heat conduction may best be illustrated by considering the simple, idealized situation shown in Fig. 1.1. Consider a plate of material having a surface area A and a thickness Δx. Let one side be

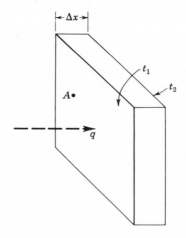

Fig. 1.1

maintained at a temperature t_1, uniformly over the surface, and the other side at temperature t_2. Let q denote the rate of heat flow (i.e., energy per unit time) through the plate, neglecting any edge effects. Experiment has shown that the rate of heat flow is directly proportional to the area A and the temperature difference $(t_1 - t_2)$ but inversely proportional to the thickness Δx. This proportionality is made an equality by the definition of a constant k. Thus,

$$q = kA \frac{t_1 - t_2}{\Delta x}. \tag{1.1}$$

The constant of proportionality, k, is called the *thermal conductivity* of the material of which the plate is composed. It is a property dependent only on the composition of the material, not on its geometrical configuration. Sometimes a gross quantity, *unit thermal conductance*, is used to express the heat-conducting capacity of a given physical configuration, so that if C denotes the unit thermal conductance,

$$q = CA(t_1 - t_2).$$

Thus, it is seen that thermal conductance is the conductivity of a substance divided by its thickness. It is no longer a physical property but depends, as well, on the geometrical configuration at hand and, thus, is a less general quantity than is thermal conductivity.

Equation (1.1) forms the basis for the fundamental relation of heat conduction. Consider now a homogeneous, isotropic solid as depicted in Fig. 1.2. If the solid is subjected to certain known boundary temperatures, what is the rate at which heat is conducted across a surface, S, within the solid?

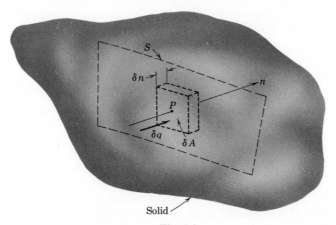

Fig. 1.2

Selecting a point P on the surface S, one can select a wafer of material having an area δA, which is part of the surface S containing P, and having a thickness δn in the direction of the normal drawn to the surface at P. If the difference between the temperature of the back face of the wafer and its front face is δt, and if δA is chosen small enough so that δt is essentially uniform over it, the rate of heat flow across the wafer, δq, is, by Eq. (1.1),

$$\delta q = -k\,\delta A\,\frac{\delta t}{\delta n}.$$

The minus sign is due to the convention that the heat flow is taken to be positive if δt is negative in the direction of increasing n, the normal displacement. Forming the ratio $\delta q/\delta A$ and allowing the area $\delta A \to 0$, one obtains what is termed the *flux* of heat conducted through the thickness δn at the point P,

$$f = \frac{dq}{dA} = -k\,\frac{\delta t}{\delta n}.$$

Further, allowing $\delta n \to 0$, one arrives at the flux of heat across S at the point P in terms of the *temperature gradient* at P in the n direction, dt/dn:

$$f = -k\,\frac{dt}{dn}. \tag{1.2}$$

This is called *Fourier's conduction law* after the French mathematician who first made an extensive analysis of heat conduction. It states that the flux of heat conducted (energy per unit time per unit area) across a surface is proportional to the temperature gradient taken in a direction normal to the surface at the point in question.

Upon returning to the situation pictured in Fig. 1.2, the total rate of heat transferred across the finite surface S would be

$$q = -\int_{s} k\,\frac{dt}{dn}\,dA.$$

Generally speaking, the normal gradient dt/dn may vary over the surface, but in many instances it is possible to select the surface as one on which the gradient is everywhere the same. This is the situation in the case depicted in Fig. 1.1, in which every plane normal to Δx is such a surface. In the case of a hollow cylinder with uniform outside and inside surface temperatures, every concentric interior cylindrical surface is isothermal with a uniform temperature gradient normal to it. In such cases, then,

$$q = -kA\frac{dt}{dn},$$ (1.3)

where A is the total area of the finite surface.

The General Heat Conduction Equation

The above relations may be used to develop an equation describing the distribution of the temperature throughout a heat-conducting solid. In general, a heat conduction problem consists of finding the temperature at any time and at any point within a specified solid which has been heated to a known initial temperature distribution and whose surface has been subjected to a known set of boundary conditions.

The development that follows will consider the case in which the heat-conducting solid may also have internal sources of heat generation. Such sources may be the result of chemical or nuclear reaction, electrical dissipation, etc., and may be either concentrated at certain locations or distributed throughout the solid.

To develop the differential equation governing this problem, consider a solid, as shown in Fig. 1.3, and select arbitrarily three mutually perpendicular coordinate directions, x, y, and z. Select in the solid a parallelepiped of dimensions Δx, Δy, and Δz. By making an energy balance on this element between the heat conducted in and out of its six faces, the heat stored within, and the heat generated within the element, an expression interrelating the temperatures throughout the solid will be obtained.

First, consider heat conduction in the x direction only. Let t denote the temperature at the point P which is located at the geometric center of the element. Let the subscripts 1 and 2 denote the left and right yz faces (shown shaded in Fig. 1.3), respectively. The excess rate at which heat is conducted into the element, q_1, over that conducted out, q_2, is

$$q_1 - q_2.$$

Application of Eq. (1.3) gives this excess to be

$$q_1 - q_2 = -k\,\Delta y\,\Delta z\left[\left(\frac{\partial t}{\partial x}\right)_1 - \left(\frac{\partial t}{\partial x}\right)_2\right],$$

where the partial derivative is used since t may depend on x, y, z, and time. The assumption that the thermal conductivity is independent of x and t has been made. The significance of this assumption will be discussed in more

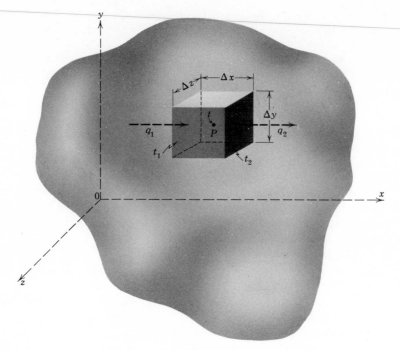

Fig. 1.3

detail later. Making a Taylor's expansion of the temperature gradients $(\partial t/\partial x)_1$ and $(\partial t/\partial x)_2$ in terms of the gradient at P, one has

$$\left(\frac{\partial t}{\partial x}\right)_2 = \frac{\partial t}{\partial x} + \frac{\partial}{\partial x}\left(\frac{\partial t}{\partial x}\right)\frac{\Delta x}{2} + \frac{\partial^2}{\partial x^2}\frac{\partial t}{\partial x}\frac{(\Delta x/2)^2}{2!} + \cdots,$$

$$\left(\frac{\partial t}{dx}\right)_1 = \frac{\partial t}{\partial x} + \frac{\partial}{\partial x}\left(\frac{\partial t}{\partial x}\right)\left(\frac{-\Delta x}{2}\right) + \frac{\partial^2}{\partial x^2}\left(\frac{\partial t}{\partial x}\right)\frac{(-\Delta x/2)^2}{2!} + \cdots,$$

where no subscript denotes derivatives evaluated at P. So the expression for the heat storage rate resulting from conduction in the x direction is

$$-k\,\Delta y\,\Delta z\left[-\frac{\partial}{\partial x}\left(\frac{\partial t}{\partial x}\right)\Delta x - \frac{\partial^3}{\partial x^3}\left(\frac{\partial t}{\partial x}\right)\frac{2}{3!}\left(\frac{\Delta x}{2}\right)^3 + \cdots\right],$$

or

$$k\,\Delta x\,\Delta y\,\Delta z\left[\frac{\partial^2 t}{\partial x^2} + \frac{1}{24}\frac{\partial^4 t}{\partial x^4}(\Delta x)^2 + \cdots\right].$$

Similar analyses in the other two coordinate directions give the following expression for the *total* rate at which heat is conducted into the element:

$$k \, \Delta x \, \Delta y \, \Delta z \left[\frac{\partial^2 t}{\partial x^2} + \frac{\partial^2 t}{\partial y^2} + \frac{\partial^2 t}{\partial z^2} + \frac{(\Delta x)^2}{24} \frac{\partial^4 t}{\partial x^4} + \frac{(\Delta y)^2}{24} \frac{\partial^4 t}{\partial y^4} + \frac{(\Delta z)^2}{24} \frac{\partial^4 t}{\partial z^4} + \cdots \right].$$

(1.4)

The rate of heat storage in the element may be expressed also in terms of the time rate of change of the average temperature of the element. Letting t_{av} denote this temperature and τ denote time, the heat storage rate is

$$\Delta x \, \Delta y \, \Delta z \, \rho c_p \frac{\partial t_{av}}{\partial \tau}.$$

(1.5)

In Equation (1.5) ρ is the density and c_p is the specific heat at constant pressure of the solid. The constant-pressure specific heat is used since most heat conduction problems in solids occur under such conditions.

If q^* is used to represent the rate at which heat is being internally generated in the element, *per unit volume*, then

$$q^*(\Delta x \, \Delta y \, \Delta z)$$

(1.6)

is the total heat released in the element. Energy conservation requires that the sum of Eqs. (1.4) and (1.6) equals Eq. (1.5). Upon making this equality; dividing by the element volume, $\Delta x \, \Delta y \, \Delta z$; allowing Δx, Δy, and $\Delta z \to 0$; and noting that under these conditions $t_{av} \to t$, one finally obtains

$$\rho c_p \frac{\partial t}{\partial \tau} = k \left(\frac{\partial^2 t}{\partial x^2} + \frac{\partial^2 t}{\partial y^2} + \frac{\partial^2 t}{\partial z^2} \right) + q^*.$$

(1.7)

Equation (1.7) represents a *volumetric* heat balance which must be satisfied at each point in an isotropic solid having constant properties. This expression, known as the *general heat conduction equation*, describes, in differential form, the dependence of the temperature in the solid on the coordinates x, y, z and on time τ. It should be noted that the heat generation term, q^*, may be a function of position and time.

Frequently Eq. (1.7) is written in the following form, which, although useful, destroys the volumetric representation of each term:

$$\frac{\partial t}{\partial \tau} = \frac{k}{\rho c_p} \left(\frac{\partial^2 t}{\partial x^2} + \frac{\partial^2 t}{\partial y^2} + \frac{\partial^2 t}{\partial z^2} \right) + \frac{q^*}{\rho c_p},$$

or

$$\frac{\partial t}{\partial \tau} = \alpha \left(\frac{\partial^2 t}{\partial x^2} + \frac{\partial^2 t}{\partial y^2} + \frac{\partial^2 t}{\partial z^2} \right) + \frac{q^*}{\rho c_p}. \tag{1.8}$$

The quantity

$$\alpha = \frac{k}{\rho c_p} \tag{1.9}$$

is called the *thermal diffusivity* and is seen to be a physical property of the material of which the solid is composed.

As was mentioned, the above development involved the assumptions that the thermal conductivity is the same in all directions and, furthermore, independent of temperature. The first of these assumptions, isotropy, is satisfied by many homogeneous materials; however, there are many sub-stances, such as wood, for which this is not true. The second assumption, the independence of k from temperature, is found to be satisfied in only a very few materials under special circumstances. More complex forms of Eq. (1.7) must be employed to account for these effects accurately and References 1 and 2 may be consulted in.this regard. However, these special forms will not be considered here and the effects in question will be treated by special means—as illustrated by certain solutions in Chapters 3 through 5.

Other Coordinate Systems

The above discussion was carried out in terms of rectangular coordinates. It is often useful to write the equation in cylindrical or spherical coordinates. The results are:

Cylindrical Coordinates, r (radius), z (axis), θ (longitude):

$$\frac{\partial t}{\partial \tau} = \alpha \left(\frac{\partial^2 t}{\partial r^2} + \frac{1}{r} \frac{\partial t}{\partial r} + \frac{1}{r^2} \frac{\partial^2 t}{\partial \theta^2} + \frac{\partial^2 t}{\partial z^2} \right) + \frac{q^*}{\rho c_p}. \tag{1.10}$$

Spherical Coordinates, r (radius), φ (longitude), θ (colatitude):

$$\frac{\partial t}{\partial \tau} = \alpha \left[\frac{1}{r^2} \frac{\partial}{\partial r} \left(r^2 \frac{\partial t}{\partial r} \right) + \frac{1}{r^2 \sin \theta} \frac{\partial}{\partial \theta} \left(\sin \theta \frac{\partial t}{\partial \theta} \right) + \frac{1}{r^2 \sin^2 \theta} \frac{\partial^2 t}{\partial \varphi^2} \right] + \frac{q^*}{\rho c_p}. \tag{1.11}$$

The proof of these equations will be left as exercises for the reader.

The Steady State

A particularly useful special case of Eq. (1.8) is one which has a very wide range of application in engineering. This is the *steady state*, in which there is no dependence on time. The heat conduction equation then reduces to Poisson's equation: In rectangular coordinates this is

$$\alpha\left(\frac{\partial^2 t}{\partial x^2} + \frac{\partial^2 t}{\partial y^2} + \frac{\partial^2 t}{\partial z^2}\right) + \frac{q^*}{\rho c_p} = 0.$$

$$\frac{\partial^2 t}{\partial x^2} + \frac{\partial^2 t}{\partial y^2} + \frac{\partial^2 t}{\partial z^2} + \frac{q^*}{k} = 0. \tag{1.12}$$

In the absence of internal heat generation, Laplace's equation is obtained:

$$\frac{\partial^2 t}{\partial x^2} + \frac{\partial^2 t}{\partial y^2} + \frac{\partial^2 t}{\partial z^2} = 0. \tag{1.13}$$

1.5   The Fundamental Laws of Convection

The Boundary Layer Concept

The discussion of Sec. 1.3 defined "convection" as the term applied to the heat transfer mechanism which takes place in a fluid because of a combination of conduction within the fluid and energy transport which is due to the fluid motion itself—the fluid motion being produced either by artificial means or by density currents.

Since fluid motion is the distinguishing feature of heat convection, it is necessary to understand some of the principles of fluid dynamics in order to describe adequately the processes of convection. When any real fluid moves past a solid surface, it is observed that the fluid velocity varies from a zero value immediately adjacent to the wall to a finite value at a point some distance away.

Considering the simplest case of flow past a plane surface as shown in Fig. 1.4(a), the fluid velocity will vary from a uniform, "free stream" value at points far away from the wall to zero at the wall—somewhat in the way shown in the figure. For fluids of low viscosity, such as air or water, the region near the surface, in which most of the velocity variation occurs, may be quite thin—depending on the free stream velocity of the fluid. In many applications—such as low speed aerodynamics, hydraulics, etc.—it is possible to obtain satisfactory results by assuming that the fluid is inviscid (i.e., without viscosity). Hence, the flow may be treated as though it slips past the surface with no viscous retardation. However, since the process of convection of heat away from the wall (if the wall is at a temperature different from the free stream of the fluid) is intimately concerned with thermal conduction and

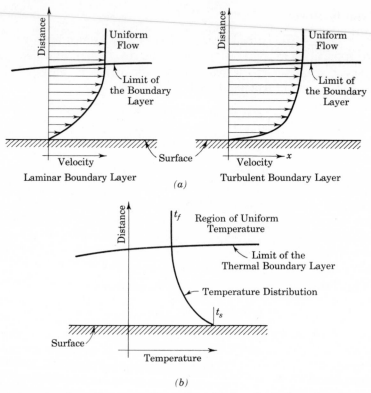

Fig. 1.4. Boundary layer flow past a flat surface: (a) the velocity boundary layer; (b) the thermal boundary layer.

energy transport due to motion in the fluid layers in the immediate vicinity of the wall, the simplification of assuming the fluid to be inviscid may not be made when analyses of heat convection are undertaken.

Since the region in which the retarding effect of the fluid viscosity plays a dominant role will often be a very thin layer near the wall, it is possible to simplify the description of the convection process by introducing the concept of the *velocity boundary layer*. The velocity boundary layer is defined as the thin layer near the wall in which one assumes that viscous effects are important. Within this region the effect of the wall on the motion of the fluid is significant. Outside the boundary layer it is assumed that the effect of the wall may be neglected. The exact limit of the boundary layer cannot be precisely defined because of the asymptotic nature of the velocity variation. The limit of the boundary layer is usually taken to be at the distance from the wall at which the fluid velocity is equal to a predetermined percentage of the free stream value. This percentage depends on the accuracy desired, 99 or 95 per cent being customary.

Outside the boundary layer region the flow is assumed to be inviscid. Inside the boundary layer the viscous flow may be either *laminar* or *turbulent*. In the case of laminar boundary layer flow, adjacent fluid layers slide relative to one another but do not mix in the direction normal to the fluid streamlines. Thus, any heat that flows from the surface to the free stream fluid does so mostly by conduction—although a transport of energy is also accomplished by virtue of the fact that the fluid has a velocity component normal to the surface. This normal velocity component is caused by the fact that the boundary layer must become progressively thicker as it moves along the surface.

In the event that the fluid motion in the boundary layer is turbulent, the mean flow is essentially parallel to the surface, but it has superimposed upon it a fluctuating motion—in directions both parallel and normal to the surface. The transverse fluctuations cause additional fluid mixing, which increases the rate at which heat is transported in the direction perpendicular to the surface. Typical velocity variations through laminar and turbulent boundary layers are illustrated in Fig. 1.4(a).

If the solid surface is maintained at a temperature, say t_s, which is different from the fluid temperature, t_f, measured at a point far removed from the surface, a variation of the temperature of the fluid is observed which is similar to the velocity variation described above. That is, the fluid temperature varies from t_s at the wall to t_f far away from the wall—with most of the variation occurring close to the surface. This is illustrated in Fig. 1.4(b), where it has been assumed that the surface is hotter than the fluid. The fluid temperature approaches t_f asymptotically. However, a *thermal boundary layer* may be defined (in the same sense that the velocity boundary layer was defined above) as the region between the surface and the point at which the fluid temperature has reached a certain percentage of t_f. Outside the thermal boundary layer the fluid is assumed to be a heat sink at a uniform temperature of t_f. The thermal boundary layer is generally not coincident with the velocity boundary layer, although it is certainly dependent on it—i.e., the velocity boundary layer thickness, the variation of the velocity, whether the flow is laminar or turbulent, etc., are all factors which determine the temperature variation in the thermal boundary layer.

Newton's Law of Cooling

Chapters 6 through 10 will consider, in some detail, the present state of knowledge of convective heat transfer—from both the theoretical and empirical points of view. As should be apparent from the discussion, the prediction of the rates at which heat is convected away from a solid surface by an ambient fluid involves a thorough understanding of the principles of heat conduction, fluid dynamics, and boundary layer theory. All the complexities involved in such an analytical approach may be lumped together in terms of a single parameter by introduction of Newton's law of cooling:

$$\frac{q}{A} = h(t_s - t_f). \qquad (1.14)$$

The quantity h in this equation is variously known as the *heat transfer coefficient, film coefficient, or unit film conductance.* It is seen to be (refer to Sec. 1.4) a unit conductance, not a material property as is thermal conductivity. It would necessarily depend on the composition of the fluid *and* on the nature and geometry of the fluid motion past the surface. Thus, h is a gross quantity which attempts only to represent the over-all effect and therefore does not attempt to explain the actual mechanism of heat transfer; more detailed analysis of the fluid motion and methods of predicting values of h will be found in Chapters 6 through 10. For the purposes of the next few chapters the above law of cooling will be adequate. Although this discussion indicates that h is a complex and variable quantity, in many instances it is possible to obtain satisfactory results by treating it as a constant—simplifying the associated analysis considerably.

Newton's law of cooling enables one to state, very simply, the boundary condition for the general heat conduction equation, Eq. (1.8), for cases when one has a solid bounded by a convecting fluid. At the surface of the solid, Fourier's conduction law, Eq. (1.3), gives

$$\frac{q}{A} = -k\left(\frac{\partial t}{\partial n}\right)_s$$

for the heat conducted to the surface. The symbol n denotes the normal direction to the surface, and the subscript s means the derivative is evaluated at the surface. Equating this to the heat convected away from the surface, one finds that Eq. (1.14) gives the following boundary condition:

$$\left(\frac{\partial t}{\partial n}\right)_s = -\frac{h}{k}(t_s - t_f). \qquad (1.15)$$

1.6   The Fundamental Laws of Radiation

As mentioned in Sec. 1.3, it has been observed that a body may lose or gain thermal energy without the need of a physical medium of transport. That is, a heated body placed in the presence of cooler surroundings but having no physical contact with them (imagine it suspended in a vacuum) is observed to lose energy. This loss is due to the electromagnetic emission known as thermal radiation. It has been pointed out that this absence of a medium of transport is a distinguishing feature of the thermal radiation mode of heat transfer.

A second distinguishing feature of thermal radiation is the effect of the level of the temperature of the emitting bodies. As the discussions of Secs. 1.4 and 1.5 have shown, the rate of heat transfer by the modes of conduction and convection is proportional to the difference of the temperature between the heat source and the heat sink. Thus, regarding physical properties to be constant for the moment, the amount of heat transferred is independent of the absolute magnitude of the temperature as long as the difference is the same. This is not the case for thermal radiation. The quantity of heat exchanged by radiation is proportional to the difference of the fourth power of the absolute temperatures of the radiating bodies. Thus, for a given temperature difference, the heat transferred is much greater at high temperatures than at low temperatures.

The exact character of radiant heat emission is not completely agreed upon, but it is known that the rate at which energy is emitted per unit area of emitting surface is given by the Stefan–Boltzmann law:

$$W = \epsilon \sigma T^4.$$

In this equation W is the rate of energy emission per unit area, T is the absolute temperature of the body, σ is a universal physical constant, and ϵ is a property of the particular emitting surface known as the *emissivity*. This law was originally proposed by Stefan based on experimental evidence. Boltzmann showed later that the law is derivable from the laws of thermodynamics and hence is not based just on experimental data as are Fourier's law [Eq. (1.3)] and Newton's law of cooling [Eq. (1.14)].

Because of the extremely difficult character of the mode of heat transfer, further discussion will be reserved for Chapter 11.

1.7   Dimensions and Units

Careful attention to dimensions and units is necessary if one is to obtain correct results in any engineering problem. Current engineering practice in the United States is still based on the English engineering system although the use of SI units is increasing rapidly in view of the almost universal adoption of these units elsewhere. This and the following section will be used to describe the English units to be used throughout the text; however, Appendix F presents conversion factors by which the quantities encountered in heat transfer may be converted from English to SI units. Also, answers in the example problems will be expressed in both English and SI units.

The main source of difficulty in dealing with systems of units results from the interrelation of *force* and *mass* through Newton's second law of motion.

It is convenient to express Newton's law as

$$F = \frac{1}{g_c} ma,$$

where F denotes force, m denotes mass, a is the acceleration of the mass, and g_c is a dimensional conversion constant. A system of units in which g_c equals 1 with no dimensions is called a "consistent" system of units. Two consistent systems exist and are described below.

The Mass–Length–Time System. This system takes as the fundamental dimensions: mass, length, and time. The units used are

Dimension	Unit
Mass	1 pound mass, lb_m
Length	1 foot
Time	1 second

All other dimensions are derived from these. By application of Newton's law and making g_c dimensionless and equal to 1, one has as the dimensions of force:

$$\text{Force} = \frac{\text{mass} \times \text{length}}{(\text{time})^2}.$$

The unit force is taken to be

$$1 \text{ poundal} = \frac{1 \ lb_m\text{-ft}}{\sec^2}.$$

The Force–Length–Time System. One could, as well, take force as a fundamental dimension, deriving the dimension of mass. In this system the units used are

Dimension	Unit
Force	1 pound force, lb_f
Length	1 foot
Time	1 second

The derived dimensions of mass are, for a consistent system,

$$\text{Mass} = \frac{\text{force} \times (\text{time})^2}{\text{length}},$$

and the standard unit of mass is

$$1 \text{ slug} = \frac{1 \text{ lb}_f\text{-sec}^2}{\text{ft}}.$$

The Force–Mass–Length–Time System. One could take all four dimensions in Newton's law as fundamental, making it necessary to give dimensions to g_c. This, then, is not a consistent system of units. In this system one takes the following fundamental dimensions with the units indicated:

Dimension	Unit
Force	1 pound force, lb_f
Mass	1 pound mass, lb_m
Length	1 foot
Time	1 second

The derived dimensions of g_c are then

$$g_c = \frac{\text{mass} \times \text{length}}{\text{force} \times (\text{time})^2}.$$

If one, in addition, makes the standard arbitrary definition

$$1 \text{ lb}_f \text{ will accelerate } 1 \text{ lb}_m \ 32.1739 \text{ ft/sec}^2,$$

the units of g_c become

$$g_c = 32.1739 \frac{\text{lb}_m\text{-ft}}{\text{lb}_f\text{-sec}^2}.$$

Usually a value of $g_c = 32.2 (\text{lb}_m\text{-ft})/(\text{lb}_f\text{-sec}^2)$ is accurate enough—particularly in the case of slide rule computations. This latter system is apparently the most commonly used system among English-speaking engineers today. One of the main difficulties encountered is the fact that when objects are weighed, the weight is a force and the acceleration is that of gravity. At sea level the acceleration of gravity is $g = 32.2 \text{ ft/sec}^2$. This is the same magnitude as g_c but certainly does not have the same dimensions. This means that, at sea level at least, 1 lb_m will weigh 1 lb_f—numerically the same, but with different dimensions!

The latter fact produces confusion in the dimensions of certain "specific" quantities, such as density, ρ, and specific weight, γ. Density has the dimensions of lb_m/ft^3, whereas specific weight has the dimensions of lb_f/ft^3. Again, one has the situation of two quantities being numerically equal but having different dimensions. The relation between the two would be expressed as

$$\gamma = \frac{g}{g_c} \rho.$$

The same would be true for other quantities based on mass or weight, such as specific heat, specific enthalpy, etc.

1.8 The Dimensions and Units of Conductivity, Conductance, and Diffusivity

When dealing with problems involving heat, it is customary to utilize two additional dimensions, temperature and heat. The latter quantity is actually an energy and, hence, could be represented by the dimensions discussed above (i.e., as force times length). However, it is more convenient to think of heat as a separate dimension. This entails the need of another dimensional constant, Joule's mechanical equivalent of heat, J, which has the dimensions of heat/force × length.

Throughout this book the British thermal unit (Btu) will be used as the standard unit of heat. Fahrenheit or Rankine temperature will be used. Thus, the rate of heat flow, q, and heat flux, q/A, will have the following units:

$$q = \text{Btu/hr, Btu/sec, etc.}$$

$$\frac{q}{A} = \text{Btu/hr-ft}^2, \text{Btu/sec-ft}^2, \text{etc.}$$

Referring to Eq. (1.1), one notes that the thermal conductivity, k, will have the following units:

$$k = \text{Btu/hr-ft-}°\text{F}.$$

A unit conductance, most notably the film coefficient h, will be expressed as

$$h = \text{Btu/hr-ft}^2\text{-}°\text{F}.$$

Equation (1.9) defines the thermal diffusivity as

$$\alpha = \frac{k}{\rho c_p},$$

where, since ρ is the *mass* density, the specific heat c_p must have the dimensions of heat per unit *mass* per unit temperature difference. If the specific heat is based on a unit weight, then γ, the specific weight, must be used in place of ρ. Either way gives the dimensions of diffusivitiy to be

$$\alpha = \frac{(\text{length})^2}{\text{time}}$$

$$= \text{ft}^2/\text{hr}, \text{ft}^2/\text{sec}, \text{etc}.$$

References

1. CARSLAW, H. S., and J. C. JAEGER, *Conduction of Heat in Solids*, 2nd ed., New York, Oxford U.P., 1959.
2. ARPACI, V., *Conduction Heat Transfer*, Reading, Mass., Addison-Wesley, 1966.

Problems

1.1. Obtain Eq. (1.10) by making a coordinate transformation of Eq. (1.8).

1.2. Obtain Eq. (1.11) by making a coordinate transformation of Eq. (1.8).

1.3. For steady state heat flow in a plane (i.e., the temperature depends only on two coordinates), show that Eq. (1.13) is satisfied by

$$t = (c_1 e^{-\lambda y} + c_2 e^{\lambda y})(c_3 \sin \lambda x + c_4 \cos \lambda x).$$

1.4. If steady heat conduction without internal heat generation in a hollow cylinder of very long length can be considered as radial only [i.e., if t in Eq. (1.10) is independent of τ, θ, or z], show that the solution of Eq. (1.10) is

$$t = c_1 \ln r + c_2.$$

1.5. Show that for nonsteady heat flow in one dimension only, Eq. (1.8) is satisfied by

$$t = e^{-\lambda^2 \alpha \tau}(c_1 \sin \lambda x + c_2 \cos \lambda x)$$

if there is no internal heat generation.

CHAPTER 2

Material Properties of Importance in Heat Transfer

2.1 Introductory Remarks

Any application to real problems of the results of the physical laws and mathematical analyses related to heat transfer will require that one have available numerical values of the necessary physical properties of the substance under consideration. This chapter will be devoted to a discussion of these important properties. The effects of various factors, such as pressure, temperature, density, porosity, etc., will also be discussed.

As shown in Chapter 1, in the derivation of the general heat conduction equation, Eq. (1.8), the properties of importance in conduction are thermal conductivity, density, and specific heat. For problems involving convection these properties are still important, but, due to the fluid motion involved, the additional property of the fluid viscosity becomes extremely important.

In the mechanism of free convection, the fluid motion, being caused by density differences, is very much influenced by the coefficient of the thermal expansion of the fluid.

2.2 The Thermal Conductivity of Homogeneous Materials

As discussed in Sec. 1.4, heat conduction is, basically, the transmission of energy by molecular motion. Thermal conductivity is, then, the physical property denoting the ease with which a particular substance can accomplish this transmission. The thermal conductivity of a material is found to depend on the chemical composition of the substance, or substances, of which it is composed, the phase (i.e., gas, liquid, or solid) in which it exists, its crystalline structure if a solid, the temperature and pressure to which it is subjected, and whether or not it is a homogeneous material.

These factors will be discussed in the following sections, with attention first directed to homogeneous materials. The nonhomogeneous materials of engineering importance are, principally, those used for insulating purposes and will be described after the general principles have been established for homogeneous materials.

The discussion of the relative order of magnitude of the thermal conductivities of various homogeneous substances can best be illustrated by examination of Table 2.1, in which some typical values are tabulated.

Table 2.1 The Thermal Conductivity of Various Substances*

Substance	Thermal Conductivity Btu/hr-ft-°F
Gases	
Sulfur dioxide (32°F, 1 atm)	0.0050
Air (32°F, 1 atm)	0.0137
Liquids	
Carbon dioxide (sat. liq., 32°F)	0.0604
Glycerine, pure (32°F)	0.163
Water (sat. liq., 32°F)	0.327
Solids	
Glass, plate (68°F)	0.44
Ice (32°F)	1.28
Magnesite brick (400°F)	2.20
Quartz (68°F)	4.4
Stainless steel (18% Cr, 8% Ni) (32°F)	9.4
Iron, pure (32°F)	42
Zinc, pure (32°F)	65
Aluminum, pure (32°F)	132
Copper, pure (32°F)	223
Silver, pure (32°F)	241

* Abstracted mainly from tables in Appendix A.

One immediately observes a vast range of possible values of thermal conductivity, with silver having a conductivity almost 50,000 times as great as that of sulfur dioxide.

One also sees that, generally speaking, a liquid is a better conductor than a gas and that a solid is a better conductor than a liquid. These facts are best illustrated by considering the three phases of a single substance, such as mercury. As a solid at $-315°F$, mercury has a thermal conductivity of 28 Btu/hr-ft-°F, as a liquid at 32°F the conductivity has dropped to 4.6 Btu/hr-ft-°F, whereas as a gas at 392°F its thermal conductivity is reported to be 0.0197 Btu/hr-ft-°F. These differences can be explained partially by the fact that while in a gaseous state, the molecules of a substance are spaced relatively far apart and their motion is random. This means that energy transfer by molecular impact is much more slow than in the case of a liquid, in which the motion is still random but in which the molecules are more closely packed. The same is true concerning the difference between the thermal conductivity of the liquid and solid phases; however, other factors become important when the solid state is formed.

Referring again to Table 2.1, one sees that a solid having a crystalline structure, such as quartz, has a higher thermal conductivity than a substance in an amorphous solid state, such as glass. Also, metals, crystalline in structure, are seen to have greater thermal conductivities than do nonmetals. In the case of the amorphous solids, the irregular arrangement of the molecules inhibits the effectiveness of the transfer of the energy by molecular impact, and, hence, the thermal conductivity is of the same order of magnitude as that observed for liquids. On the other hand, in a solid having a crystalline structure, there is an additional transfer of heat energy, as a result of a vibratory motion of the crystal lattice as a whole, in the direction of decreasing temperature. Imperfections in the lattice structure tend to distort and scatter these "thermoelastic" vibrations and, hence, tend to decrease their intensity.

In the case of metallic conduction, still a third mechanism of energy transfer, in addition to the molecular communication and lattice vibration mentioned above, comes into play. When the crystal of a nonmetallic substance is formed, the valence electrons (i.e., the outermost electrons) are shared among atoms to form the chemical bond which holds the atoms together as a molecule. In a metal crystal, however, these valence electrons become detached and are free to move within the lattice formed by the remaining positive ions of the metal atoms. When a difference exists between the temperatures of different parts of the metal, a general drift of these free electrons occurs in the direction of the decreasing temperature. It is this drift of free electrons which makes the metals so much better as conductors than other solids. These free electrons account for the observed proportionality between the thermal and electrical conductivities of pure metals.

With these brief remarks concerning the mechanism of conduction in gases, liquids, and solids, the following three sections will consider each phase separately. The effects of pressure and temperature will also be discussed, as will tabulations in Appendix A of the thermal conductivity of various substances of importance in engineering.

The Thermal Conductivity of Homogeneous Solids

Table A.1 of Appendix A tabulates, among other properties, the thermal conductivity of various pure metals and metal alloys. Many factors are known to influence the thermal conductivity of metals, such as chemical composition, atomic structure, phase changes, grain size, temperature, pressure, and deformation. The factors with the greatest influence are the chemical composition, phase changes, and temperature. Usually the first two of these do not enter a case in which one is interested in a particular material, and, hence, only the temperature effect has to be accounted for.

It is known (Ref. 1) that the thermal conductivity of metals is directly proportional to the absolute temperature and the mean free path of the molecules. The mean free path tends to decrease with increasing temperature so that the net variation is the result of opposing influences. Pure metals generally have thermal conductivities which decrease with temperature, but the presence of impurities or alloying elements, even in minute amounts, may reverse this trend. These effects may be observed by examination of Table A.1. The data of Table A.1 for chrome steels are plotted in Fig. 2.1, which illustrates typical effects of temperature and composition on the thermal conductivity of a metal.

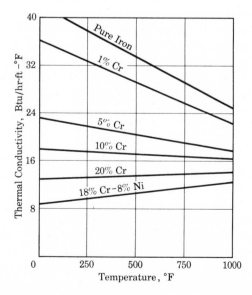

Fig. 2.1. The effect of temperature and composition on the thermal conductivity of steel.

It is usually possible to represent the temperature dependence of the thermal conductivity of a metal by a linear relation of the form $k = k_0(1 + bt)$, in which k_0 is the thermal conductivity at $t = 0°F$ and b is a constant.

The thermal conductivities of other homogeneous solids are presented in Table A.2, along with many nonhomogeneous solids. Generally speaking, the thermal conductivities of these materials do not vary with pressure, only with temperature. The temperature variation is usually approximately linear, with the conductivity increasing with temperature.

The Thermal Conductivity of Liquids

Table A.4 and A.5 present the thermal conductivities of various saturated liquids as functions of temperature. Other properties to be discussed later are given also. In general, the thermal conductivities of liquids are relatively insensitive to pressure, particularly at pressures not too close to the critical pressure. For this reason the temperature variation is usually the only influence which is taken into account, and the saturated state is the condition at which the thermal conductivity is reported because of the uniqueness of this state.

Most liquids exhibit a decreasing thermal conductivity with temperature, although water is, as usual, a notable exception.

Because of its importance as an engineering fluid, the properties of water are presented separately in the more extensive tabulation of Table A.4. Table A.5 presents data for several other liquids of importance.

Of the nonmetallic liquids, water has the highest thermal conductivity, with a maximum value occurring at about 300°F. Some of the data of Tables A.4 and A.5 are plotted in Fig. 2.2 for purposes of comparison. Table A.11 gives the thermal properties of some liquid metals. These data are of current interest because of their use as heat transfer media in nuclear power plants. One can note that the thermal conductivities of liquid metals are significantly greater than for other liquid substances.

The Thermal Conductivity of Gases

The thermal conductivities of several gases are given in Table A.9 for low pressures. In general, these values may be interpreted as applying at a pressure of about 1 atmosphere except in the cases in which the gaseous phase cannot exist at the temperature quoted. In such cases the values should be interpreted as applying at pressures near the saturation pressure for the temperature quoted. Some of the values have been plotted in Fig. 2.3 for purposes of discussion. It can be noted that in general the conductivity increases with increasing temperature and decreases with increasing molecular weight.

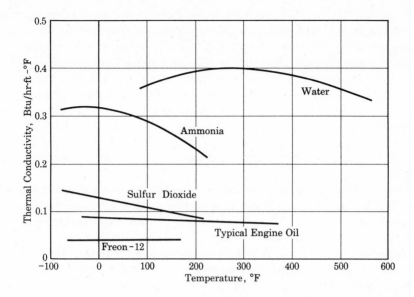

Fig. 2.2. The effect of temperature on the thermal conductivity of several saturated liquids.

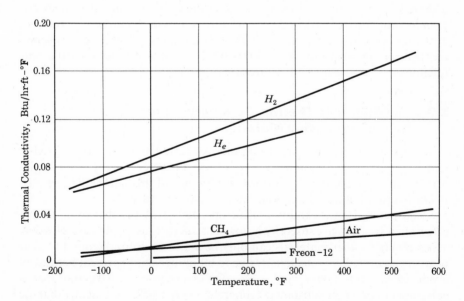

Fig. 2.3. The effect of temperature on the thermal conductivity of several gases at atmospheric pressure.

In general, the thermal conductivity of a gas is relatively independent of pressure if the pressure is near atmospheric pressure. Vapors near the saturation point show a strong pressure dependence. In the absence of other information, the pressure effect on the thermal conductivity may be approximated at high pressures by use of the generalized chart shown in Fig. A.1 of Appendix A. This chart plots the ratio of the thermal conductivity at high pressure to that at 1 atmosphere pressure as a function of the *reduced pressure* and *reduced temperature*. The latter quantities are the ratios of the pressure and temperature, respectively, to their values at the critical point. Table A.10 tabulates the critical constants for various substances.

Two particular gases of great engineering importance are steam and air. For this reason, more extensive presentations are made for their thermal conductivities in Tables A.6, A.7, and A.8. Other important properties, to be discussed later, are given also.

As usual, steam (Table A.6) proves to be an irregularly behaving gas, showing a rather strong pressure dependence for the thermal conductivity as well as a temperature dependence.

Table A.7 gives the temperature dependence of air at atmospheric pressure, and Table A.8 presents the data for higher pressures. It is apparent from the latter table that there is little pressure effect on the thermal conductivity of air until elevated temperatures are reached. Some small discrepancies between Tables A.7 and A.8 in the values of the thermal conductivities at atmospheric pressure are apparent. This is typical of the status of the present knowledge of the thermal conductivity of air. The more complete Table A.7 is recommended for use at atmospheric pressure, but Table A.8 presents the only reliable information for higher pressures.

2.3 The Apparent Thermal Conductivity of Nonhomogeneous Materials

The above discussions concerning thermal conductivity were restricted to materials composed of homogeneous or pure substances. Many of the engineering materials encountered in practice are not of this nature. This is particularly true of building materials and insulating materials.

Some materials may exhibit nonisotropic conductivities that result from a directional preference caused by a fibrous structure (as in the case of wood, asbestos, etc.). Other materials can only be discussed from the point of view of an *apparent thermal conductivity* due to inhomogeneities present because of a porous structure (glass wool, cork, etc.) or because of a structure that is composed of different substances (concrete, stone, brick, etc.). In any of these instances the thermal conductivity may vary from sample to sample due to variations in structure, composition, density, or porosity. The values reported in Table A.2 are only typical values encountered for such substances, and a considerable deviation from these values in practice should be expected.

Insulating Materials

Insulating materials are used in cases in which one wishes to obstruct the flow of heat between an enclosure and its surroundings. Low temperature insulations are used in instances in which the enclosure in question is at a temperature lower than the ambient temperatures and when one wishes to prevent the enclosure from gaining heat. High temperature insulations are used in the reverse case when one wishes to prevent an enclosure at a temperature higher than the ambient from losing heat to its surroundings.

Cork, rock wool, glass wool, etc., are examples of low temperature insulations. Asbestos, diatomaceous earth, magnesia, etc., are examples of high temperature insulations. The thermal conductivities of these materials may be found in Table A.2. One observes that for a given material there is a strong dependence of the thermal conductivity with density as well as temperature. The term *apparent bulk density* should actually be used. This is the mass of the substance divided by its total volume (including the volume of pores in the case of a porous material). The low conductivity of these insulating materials is due primarily to the air (a poorly conducting gas) that is contained in the pores rather than to a low conductivity of the solid substance itself. Generally speaking, the apparent thermal conductivity varies directly with the apparent bulk density, although a limit exists at which the temperature of the material becomes great enough that convection and radiation within the pore may increase the transfer of heat through the material, reversing the effect of the apparent bulk density. Figure 2.4 is a plot of the data of Table A.2 for rock

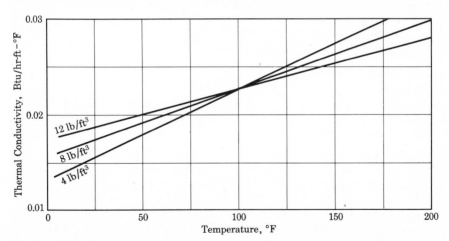

Fig. 2.4. The effect of temperature and apparent bulk density on the thermal conductivity of rock wool.

wool. The effect of the apparent density and temperature is easily seen. The increase of the conductivity with temperature is typical of insulating materials.

Refractory Materials

Refractory materials are used in instances in which it is desired to have a material capable of withstanding high temperatures without physical deterioration. The various fire-clay bricks (kaolin, magnesite, chrome, etc.) are usually encountered in this application. Some of these are included in Table A.2.

The refractories, as may be seen by consulting Table A.2, are not necessarily insulating materials as well. Further examination of this table will show the dependence on apparent density, as in the case of the insulations, as well as a dependence on the temperature at which the brick is fired. The variation of the thermal conductivity with temperature depends on whether or not a predominantly crystalline structure is formed. Usually the thermal conductivity increases with temperature for such substances, although an exception may be noted in the case of magnesite brick, which, like a pure metal, shows a decrease in conductivity with temperature.

2.4   Specific Heat

The thermal conductivity discussed above denotes the facility with which heat energy is propagated through a substance due to a temperature gradient. The variation of the temperature of a substance with the amount of energy, heat, stored within it is expressed in terms of the *specific heat* of the substance. Because of the different ways, thermodynamically speaking, in which energy may be stored in a substance, the definition of specific heat depends upon the nature of the process of heat addition. Two particularly important specific heats are defined:

Specific heat at constant pressure:

$$c_p = \left(\frac{\partial h}{\partial T}\right)_p.$$ (2.1)

Specific heat at constant volume:

$$c_v = \left(\frac{\partial u}{\partial T}\right)_v.$$ (2.2)

In Eqs. (2.1) and (2.2) h denotes specific enthalpy, u denotes specific internal energy, and T denotes the temperature. The subscripts p and v mean that the indicated differentiations are performed at constant pressure and constant volume, respectively. The first law of thermodynamics shows that, for reversible processes, changes in enthalpy at constant pressure and changes in internal energy at constant volume represent the heat transferred to or from a system during such processes. Thus, c_p represents the quantity of heat transferred per unit temperature difference during a reversible constant pressure process, and c_v represents the same during a reversible constant volume process; hence the name "specific *heat*."

The dimensions of specific heat are heat per unit mass per unit temperature difference. The units Btu/lb_m-°F will be used in this book.

In the heat transfer processes with which this book is concerned, conditions of constant, or virtually constant, pressure will prevail. Hence, only tabulated values of c_p will be needed.

The partial differentiation indicated in Eq. (2.1) implies that, in general, the specific heat of a substance is dependent on pressure and temperature. Some simplifications may be made of this fact, particularly in the cases of solids and liquids.

Specific Heats of Solids and Liquids

The theory of the specific heat of solids is a highly developed one, and a comprehensive summary may be found in Ref. 2. The same is true for liquids, and one may consult Ref. 3 in this connection.

These theories show, and experiment verifies, the fact that there is very little pressure dependence of the specific heat of solids and liquids until extremely high pressures are encountered. For this reason only the temperature dependence for various solids and liquids is noted in the tables of Appendix A. Even this temperature dependence is slight, particularly for liquids, for moderate changes in temperature.

Table A.1 gives the specific heat at constant pressure for the various metals and alloys listed but only at the single temperature of 68°F. In the absence of any other information, these values may be used at other temperatures without grave errors. Table A.3 shows the temperature dependence of c_p for various metals. Some of the other solids given in Table A.2 have values of c_p tabulated. These are apparent values for those substances which are not homogeneous.

Tables A.4 and A.5 give the c_p values for saturated water and other liquids, and the relative lack of dependence upon temperature is obvious. This is also true of the liquid metals listed in Table A.11.

Specific Heats of Gases

In the case of gases, the temperature dependence of the specific heat is much more pronounced than in the case of solids and liquids. A dependence on pressure also enters, but for gases far removed from saturation, this may often be neglected. The higher the temperature, the less the effect of pressure. Reference 3 may be consulted for a very comprehensive review of the molecular theory of the specific heat of gases.

Again, steam is an ill-behaved substance, and the pressure and temperature dependence of the specific heat at constant pressure may be seen in Table A.6.

Tables A.7 and A.8 give the specific heat data for air at atmospheric pressure and at high pressures. Because of a high degree of ideality, air does not show a very strong pressure dependence of specific heat.

For many other gases, the specific heat–temperature relations may be expressed by relatively simple empirical equations. Such equations for several useful gases are given in Table A.10. These equations apply only for pressures near atmospheric but may be applied with fair accuracy for higher pressures, say up to 200 psia.

In the absence of any other information, the results of the elementary kinetic theory of gases may be used to obtain at least an order-of-magnitude estimate of the specific heat. The kinetic theory gives

$$\text{Monatomic gases:} \quad c_p = \frac{5}{2}\frac{R}{M},$$

$$\text{Diatomic gases:} \quad c_p = \frac{7}{2}\frac{R}{M},$$

$$\text{Polyatomic gases:} \quad c_p = \frac{8}{2}\frac{R}{M}.$$

R is the universal gas constant equal to 1.986 Btu/mole-R°, and M is the molecular weight of the gas.

2.5 Thermal Diffusivity

In Chapter 1 the derivation of the general heat conduction equation introduced the defined property of *thermal diffusivity*, α. [See Eq. (1.9).] This property is defined in terms of those already discussed:

$$\alpha = \frac{k}{\rho c_p}.$$

As noted in Sec. 1.8, the units ft^2/hr will be used in this work.

Since the thermal diffusivity involves the properties of thermal conductivity and specific heat discussed above, no further discussion here is necessary except to note that values of this property are included in the tables of Appendix A for those substances (solid, liquid, and gaseous) for which sufficient data exist and for which tabulated values of thermal diffusivity may be desired.

2.6 The Coefficient of Thermal Expansion

The driving force in the process of free convection is that of gravity acting on regions of different densities in a fluid. These density differences come about because of the thermal expansion of a heated portion of the fluid. The physical property denoting this effect is the *coefficient of thermal expansion.*

This discussion will not consider the solid state, as it has little importance in heat transfer. Only the liquid and gaseous states will be discussed.

If a substance has an initial volume of V_i and is heated so that its volume increases in an amount ΔV while its temperature increases by Δt, the average coefficient of thermal expansion, β_{av}, for that temperature range is defined:

$$\beta_{av} = \frac{\dfrac{\Delta V}{V_i}}{\Delta t}.$$

The instantaneous value of the coefficient of expansion is found by taking the limit of the above quotient as $\Delta t \to 0$.

Thus,

$$\beta = \frac{1}{V}\left(\frac{\partial V}{\partial t}\right)_P.$$

The subscript P which suddenly appeared in the above expression denotes that β is thus defined when the heating process occurs at constant pressure. This specification is required to make the definition thermodynamically unique, particularly for the case of gases.

One could, as well, make the definition of β in terms of the specific volume, v, of the substance:

$$\beta = \frac{1}{v}\left(\frac{\partial v}{\partial t}\right)_P. \tag{2.3}$$

This leads to an alternative, and perhaps more useful, expression in terms of density since the density, ρ, is related to the specific volume:

$$\rho = \frac{1}{v}.$$

Then the expression for β may be taken as

$$\beta = -\frac{1}{\rho}\left(\frac{\partial \rho}{\partial t}\right)_P.$$ (2.4)

The dimensions of β are seen to be the reciprocal of temperature. This book will use as units: $1/^\circ$Rankine.

The Coefficient of Expansion of Gases

The coefficient of thermal expansion of a gas may be readily evaluated from Eq. (2.4) with the appropriate equation of state of the gas, i.e., the known relation among the pressure P, the density ρ, and the absolute temperature T.

For gases that may be considered to behave as ideal gases as far as their P-ρ-T behavior is concerned, one has

$$P = \rho RT,$$

from which the coefficient of expansion is evaluated as

$$\beta = \frac{1}{T}.$$ (2.5)

Thus, for ideal gases, or gases that may be approximated as ideal, one needs only the *absolute* temperature to evaluate β.

The ideal gas approximation is frequently adequate for many gases; if not, the more complex equations of state may be used. For example, for van der Waals' equation of state of a gas:

$$P = \frac{\rho RT}{1 - b\rho} - a\rho^2,$$

$$\beta = \frac{1}{T}\frac{1}{1 + \dfrac{b\rho}{1 - b\rho} - \dfrac{2a\rho^2}{P + a\rho^2}}.$$ (2.6)

The constants a and b are van der Waals' constants. Tabulated values of these constants may be found for several gases in Table A.10. These values may also be of use for the determination of densities when the ideal gas law does not apply.

For gases which cannot be represented by the above equation of state, β may be obtained from some of the other well-known equations of state (see Ref. 2), or if tabulated values of density or specific volume are available, as they are for steam, average values may be obtained over a finite temperature change by applying Eq. (2.3) or (2.4) in a finite difference form.

The Coefficient of Expansion of Liquids

The case of liquids is somewhat more complex than that of gases because of the lack of equations of state of sufficient generality to permit the evaluation of β in general forms.

For a great many liquids the approximate rule (Ref. 13)

$$\beta \approx 0.06284(T_c - T)^{-0.641} \tag{2.7}$$

may be found to apply rather well. T_c is the critical temperature of the substance in °R. This relation is an empirical rule that may become inadequate at extreme pressures and temperatures, but it is useful when other data are lacking.

In general, as long as extreme pressures are avoided, one may consider that the coefficient of expansion of liquids is dependent on temperature only. Table A.4 tabulates values of β for saturated water at various temperatures. In Table A.5 the values of β for various liquids are given at the single temperature of 68°F.

2.7   Fluid Viscosity

All real substances are known to offer a certain amount of resistance to deformation. Generally speaking, solids in the so-called "elastic" range offer a resistance which is proportional to the amount of deformation. Fluids, both gases and liquids, are observed to resist deformation in a manner which is proportional to the time rate of deformation.

When, as in the processes of heat convection (free or forced), a fluid is caused to move past solid bodies, velocity gradients are set up within the fluid because of the relative motion of various parts of the fluid. This relative motion may be a sliding motion of adjacent fluid elements producing shearing resistance, or it may be a relative expanding or contracting motion producing tensile or compressive resistance.

In general, a fluid motion will be a complex combination of such movements with a correspondingly complex set of resisting forces acting within the fluid. The resistive forces, the inertia force due to the motion, and the applied external forces (pressure, gravity, etc.) determine the movements of the various portions of the fluid.

In order to analyze this motion and to apply the results to cases of convective heat transfer, it is necessary to be able to express the forces of resistance in terms of the velocity field of the fluid. The resistance to the shearing motion is the dominant resistive force and is expressed in terms of the fluid property known as *viscosity*.

Various concepts and definitions exist for the viscosity of a fluid, but the most useful is that of Newton, as discussed in the next section.

Dynamic Viscosity

The fundamental definition of viscosity is best illustrated by considering "parallel," laminar fluid motion such as that past a plane wall as depicted in Fig. 2.5. Some distribution in the velocity (parallel to the wall) is presumed to exist, varying from zero at the wall to some value far removed from the wall.

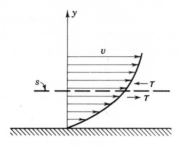

Fig. 2.5

By selecting a plane, S, parallel to the wall, experiment has shown that the fluid layers on either side of S experience a shearing force or viscous drag which is due to their relative motion. Newton postulated that the shearing stress, τ, produced by this relative motion is directly proportional to the velocity gradient taken in a direction normal to the plane S. The constant of proportionality, μ, is termed the *coefficient of dynamic viscosity*. Thus,

$$\tau = \mu \frac{dv}{dy}, \tag{2.8}$$

in which v is the fluid velocity, parallel to the wall, and y is the coordinate measured normal to the wall.

The physical concept behind this definition of viscosity is that of a transfer of momentum across S due to an exchange of molecules between the fluid layers on either side. As a molecule (moving because of thermal excitation) crosses the plane S, it will transfer an amount of momentum which is proportional to the difference in the flow velocity between the point on one side of S from which the molecule started and the point on the other side of S where it was stopped by the next molecular collision. The velocity difference, and hence the momentum exchanged, is directed parallel to S. According to

Newton's law of motion, this change in momentum is accompanied by a resultant force in the same direction. It is this latter shearing force which is the viscous resistance to deformation offered by the fluid. This physical interpretation requires the existence of a velocity gradient in the fluid, and Newton's definition of the dynamic viscosity coefficient in Eq. (2.8) relates the viscous shear to the velocity gradient. This concept of viscosity is based on a molecular exchange. The existence of any transport of mass in a direction normal to the velocity on a scale larger than a molecular one is excluded in this definition. Thus, the above definition of viscosity applies only to laminar flow—not to turbulent flow, in which gross transverse fluctuations and mixing occur.

Extensive experiments have confirmed the law stated in Eq. (2.8), and most of the commonly encountered fluids are known to be very nearly Newtonian. The coefficient of dynamic viscosity is then a characteristic molecular property of the fluid and is not dependent upon the fluid motion. Reference 3 presents an extensive summary of the theoretical molecular theory of the dynamic viscosity.

The Dimensions and Units of Dynamic Viscosity

The defining relation for the dynamic viscosity, Eq. (2.8), gives its dimensions as

$$\frac{\text{force} \times \text{time}}{(\text{length})^2}.$$

Generally, the units used are

$$\text{English}: \quad \frac{\text{lb}_f\text{-sec}}{\text{in.}^2}, \quad \frac{\text{lb}_f\text{-hr}}{\text{ft}^2},$$

$$\text{Metric}: \quad \frac{\text{dyne-sec}}{\text{cm}^2}, \quad \frac{\text{g}_f\text{-sec}}{\text{cm}^2}.$$

These units are convenient in many cases; however, it is often desirable to express the dimensions of viscosity in terms of the fundamental dimensions of mass rather than force. The dimensional constant g_c (see Sec. 1.7) gives the following relation between force and mass:

$$\text{Force} = \text{mass} \times \frac{\text{length}}{(\text{time})^2}.$$

Thus, the dimensions of dynamic viscosity may also be

$$\frac{\text{mass}}{\text{length} \times \text{time}},$$

and the units used most often are

$$\text{English}: \quad \frac{\text{lb}_m}{\text{ft-sec}}, \quad \frac{\text{lb}_m}{\text{ft-hr}}, \quad \frac{\text{slugs}}{\text{ft-hr}}.$$

$$\text{Metric}: \quad \text{poise} = \frac{\text{g}_m}{\text{cm-sec}}.$$

$$\text{centipoise} = \tfrac{1}{100} \text{ poise}.$$

As is rather apparent from the above, there are a great number of viscosity units in use. The conversion from one to another can be facilitated by use of Table B.1 in Appendix B or the conversion factors of Appendix F.

Kinematic Viscosity

Since the forces of viscosity act directly on the fluid, and since the fluid inertia resists these forces, the ratio of the viscous force to the inertia force would be expected to be an important parameter of the fluid motion. Thus, the *kinematic viscosity*, defined as the ratio of the dynamic viscosity to the fluid mass density, becomes an important fluid property. The kinematic viscosity is denoted by v:

$$v = \frac{\mu}{\rho}. \tag{2.9}$$

This gives the dimensions of kinematic viscosity to be

$$\frac{(\text{length})^2}{\text{time}},$$

with the usual units being:

$$\text{English}: \quad \text{ft}^2/\text{hr, ft}^2/\text{sec}.$$

$$\text{Metric}: \quad \text{stoke} = \text{cm}^2/\text{sec},$$

$$\text{centistoke} = \tfrac{1}{100} \text{ stoke}.$$

The Dynamic and Kinematic Viscosity of Various Gases and Liquids

The molecular theory of the viscosity of gases and liquids (Ref. 3) predicts that the dynamic viscosity depends primarily on temperature and, to a lesser degree, on pressure. These facts are borne out by experimental measurements.

In liquids, as in the case of thermal conductivity and specific heat, the pressure dependence of the dynamic viscosity is found to be quite slight unless extreme pressures are reached. Usually, then, this effect may be ignored. The same is true of the kinematic viscosity. Tables A.4 and A.5 present, as functions of temperature, the dynamic and kinematic viscosities of saturated water and several other saturated liquids. Table A.11 contains corresponding data for liquid metals.

The Dynamic and Kinematic Viscosity of Gases

For gases the picture is somewhat different from that noted above for liquids. Generally speaking, the dynamic viscosity of gases is quite a bit smaller than that of liquids, although the kinematic viscosity may be greater because of the lower gas densities. In most gases the temperature is again the most significant factor influencing the dynamic viscosity. The pressure effect is quite small for most gases—as long as the saturation state is not approached. The kinematic viscosity of gases is strongly dependent on pressure, owing to their high compressibilities.

Table A.6 includes values of the dynamic viscosity of steam. These values show that the viscosity of steam has a significant pressure dependence.

Table A.7 gives values of both the dynamic and kinematic viscosity of air as functions of temperature, at atmospheric pressure. No values of the dynamic viscosity of air are included in Table A.8 for higher pressures because the pressure dependence is so slight.

The dynamic viscosity of a gas can be expressed in terms of rather generalized expressions. For most gases the temperature variation of the dynamic viscosity can be rather accurately determined from the Sutherland formula,

$$\mu = C_1 \frac{T^{3/2}}{T + C_2}, \tag{2.10}$$

in which T is the absolute temperature and C_1 and C_2 are constants. Tabulated values of the Sutherland constants for several gases may be found in Table A.10.

The values of μ obtained from the Sutherland formula may be interpreted as applying at pressures of 1 atmosphere. As in the case of thermal conductivity, the behavior at high pressures may be estimated rather well from the generalized correlation in Fig. A.2. This chart relates the ratio of the viscosity at high pressures to that at 1 atmosphere in terms of the reduced pressure, P_r, and reduced temperature, T_r, defined in Sec. 2.2.

2.8   The Prandtl Number

In problems of convection in which one encounters fluid motion and heat conduction occurring simultaneously, the ratio of the kinematic viscosity, v, to the thermal diffusivity, α, is found to play an important role. The physical significance and usefulness of this parameter will be discussed in Chapter 7, but it seems appropriate to mention it here since it is purely a fluid property.

This ratio is called the *Prandtl number* in honor of the German scientist who contributed so much to the knowledge of fluid motion and heat transfer. The Prandtl number is, then,

$$N_{PR} = \frac{v}{\alpha}$$

$$= \frac{\mu c_p}{k}. \tag{2.11}$$

Since both v and α have dimensions of (length)^2/time, N_{PR} has no dimensions, although it is a fluid property.

Tables A.4, A.5, A.6, A.7, A.8, and A.11 tabulate values of the Prandtl number for water, steam, air, and various liquids. It may be computed for other substances from the known values of μ, c_p, and k.

2.9   Closure

This chapter has discussed, in brief, the physical properties of substances that have significance in convective and conductive heat transfer. These properties are thermal conductivity, specific heat, thermal diffusivity, thermal expansion, and fluid viscosity.

Generally speaking, all these properties are dependent on temperature, but such is not the case with pressure. They are practically independent of pressure in the case of solids or liquids unless the pressure is extreme. Gases, on the other hand, show a strong pressure dependence in the properties, particularly if the saturation state or critical state is approached.

References

1. AUSTIN, J. B., *The Flow of Heat in Metals*, Cleveland, American Society for Metals, 1942.

2. ZEMANSKY, M. W., *Heat and Thermodynamics*, 4th ed., New York, McGraw-Hill, 1957.
3. HIRSCHFELDER, J. O., C. F. CURTISS, and R. B. BIRD, *Molecular Theory of Gases and Liquids*, New York, Wiley, 1954.
4. ECKERT, E. R. G., *Introduction to the Transfer of Heat and Mass*, New York, McGraw-Hill, 1950.
5. BROWN, A. I., and S. M. MARCO, *Introduction to Heat Transfer*, 3rd ed., New York, McGraw-Hill, 1958.
6. MCADAMS, W. H., *Heat Transmission*, 3rd ed., New York, McGraw-Hill, 1954.
7. HALL, N. A., *Thermodynamics of Fluid Flow*, Englewood Cliffs, N.J., Prentice-Hall, 1951.
8. JAKOB, M., and G. A. HAWKINS, *Elements of Heat Transfer and Insulation*, 3rd ed., New York, Wiley, 1957.
9. FANO, L., J. H. HUBBELL, and C. W. BECKETT, "Compressibility Factor, Density, Specific Heat Enthalpy, Entropy, Free Energy Function, Viscosity, and Thermal Conductivity of Steam," *N.A.C.A. Tech. Note 3273*, Washington, D.C., 1956.
10. *Handbook of Supersonic Aerodynamics*, Vol. 5, Washington, D.C., Bureau of Ordnance, Department of the Navy, 1953.
11. *International Critical Tables*, New York, McGraw-Hill, 1929.
12. MARKS, L. S., *Mechanical Engineers' Handbook*, 6th ed., New York, McGraw-Hill, 1958.
13. SMITH, W. T., S. GREENBAUM, and G. P. RUTLEDGE, "Correlation of Critical Constants with the Thermal Expansion Coefficient of Organic Liquids," *J. Phys. Chem.*, Vol. 58, 1954, p. 443.

Problems

2.1. With the aid of Table A.9 and Fig. A.1, find the thermal conductivity of the following gases at the pressure and temperature indicated:
   (a) nitrogen at 1500 psia, 32°F;
   (b) hydrogen at 900 psia, 572°F;
   (c) carbon dioxide at 3300 psia, 100°F.

2.2. Using Table A.7 and Fig. A.1, determine the thermal conductivity of dry air at 80°F, 100 atm pressure. Compare this result with that found in Table A.8.

2.3. Treating carbon dioxide as
   (a) an ideal gas,
   (b) a van der Waals gas,
      find the coefficient of volume expansion at 1000 psia, 100°F, and compare the two results.

2.4. Repeat Prob. 2.3 for dry air at 1000 psia, 100°F.

2.5. Using Sutherland's formula, determine the dynamic viscosity of the following gases at atmospheric pressure and at the indicated temperature:
   (a) methane at 500°F;
   (b) carbon dioxide at 100°F;

(c) freon-12 at 200°F;

(d) helium at 500°F.

Express answers in the following units:

$$\text{lb}_m/\text{ft-hr}, \frac{\text{lb}_f\text{-sec}}{\text{in}^2.}, \text{centipoise}, \frac{\text{dyne-sec}}{\text{cm}}, \frac{\text{kg}}{\text{m-sec}}$$

2.6. Using Sutherland's formula and Fig. A.2, determine the dynamic viscosity of the following gases at the indicated pressure and temperature:

(a) nitrogen at 1500 psia, 100°F;

(b) freon-12 at 1200 psia, 250°F;

(c) helium at 65 psia, 1400°F.

Express the answers in $\text{lb}_m/\text{ft-hr}$.

2.7. Determine the thermal diffusivity of the following gases at the indicated conditions:

(a) methane at 14.7 psia, 500°F;

(b) carbon monoxide at 14.7 psia, 250°F;

(c) oxygen at 1500 psia, 150°F.

2.8. Determine the value of the Prandtl number for the gases listed in Prob. 2.7.

CHAPTER 3

Steady State Heat Conduction in One Dimension

3.1 The Meaning of "One-Dimensional" Conduction

The term "steady state conduction" was defined in Sec. 1.4 as the condition which prevails when the temperatures of fixed points within a heat-conducting body do not change with time. This also implies that the time rate of heat flow between any two selected points is constant with time. This, as was shown, leads to the following differential equation as the relation governing the distribution of temperature throughout the heat-conducting body:

$$\frac{\partial^2 t}{\partial x^2} + \frac{\partial^2 t}{\partial y^2} + \frac{\partial^2 t}{\partial z^2} = \frac{q^*}{k}. \tag{3.1}$$

In this equation t symbolizes the temperature, and x, y, and z represent the space coordinates.

The term "one-dimensional" is applied to a heat conduction problem when only one space coordinate is required to describe the distribution of temperature within a heat-conducting body. Such a situation rarely exists precisely in real problems, but a very great number of problems of practical engineering importance may be approximated quite well by assuming a one-dimensional condition.

The flow of heat through a plane wall, at regions sufficiently removed from the edges, depends only on the coordinate measured normal to the plane of the wall. Hence, neglecting edge effects in a wall leads one to a one-dimensional heat conduction problem. This is sometimes referred to as conduction through an "infinite slab" (i.e., a wall with no edge effects).

Similarly, conduction through a very long, hollow cylinder (such as pipe) which is maintained at uniform temperatures on its inner and outer surfaces may be accurately considered to depend only on a single coordinate, the radial distance.

A very thin rod, or a wire, maintained at different temperatures on its ends may be considered to conduct heat one-dimensionally if it is sufficiently thin so that the temperature can be taken as uniform over any cross section.

Thus, it is seen that the term "one-dimensional" applies only to the number of coordinates needed to describe the temperature distribution in the body —not the number of space dimensions occupied by the conducting body. Also this one-dimensional coordinate need not be straight, for the wire described in the above paragraph might be in the form of a helical coil with the one-dimensional coordinate being the distance along the wire. This chapter will be devoted to the analysis of such one-dimensional situations.

The solution of heat conduction problems involves, in general, the writing of the solution of the general heat conduction equation in terms of the appropriate number of arbitrary constants and then the evaluating of these constants by use of the imposed boundary conditions.

The simplest type of boundary condition in heat conduction is that of specified boundary temperatures, and certain examples will be presented in the next few sections. Cases with and without internal heat generation will be treated. Problems in which the solid boundaries are in contact with convecting fluids of specified temperature will then be discussed.

3.2   The Plane Wall with Specified Boundary Temperatures

The most elementary of all heat conduction problems is that of a plane wall of finite thickness but infinite in extent (so that the conduction may be taken as one-dimensional) with each face maintained at a uniform temperature.

The expressions for the temperature distribution (linear) and the rate of heat flow through the wall may be immediately deduced from Fourier's conduction law, Eq. (1.3), without going into an elaborate mathematical analysis. However, because of the generality of the method and the simplicity of this problem, the mathematical approach will be used to illustrate the principles to be applied in the more complex problems of later chapters.

Figure 3.1 illustrates such a plane wall of thickness Δx and composed of a material having a thermal conductivity k. Its two faces are maintained at the uniform temperatures t_1 and t_2. The single coordinate required is the displacement in the direction normal to the wall and is denoted by x. The

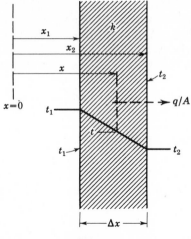

Fig. 3.1

origin of the x coordinate is arbitrarily placed outside the wall as shown. The symbol x_1 denotes the face at temperature t_1, and x_2 denotes the face at temperature t_2.

Since there is no internal heat generation, the differential equation for steady state heat conduction, Eq. (3.1), reduces to

$$\frac{d^2t}{dx^2} = 0$$

with the boundary conditions:

$$\text{At } x = x_1: \quad t = t_1.$$

$$\text{At } x = x_2: \quad t = t_2.$$

The solution of the rather elementary equation is

$$t = Bx + C. \tag{3.2}$$

The symbols B and C denote arbitrary constants. Application of the two boundary conditions yields two equations to be solved for B and C, which, upon substitution into Eq. (3.2), give the following linear expression for the temperature distribution in the wall:

$$t = t_1 + \frac{t_2 - t_1}{x_2 - x_1}(x - x_1). \tag{3.3}$$

The rate of heat flow through the wall may be found from Fourier's law. Since this is a situation in which the temperature gradient is uniform over every plane drawn normal to the direction of heat flow, the form of Fourier's law expressed in Eq. (1.3) is applicable:

$$q = -kA\frac{dt}{dx}.$$

Introducing dt/dx from Eq. (3.3), one has

$$\frac{q}{A} = k\frac{t_1 - t_2}{x_2 - x_1}, \quad \text{or} \quad \frac{q}{A} = k\frac{t_1 - t_2}{\Delta x}. \tag{3.4}$$

The above expression is written in terms of q/A, the heat flow per unit area, since a wall of infinite extent was assumed in order that edge effects could be neglected.

The derivations of the general heat conduction equation and the form of Fourier's law used here involved the assumption of a constant thermal conductivity, k. As was apparent in Chapter 2, this is generally not true for actual substances. The simplification of this assumption will be discussed in Sec. 3.6, in which a variable conductivity is considered.

3.3   The Multilayer Wall with Specified Boundary Temperatures

Consider now a plane wall composed of layers of materials having different thicknesses and thermal conductivities. Figure 3.2 illustrates a wall of three layers for the purposes of discussion, although any number of layers may be considered.

Denote each juncture of two different materials by a number (1, 2, 3, 4 in Fig. 3.2) and adopt the following notational convention:

Δx_{mn} = thickness of material between plane m and plane n,

k_{mn} = thermal conductivity of material between plane m and plane n.

If the outside temperatures are specified (i.e., t_1 and t_4) and if one knows the thickness and conductivities of all the materials between the planes where these temperatures are specified, what is the distribution of the temperature through the wall and what is the rate of heat flow through the wall?

Since the steady state is assumed to exist, Eq. (3.4) may be applied to obtain the following set of equations since the rate of heat flow per unit area, q/A, is

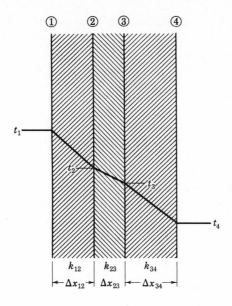

Fig. 3.2

the same for each layer:

$$\frac{q}{A} = \frac{t_1 - t_2}{\Delta x_{12}/k_{12}} = \frac{t_2 - t_3}{\Delta x_{23}/k_{23}} = \frac{t_3 - t_4}{\Delta x_{34}/k_{34}}.$$

Solving for each of the temperature differences, one obtains

$$t_1 - t_2 = \frac{q}{A}\frac{\Delta x_{12}}{k_{12}},$$

$$t_2 - t_3 = \frac{q}{A}\frac{\Delta x_{23}}{k_{23}}, \qquad (3.5)$$

$$t_3 - t_4 = \frac{q}{A}\frac{\Delta x_{34}}{k_{34}}.$$

Upon addition of these equations the unknown temperatures are eliminated, and the following expression for the heat flow through the wall in terms of the over-all temperature difference may be obtained:

$$\frac{q}{A} = \frac{t_1 - t_4}{\dfrac{\Delta x_{12}}{k_{12}} + \dfrac{\Delta x_{23}}{k_{23}} + \dfrac{\Delta x_{34}}{k_{34}}}. \tag{3.6}$$

Once the heat flow rate is obtained from Eq. (3.6), one may determine the temperatures t_2 and t_3 from Eq. (3.5). The distribution of temperature is then known to be linear between the known values.

Thermal Resistance and Thermal Conductance. In Sec. 1.4 a unit thermal conductance was defined:

$$C = \frac{k}{\Delta x}.$$

A unit thermal resistance R could be defined as the reciprocal of the unit conductance:

$$R = \frac{1}{C} = \frac{\Delta x}{k},$$

which would lead to an analogy between electrical and thermal conduction. However, a clearer analogy may be obtained if one defines a *total* thermal resistance

$$\mathscr{R}_k = \frac{1}{CA} = \frac{\Delta x}{kA},$$

the subscript k being used to emphasize the fact that the resistance defined is a *conduction* resistance. On the basis of this definition, the total heat flow through a multilayer wall is

$$q = \frac{\Delta t_{\text{over-all}}}{\mathscr{R}_{k_{12}} + \mathscr{R}_{k_{23}} + \cdots},$$

$$\frac{q}{A} = \frac{\Delta t_{\text{over-all}}}{A(\mathscr{R}_{k_{12}} + \mathscr{R}_{k_{23}} + \cdots)}.$$

Introduction of the definition of $\mathscr{R}$ will yield Eq. (3.6).

Figure 3.2 depicts the equivalent electrical network implied by the above analysis. One sees that thermal resistance in series should be added—as in the electrical equivalent. For the plane wall, the area, A, is the same for each layer, and the unit resistance could have been employed. However, for instances in which the areas might not be the same (in the case of a parallel circuit, for instance), the formulation given above is necessary.

EXAMPLE 3.1:   A house wall consists of an outer layer of common brick 4 in. thick ($k = 0.40$ Btu/hr-ft-°F), followed by a $\frac{1}{2}$-in. layer of Celotex sheathing ($k = 0.028$ Btu/hr-ft-°F). A $\frac{1}{2}$-in. layer of sheetrock ($k = 0.43$ Btu/hr-ft-°F) forms the inner surface and is separated from the sheathing by $3\frac{3}{4}$ in. of air space—as provided by the wall studs. The air space has a *conductance* of 1.10 Btu/hr-ft²-°F. The outside brick surface temperature is 40°F; the inner wall surface is maintained at 70°F. What is the rate of heat loss from the house, per square foot of wall area? What is the temperature at a point midway through the Celotex layer?

Solution: From Eq. (3.6) the heat loss is

$$\frac{q}{A} = \frac{70 - 40}{\dfrac{0.5}{12 \times 0.43} + \dfrac{1}{1.10} + \dfrac{0.5}{12 \times 0.028} + \dfrac{4}{12 \times 0.40}} = 9.05 \text{ Btu/hr-ft}^2 \ (28.5 \text{ W/m}^2).$$

A point midway through the Celotex is 4.25 in. from the outside surface. The heat flow between these two points is the same as the above since the steady state exists. Hence the desired temperature is

$$\frac{q}{A} = 9.05 = \frac{t - 40}{\dfrac{0.25}{12 \times 0.028} + \dfrac{4}{12 \times 0.4}},$$

$$t = 54.2°\text{F} \ (12.3°\text{C}).$$

3.4   The Single-Layer Cylinder with Specified Boundary Temperatures

A geometrical configuration which is mathematically simple and also of great engineering importance is that of a hollow cylinder, as pictured in Fig. 3.3. The cylindrical system of coordinates shown is the natural one to use.

Fig. 3.3

If the inner surface of radius r_1 and the outer surface of radius r_2 are maintained at uniform temperatures t_1 and t_2, respectively, then for a sufficiently long cylinder the end effects may be ignored. One may thus eliminate any dependence of the temperature on the axial coordinate, z, or the circumferential coordinate, θ. Thus, the problem is reduced, in the steady state, to a one-dimensional situation, with the radial distance r as the coordinate. Under these conditions the general heat conduction equation in cylindrical coordinates [Eq. (1.10)] reduces to the following expression if the generation term is omitted:

$$\frac{d^2t}{dr^2} + \frac{1}{r}\frac{dt}{dr} = 0 \quad \text{or} \quad \frac{d}{dr}\left(r\frac{dt}{dr}\right) = 0. \tag{3.7}$$

The boundary conditions are

$$\text{At } r = r_1: \quad t = t_1.$$
$$\text{At } r = r_2: \quad t = t_2. \tag{3.8}$$

Equation (3.7) has the solution

$$t = B \ln r + C, \tag{3.9}$$

where the arbitrary constants B and C are found from applying Eq. (3.8) to Eq. (3.9). Thus,

$$B = \frac{t_2 - t_1}{\ln (r_2/r_1)},$$

$$C = t_1 - \ln r_1 \frac{t_2 - t_1}{\ln (r_2/r_1)},$$

so that Eq. (3.9) gives

$$t = t_1 + \frac{t_2 - t_1}{\ln (r_2/r_1)} \ln (r/r_1). \tag{3.10}$$

The temperature, then, is seen to vary logarithmically through the cylinder wall in contrast to the linear variation in the plane wall.

The rate of heat flow through the cylindrical wall is, since the steady state exists, the heat flow across any arbitrary cylindrical surface of radius r between r_1 and r_2. With the temperature gradient uniform on such a surface, Fourier's law expressed in Eq. (1.3) is applicable:

$$q = -kA_r \frac{dt}{dr}.$$

A_r is the area of the cylindrical surface of radius r and is equal to $2\pi rL$, L being the length of the cylinder. Since the cylinder was assumed to be infinite in length in order to obtain the one-dimensional condition, it is more appropriate to speak in terms of heat flow per unit length. Thus,

$$\frac{q}{L} = -k2\pi r\frac{dt}{dr}.$$

Equation (3.9) gives

$$\frac{dt}{dr} = \frac{B}{r}$$

$$= \frac{1}{r}\frac{t_2 - t_1}{\ln(r_2/r_1)},$$

so that

$$\frac{q}{L} = \frac{2\pi k(t_1 - t_2)}{\ln(r_2/r_1)}. \tag{3.11}$$

3.5   The Multilayer Cylinder with Specified Boundary Temperatures

A pipe covered with insulation is a perfect example of the next configuration to be discussed—that of a long cylinder composed of two or more layers of materials having different thermal conductivities. For purposes of discussion consider the three-layer cylinder shown in Fig. 3.4. The same notational scheme employed in the case of the multilayer wall is used here.

The same procedure used for the plane wall may be applied in this case. Writing three expressions for the heat flow through each of the layers, in terms of unspecified temperatures t_2 and t_3:

$$\frac{q}{L} = \frac{2\pi k_{12}(t_1 - t_2)}{\ln(r_2/r_1)} = \frac{2\pi k_{23}(t_2 - t_3)}{\ln(r_3/r_2)} = \frac{2\pi k_{34}(t_3 - t_4)}{\ln(r_4/r_3)}.$$

Elimination of the unknown t_2 and t_3 gives

$$\frac{q}{L} = \frac{2\pi(t_1 - t_4)}{\dfrac{\ln(r_2/r_1)}{k_{12}} + \dfrac{\ln(r_3/r_2)}{k_{23}} + \dfrac{\ln(r_4/r_3)}{k_{34}}} \tag{3.12}$$

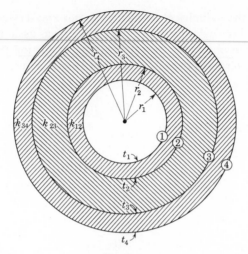

Fig. 3.4

as the heat flow per unit length in terms of the over-all temperature difference. Once q/L is known, the temperatures t_2 and t_3 may be found from the expressions for the heat flow through each layer.

EXAMPLE 3.2:   A 4-in. schedule 40 wrought iron pipe ($k = 32$ Btu/hr-ft-°F) is covered with 1 in. of 85 % magnesia insulation ($k = 0.041$ Btu/hr-ft-°F). For an inside pipe wall temperature of 300°F and an outer insulation surface temperature of 85°F, find the heat loss per foot of length of the pipe.

Solution: Appendix B gives the inside radius of a 4-in. schedule 40 pipe to be $4.03/2 =$ 2.015 in., and the outside radius to be $4.500/2 = 2.250$ in. Hence, Eq. (3.12) gives the loss to be

$$\frac{q}{L} = \frac{2\pi(300 - 85)}{\dfrac{\ln\left(\dfrac{2.250}{2.015}\right)}{32} + \dfrac{\ln\left(\dfrac{2.250 + 1.0}{2.250}\right)}{0.041}}$$

$$= 150.7 \text{ Btu/hr-ft } (144.9 \text{ W/m}).$$

Problem 3.39 shows that for a cylindrical wall a "log-mean" area may be defined:

$$A_m = \frac{A_2 - A_1}{\ln\,(A_2/A_1)} = \frac{2\pi L(r_2 - r_1)}{\ln\,(r_2/r_1)}.$$

If a thermal resistance is defined on the basis

$$\mathscr{R}_k = \frac{\Delta r}{kA_m},$$

the electrical network analogy gives

$$q = \frac{\Delta t_{\text{over-all}}}{\mathscr{R}_{k_{12}} + \mathscr{R}_{k_{23}} + \cdots}.$$

This is seen to be the same form as the result given in Eq. (3.12).

3.6 The Effect of Variable Thermal Conductivity

All the examples discussed above involve the assumption of constant thermal conductivity. Chapter 2 noted the fact that almost all substances show some dependence of thermal conductivity on temperature. It is the purpose of this section to illustrate the effect of this dependence on the temperature distribution and the heat flow rate.

Examination of the tables of Appendix A or Figs. 2.1 through 2.4 will show that for most substances, particularly solids, a *linear* dependence of the thermal conductivity on temperature exists, at least for limited ranges of temperatures. Thus, let the thermal conductivity be expressed as a function of temperature in the following way:

$$k = k_0(1 + bt). \tag{3.13}$$

The symbol k_0 is the value of the conductivity at $t = 0$, and b is a constant property of the material.

For simplicity of discussion, consider conduction through a plane wall (depicted in Fig. 3.5) of thickness Δx and with uniform surface temperatures t_1 and t_2.

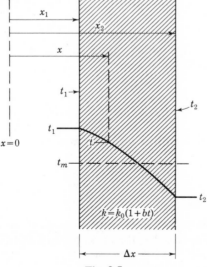

Fig. 3.5

Again using Fourier's law, the heat flow per unit area of wall is

$$q = -kA\frac{dt}{dx}$$

$$= -k_0(1 + bt)\frac{dt}{dx}A.$$

Since the steady state exists, q/A is a constant, and the equation above may be separated and integrated directly:

$$\frac{q}{A}\int_{x_1}^{x_2} dx = -k_0\int_{t_1}^{t_2}(1 + bt)\,dt,$$

$$\frac{q}{A}(x_2 - x_1) = -k_0\left[\left(t_2 + \frac{b}{2}t_2^2\right) - \left(t_1 + \frac{b}{2}t_1^2\right)\right]. \tag{3.14}$$

Equation (3.14) may be factored and written

$$\frac{q}{A} = -k_0\left[1 + \frac{b}{2}(t_1 + t_2)\right]\frac{t_2 - t_1}{\Delta x}. \tag{3.15}$$

If one now evaluates the thermal conductivity at the arithmetic mean of the surface temperature,

$$t_m = \frac{t_1 + t_2}{2},$$

Eq. (3.13) gives the mean conductivity to be

$$k_m = k_0\left[1 + \frac{b}{2}(t_1 + t_2)\right].$$

Equation (3.15) for the heat flux then becomes

$$\frac{q}{A} = k_m\frac{t_1 - t_2}{\Delta x}. \tag{3.16}$$

This is identical in form to Eq. (3.4) for the wall of constant thermal conductivity except for the fact that the mean thermal conductivity replaces the constant value.

Thus, as far as the rate of heat transfer is concerned, it is adequate to employ the relations developed for constant thermal conductivity if the conductivity is evaluated at the arithmetic mean temperature. This is true, of course, only if the linear variation assumed adequately represents the actual temperature dependence of the conductivity.

The temperature distribution is, in the case of the plane wall, no longer linear. If in Eq. (3.14) the dimension x_2 is replaced by the unspecified coordinate x and t_2 by t, the temperature at x, the distribution of the temperature in the wall is given by

$$\frac{q}{A}(x - x_1) = -k_0\left[\left(t + \frac{b}{2}t^2\right) - \left(t_1 + \frac{b}{2}t_1^2\right)\right].$$

The term q/A can be eliminated by use of Eq. (3.16), and the temperature distribution is given by the following quadratic expression:

$$\frac{b}{2}t^2 + t = \left(t_1 + \frac{b}{2}t_1^2\right) - \frac{k_m}{k_0}\frac{x - x_1}{x_2 - x_1}(t_1 - t_2).$$

In some instances it may not be possible to know in advance the mean temperature at which the thermal conductivity should be evaluated. This would be the case when plane or cylindrical walls of multiple layers are encountered. Only the over-all temperature range for the entire structure is usually known—not the surface temperature of each layer of material. In such cases one would have to assume reasonable values of the thermal conductivities of the materials involved and use these values to determine the temperature of the boundaries of each layer. Then the mean temperature of each layer may be computed, and the correctness of the assumed conductivity may be determined. This would then be repeated until a satisfactory check is obtained. However, unless there is a strong dependence of the thermal conductivity on temperature, and unless the temperature differences are large, such a procedure is usually not worthwhile since thermal conductivities are not generally known accurately enough to warrant such precision. Furthermore, in many problems of practical engineering importance, the accuracy of the given data (e.g., the temperature) does not justify such procedures.

3.7   Cases Involving Internal Heat Generation

The cases considered thus far have been those in which the conducting solid is free of internal heat generation. In this section three instances in which internal generation is present will be considered, and all will treat cases in which the heat generation is distributed throughout the solid.

Numerous practical instances exist in which such generation must be accounted for. The dissipative processes in current-carrying electrical conductors result in heat generation. Induction heating produces distributed heat additions as well as certain exothermic chemical reactions such as in the curing of concrete. Although the processes just described will produce distributed heat generation, the generation rate need not be uniform throughout the body. Both uniform and temperature-dependent generation rate cases will be discussed in the following paragraphs.

Uniformly Distributed Generation in a Plane Wall

Consider the instance in which a plane wall, for which the surface temperatures are specified, has a uniformly distributed heat generation rate of q^* (per unit volume) throughout its interior. The steady state temperature distribution and heat flow rates will be found and compared with the results of Sec. 3.2, in which generation is absent. For convenience, then, use the geometry and boundary conditions of that section as depicted in Fig. 3.1. In this instance the governing differential equation, Eq. (3.1), and the associated boundary conditions become

$$\frac{d^2t}{dx^2} + \frac{q^*}{k} = 0.$$

$$\text{At } x = x_1: \quad t = t_1.$$ (3.17)

$$\text{At } x = x_2: \quad t = t_2.$$

In this instance the generation rate q^* is taken as constant. Twice integrating Eq. (3.17) one obtains

$$t + \frac{q^*}{k}\frac{x^2}{2} = Bx + C.$$

Application of the two boundary conditions at x_1 and x_2 to the above expression yields two expressions for the integration constants B and C which may then be solved to find

$$B = \frac{t_2 - t_1}{x_2 - x_1} + \frac{q^*}{2k}(x_2 + x_1),$$

$$C = \frac{x_2 t_1 - x_1 t_2}{x_2 - x_1} - \frac{q^*}{2k}x_1 x_2.$$

Substitution of these results into the original equation and subsequent re-arrangement gives the following equation for the temperature distribution:

$$t = \left[t_1 + \frac{t_2 - t_1}{x_2 - x_1}(x - x_1) \right] + \left[\frac{q^*(x_2 - x_1)^2}{2k} \right] \left[\frac{x - x_1}{x_2 - x_1} - \left(\frac{x - x_1}{x_2 - x_1} \right)^2 \right]$$

$$(3.18)$$

One notes when comparing this result with that for the wall without genera-tion [Eq. (3.3)] that the linear temperature variation is altered by the presence of the second term involving the internal heat generation rate q^*. If one were to consider as an example, the instance in which $t_1 > t_2$, it may be noted that if the parameter

$$\frac{q^*(x_2 - x_1)^2}{2k(t_1 - t_2)} = \frac{q^*(\Delta x)^2}{2k(t_1 - t_2)} > 1, \qquad (3.19)$$

temperatures in excess of either t_1 or t_2 will occur within the wall. This is more easily seen, perhaps, by considering the heat flux. Since

$$\frac{q}{A} = -k\frac{dt}{dx},$$

Eq. (3.18) gives

$$\frac{q}{A} = k\frac{t_1 t_2}{x_2 - x_1} - q^*(x_2 - x_1)\left(\frac{1}{2} - \frac{x - x_1}{x_2 - x_1} \right). \qquad (3.20)$$

Once again one may note how the heat generation term modifies the uniform heat flux of the nongeneration case. It is instructive to observe the heat flux evaluated at each surface ($x = x_1$ and $x = x_2$):

$$\left(\frac{q}{A} \right)_{x_1} = k\frac{t_1 - t_2}{\Delta x} - \frac{q^* \Delta x}{2},$$

$$(3.21)$$

$$\left(\frac{q}{A} \right)_{x_2} = k\frac{t_1 - t_2}{\Delta x} + \frac{q^* \Delta x}{2}.$$

One sees that one-half of the total generated heat flows out each face of the wall—appropriately adding to or subtracting from the nongeneration heat flow, depending on the surface in question. It is also apparent that, for example, in the case when $t_1 > t_2$, the condition stated in Eq. (3.19) reverses the direction of heat flow at the face, $x = x_1$.

Uniformly Distributed Generation in a Cylinder

The case of uniformly distributed generation may be treated for the cylindrical geometry much in the same fashion as was just done for the plane wall. Using the same geometry and boundary conditions as for the case without generation shown in Fig. 3.3, Eq. (1.10) becomes

$$\frac{d^2t}{dr^2} + \frac{1}{r}\frac{dt}{dr} + \frac{q^*}{k} = 0,$$

$$\frac{d}{dr}\left(r\frac{dt}{dr}\right) + r\frac{q^*}{k} = 0. \tag{3.22}$$

The boundary conditions are

$$\text{At } r = r_1: \quad t = t_1.$$

$$\text{At } r = r_2: \quad t = t_2.$$

Twice integrating Eq. (3.22), one obtains

$$t = -\frac{r^2}{4}\frac{q^*}{k} + B\ln r + C. \tag{3.23}$$

Application of the boundary conditions yields the temperature distribution:

$$t = t_1 + (t_2 - t_1)\frac{\ln(r/r_1)}{\ln(r_2/r_1)} + \frac{q^*}{4k}\left[(r_2^2 - r_1^2)\frac{\ln(r/r_1)}{\ln(r_2/r_1)} - (r^2 - r_1^2)\right]. \tag{3.24}$$

Although the above expression might be written more concisely, it is left in this form in order that it may be compared to the result in which generation is absent [see Eq. (3.10)].

The heat flow, per unit length, may be found at the two surfaces by using

$$\frac{q}{L} = -k2\pi r\frac{dt}{dr}.$$

The results may be found to be

$$\left(\frac{q}{L}\right)_{r_1} = \frac{2\pi k(t_1 - t_2)}{\ln(r_2/r_1)} - \pi r_1^2 q^*\left[\frac{\frac{1}{2}\left(\frac{r_2^2}{r_1^2} - 1\right)}{\ln(r_2/r_1)} - 1\right], \tag{3.25}$$

$$\left(\frac{q}{L}\right)_{r_2} = \frac{2\pi k(t_1 - t_2)}{\ln(r_2/r_1)} + \pi r_1^2 q^*\left[1 - \frac{\frac{1}{2}\left(1 - \frac{r_1^2}{r_2^2}\right)}{\ln(r_2/r_1)}\right]. \tag{3.26}$$

As in the plane case, one may ascertain the magnitude of the generation rate which reverses the heat flow at r_1 when $t_1 > t_2$.

The special case that results when the cylinder is a solid bar is most easily found by returning to Eq. (3.23) and noting that t must remain finite as $r \to 0$. Hence $B = 0$. The remaining boundary condition of $t = t_2$ at $r = r_2$ gives

$$t = t_2 + \frac{q^* r_2^2}{4k} \left(1 - \frac{r^2}{r_2^2} \right). \tag{3.27}$$

The centerline temperature, call it t_0, is then

$$t_0 = t_2 + \frac{q^* r_2^2}{4k}. \tag{3.28}$$

The heat flow at the surface must necessarily be the total heat generated in the bar:

$$\left(\frac{q}{L} \right)_{\text{surf}} = \pi r_2^2 q^*. \tag{3.29}$$

Temperature-Dependent Generation in a Plane Wall

As a final example of conduction problems in which internal heat generation is present, the case of a plane wall in which the internally generated heat varies linearly with temperature will be considered. Figure 3.6(a) depicts the

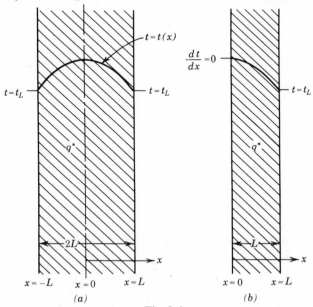

Fig. 3.6

case to be studied. The wall has a thickness $2L$ and symmetrically imposed boundary temperatures, t_L, at the two faces. If the x coordinate is measured from the centerline of the wall, the resulting temperature distribution will be symmetrical, and the problem may be simplified by the equivalent situation shown in Fig. 3.6(b), in which a wall of thickness L has a temperature t_L imposed at $x = L$ with the surface at $x = 0$ insulated ($dt/dx = 0$).

Let the linearly dependent generation follow the relation

$$q^* = q_L^*[1 + b(t - t_L)], \tag{3.30}$$

in which q_L^* represents the volumetric generation rate evaluated at the specified surface temperature, and b is a linear proportionality constant. Incorporating this into Eq. (3.17), one finds that the resulting differential equation and boundary conditions are

$$\frac{d^2t}{dx^2} + \frac{q_L^*}{k}[1 + b(t - t_L)] = 0.$$

$$\text{At } x = 0: \quad \frac{dt}{dx} = 0. \tag{3.31}$$

$$\text{At } x = L: \quad t = t_L.$$

This system of equations can be simplified by introducing the temperature difference variable

$$\theta = t - t_L, \tag{3.32}$$

so that

$$\frac{d^2\theta}{dx^2} + \frac{q_L^* b}{k}\theta + \frac{q_L^*}{k} = 0.$$

$$\text{At } x = 0: \quad \frac{d\theta}{dx} = 0. \tag{3.33}$$

$$\text{At } x = L: \quad \theta = 0.$$

The solution to the system of equations in Eq. (3.33) is

$$\theta = t - t_L = \frac{1}{b}\left(\frac{\cos mx}{\cos mL} - 1\right), \tag{3.34}$$

in which

$$m = \sqrt{b\frac{q_L^*}{k}}. \tag{3.35}$$

Equation (3.34) gives the steady temperature in the wall—varying from t_L at the surface to

$$t_0 = t_L + \frac{1}{b}\left(\frac{1}{\cos mL} - 1\right) \tag{3.36}$$

at the center. However, Eq. (3.34) makes sense only if $mL < \pi/2$, for when $mL = \pi/2$, infinite temperatures are indicated for all x. The physical interpretation of this fact is that there may be a temperature dependence of the generation rate (as given by the constant b) sufficiently large that the temperatures required to conduct the heat out the surface result in generation rates too great to be conducted away. Consequently, the temperatures and generation rates increase without bound unless

$$mL < \frac{\pi}{2} \quad \text{or} \quad bq_L^* < \left(\frac{\pi}{2L}\right)^2 k. \tag{3.37}$$

References 2 and 3 present additional examples involving internal heat generation for other geometries and other generation rate laws.

EXAMPLE 3.3: A copper rod (k = 220 Btu/hr-ft-°F) 0.2 in. in diameter and 1 ft long has its two ends maintained at 70°F. The lateral surface of the rod is perfectly insulated, so conduction may be taken as one-dimensional down the length of the rod. Find the maximum electrical current that the rod may carry if the temperature is not to exceed 250°F at any point and the electrical resistivity of the rod is
(a) constant at 1.73×10^{-6} ohm-cm;
(b) equal to $1.73[1 + 0.002(t - 70)] \times 10^{-6}$ ohm-cm.

Solution: (a) Since the resistance of a wire is the product of the resistivity and length divided by the cross-sectional area, one has

$$R = 1.73 \times 10^{-6} \times \frac{1}{2.54} \times \frac{12}{\pi(0.2)^2/4} = 2.60 \times 10^{-4} \text{ ohm.}$$

The conditions imposed meet the requirements of the first case considered above and as given by Eq. (3.18). Since the maximum temperature will occur at the midpoint of the rod, this equation gives

$$250 = \left[70 + \frac{70 - 70}{1}(0.5)\right] + \frac{q^*(1)^2}{2 \times 220}\left[\frac{0.5}{1} - \left(\frac{0.5}{1}\right)^2\right],$$

so that

$$q^* = 3.168 \times 10^5 \text{ Btu/hr-ft}^3.$$

Since

$$q^* = \frac{RI^2}{\text{volume}},$$

then

$$3.168 \times 10^5 = \frac{2.60 \times 10^{-4} \times I^2 \times 3.413}{[12 \times \pi \times (0.2)^2/4]/1728},$$

$$I = 279 \text{ A}.$$

(b) The given linear resistivity law yields a linear resistance:

$$R = 2.60 \times 10^{-4}[1 + 0.002(t - 70)].$$

The resulting heat generation will also depend linearly on temperature with the same rate constant, $b = 0.002$ 1/°F. Thus, the conditions of the last case considered above are met. The maximum temperature in that case is given by Eq. (3.36). Since L in that equation is the half-length of the rod, one has

$$250 = 70 + \frac{1}{0.002}\left[\frac{1}{\cos(0.5m)} - 1\right],$$

$$m = 1.489 \text{ 1/ft},$$

since

$$m = \sqrt{bq_L^*/k},$$

$$q_L^* = (1.489)^2\frac{220}{0.002} = 2.44 \times 10^5 \text{ Btu/hr-ft}^3.$$

The quantity q_L^* is the dissipated energy corresponding to the coefficient in the above resistance law. Thus, as in part (a),

$$q_L^* = 2.44 \times 10^5 = \frac{2.60 \times 10^{-4}I^2 \times 3.413}{[12 \times \pi \times (0.2)^2/4]/1728},$$

$$I = 245 \text{ A}.$$

3.8   Boundaries Surrounded by Fluids of Specified Temperatures

The examples presented thus far have been restricted to cases in which certain boundary temperatures of the bodies in question were assumed to be known. In many problems of practical importance this is not the case. Usually the configurations mentioned above are encountered in situations in which they separate fluids of different temperatures—these fluid temperatures being specified or known. If there is no fluid motion, the transfer of heat is done purely by conduction, and the methods discussed above may be employed while treating the fluid as a solid body.

If fluid motion does exist, as it invariably will because of either free or forced convection, the resulting thermal and velocity boundary layers cause a temperature difference to exist between the main bulk of the fluid (essentially at a uniform temperature) and the surface. This was discussed in Sec. 1.5. In that discussion the film coefficient, h, was defined in Eq. (1.14) in order to relate the rate of heat flow per unit area to the difference between the surface and fluid temperatures:

$$\frac{q}{A} = h(t_s - t_f). \tag{1.14}$$

Application of this definition will be made in the following sections for various geometrical configurations.

Since this chapter is meant to deal primarily with conduction, the mechanism of convection at the boundary of a solid is introduced mainly to illustrate the application of this important boundary condition to a conduction problem. The actual mechanism of heat convection is not of interest at this point. Hence, it will be assumed that the film coefficient h is a known quantity. Chapters 6 through 10 are devoted to a detailed analysis of convection and the determination of the coefficient h. In Chapters 12 and 13 the problems of combined conduction and convection will be reconsidered.

The Plane Wall Bounded by Fluids of Different Temperature

Figure 3.7 shows a plane wall (composed of two solid layers for purposes of discussion) bounded on each side by convecting fluids. By denoting the fluid regions of uniform temperature and the various junctures between different materials by numbers, the subscript notation introduced in Sec. 3.3 may be used to denote the thermal conductances and conductivities separating the numbered regions: $h_{12}, k_{23}, k_{34},$ and h_{45}.

The heat flow per unit wall area may be written for each "layer" by application of the defining relation for h given above and the equations obtained

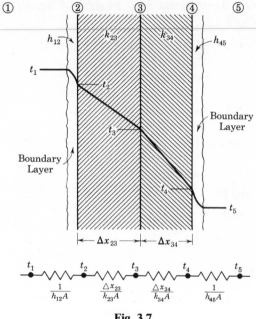

Fig. 3.7

in Sec. 3.3. Thus,

$$\frac{q}{A} = h_{12}(t_1 - t_2),$$

$$\frac{q}{A} = \frac{t_2 - t_4}{\dfrac{\Delta x_{23}}{k_{23}} + \dfrac{\Delta x_{34}}{k_{34}}},$$

$$\frac{q}{A} = h_{45}(t_4 - t_5).$$

Again, by combining these relations to eliminate t_2 and t_4, the following expression for the heat flux is obtained in terms of the over-all temperature difference and the thermal properties of the matter in between:

$$\frac{q}{A} = \frac{t_1 - t_5}{\dfrac{1}{h_{12}} + \dfrac{\Delta x_{23}}{k_{23}} + \dfrac{\Delta x_{34}}{k_{34}} + \dfrac{1}{h_{45}}}. \tag{3.38}$$

In keeping with the thermal resistance defined above,

$$\mathscr{R} = \frac{1}{CA},$$

a *convective* thermal resistance may be defined:

$$\mathscr{R}_c = \frac{1}{hA}.$$

The application of this definition, along with that already used for conduction resistances, to the network shown in Fig. 3.7 leads to the result given by Eq. (3.38).

EXAMPLE 3.4: The masonry wall of a building consists of an outer layer of facing brick ($k = 0.76$ Btu/hr-ft-°F) 4 in. thick, followed by a 6-in.-thick layer of common brick ($k = 0.40$ Btu/hr-ft-°F), followed by a $\frac{1}{2}$-in. layer of gypsum plaster ($k = 0.28$ Btu/hr-ft-°F). An outside coefficient of 5.5 Btu/hr-ft²-°F may be expected, and a coefficient of 1.4 Btu/hr-ft²-°F is a reasonable value to use for the inner surface of a ventilated room. Under these conditions, what will be the rate of heat gain per square foot of wall surface when the outside air is at 95°F and the inside air is conditioned to 72°F? What will be the temperature of the surface of the plaster wall?

Solution: From Eq. (3.38), the heat flow is

$$\frac{q}{A} = \frac{95 - 72}{\dfrac{1}{5.5} + \dfrac{4}{12 \times 0.76} + \dfrac{6}{12 \times 0.40} + \dfrac{0.5}{12 \times 0.28} + \dfrac{1}{1.4}} = 8.42 \text{ Btu/hr-ft}^2 \ (26.6 \text{ W/m}^2).$$

From the definition of h in Eq. (1.14), the inside surface temperature may be found from

$$\frac{q}{A} = 1.4(t_2 - 72) = 8.42,$$

$$t_2 = 78°F \ (25.6°C).$$

Cylindrical Surfaces Bounded by Fluids of Fixed Temperatures

A cylindrical surface separating fluids of different temperatures is one of immense practical importance since fluids are transported, heated, cooled, evaporated, and condensed in cylindrical pipes, tubes, and vessels. Such processes are encountered in almost every phase of engineering work involving fluids or heat transfer.

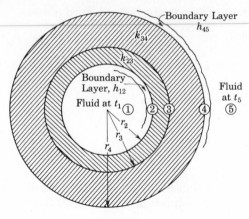

Fig. 3.8

The notation and numbering scheme introduced above is used in Fig. 3.8, which shows a double-layer pipe separating fluids of different fixed temperatures.

The rate of heat flow through the various layers of the configuration shown in Fig. 3.8 may be written by use of the equations of Sec. 3.5 and Eq. (1.14), remembering that the area through which the heat is convected is different at the two surfaces. Thus,

$$q = 2\pi L r_2 h_{12}(t_1 - t_2) = \frac{2\pi L(t_2 - t_4)}{\dfrac{\ln (r_3/r_2)}{k_{23}} + \dfrac{\ln (r_4/r_3)}{k_{34}}} = 2\pi L r_4(t_4 - t_5)h_{45}.$$

The above set of equations yields the following expression for the rate of heat flow through the cylinder per unit length:

$$\frac{q}{L} = \frac{2\pi(t_1 - t_5)}{\dfrac{1}{r_2 h_{12}} + \dfrac{\ln (r_3/r_2)}{k_{23}} + \dfrac{\ln (r_4/r_3)}{k_{34}} + \dfrac{1}{r_4 h_{45}}}. \tag{3.39}$$

EXAMPLE 3.5: A $\frac{3}{4}$-in., 18-gage brass condenser tube ($k = 66$ Btu/hr-ft-°F) is used to condense steam on its outer surface at 3 in. Hg abs. pressure. Circulating water at 65°F passes through the tube. If the inside and outside film coefficients are 300 Btu/hr-ft²-°F and 1500 Btu/hr-ft²-°F, respectively, find the pounds of steam condensed per hour per foot of tube length.

Solution: Appendix B gives the dimensions of a $\frac{3}{4}$-in., 18-gage tube to be

$$\text{O.D.} = 0.75 \text{ in., I.D.} = 0.652 \text{ in.}$$

The steam tables give, at 3 in. Hg abs. pressure:

$$\text{Saturation temperature} = 115.1°F.$$

$$\text{Latent heat of vaporization} = h_{fg} = 1028.6 \text{ Btu/lb}_m.$$

Equation (3.39) gives the heat removed per foot of length to be

$$\frac{q}{L} = \frac{2\pi(115.1 - 65)}{\dfrac{1}{300\dfrac{0.652}{2 \times 12}} + \dfrac{\ln\left(\dfrac{0.750}{0.652}\right)}{66} + \dfrac{1}{1500\dfrac{0.75}{2 \times 12}}}$$

$$= 2150 \text{ Btu/hr-ft (2067 W/m)}.$$

Thus, the hourly rate of steam condensed is

$$\frac{2150}{1028.6} = 2.09 \text{ lb}_m/\text{hr-ft} (8.64 \times 10^{-4} \text{ kg/s-m}).$$

3.9   The Critical Thickness of Pipe Insulation

An application of the above formulas having some practical significance is found in the case of insulation of small pipes or electrical wires. Given a pipe of fixed outside radius, let it be desired to find the thickness of an insulation which yields the optimum insulating effect. As insulation is added to the pipe, the outer surface will decrease in temperature, but at the same time the surface area for convective heat dissipation will be increased. It is possible that some optimum thickness of insulation exists due to these opposing effects.

For ease of analysis some simplifying assumptions will be made. Denote the fixed outside pipe radius by R, as shown in Fig. 3.9; r will be used to symbolize the insulation radius, so the thickness of insulation is $(r - R)$.

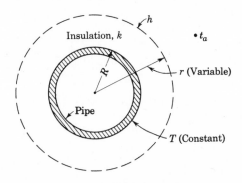

Fig. 3.9

As insulation is added, the temperature of the pipe surface (i.e., the temperature of the inner surface of the insulation) will be affected. However, if the thermal conductivity of the insulation is small compared to that of the pipe metal, as is often the case, the relative thermal resistance of the pipe will be so small that the pipe surface temperature will be essentially constant as r varies. Denote this constant by T.

The thermal conductivity of the insulation will be denoted by k, and the film coefficient due to convection at the outer boundary will be denoted by h. This latter quantity is dependent on the outer radius r as well as the outer surface temperature. For air, however, the value of h will not vary much as r is increased, so it may be treated as approximately constant.

The heat flow, per unit length of pipe, through the insulation is then

$$\frac{q}{L} = \frac{2\pi(T - t_a)}{\dfrac{1}{hr} + \dfrac{\ln(r/R)}{k}}, \qquad (3.40)$$

where t_a is the temperature of the surrounding air.

An optimum value of the heat loss may be found by setting the first derivative of q/L with respect to r equal to zero. When this is done, one finds that (ignoring the degenerate cases of $r = 0$ and $r = \infty$) this condition is satisfied when the radius r is equal to

$$r = r_c = \frac{k}{h}. \qquad (3.41)$$

The symbol r_c will be used to denote this "critical radius," and it should be noted that this "radius" is a quantity dependent only on the thermal quantities k and h.

However, if one now evaluates the second derivative of q/L at $r = r_c$, the result is

$$\left[\frac{d^2}{dr^2}\left(\frac{q}{L}\right)\right]_{r=r_c} = -2\pi(T - t_a)\left[\frac{\left(\dfrac{k}{hr} + \ln\dfrac{r}{R}\right)\left(\dfrac{2k}{hr} - 1\right) - 2\left(1 - \dfrac{k}{hr}\right)^2}{\dfrac{1}{rk}\left(\dfrac{k}{h} + r\ln\dfrac{r}{R}\right)^3}\right]_{r=r_c}$$

$$= -2\pi(T - t_a)\frac{\dfrac{h^2}{k}}{\left(1 + \ln\dfrac{r_c}{R}\right)^2}. \qquad (3.42)$$

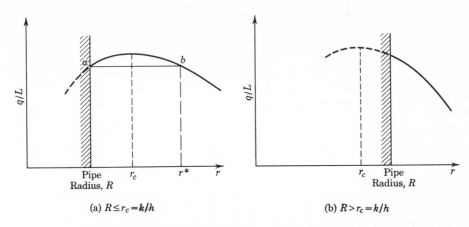

Fig. 3.10. The critical thickness of pipe insulation.

This is seen to be always negative. Hence, the optimum radius denoted by Eq. (3.41) is one of *maximum* heat loss, not minimum!

Thus, it appears possible to *increase* the heat loss from a pipe by the addition of insulation when the critical radius, k/h, is larger than the bare pipe radius. This is illustrated by the sketch in Fig. 3.10, wherein the heat loss is plotted against the insulation radius for two different cases.

Since the critical radius, $r_c = k/h$, is fixed by the thermal properties involved, it may be possible that for small pipes or wires the pipe radius, R, might be smaller than r_c. This is shown in Fig. 3.10(a). In such an instance, addition of insulation to the bare pipe (point a in the figure) causes the heat loss rate to increase until the critical radius is reached. Further increase in the insulating thickness will cause the heat loss rate to decrease from this peak value, but until a certain amount of insulation, denoted by r^* at b in Fig. 3.10(a), is added, the heat loss rate is still greater than that for the bare pipe. Thus, an insulation thickness in excess of $(r^* - R)$ must be added to reduce the heat loss below the uninsulated rate.

Figure 3.10(b) illustrates the situation typical of large pipes, in which the outside pipe radius, R, is larger than the critical radius, r_c, and any insulation added will decrease the heat loss.

3.10   The Over-all Heat Transfer Coefficient

In many instances it is customary to express the heat flow rate in the cases of the plane wall and cylinder (single or multilayered) with convection at the boundaries in terms of an "over-all conductance" or "over-all heat transfer coefficient." This over-all heat transfer coefficient, symbolized by U, is simply defined as a quantity such that the rate of heat flow through a configuration is given by taking a product of U, the surface area, and the over-all temperature difference:

$$q = UA(\Delta t)_{\text{over-all}}. \tag{3.43}$$

The dimensions of U are seen to be those of a conductance, and Btu/hr-ft²-°F are the units commonly used.

If one utilizes the equivalent resistance concept of the electrical networks introduced earlier, the over-all conductance U is simply related to the total resistance between the points at which the over-all potential is applied:

$$U = \frac{1}{\mathscr{R}_{\text{total}} A}.$$

The Plane Wall

The case of the plane wall is easily deduced from the discussion of Sec. 3.8. For a wall of n layers bounded on either side by fluids of temperatures t_1 and t_{n+3},

$$q = UA(t_1 - t_{n+3}),$$

$$U = \frac{1}{\dfrac{1}{h_{12}} + \dfrac{\Delta x_{23}}{k_{23}} + \dfrac{\Delta x_{34}}{k_{34}} + \cdots + \dfrac{1}{h_{(n+2)(n+3)}}}. \tag{3.44}$$

This representation is not essentially different from that already discussed, but it is found to be useful in the calculation of heat flow rates in the determination of heating or cooling needs of buildings. In such cases the over-all heat transfer coefficient of standard structural walls, floors, roofs, etc., may be tabulated for ready reference. This has been done, and the results are available in various handbooks.

EXAMPLE 3.6: The over-all heat transfer coefficient for the configuration of Example 3.4 is

$$U = \frac{1}{\dfrac{1}{5.5} + \dfrac{4}{12 \times 0.76} + \dfrac{6}{12 \times 0.40} + \dfrac{0.5}{12 \times 0.28} + \dfrac{1}{1.4}}$$

$$= 0.366 \text{ Btu/hr-ft}^2\text{-}°\text{F} \ (2.08 \text{ W/m}^2\text{-}°\text{C}).$$

The Cylinder

Much use is made of the over-all heat transfer coefficient for the cylindrical case in expressing the heat transfer capacity of heat exchangers. Frequently heat exchangers use a bundle of cylindrical tubes to provide the surface area for the transfer of heat between two fluids.

In the case of the cylindrical configuration, the over-all heat transfer coefficient depends on what surface area (inside or outside) is used in the defining relation given by Eq. (3.43), although the product UA is always the same. It is customary to use the outside surface area, and the associated over-all coefficient is sometimes called the "outside over-all heat transfer coefficient." This is best illustrated by use of the double-layered cylinder discussed in Sec. 3.8 and pictured in Fig. 3.8. If U_4 is used to denote the over-all heat transfer coefficient based on the outside area A_4 of radius r_4, then (see Fig. 3.8)

$$q = U_4 A_4 (t_1 - t_5). \tag{3.45}$$

Since $A_4 = 2\pi r_4 L$, Eqs. (3.39) and (3.45) combine to give the following expression for U_4:

$$U_4 = \frac{1}{\dfrac{r_4}{r_4 h_{12}} + \dfrac{r_4 \ln (r_3/r_2)}{k_{23}} + \dfrac{r_4 \ln (r_4/r_3)}{k_{34}} + \dfrac{1}{h_{45}}}. \tag{3.46}$$

It is not difficult to deduce the following general expression for a cylinder of n layers, having a fluid at temperature t_1 flowing inside and a fluid of temperature t_{n+3} on the outside:

$$U_{n+2} = \frac{1}{\dfrac{r_{n+2}}{r_2 h_{12}} + \dfrac{r_{n+2} \ln (r_3/r_2)}{k_{23}} + \dfrac{r_{n+2} \ln (r_4/r_3)}{k_{34}} + \cdots + \dfrac{1}{h_{(n+2)(n+3)}}}.$$

EXAMPLE 3.7: The over-all heat transfer coefficient, based on the outside surface area, of the condenser tube illustrated in Example 3.5 is

$$U_3 = \frac{1}{\dfrac{0.750}{0.652 \times 300} + \dfrac{0.750}{2 \times 12 \times 66} \ln \left(\dfrac{0.750}{0.652}\right) + \dfrac{1}{1500}}$$

$$= 218 \text{ Btu/hr-ft}^2\text{-}°\text{F} \ (1238 \text{ W/m}^2\text{-}°\text{C}).$$

3.11   Extended Surfaces

When it is desired to increase the heat removal between a structure and a surrounding ambient fluid, it is common practice to utilize "extended surfaces" attached to the primary surface. In such instances the extended surfaces are provided to increase artificially the surface area of heat transmission, although the average surface temperature may be decreased by so doing. If the surface is proportioned properly, the net result will be an increase in the heat transmission rate between the structure and the ambient fluid.

The uses of extended surfaces in applications of practical importance are numerous. Examples may be found in the cooling fins of air-cooled engines, the fin extensions to the tubes of radiators and other heat exchangers, the "pins" or "studs" attached to boiler tubes, etc. The extended surface applications noted above are all cases in which one purposely wishes to increase the rate of heat exchange between a source and an ambient fluid. Similar extended surface configurations may occur in other instances where the exchange of heat with the ambient fluid may be a disadvantage. Such instances are encountered in the measurement of temperature—the conduction of heat along thermocouple wires attached to a heated surface being an example.

An extended surface configuration is generally classed as a straight fin, an annular fin, or a spine. The term *straight fin* is applied to the extended surface attached to a wall which is otherwise plane, whereas an *annular fin* is one attached, circumferentially, to a cylindrical surface. A *spine* or *pin fin* is an extended surface of cylindrical or conical shape. These definitions are illustrated in Figs. 3.11 and 3.12.

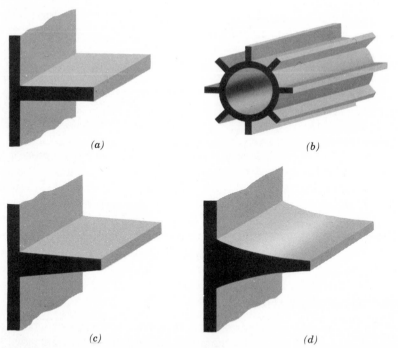

(a)   *(b)*

(c)   *(d)*

Fig. 3.11. Examples of extended surfaces: (a) and (b), straight fins of uniform thickness; (c) and (d), straight fins of nonuniform thickness.

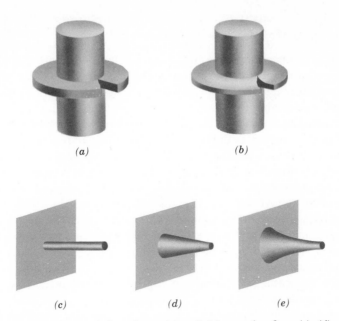

(a)   (b)

(c)   (d)   (e)

Fig. 3.12. Examples of extended surfaces: (a) and (b), annular fins; (c), (d), and (e), spines.

The basic problem with which the designer is faced is: Given a fin of a certain configuration and size attached to a surface of a fixed temperature (i.e., the fin base temperature) and surrounded by a fluid of fixed temperature, what is the rate of heat dissipated by the fin, and what is the variation in the temperature of the fin as one proceeds from the base to the tip?

In most instances the fin proportions used in practice are such that the length of the fin (its dimension measured normal to the primary surface to which it is attached) is large compared to its maximum thickness. When this is the case, one may assume that the temperature of the fin depends on only the single coordinate measured in the direction of the fin's length. For example, if a cylindrical rod protruding from a heated wall is very long compared to its diameter, one may assume that the temperature is uniform over any cross section taken normal to the axis. Hence, the temperature in the rod depends only on the axial coordinate measured along the rod. Similarly, if an annular fin is sufficiently slender, the distribution of the temperature may be taken to depend only on the radial coordinate.

If the above simplification can be made, the problem for the solution of the temperature distribution and heat flux rate becomes a one-dimensional conduction problem. A few shapes of practical importance will be considered in detail in this chapter. Only the steady state case will be considered.

3.12   The Straight Fin of Uniform Thickness and the Spine of Uniform Cross Section

The straight fin of uniform thickness and the spine of uniform cross section may be treated identically since in either case the cross-sectional area for heat flow (normal to the length coordinate) is constant. Also, in either case, the exposed surface area for heat convection is a linear function of the distance measured along the length; i.e., the perimeter of the cross section is constant.

Figure 3.13 illustrates these two configurations and shows the notation to be used in the following analysis. The symbol L will denote the length of the fin, k the thermal conductivity, h the film coefficient at the exposed surface, t_0 the fixed temperature at the fin base, and t_f the temperature of the bulk of the ambient fluid. The coordinate distance along the fin length will be symbolized by x. The symbols A and C will be used to denote the area of a cross section normal to x and the perimeter of this section, respectively.

Usually when dealing with straight fins it is customary to neglect heat losses from the side edges of the fin and to express the heat flow, etc., per unit width. In this case, as illustrated in Fig. 3.13, the perimeter C will be 2 and the cross-sectional area will equal the thickness, w. If the one-dimensional condition prevails, one does not need to neglect edge heat losses, in which case one would have $C = 2$ (width + thickness).

If one now considers an element of the fin at a distance x (thickness δx), an energy balance may be made on this element (see Fig. 3.13). Adopt the following notation:

$q_1 =$ heat conducted *into* the element at x,

$q_2 =$ heat conducted *out of* the element at $x + \delta x$,

$q_3 =$ heat convected *out of* the element between x and $x + \delta x$.

Conservation of energy requires, in the steady state, that

$$q_1 = q_2 + q_3.$$

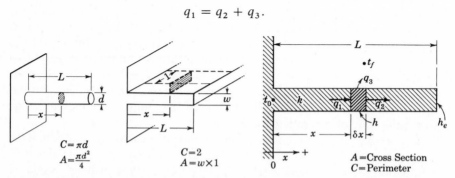

Fig. 3.13. The straight fin of uniform thickness or spine of constant cross section.

From Fourier's law and the definition of the film coefficient h, the equation above is

$$-kA\left(\frac{dt}{dx}\right)_x = -kA\left(\frac{dt}{dx}\right)_{x+\delta x} + h(C \cdot \delta x)(t_{av} - t_f), \qquad (3.47)$$

where t_{av} denotes the average temperature of the element. Making a Taylor's expansion of the temperature gradient at $(x + \delta x)$ in terms of that at x, one finds that

$$\left(\frac{dt}{dx}\right)_{x+\delta x} = \left(\frac{dt}{dx}\right)_x + \frac{d}{dx}\left(\frac{dt}{dx}\right)_x \delta x + \frac{d^2}{dx^2}\left(\frac{dt}{dx}\right)_x \frac{(\delta x)^2}{2!} + \cdots.$$

By making this substitution in Eq. (3.47) and dividing by the element volume, $\delta x \cdot A$, one obtains the following expression of the energy balance per unit volume:

$$-\frac{k}{\delta x}\left(\frac{dt}{dx}\right)_x = -\frac{k}{\delta x}\left(\frac{dt}{dx}\right)_x - k\left(\frac{d^2t}{dx^2}\right)_x - k\frac{d^3t}{dx^3}\frac{\delta x}{2} + \cdots + \frac{hC}{A}(t_{av} - t_f).$$

This conservation of energy must be satisfied for all elements of the fin, so upon allowing $\delta x \to 0$ (noting that $t_{av} \to t$, the temperature at x), one obtains the following differential equation for the distribution of temperature in the fin:

$$\frac{d^2t}{dx^2} - \frac{hC}{kA}(t - t_f) = 0.$$

This equation must be satisfied at every point in the fin, and if the film coefficient and the thermal conductivity may be assumed constant, it can be easily solved.

The equation may be put in a more concise form by defining a temperature difference variable θ:

$$\theta = t - t_f$$

and a parameter m:

$$m = \sqrt{\frac{hC}{kA}}. \qquad (3.48)$$

Then the differential equation is

$$\frac{d^2\theta}{dx^2} - m^2\theta = 0.$$

The general solution is

$$\theta = B e^{-mx} + D e^{mx}. \tag{3.49}$$

The B and D in Eq. (3.49) are arbitrary constants determined by the boundary conditions imposed at the ends of the fin. At the base of the fin, the temperature is fixed at t_0 if the fin material is integral with the material of the primary surface. If, as is sometimes the case, the fin is a separate piece attached to the primary surface, some contact resistance may exist at the fin base and its temperature may be different from that of the primary surface. In such a case t_0 should be interpreted as the fin base temperature, not the temperature of the primary surface. At the outer end of the fin, convection is taking place. In other instances some other condition may be imposed—such as another fixed temperature if the fin extends between two heat sources.

For the present analysis let the conditions be those indicated by Fig. 3.13—an imposed temperature at the base and convection at the free end. Then the conditions determining B and D are [with reference to Eq. (1.15)],

$$At\ x = 0: \quad t = t_0.$$

$$At\ x = L: \quad -k\frac{dt}{dx} = h_e(t - t_f).$$

It should be noted that the boundary condition of convection at the end of the fin as stated above assumes that the film coefficient at the end, h_e, is different from that at the other surface, h. This is done for two reasons. First, as will be shown later, the convective film coefficient depends on the orientation of the surface, and it is quite likely that the coefficient at the find end will differ from that on the other surfaces. Second, this distinction between the film coefficient will permit easy simplification of future results to the special case in which the heat convected out the end is considered negligible by merely setting $h_e = 0$.

Written in terms of the temperature difference variable, θ, defined above, these boundary conditions are:

$$At\ x = 0: \quad \theta = \theta_0 = t_0 - t_f.$$

$$At\ x = L: \quad \frac{d\theta}{dx} = -\frac{h_e}{k}\theta. \tag{3.50}$$

The constants B and D of Eq. (3.49) may be found by use of the conditions of Eq. (3.50). The application of these conditions gives the following two equations:

$$At \; x = 0: \quad \theta_0 = B + D.$$

$$At \; x = L: \quad m(-B \, e^{-mL} + D \, e^{mL}) = -\frac{h_e}{k}(B \, e^{-mL} + D \, e^{mL}).$$

Since m and L are constants, B and D may be found from these two expressions. The results are

$$D = \theta_0 - B,$$

$$B = \frac{\theta_0 \left(1 + \dfrac{h_e}{km}\right) e^{mL}}{(e^{mL} + e^{-mL}) + \dfrac{h_e}{km}(e^{mL} - e^{-mL})}.$$

Substitution of these results into Eq. (3.49) gives the following expression for the distribution of the temperature along the length of the fin:

$$\frac{\theta}{\theta_0} = \frac{t - t_f}{t_0 - t_f} = \frac{[e^{m(L-x)} + e^{-m(L-x)}] + \dfrac{h_e}{km}[e^{m(L-x)} - e^{-m(L-x)}]}{(e^{mL} + e^{-mL}) + \dfrac{h_e}{km}(e^{mL} - e^{-mL})}.$$

$$(3.51)$$

This is, perhaps, more conveniently expressed in terms of the hyperbolic functions:

$$\frac{\theta}{\theta_0} = \frac{t - t_f}{t_0 - t_f} = \frac{\cosh m(L - x) + H \sinh m(L - x)}{\cosh mL + H \sinh mL}, \qquad (3.52)$$

where

$$H = \frac{h_e}{km} \quad \text{and} \quad m = \sqrt{\frac{hC}{kA}}.$$

The rate of heat flow from the fin could be evaluated by integrating the convected heat [i.e., q_3 in Eq. (3.47)] over the fin surface. However, the result is obtained more directly by noting the fact that all the heat dissipated by the fin must be *conducted* past the point where $x = 0$. Thus, if q symbolizes the heat flow rate,

$$q = -kA \left(\frac{dt}{dx} \right)_{x=0}.$$

Introduction of the temperature distribution from Eq. (3.52) yields

$$q = kmA\theta_0 \frac{\sinh mL + H \cosh mL}{\cosh mL + H \sinh mL}. \tag{3.53}$$

In the development of these relations for the temperature distribution and heat dissipation, it was assumed that the fin was slender enough for a one-dimensional condition to prevail. If this condition is met, it is very likely that the amount of heat convected out the end of the fin is a small fraction of the heat convected out the other surfaces. If this is the case, the above relations may be simplified by neglecting this heat loss at the end. This implies that $h_e = 0$, or $H = 0$. Then Eqs. (3.52) and (3.53) reduce to

$$\frac{\theta}{\theta_0} = \frac{t - t_f}{t_0 - t_f} = \frac{\cosh m(L - x)}{\cosh mL}, \tag{3.54}$$

$$q = kmA\theta_0 \tanh mL. \tag{3.55}$$

An empirical correction that is often made to these simplified relations to allow for the heat loss from the end is to use, instead of the true fin length, an apparent length which gives a lateral surface area equal to the true lateral surface area plus the cross-sectional area of the fin end. That is, for cylindric rods and straight fins of uniform thickness, the apparent fin lengths, L', to use are

Cylindric rod: $L' = $ (true length) $+ \frac{1}{2}$(radius).

Straight fin: $L' = $ (true length) $+ \frac{1}{2}$(fin width).

It should be pointed out that the simplified approach of using this apparent length L' applies only to the evaluation of the heat flow rate, q.

EXAMPLE 3.8: Three rods, one made of glass ($k = 0.63$ Btu/hr-ft-°F), one pure aluminum ($k = 132$ Btu/hr-ft-°F), and one wrought iron ($k = 33$ Btu/hr-ft-°F) all $\frac{1}{2}$ in. in diameter, 12 in. long, and are heated to a temperature of 250°F on one end. If the rods extend into air at 70°F, and if the surface film coefficient is known to be 1.6 Btu/hr-ft²-°F, find (a) the distribution of temperature in the rods if the heat flux from the ends may be neglected; (b) the total heat flow from the rods under the same conditions; and (c) the heat flow rate from the rods, if the heat flux from the ends is not neglected. For purposes of comparison, take the film coefficient at the ends to be equal to that on the lateral surface, 1.6 Btu/hr-ft^2-°F.

Solution:

(a) For a cylindrical rod, the parameter $m = \sqrt{hC/kA} = \sqrt{4h/kd}$.
For glass:

$$m = \sqrt{\frac{4 \times 1.6}{0.63 \times \dfrac{1}{2 \times 12}}} = 15.61 \frac{1}{\text{ft}}.$$

For iron:

$$m = 2.158 \frac{1}{\text{ft}}.$$

For aluminum:

$$m = 1.079 \frac{1}{\text{ft}}.$$

Since $L = 12$ in. $= 1$ ft, the temperature distributions are given by Eq. (3.54).
For glass:

$$\frac{t - 70}{250 - 70} = \frac{\cosh\,[15.61(1 - x)]}{\cosh\,(15.61 \times 1)}.$$

For iron:

$$\frac{t - 70}{250 - 70} = \frac{\cosh\,[2.158(1 - x)]}{\cosh\,(2.158 \times 1)}.$$

For aluminum:

$$\frac{t - 70}{250 - 70} = \frac{\cosh\,[1.079(1 - x)]}{\cosh\,(1.079 \times 1)}.$$

These distributions are plotted in Fig. 3.14 for the sake of comparison.

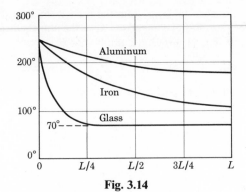

Fig. 3.14

(b) Equation (3.55) gives the heat flow rates to be

For glass:

$$q = (0.63)(15.61)\frac{\pi(1/2)^2}{4 \times 144}(250 - 70)\tanh(15.61 \times 1)$$

$$= 2.413 \tanh(15.61)$$

$$= 2.413 \text{ Btu/hr } (0.707 \text{ W}).$$

For iron:

$$q = (33)(2.158)\frac{\pi(1/2)^2}{4 \times 144}(250 - 70)\tanh(2.158 \times 1)$$

$$= 17.48 \tanh(2.158)$$

$$= 17.01 \text{ Btu/hr } (4.99 \text{ W}).$$

For aluminum:

$$q = (132)(1.079)\frac{\pi(1/2)^2}{4 \times 144}(250 - 70)\tanh(1.079 \times 1)$$

$$= 34.95 \tanh(1.079)$$

$$= 27.71 \text{ Btu/hr } (8.12 \text{ W}).$$

(c) If the heat loss out of the end is to be accounted for, one needs the parameter H defined in Eq. (3.52):

$$H = \frac{h_e}{km}.$$

Hence, since $h_e = h = 1.6 \text{ Btu/hr-ft}^2\text{-}°\text{F}$:

For glass:

$$H = \frac{1.6}{0.63 \times 15.61} = 0.1627 \text{ (dimensionless)}.$$

For iron:

$$H = \frac{1.6}{33 \times 2.158} = 0.0225.$$

For aluminum:

$$H = \frac{1.6}{132 \times 1.079} = 0.0112.$$

Thus, Eq. (3.53) for the heat flow gives

For glass:

$$q = 2.413 \frac{\sinh(15.61 \times 1) + 0.1627 \cosh(15.61 \times 1)}{\cosh(15.61 \times 1) + 0.1627 \sinh(15.61 \times 1)}$$

$$= 2.413 \text{ Btu/hr } (0.707 \text{ W}).$$

For iron:

$$q = 17.48 \frac{\sinh(2.158 \times 1) + 0.0225 \cosh(2.158 \times 1)}{\cosh(2.158 \times 1) + 0.0225 \sinh(2.158 \times 1)}$$

$$= 17.14 \text{ Btu/hr } (5.02 \text{ W}).$$

For aluminum:

$$q = 34.95 \frac{\sinh(1.079 \times 1) + 0.0112 \cosh(1.079 \times 1)}{\cosh(1.079 \times 1) + 0.0112 \sinh(1.079 \times 1)}$$

$$= 27.85 \text{ Btu/hr } (8.16 \text{ W}).$$

3.13   Extended Surfaces of Nonuniform Cross Section— General Considerations

Some of the other extended surfaces shown in Figs. 3.11 and 3.12 differ from the straight fin of uniform thickness or the spine of uniform cross section in that the cross-sectional area for heat conduction will not be constant, nor will the surface area for convection to an ambient fluid vary linearly with the distance from the fin base. For example, the annular fin of uniform thickness (for radial conduction) has a cross-sectional area which varies linearly with the radius and a surface area which varies with the square of the radius.

The differential equation of the temperature distribution in these extended surfaces may be deduced in the same manner as was done above (i.e., by making an energy balance on an element of the fin), but certain general facts may be deduced before treating specific configurations. As before, a one-dimensional problem will result if the analysis is limited to those fins in which the thickness is quite small compared to the fin length. Imagine, then, an extended surface (geometry unspecified, for the present) so proportioned that the heat conduction within may be treated as one-dimensional. That is, let the temperature be expressible as a function of a single coordinate, say x. Such an arrangement is depicted in Fig. 3.15. In general, the cross-sectional area (normal to x) for heat conduction will be a function of x, as will be the surface area for heat convection to the surroundings. Let $A(x)$ represent the cross-sectional area at any value of x and let $S(x)$ represent the surface area exposed for convection between $x = 0$ and $x = x$.

Select now an element of the fin contained between the cross sections at x, and $x + \delta x$. Thus, δx is the thickness of this finite element. Denote, as before, the heat conducted into the element across the area at x by q_1, the heat conducted out of the element across the area at $x + \delta x$ by q_2 and the heat convected out of the surface of the element and into the surrounding fluid by q_3. Then the requirement of the steady state gives

$$q_1 = q_2 + q_3.$$

The conducted heat, q_1 or q_2, depends on the temperature gradient in the x direction and on the cross-sectional area. The area is itself a function of x, and hence the conducted heat is purely a function of x. Thus, one can express q_2 in terms of q_1 by use of Taylor's expansion:

$$q_2 = q_1 + \frac{d}{dx}(q_1)\,\delta x + \frac{d^2}{dx^2}(q_1)\frac{(\delta x)^2}{2!} + \cdots.$$

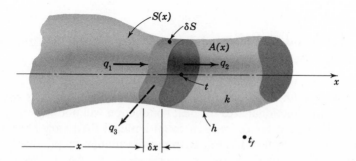

Fig. 3.15. A generalized extended surface.

Thus, the heat balance above becomes

$$\frac{d}{dx}(q_1)\,\delta x + \frac{d^2}{dx^2}(q_1)\frac{(\delta x)^2}{2!} + \cdots + q_3 = 0. \tag{3.56}$$

By again using θ to denote the difference between the fin temperature and the surrounding fluid temperature, Eq. (3.56) is expressible as

$$\frac{d}{dx}\left(-kA\frac{d\theta}{dx}\right)\delta x + \frac{d^2}{dx^2}\left(-kA\frac{d\theta}{dx}\right)\frac{(\delta x)^2}{2!} + \cdots + h\theta(\delta S) = 0,$$

where δS denotes the exposed surface area of the chosen element.

In the above expression h and k, as usual, denote the convective film coefficient and the thermal conductivity. It should be noted that A and S are to be treated as functions of the coordinate x. Rewriting the above heat balance on a per-unit-volume basis, one finds that

$$\frac{1}{A}\frac{d}{dx}\left(-kA\frac{d\theta}{dx}\right) + \frac{1}{A}\frac{d^2}{dx^2}\left(-kA\frac{d\theta}{dx}\right)\frac{\delta x}{2} + \cdots + \frac{h\theta}{A}\frac{\delta S}{\delta x} = 0.$$

Now, this expression must be satisfied for *all* elements, no matter how large or small. Hence, it must be satisfied as $\delta x \to 0$. Allowing, then, $\delta x \to 0$ in this equation, one obtains

$$\frac{1}{A}\frac{d}{dx}\left(-kA\frac{d\theta}{dx}\right) + \frac{h\theta}{A}\frac{dS}{dx} = 0,$$

or

$$\frac{d^2\theta}{dx^2} + \frac{1}{A}\frac{dA}{dx}\frac{d\theta}{dx} - \frac{h}{k}\frac{1}{A}\frac{dS}{dx}\theta = 0. \tag{3.57}$$

This equation may be applied generally to all extended surface configurations for which the one-dimensional assumption is valid. For example, if the special case of the straight fin or uniform spine treated in Sec. 3.12 is considered, $A = $ constant and $S = C \cdot x$. In this instance Eq. (3.57) becomes

$$\frac{d^2\theta}{dx^2} - \frac{hC}{kA}\theta = 0.$$

This equation is identical to the equation obtained earlier.

This expression for the general extended surface [Eq. (3.57)] will now be applied to certain other configurations of practical importance.

3.14   The Annular Fin of Uniform Thickness

As mentioned earlier, a second extended surface configuration of considerable engineering importance is that of an annular fin of constant thickness attached circumferentially to a circular cylinder. Such a configuration is found to be utilized on certain liquid-to-gas heat exchanger tubes and on the cylinders of air-cooled engines.

Figure 3.16 shows the notation to be used in the analysis of an annular fin. The symbols r_b and r_e denote the radii of the base and end of the fin, respectively, whereas w denotes the constant fin thickness. By assuming that there is symmetry of the temperature distribution with respect to a circumferential coordinate and that $w \ll (r_e - r_b)$, the heat conduction within the fin may be taken to depend on the radial coordinate r only.

The cross-sectional area and surface area are functions of this radial coordinate:

$$A = 2\pi r w,$$

$$S = 2[\pi(r^2 - r_b^2)].$$

Thus, Eq. (3.57) leads to the following expression for the temperature distribution as a function of r:

$$\frac{d^2\theta}{dr^2} + \left(\frac{1}{2\pi r w} \times 2\pi w\right)\frac{d\theta}{dr} - \frac{h}{k}\left(\frac{1}{2\pi r w} \times 4\pi r\right)\theta = 0,$$

$$\frac{d^2\theta}{dr^2} + \frac{1}{r}\frac{d\theta}{dr} - \frac{2h}{kw}\theta = 0. \qquad (3.58)$$

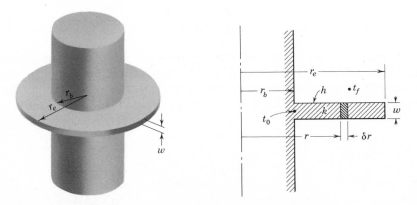

Fig. 3.16. The annular fin of uniform thickness.

This is recognized as Bessel's equation of zero order. A summary of the essential features of this equation and of the properties of the functions which are its solutions is given in Appendix C. Appendix C shows that the solution to Eq. (3.58) is

$$\theta = BI_0(nr) + CK_0(nr),\tag{3.59}$$

in which

$$n = \sqrt{\frac{2h}{kw}},$$

I_0 = modified Bessel function, first kind,

K_0 = modified Bessel function, second kind,

B, C = constants.

The constants B and C are determined, as usual, by the conditions imposed at r_b and r_e. For the case of a fin heated uniformly at its base to a temperature t_0 and for which the heat loss out its end may be neglected, one has

$$At\ r = r_b: \quad t = t_0 \quad or \quad \theta = \theta_0 = t_0 - t_f.$$

$$At\ r = r_e: \quad \frac{dt}{dr} = \frac{d\theta}{dr} = 0.$$

Referring again to Appendix C., one finds that the application of these boundary conditions to Eq. (3.59) leads to the following two expressions for the constants B and C:

$$\theta_0 = BI_0(nr_b) + CK_0(nr_b),$$

$$\left(\frac{d\theta}{dr}\right)_{r_e} = 0 = nBI_1(nr_e) - nCK_1(nr_e).$$

When these are solved for B and C, one obtains

$$B = \frac{\theta_0 K_1(nr_e)}{I_0(nr_b)K_1(nr_e) + I_1(nr_e)K_0(nr_b)},$$

$$C = \frac{\theta_0 I_1(nr_e)}{I_0(nr_b)K_1(nr_e) + I_1(nr_e)K_0(nr_b)}.$$

Substitution of these expressions into the general solution, Eq. (3.59), leads to the following equation for the distribution of the fin temperature as a function of the radius r:

$$\frac{\theta}{\theta_0} = \frac{I_0(nr)K_1(nr_e) + K_0(nr)I_1(nr_e)}{I_0(nr_b)K_1(nr_e) + K_0(nr_b)I_1(nr_e)}. \tag{3.60}$$

The rate of heat dissipation by the fin may be found by evaluating the rate at which heat is conducted past the base radius:

$$q = -k\left(A\frac{d\theta}{dr}\right)_{r=r_b}.$$

From Eq. (3.60) and the rules of differentiation of Bessel functions as outlined in Appendix C, this becomes

$$q = 2\pi knw\theta_0 r_b \frac{K_1(nr_b)I_1(nr_e) - I_1(nr_b)K_1(nr_e)}{K_0(nr_b)I_1(nr_e) + I_0(nr_b)K_1(nr_e)}. \tag{3.61}$$

These equations for the temperature distribution and the heat flow rate may be written in a dimensionless form for a more general presentation. Since the problem is a one-dimensional one, these dimensionless expressions for the temperature and heat flow can be obtained in terms of only a dimensionless film coefficient parameter, a dimensionless size parameter, and a dimensionless coordinate parameter. For this purpose the following symbols will be used for these parameters for the present problem of the annular fin and for the other configurations to be treated in this chapter:

β = a dimensionless film coefficient parameter

α = a dimensionless size parameter

η = a dimensionless coordinate parameter

In particular, for the annular fin, define these parameters in the following way:

$$\beta_{an} = nr_e = \sqrt{\frac{2hr_e^2}{kw}},$$

$$\alpha_{an} = \frac{r_b}{r_e}, \tag{3.62}$$

$$\eta_{an} = \frac{r}{r_e}.$$

where the subscript "an" denotes the annular fin.

Equation (3.60) may now be expressed dimensionlessly in terms of these parameters:

$$\frac{\theta}{\theta_0} = \frac{I_0(\beta_{an}\eta_{an})K_1(\beta_{an}) + K_0(\beta_{an}\eta_{an})I_1(\beta_{an})}{I_0(\beta_{an}\alpha_{an})K_1(\beta_{an}) + K_0(\beta_{an}\alpha_{an})I_1(\beta_{an})}. \tag{3.63}$$

For specific problems the parameters β_{an}, α_{an}, and η_{an} may be calculated from the given fin geometry and the imposed temperatures. Then the temperature distribution and heat flow may be evaluated by use of the equations above and the tables of Bessel functions given in Appendix C.

For rapid calculations, sufficiently accurate for most engineering work, the temperature distribution may be obtained from charts which represent Eq. (3.63). At first examination one would imagine that a graphical presentation of Eq. (3.63) would involve the four parameters θ/θ_0, β_{an}, α_{an}, and η_{an}, requiring a rather extensive number of charts. However, one parameter may be eliminated (thus reducing the graphical presentation to a single chart) in the following way. A characteristic temperature of the fin is that at the free end and may be denoted by θ_e. Then $\theta = \theta_e$ at $r = r_e$ or $\eta_{an} = 1$. Then Eq. (3.63) gives

$$\frac{\theta_e}{\theta_0} = \frac{I_0(\beta_{an})K_1(\beta_{an}) + K_0(\beta_{an})I_1(\beta_{an})}{I_0(\beta_{an}\alpha_{an})K_1(\beta_{an}) + K_0(\beta_{an}\alpha_{an})I_1(\beta_{an})}. \tag{3.64}$$

Dividing this by Eq. (3.63), one obtains

$$\frac{\theta_e}{\theta} = \frac{I_0(\beta_{an})K_1(\beta_{an}) + K_0(\beta_{an})I_1(\beta_{an})}{I_0(\beta_{an}\eta_{an})K_1(\beta_{an}) + K_0(\beta_{an}\eta_{an})I_1(\beta_{an})}. \tag{3.65}$$

Now, Eqs. (3.64) and (3.65) are of identical form—θ replacing θ_0 and η replacing α. By defining a function $G_1(\gamma, \beta)$,

$$G_1(\gamma, \beta) = \frac{I_0(\beta)K_1(\beta) + K_0(\beta)I_1(\beta)}{I_0(\beta\gamma)K_1(\beta) + K_0(\beta\gamma)I_1(\beta)},$$

the two equation above are

$$\frac{\theta_e}{\theta_0} = G_1(\alpha_{an}, \beta_{an}) \tag{3.66}$$

and

$$\frac{\theta_e}{\theta} = G_1(\eta_{an}, \beta_{an}), \qquad \alpha < \eta < 1. \tag{3.67}$$

The function $G_1(\gamma, \beta)$ is plotted in Fig. 3.17. The temperature distribution in a given annular fin can be determined from this chart in the following way:

(a) With the known geometry, thermal properties, and imposed temperatures, θ_0, α_{an}, and β_{an} may be calculated from their definitions in Eq. (3.62).

(b) Interpreting γ as the parameter α_{an}, one can use Fig. 3.17 to obtain the end temperature θ_e from Eq. (3.66).

(c) With θ_e known, Eq. (3.67) shows that Fig. 3.17 may be used to evaluate θ at any radius between r_b and r_e by interpreting γ to represent $\eta_{an} = r/r_e$, allowing η_{an} to vary between α_{an} and 1.

Equation (3.61) for the heat flow rate may also be conveniently expressed in terms of α_{an}, β_{an}, and η_{an} by nondimensionalizing q in the following way:

$$\frac{q}{\pi(1 - \alpha_{an}^2)kw\theta_0}.$$

Thus, by employing this and the definitions given in Eq. (3.62), Eq. (3.61) may be written

$$\frac{q}{\pi(1 - \alpha_{an}^2)kw\theta_0} = \beta_{an}^2 \left[\frac{2\alpha_{an}}{\beta_{an}(1 - \alpha_{an}^2)} \frac{K_1(\beta_{an}\alpha_{an})I_1(\beta_{an}) - I_1(\beta_{an}\alpha_{an})K_1(\beta_{an})}{K_1(\beta_{an})I_0(\beta_{an}\alpha_{an}) + I_1(\beta_{an})K_0(\beta_{an}\alpha_{an})} \right]$$

$$= \beta_{an}^2 \times G_2(\alpha_{an}, \beta_{an}), \tag{3.68}$$

in which the function $G_2(\alpha, \beta)$ is defined as

$$G_2(\alpha, \beta) = \frac{2\alpha}{\beta(1 - \alpha^2)} \frac{K_1(\beta\alpha)I_1(\beta) - I_1(\beta\alpha)K_1(\beta)}{K_1(\beta)I_0(\beta\alpha) + I_1(\beta)K_0(\beta\alpha)}. \tag{3.69}$$

This function, $G_2(\alpha, \beta)$, is plotted in Fig. 3.18, from which the heat flow rate may be readily determined using Eq. (3.68).

3.15 The Straight Fin of Triangular Profile

As a second example of a fin with a nonuniform cross section, consider a straight fin having a tapered or trapezoidal profile. This is illustrated in Fig. 3.19, wherein the dimensions and coordinate system are shown. It is more convenient to place the origin of the distance coordinate at the point of intersection of the extension of the sides of the fin. L and x_e represent the distances of the base and fin end, respectively, from this origin, and w is the fin thickness at the base.

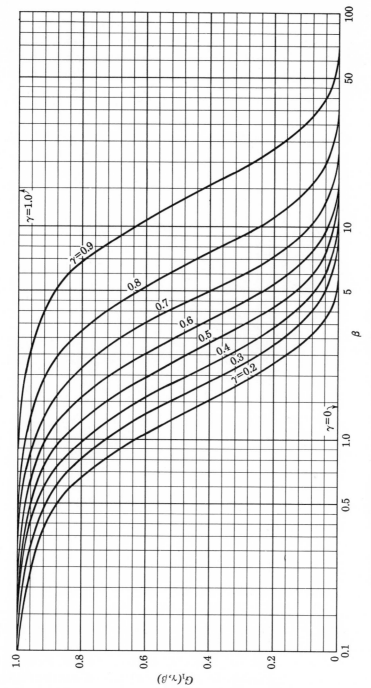

Fig. 3.17. The function G_1 for the temperature distribution in the annular fin of uniform thickness.

$\gamma=1.0$

$\gamma=0.9$

0.8

0.7

0.6

0.5

0.4

0.3

$\gamma=0.2$

$\gamma=0$

$G_1(\gamma,\beta)$

β

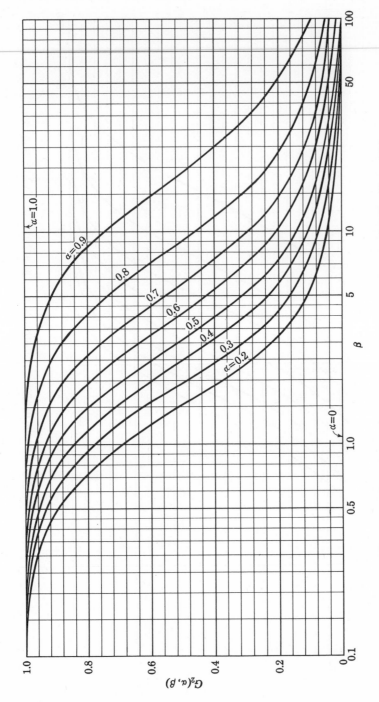

Fig. 3.18. The function G_2 for the heat flow rate from the annular fin of uniform thickness.

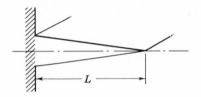

Full Triangular Fin, $x_e = 0$

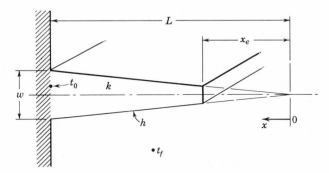

Fig. 3.19. The straight fin of trapezoidal profile.

Again one assumes that the fin is sufficiently thin [i.e., $w \ll (L - x_e)$] so that a one-dimensional situation exists. Now, for a *unit width* of the fin (neglecting side losses) the cross-sectional area A and surface area S vary with x in the following ways:

$$A = \left(w\frac{x}{L}\right) \times 1, \tag{3.70}$$

$$S = 2\sqrt{(x - x_e)^2 + \left(\frac{w}{2}\frac{x - x_e}{L}\right)^2}$$

$$= 2(x - x_e)\sqrt{1 + \left(\frac{w}{2L}\right)^2}$$

$$= 2(x - x_e)f, \tag{3.71}$$

where $f = \sqrt{1 + (w/2L)^2}$ is simply a geometrical property of the fin. Usually it is sufficiently accurate for engineering applications to let $f \approx 1$ if the condition for one-dimensional conduction is to be satisfied.

Equation (3.57) for the general extended surface may now be applied, and Eqs. (3.70) and (3.71) give

$$\frac{d^2\theta}{dx^2} + \left(\frac{L}{wx}\frac{w}{L}\right)\frac{d\theta}{dx} - \frac{h}{k}\left(\frac{L}{wx}2f\right)\theta = 0.$$

This may also be written as

$$\frac{d^2\theta}{dx^2} + \frac{1}{x}\frac{d\theta}{dx} - p^2\frac{\theta}{x} = 0, \tag{3.72}$$

in which

$$p = \sqrt{\frac{2fhL}{kw}}. \tag{3.73}$$

By comparing this to the generalized Bessel equations in Appendix C, the solution to Eq. (3.72) is seen to be

$$\theta = BI_0(2px^{1/2}) + CK_0(2px^{1/2}). \tag{3.74}$$

For the special case in which the fin comes to a point (i.e., $x_e = 0$ as noted in Fig. 3.19), one notes that the constant C must be zero, since the modified Bessel function of the second kind, K_0, approaches infinity as the argument goes to zero (see Appendix C). If C were not zero, an undefined temperature would result. Thus, for the *complete triangular fin*,

$$\theta = BI_0(2px^{1/2}),$$

and for a fixed base temperature t_0,

$$\theta = \theta_0 \frac{I_0(2px^{1/2})}{I_0(2pL^{1/2})}.$$

The heat transfer rate is then, per unit width,

$$q = -k\left(A\frac{d\theta}{dx}\right)_{x=L}$$

$$= -k\left[w\theta_0\frac{2p(\frac{1}{2}L^{-1/2})I_1(2pL^{1/2})}{I_0(2pL^{1/2})}\right]$$

$$= -\frac{kw\theta_0 p}{L^{1/2}}\frac{I_1(2pL^{1/2})}{I_0(2pL^{1/2})}. \tag{3.75}$$

The minus sign in this equation is due to the choice of the direction of the coordinate x.

These equations for θ and q may, as in the above case of the annular fin, be expressed in dimensionless forms for ease of presentation. Define the β and η parameters in the following way:

$$\beta_t = 2pL^{1/2} = \sqrt{\frac{8fhL^2}{kw}},$$

$$\eta_t = \left(\frac{x}{L}\right)^{1/2}.$$

(3.76)

The subscript t is used to denote the triangular fin. Thus,

$$\frac{\theta}{\theta_0} = \frac{I_0(\beta_t \eta_t)}{I_0(\beta_t)},$$

$$\frac{qL}{\theta_0 kw} = -\frac{\beta_t}{2}\frac{I_1(\beta_t)}{I_0(\beta_t)}.$$

(3.77)

If the functions G_3 and G_4 are defined as

$$G_3(\eta, \beta) = \frac{I_0(\beta\eta)}{I_0(\beta)},$$

(3.78)

$$G_4(\beta) = \frac{I_1(\beta)}{I_0(\beta)},$$

(3.79)

then

$$\frac{\theta}{\theta_0} = G_3(\eta_t, \beta_t),$$

(3.80)

$$\frac{qL}{\theta_0 kw} = -\frac{\beta_t}{2}G_4(\beta_t).$$

(3.81)

The functions G_3 and G_4 are plotted in Figs. 3.20 and 3.21 for rapid calculation, although more precise results may be obtained by using the tables of Appendix C.

3.16 Other Shapes of Nonuniform Cross Section

The cases discussed above (i.e., the uniform thickness straight fin, the triangular straight fin, the uniform annular fin) are only a few of the possible extended surface configurations. Extended surfaces, in which the thickness

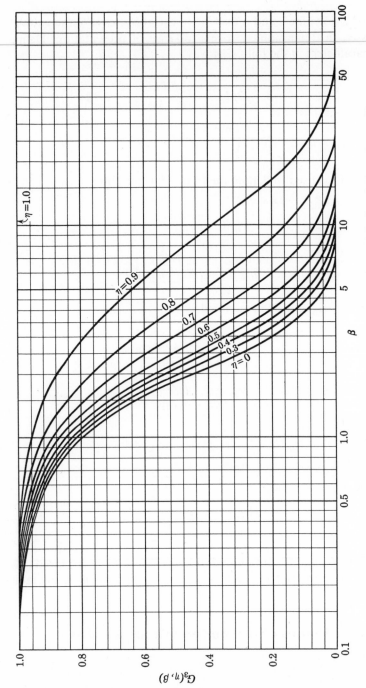

Fig. 3.20. The function G_3 for the temperature distribution in the straight fin of triangular profile.

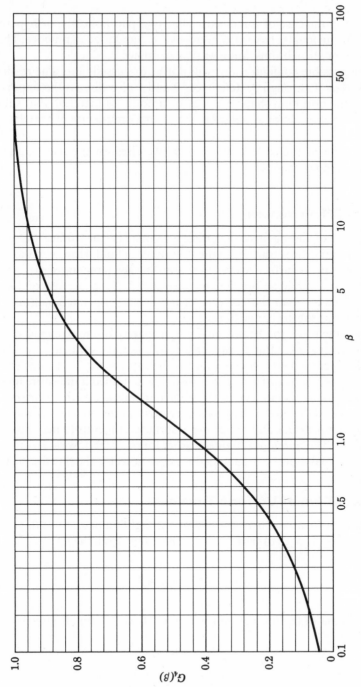

Fig. 3.21. The function G_4 for the heat flow rate from the straight fin of triangular profile.

(or diameter in the case of spine) varies linearly, parabolically, or hyperbolically with the fin length, have also been analyzed and used in practice. This is equally true for straight fins, annular fins, or spines.

All these possible configurations will not be discussed here. The reader is referred to Refs. 1 through 3. References 2 and 3 also discuss the important problem of the optimum shape, for a given amount of material, that should be employed in straight or annular fins for the maximum rate of heat dissipation.

The problems at the end of the chapter, exercises for the reader, include some of these other shapes.

3.17   Fin Effectiveness

The purpose of adding fins to a surface is to increase the surface area available for convective heat transfer to the surrounding fluid. However, the addition of the extended surface lowers the mean surface temperature to a value below that which it had before the fins were added. If the increase of the surface area is greater than the decrease of the mean surface temperature, the fins will enhance the exchange of heat.

In order to express the heat-exchanging capacity of an extended surface relative to the heat-exchanging capacity of the primary surface with no fins, it is useful to define the *fin effectiveness* as the ratio of the heat transfer rate from a fin to the heat transfer rate that would be obtained if the entire fin surface area were to be maintained at the same temperature as the primary surface. It is assumed that no contact resistance exists at the fin base in order that the fin base temperature and the primary surface temperature may be taken to be the same. The relations obtained earlier for the heat flow rate from various fin shapes may now be used to deduce equations for the fin effectiveness.

The Straight Fin of Uniform Thickness

Neglecting the heat loss from the end of a straight fin of uniform thickness, one may use Eq. (3.55) to obtain the rate of heat flow from the fin:

$$q = kmA\theta_0 \tanh mL.$$

The surface area of the fin, per unit width, is $2L$, so the fin effectiveness, κ_u, is given as

$$\kappa_u = \frac{kmA\theta_0 \tanh mL}{2Lh\theta_0}.$$

Since Eq. (3.48) gives $m^2 = 2h/kw$,

$$\kappa_u = \frac{1}{mL} \tanh mL. \tag{3.82}$$

The subscript u is used to denote a straight fin of uniform thickness.

The Straight Fin of Triangular Profile

If the same procedure is followed, Eq. (3.75) leads to the following expression for the fin effectiveness, κ_t, for a triangular fin in which f [see Eq. (3.71)] is taken as unity:

$$\kappa_t = \frac{kw\theta_0 p}{L^{1/2}} \frac{I_1(2pL^{1/2})}{I_0(2pL^{1/2})} \times \frac{1}{2hL\theta_0}$$

$$= \frac{1}{pL^{1/2}} \frac{I_1(2pL^{1/2})}{I_0(2pL^{1/2})}. \tag{3.83}$$

In terms of the dimensionless parameter β_t defined in Eq. (3.76), this equation becomes

$$\kappa_t = \frac{2}{\beta_t} \frac{I_1(\beta_t)}{I_0(\beta_t)} = \frac{2}{\beta_t} G_4(\beta_t). \tag{3.84}$$

The function $G_4(\beta_t)$ was encountered earlier and is plotted in Fig. 3.21.

The Annular Fin of Uniform Thickness

The effectiveness of the annular fin of uniform thickness is similarly deduced from the relations of Sec. 3.14. The result in terms of the parameters α_{an} and β_{an} defined in Eq. (3.52) is

$$\kappa_{an} = G_2(\alpha_{an}, \beta_{an}). \tag{3.85}$$

The proof of this equation will be left as an exercise for the reader. The function G_2 was defined in Eq. (3.69) and is plotted in Fig. 3.18.

3.18   Total Surface Temperature Effectiveness

The *fin effectiveness*, κ, discussed in Sec. 3.17, is concerned with expressing the performance of the fin itself. However, most applications employing extended surfaces involve the use of an array of fins attached to the primary

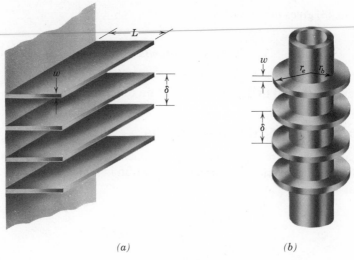

(a)   *(b)*

Fig. 3.22

surface. Figure 3.22 depicts such arrays for straight and annular fins. In such applications it proves useful to define a *total surface temperature effectiveness*, which gives a measure of the performance of an entire array. Let the following notation be adopted:

A_f = surface area of the fins only,

A = total exposed surface area, including the fins and the unfinned primary surface,

κ = fin effectiveness defined in Sec. 3.17.

If the total surface temperature effectiveness, η, is defined as the heat rejected by an array divided by the heat the array would reject if the *entire* surface were maintained at the base temperature, then

$$\eta = \frac{(A - A_f)h\theta_0 + \kappa A_f h\theta_0}{Ah\theta_0}.$$

The first term in the numerator of this equation represents the heat given up by the exposed portions of the primary surface; the second term is that given up by the fins. The equation reduces to

$$\eta = 1 - \frac{A_f}{A}(1 - \kappa). \tag{3.86}$$

Since $A_f/A < 1$ and $\kappa \le 1$, one deduces that $\eta \le 1$.

For the straight fin array shown in Fig. 3.22(a), when the effect of the fin ends is neglected,

$$\eta = 1 - \frac{2L}{2L + \delta - w}\left(1 - \frac{1}{mL} \tanh mL\right). \tag{3.87}$$

A similar expression may be written for the annular array shown in Fig. 3.22(b).

The principal application of the total surface temperature effectiveness, just defined, will be made in Chapter 12 in connection with the analysis of heat exchangers.

References

1. GARDNER, K. A., "Efficiency of Extended Surface," *Trans. A.S.M.E.*, Vol. 67, No. 8, 1945, pp. 621–631.
2. JAKOB, M., *Heat Transfer*, Vol. I, New York, Wiley, 1949.
3. SCHNEIDER, P. J., *Conduction Heat Transfer*, Reading, Mass., Addison-Wesley, 1955.

Problems

3.1. It is desired to limit the heat loss through a boiler furnace wall to 700 Btu/hr-ft². The wall is composed of a material having a thermal conductivity of 0.65 Btu/hr-ft-°F. If the inner surface temperature is 2000°F and the outer surface temperature is 700°F, what thickness should be used?

3.2. If the wall of Prob. 3.1 has 2 in. of insulation ($k = 0.05$ Btu/hr-ft-°F) added to its outer surface to reduce the outer surface temperature to 350°F, what will be the rate of heat loss through each square foot of the wall?

3.3. A composite wall is made of 8 in. of fire-clay brick (burnt at 2426°F), 6 in. of fired diatomaceous earth brick, and an outer 4-in. layer of common brick. If the inside surface temperature is 1900°F and the outside surface is held at 300°F, find (a) the heat loss rate per square foot of wall area; (b) the temperature at the junction between the different layers of brick; and (c) the temperature at a point 6 in. from the outer surface.

3.4. A wall is composed of a 5-in.-thick layer of material A and an 8-in.-thick layer of material B. The temperature of the outer surface of layer A is known to be 500°F when the outer surface of layer B is at 90°F. A layer of 1-in.-thick insulation ($k = 0.05$ Btu/hr-ft-°F) is added to the outer surface of layer B. Under these conditions it is observed that the surface temperature of layer A rises to 600°F, and the junction between layer B and the insulation (formerly the outer surface of layer B) becomes 425°F. If the insulation surface temperature is measured to be 80°F, what is the rate of heat flow per square foot of wall area *before* and *after* the addition of insulation?

3.5. A wall of a house is composed of 4 in. of common brick, $\frac{1}{2}$ in. of Celotex, a $3\frac{5}{8}$-in. air space created by 2-in. × 4-in. studs, and $\frac{1}{2}$ in. of asbestos cement board. If the outside brick wall is 90°F, and the inner surface is at 70°F, what is the heat flow rate through each square foot of wall surface if (a) the air space *conductance* is 1.17 Btu/hr-ft²-°F and (b) the air space is filled with glass wool at a density of 4.0 $\mathrm{lb_m/ft^3}$?

3.6. A nominal 5-in., wrought iron, schedule 40 pipe is covered with a 2-in. layer of 85% magnesia insulation. If the pipe carries steam so that the inner surface is at 800°F and the outer insulation surface is 100°F, what is the heat loss per hour from 100 ft of this pipe?

3.7. If an 8-in. nominal, schedule 80, steel pipe is covered with first a 3-in. layer of 85% magnesia insulation and then a 1-in. layer of air-cell insulation ($k = 0.037$ Btu/hr-ft-°F), find the heat loss per square foot of outside surface area if the inside pipe surface is at 1000°F and the outside insulation surface is at 200°F.

3.8. Find the temperature at the junction between the two insulations in Prob. 3.7.

3.9. Considering two pipes of different diameters, in which case is the maximum insulating effect obtained if equal thickness of an insulation is placed on each? Assume that the pipe surface and insulation surface temperatures are the same in both cases.

3.10. A 2-in. nominal, schedule 40, steel pipe carrying chilled water is to be covered with a 1-in. thickness of expanded scrap cork and a 1-in. thickness of rock wool (4.0 $\mathrm{lb_m/ft^3}$). To achieve the maximum insulating effect, which insulation should be placed next to the pipe surface? The pipe surface is 35°F and the outer insulation surface is 95°F in either case.

3.11. Derive the following expression for the heat flow per unit surface area from a sphere of inside radius r_1 and outside radius r_2. The sphere is heated to uniform inside and outside surface temperatures of t_1 and t_2, respectively.

$$\frac{q}{A} = \frac{k(t_1 - t_2)}{\frac{r_2}{r_1}(r_2 - r_1)}.$$

3.12. A hollow cylinder of inside and outside radii r_1 and r_2, respectively, is heated such that its inner and outer surfaces are at uniform temperatures t_1 and t_2. If the material of which the cylinder is composed has a thermal conductivity which varies with temperature in the following way:

$$k = k_0(1 + bt),$$

find the rate of heat flow through the cylinder. At what mean temperature should one evaluate k in order to use the constant-thermal-conductivity formula given in Eq. (3.11)?

3.13. Find the relation for the rate of heat flow through a single-layered plane wall, the thermal conductivity of which varies quadratically:

$$k = k_0(1 + bt + ct^2).$$

3.14. Determine, from the data of Table A.2, k_0 and b for glass wool (density $1.5 \, \text{lb}_m/\text{ft}^3$) for use in the linear relation $k = k_0(1 + bt)$. If a layer of this material 6 in. thick is heated so that its surfaces are at $100°F$ and $0°F$, respectively, find the rate of heat flow per unit area and the temperature midway through the wall.

3.15. Verify the algebra leading to Eqs. (3.18) and (3.20).

3.16. Verify the algebra leading to Eq. (3.24).

3.17. Verify the algebra leading to Eq. (3.34).

3.18. A large slab of concrete, 3 ft thick, has both surfaces maintained at $70°F$. During the curing process a uniform internal heat generation of $6 \, \text{Btu/ft}^3$ occurs. If the concrete thermal conductivity is $0.65 \, \text{Btu/hr-ft-}°F$, find the steady temperature which results at the center of the slab.

3.19. For the rod considered in Example 3.3(b), what is the maximum permissible temperature at which the ends may be maintained?

3.20. Repeat Example 3.3(b) for a rod 0.1 in. in diameter.

3.21. A bare copper wire, $\frac{1}{8}$ in. in diameter, has its outer surface maintained at $80°F$. It has a constant resistivity of 1.73×10^{-6} ohm-cm. If the centerline temperature is not to exceed $250°F$, what is the maximum current it will carry?

3.22. Imagine that the wall described in Prob. 3.3, rather than having its surface temperatures specified, is surrounded on its fire-brick side by hot gases at $2100°F$ with a film coefficient of $5.6 \, \text{Btu/hr-ft}^2\text{-}°F$ and is surrounded on its other side by air at $100°F$ with a film coefficient of $1.5 \, \text{Btu/hr-ft}^2\text{-}°F$. Find (a) the rate of heat flow through each square foot of wall, and (b) the temperature of the two surfaces.

3.23. Let the wall of the house described in Prob. 3.5 be subjected to an outdoor wind velocity of 10 mph such that a film coefficient of $5.5 \, \text{Btu/hr-ft}^2\text{-}°F$ exists at the outer surface. Assuming that the inside surface film coefficient is $1.8 \, \text{Btu/hr-ft}^2\text{-}°F$, find the rate of heat flow and the wall surface temperatures for outside and inside air temperatures of $100°F$ and $65°F$, respectively.

3.24. A large pressure vessel (large enough to treat the walls as plane) with steel walls ($k = 25 \, \text{Btu/hr-ft-}°F$) has its inside surface temperature exposed to steam at $1200°F$ through a film coefficient of $100 \, \text{Btu/hr-ft}^2\text{-}°F$. It is desired to insulate the outer surface so that the exposed insulation surface is not more than $100°F$. In order to minimize cost, two insulating materials are to be used. First, an expensive high temperature insulation ($k = 0.15 \, \text{Btu/hr-ft-}°F$) is applied next to the vessel wall, and then a less-expensive insulation ($k = 0.05 \, \text{Btu/hr-ft-}°F$) is placed on the outside. The maximum allowable temperature of the less-expensive insulation is $600°F$. The film coefficient at the outermost surface is $2.0 \, \text{Btu/hr-ft}^2\text{-}°F$, and the temperature of the ambient air is $85°F$. Find the thickness of each layer of insulation.

3.25. A nominal 4-in., schedule 40, wrought iron pipe is covered with $1\frac{1}{2}$ in. of 85% magnesia insulation. The pipe carries superheated steam at a temperature of 750°F, and the outer insulation surface is exposed to air at 80°F. If the inside and outside film coefficients are 248 Btu/hr-ft²-°F and 2.0 Btu/hr-ft²-°F, respectively, find the rate of heat loss per foot of pipe length, and find the temperatures of the inside pipe surface, outside pipe surface, and outside insulation surface.

3.26. The pipe configuration described in Prob. 3.7 carries steam at 1200°F with an inside film coefficient of 485 Btu/hr-ft²-°F. The outer insulation surface is exposed to air at 120°F with a surface film coefficient of 2.0 Btu/hr-ft²-°F. Find the heat loss per square foot of exposed surface area.

3.27. Calculate the heat loss per foot of length of a 4-in., schedule 40 pipe covered with a $\frac{1}{2}$-in. layer of insulation ($k = 0.05$ Btu/hr-ft-°F), if the inside pipe surface temperature is 400°F and the outside air temperature is 65°F. Assume an outside film coefficient of 30 Btu/hr-ft²-°F.

3.28. A bare 1-in. schedule 40 pipe has a surface temperature of 350°F and is placed in air at 85°F. The convective film coefficient between the surface and the air is 1.0 Btu/hr-ft²-°F. It is desired to reduce the heat loss to 50% of its present value by the addition of an insulation with $k = 0.10$ Btu/hr-ft-°F. Assuming that the pipe surface temperature and the exposed surface film coefficient remain unchanged as insulation is added, find the required thickness of insulation. Is this thickness an economically reasonable value?

3.29. Repeat Prob. 3.28 for a 4-in. schedule 40 pipe, all other data remaining unchanged.

3.30. Repeat Prob. 3.25 for a 10-in. nominal, schedule 40, steel pipe subjected to identical conditions. Take all conductivities and film coefficients to be the same.

3.31. A 6-in., schedule 40, steel pipe carries steam at 200 psia, 400°F. Minimum cost is to determine the selection of the thickness of magnesia insulation to be used in insulating the pipe. The air temperature in the room through which the pipe passes is 90°F and a film coefficient of 1.5 Btu/hr-ft²-°F exists at the exposed insulation surface. The cost of generating the steam is $0.50 per million Btu. The cost of the insulation, installed, depends on the thickness:

1-in.:	$ 2.50 per foot of length
2-in.:	$ 4.00
3-in.:	$ 7.00
4-in.:	$10.00
5-in.:	$15.00

Annual fixed charges for interest, repairs, etc., are 10% of the initial cost. The steam line operates 8000 hr per year. Recommend the thickness of insulation to be used based on an estimated life of 10 years.

3.32. Verify Eqs. (3.41) and (3.42).

3.33. An electrical wire has a diameter of 0.125 in. If it is to be covered with an electrical insulation having a thermal conductivity of 0.05 Btu/hr-ft-°F, find the thickness of insulation that will give the maximum rate of heat dissipation for a surface film coefficient of 4.0 Btu/hr-ft²-°F.

3.34. Find the over-all heat transfer coefficient, U, for the situations described in Probs. 3.22 and 3.23.

3.35. Find the over-all heat transfer coefficient, U, for the pipe configuration of Probs. 3.25, 3.26, and 3.27. Base the coefficient on the outside exposed surface area.

3.36. A condenser tube made of brass is a $\frac{3}{4}$-in., 18-gage tube. The film coefficient on the inside surface is 950 Btu/hr-ft²-°F, and the outside film coefficient due to the condensing steam is 1200 Btu/hr-ft²-°F. The average cooling water temperature is 85°F. Find (a) the over-all heat transfer coefficient, U, based on the outer surface area and (b) the pounds of saturated steam per hour that will be condensed at 1 psia for each foot of tube length.

3.37. A water-to-water heat exchanger is made of brass tubes, 1-in. O.D., 16 gage. The inside and outside film coefficients are 800 Btu/hr-ft²-°F and 1200 Btu/hr-ft²-°F, respectively. Find the over-all heat transfer coefficient, U.

3.38. Write an expression for the over-all heat transfer coefficient of a two-layer sphere with inside and outside film coefficients. Base the over-all coefficient on the outside surface area.

3.39. Show, for a single-layered cylinder, that the equation for the rate of heat flow through a plane wall may be used if one uses for the wall area the log mean of the inner and outer cylindrical surface areas. That is,

$$A_m = \frac{A_2 - A_1}{\ln(A_2/A_1)}.$$

3.40. Repeat Prob. 3.39 for a single-layered sphere, showing that the geometric mean area should be used:

$$A_m = \sqrt{A_2 A_1}.$$

3.41. A long, hollow, cylinder (inside diameter 2 in., outside diameter 3 in.) has a thermal conductivity of 15 Btu/hr-ft-°F. It is covered with a $\frac{1}{2}$-in. layer of insulation ($k = 3.0$ Btu/hr-ft-°F). The temperature of a fluid flowing inside the cylinder is 60°F, and the temperature of the inner wall of the cylinder is 100°F. The temperature of a fluid surrounding the outside of the cylinder is 300°F, and a surface film coefficient of 10 Btu/hr-ft²-°F exists at this outer surface. (a) What is the rate of heat flow, per foot of length, between the two fluids? (b) What is the temperature of the outer surface of the insulation? (c) What is the over-all conductance, based on the outside surface area?

3.42. Write an expression for the rate of heat flow, per unit of outside surface area, from a two-layer hollow sphere which has its inside surface temperature specified and the temperature of an external convecting fluid specified. Presume any necessary conductivities or conductances to be known.

3.43. A sphere of fixed outside radius and fixed surface temperature is to be insulated with a material of known thermal conductivity. For fixed values of an ambient fluid temperature and surface film coefficient, find if a "critical thickness" of insulation exists (as in the cylindrical case considered in Sec. 3.9) and, if so, what its value is.

3.44. Show, for a spine of constant cross section, that as $L \to \infty$, the temperature distribution and heat loss are given by

$$\theta = \theta_0 e^{-mx},$$

$$q = kmA\theta_0.$$

3.45. A cylindrical rod of $\frac{3}{4}$-in. diameter, 9 in. long protrudes from a heat source at 300°F into air at 90°F. The film coefficient of convective heat transfer is known to be 1 Btu/hr-ft²-°F on all exposed surfaces. Find the following information for the three cases of the rod being composed of copper, cast iron, and glass:
  (a) The temperature at points located 1/4, 2/4, 3/4, 4/4 the distance from the source to the rod end. Neglect end heat loss.
  (b) The rate of heat flow out of the source if the end heat loss is neglected, and if the end heat loss is not neglected.

3.46. A $\frac{1}{2}$-in. diameter rod of iron ($k = 26$ Btu/hr-ft-°F) is heated to 500°F at its base and protrudes into air at 100°F, where $h = 1.5$ Btu/hr-ft²-°F. How long must the rod be in order that its end temperature may be computed, with less than 5°F error, by use of the infinite rod equations? What error in the heat dissipation is made, at this length, by use of the infinite rod equation? Refer to Prob. 3.44.

3.47. A $\frac{1}{4}$-in.-diameter pure copper rod is 18 in. long and is heated to 200°F at each end. If the fluid which surrounds the rod is at 80°F and a film coefficient of 4.5 Btu/hr-ft²-°F exists at the surface, find the temperature at the midpoint of the rod and at a point 4 in. from one end. How much heat is dissipated to the surroundings by the first 4 in. of the rod?

3.48. A rod of diameter D, conductivity k, and length L is surrounded by a fluid at temperature t_f and a film coefficient h exists at the surface. If the right-hand end of the rod is heated to t_{oR} and the left-hand end to t_{oL}, find expressions for (a) the temperature distribution in the rod and (b) the heat flow rate in or out of each heat source.

3.49. In Prob. 3.48 let $D = \frac{1}{2}$ in., $k = 25$ Btu/hr-ft-°F, $L = 18$ in., $t_f = 70°F$, $h = 2.0$ Btu/hr-ft²-°F, $t_{oR} = 120°F$, and $t_{oL} = 350°F$. Find the heat flow rate out of each source.

3.50. A Chromel–Alumel thermocouple (wire diameters = 0.049 in.) is attached to a surface at 250°F and extends into air at 80°F ($h = 1.8$ Btu/hr-ft²-°F). Estimate the rate of heat loss from the surface due to the attachment of the thermocouple.

3.51. For free convection coefficients in air of the order of magnitude of 1 Btu/hr-ft²-°F, make a reasonable estimate for the minimum length of a wiener-roasting wire made of an old coat hanger in order to avoid an uncomfortably hot temperature on the end held by the user.

3.52. Two rods of identical size and shape are both supported between two heat sources at 212°F and are surrounded by air at 80°F. One rod is known to have a thermal conductivity of 25 Btu/hr-ft-°F and its midpoint temperature is measured to be 120°F. If the midpoint temperature of the other rod is measured to be 167°F, what is its thermal conductivity?

3.53. Two circular rods, both of diameter D and length L, are joined at one end and both heated to the same temperature, t_0, at the free ends. The film coefficient, h, is the same for all surfaces. If t_f is the temperature of the surrounding fluid and if the thermal conductivities of the two rods are k_a and k_b, show that the temperature of the junction, t_j, is generally

$$\frac{t_j - t_f}{t_0 - t_f} = \frac{\sqrt{\frac{k_a}{k_b}} \sinh m_b L + \sinh m_a L}{\sqrt{\frac{k_a}{k_b}}(\cosh m_a L)(\sinh m_b L) + (\sinh m_a L)(\cosh m_b L)}.$$

3.54. An annular fin of uniform thickness has an inner radius of 3 in. and an outer radius of 5 in. The fin has a uniform thickness of 0.2 in. and is composed of a material with $k = 25$ Btu/hr-ft-°F. The base of the fin is maintained at 395°F and the surrounding fluid is at 95°F. The film coefficient between the fin surface and the fluid is 10 Btu/hr-ft²-°F. Find
(a) the rate at which heat is dissipated by the fin;
(b) the temperature at the fin end and at a point midway between the base and the end;
(c) the heat dissipated by the last inch of fin.

3.55. An annular web connects two concentric cylinders. The web is $\frac{1}{4}$ in. thick and has a thermal conductivity of 36 Btu/hr-ft-°F. The outer cylinder has an I.D. = 6 in. and its inner surface is maintained at 730°F; the inner cylinder has an O.D. = 3 in. and a surface temperature of 216°F. The web is surrounded by a convecting fluid at 123°F with $h = 13.5$ Btu/hr-ft²-°F. Assuming that the temperature in the web is a function of the radius only, find the rate at which heat is being given up by the web to the fluid.

3.56. A plane wall has its surface maintained at 250°F and is in contact with a fluid of 80°F ($h = 3.5$ Btu/hr-ft²-°F). Find the per cent increase, per square foot of wall, in the heat dissipation if the wall is equipped with tapered straight fins (triangular), $\frac{1}{8}$ in. thick at the base, 1.5 in. long, and spaced $\frac{3}{5}$ in. apart on centers. The fins are made of steel with $k = 26$ Btu/hr-ft-°F.

3.57. A spine protruding from a wall at 375°F has the shape of a cylindrical rod ($\frac{1}{2}$-in. diameter, 7 in. long) capped by a cone 5 in. long. The entire spine is made of a material with $k = 25$ Btu/hr-ft-°F, and a film coefficient of 5.4 Btu/hr-ft²-°F exists between the spine and the ambient fluid at 125°F. Find

(a) the temperature at the cross section at which the spine changes from the cylindrical to conical shape;

(b) the rate at which heat is being given up to the ambient fluid.

3.58. A spine protruding from a wall at temperature t_0 has the shape of a circular cone. The radius of the cone base is R. The spine comes to a point at its tip, and its length is L.

(a) If t_f denotes the temperature of the ambient fluid, h the surface film coefficient, and k the thermal conductivity of the cone, show that the differential equation of the temperature distribution is as follows, where x is the distance measured from the cone tip:

$$\frac{d^2\theta}{dx^2} + \frac{2}{x}\frac{d\theta}{dx} - l^2\frac{\theta}{x} = 0,$$

$$l^2 = \frac{2hL}{kR}\sqrt{1 + (R/L)^2}.$$

(b) Show that the general solution to this equation is

$$\theta = \frac{BI_1(2lx^{1/2}) + DK_1(2lx^{1/2})}{x^{1/2}}.$$

3.59. For the conical spine described in Prob. 3.58, show that as $x \to 0$, D must be 0 and that the ratio $I_1(2lx^{1/2})/x^{1/2} \to l$. Then show, for $\theta = \theta_0$ at $x = L$, that the temperature distribution in the spine is given by

$$\frac{\theta}{\theta_0} = \left(\frac{L}{x}\right)^{1/2}\frac{I_1(2lx^{1/2})}{I_1(2lL^{1/2})}.$$

3.60. Show that the rate of heat flow from the spine in Probs. 3.58 and 3.59 is given by

$$q = -k\pi R^2\theta_0\left[\frac{l}{L^{1/2}}\frac{I_0(2lL^{1/2})}{I_1(2lL^{1/2})} - \frac{1}{L}\right].$$

3.61. A straight fin has a thickness which varies parabolically—i.e., its half thickness at the base is $w/2$, at the tip it is 0, and it varies as x^2 between the tip and the base. If its total length is L and if its base is maintained at a uniform temperature, deduce the equation of the temperature distribution and the equation for the heat loss per unit width if the fin is very thin.

3.62. A spine of circular cross section has a base radius of R and a tip radius of 0. The total length is L and the radius varies as $x^{1/2}$, where x is the coordinate measured from the tip to the wall. Write the expression for the temperature distribution in the spine if the spine is very thin.

3.63. An annular fin has a form which varies hyperbolically from its base to its tip. If its base thickness is w at r_b and if the thickness varies as $1/r$ (r = radius between r_b and r_e), deduce the equation for the temperature distribution in the fin if the fin is very thin.

3.64. For a straight fin of uniform thickness, the parameter m is, for unit width, $m^2 = 2h/kw$. If such a fin, of given k and given width, w, is to be utilized under fixed service conditions (i.e., θ_0, h), the heat loss from the fin may be expressed as a function of the fin length L. Write such an expression, including the loss of heat from the end with $h_e = h$, and show that if $(hw/2k) > 1$, $dq/dL < 0$ for *all* L; and if $hw/2k < 1$, $dq/dL > 0$. This shows that unless $hw/2k < 1$, the application of such fins will produce an insulating effect.

3.65. Consider a straight fin of uniform thickness. Let the "profile area" be that area taken in a plane parallel to the fin length and normal to the width, $A_p = w \times L$. For a fixed amount of material (i.e., A_p = constant), show that if the heat loss from the fin end is negligible, the fin dissipates the maximum amount of heat if its length, L, and thickness, w, are related by the following condition:

$$\tanh \xi = 3\xi \operatorname{sech}^2 \xi,$$
$$\xi = L\sqrt{2h/kw}.$$

3.66. For a set of circumstances similar to those given in Prob. 3.65, the profile area of a fully triangular fin is $A_p = \frac{1}{2} \times w \times L$. Show, for a fixed A_p, that the maximum heat is dissipated when w and L are related by

$$\frac{4}{3} \frac{I_1(\eta)}{I_0(\eta)} = 4\left[1 - \left(\frac{I_1(\eta)}{I_0(\eta)}\right)^2\right],$$
$$\eta = 2L\sqrt{2h/kw}.$$

3.67. In Prob. 3.65 the value of ξ which satisfies the optimum condition is $\xi = 1.4192$, and in Prob. 3.66 the optimum condition is given by $\eta = 2.6188$. With these data, show that for equal amounts of heat dissipated, the optimum proportioned triangular fin has a profile area 0.690 times that of the optimum proportioned rectangular fin. Assume that θ_0 is the same in both instances.

3.68. A cylindrical spine (radius R, length L) is maintained at a fixed base temperature, t_0, and extends into a fluid at temperature t_f. A heat transfer coefficient, h, exists at the exposed surfaces. For a fixed volume of material, find the proportions of L and R for the spine which dissipates the maximum amount of heat. Neglect end heat losses.

3.69. Write an expression for the total surface temperature effectiveness of the annular array shown in Fig. 3.22(b) in terms of the geometry shown.

3.70. Find the total surface temperature effectiveness for the fin array described in Prob. 3.56.

3.71. A heat exchanger surface is equipped with an array of straight fins as shown in Fig. 3.22(a). The fins are 0.75 in. long, 0.05 in. thick, spaced 0.75 in. apart, and made of Duralumin. The ambient fluid is at 120°F, and the surface film coefficient is 25 Btu/hr-ft²-°F. Find the total surface temperature effectiveness of this array.

3.72. Because of their very small size, transistors have little surface area for the dissipation of internally generated heat. As a consequence, they are often provided with a cap which not only protects the transistor but also provides additional heat transfer area to the ambient air. Such a cap is shown in the accompanying figure,

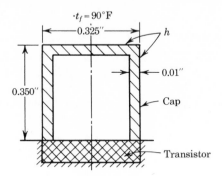

Problem 3.72

with all dimensions given. The cap has cylindrical symmetry with respect to the centerline shown—a hollow cylindrical wall topped with a circular disk. The walls of the cap are so thin that one may imagine that the heat is conducted up the cylindrical walls (one-dimensionally, but losing heat to the surroundings) and thence radially in the top disk (again one-dimensionally with heat loss to the surroundings). Presumably no heat is lost to the cap interior. The cap illustrated is made of steel with $k = 12.0$ Btu/hr-ft-°F. The ambient air is at 90°F and the heat transfer coefficient at all exposed surfaces is 4.5 Btu/hr-ft²-°F. The transistor is dissipating 375 milliwatts. Find

(a) the surface temperature of the transistor—i.e., the temperature at the juncture between the cap and the transistor;

(b) the surface temperature of the transistor if no cap were provided and the 375 milliwatts had to be transferred from the circular transistor surface with the same heat transfer coefficient and ambient air temperature.

3.73. The cross section of the core of a plate-fin heat exchanger is shown in the accompanying figure. The two primary heated surfaces are maintained at the same temperature, t_0, as shown. A fluid, of temperature t_f, flows in the open passages between the fin-matrix array shown. The fluid flows in a direction normal to the plane of the paper, and the fins are straight in that direction. The heat transfer coefficient, h, is presumed to be the same for all exposed surfaces.

(a) Find the total surface effectiveness for the primary surfaces maintained at t_0 in terms of the dimensions and parameters shown in the figure.

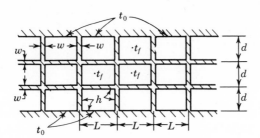

Problem 3.73

(b) If $h = 30$ Btu/hr-ft^2-°F, $k = 30$ Btu/hr-ft-°F, $t_0 = 120$°F, $t_f = 90$°F, $d = 0.4$ in., $L = 0.5$ in., and $w = 0.05$ in., find the total heat transferred to the fluid for a section of core 1 ft in depth (normal to the paper) and 1 ft in width (parallel to the primary surface).

CHAPTER 4

Heat Conduction in Two or More Independent Variables

4.1 Introductory Remarks

The cases of heat conduction discussed in Chapter 3 were all ones in which the temperature distribution throughout the bodies could be expressed in terms of a single variable. This meant that only steady state, one-dimensional systems could be discussed.

The present chapter will be devoted to discussions of conduction problems that have to be described in two or more independent variables. The major portion of the chapter will be used to describe systems in two independent variables. This means either steady conduction in two space dimensions or nonsteady conduction in one dimension (the second variable being time). The two-dimensional, steady state systems will be treated first. The particular cases to be studied will be the rectangular plate and the circular bar because of the simplicity in describing the boundary conditions. One-dimensional, transient conduction in the plane wall and circular cylinder will be treated to illustrate nonsteady conduction problems. Finally, a section will be devoted to some special cases of transient conduction in two or three space dimensions.

No attempt will be made to present a comprehensive coverage of the mathematical theory of multidimensional heat conduction. Primary attention will be directed toward geometrical shapes of frequent occurrence and subjected to boundary conditions which are significant from the point of view of practical applications. In order to establish the mathematical techniques to be used later in the chapter, Secs. 4.2 and 4.3 will be devoted to two elementary systems subjected to simple boundary conditions. Although the solutions to these first few cases have little practical value, they will prove useful to the understanding of later applications.

For the rapid application of the results of the analyses to real physical problems, the solutions of problems having practical value have been reduced to relatively simple graphical presentations.

4.2   Steady State Conduction in Rectangular Plates

If attention is now directed to steady state conduction in rectangular plates, such as shown in Fig. 4.1, it is most convenient to use cartesian co-ordinates to describe the temperature distribution in the plate. The plate will be considered to be in the x-y plane with the origin of the coordinates at one corner. No conduction will be considered in the z direction normal to the plate. This may be imagined to be the case if the plate has such a great extent in the z direction that no end effects exist or if the x-y faces of the plate are insulated so that no heat will pass in the z direction.

The heat conduction equation for the steady state [Eq. (1.13)] is, for cartesian coordinates in two dimensions,

$$\frac{\partial^2 t}{\partial x^2} + \frac{\partial^2 t}{\partial y^2} = 0. \tag{4.1}$$

Since this steady state heat conduction equation is a linear differential equation, the principle of superposition of solutions is applicable. This fact will be used later to build the solutions to more complex situations by adding the solutions of simpler problems.

The solution of Eq. (4.1) is obtained by assuming the temperature distribution to be expressible as the product of two functions, each of which involves only one of the independent variables. That is, if $X(x)$ is a function of x only and if $Y(y)$ is a function of y only, one assumes that the temperature, t, is given by

$$t = X(x) \cdot Y(y). \tag{4.2}$$

When this is substituted into Eq. (4.1) and the resulting expression is re-arranged, one obtains

$$-\frac{1}{X}\frac{d^2 X}{dx^2} = \frac{1}{Y}\frac{d^2 Y}{dy^2}.$$

Since each side of this equation involves only one of the independent variables, they may be equal only if they are both the same constant. Calling this constant λ^2, one finds that

$$-\frac{1}{X}\frac{d^2X}{dx^2} = \frac{1}{Y}\frac{d^2Y}{dy^2} = \lambda^2.$$

This equation is equivalent to the following two ordinary differential equations:

$$\frac{d^2X}{dx^2} + \lambda^2 X = 0,$$

$$\frac{d^2Y}{dy^2} - \lambda^2 Y = 0.$$

The solutions of these equations are

$$Y = B_1 \sinh \lambda y + B_2 \cosh \lambda y,$$

$$X = B_3 \sin \lambda x + B_4 \cos \lambda x.$$

When these results are introduced into the assumed form of the solution given in Eq. (4.2), one finally obtains the following general solution of Eq. (4.1):

$$t = (B_1 \sinh \lambda y + B_2 \cosh \lambda y)(B_3 \sin \lambda x + B_4 \cos \lambda x). \qquad (4.3)$$

The λ's and B's are constants to be determined by application of boundary conditions. An example of this will now be illustrated.

The Rectangular Plate with a Specified Temperature Distribution on One Edge —The Other Edges at Constant Temperature

Let the configuration under consideration be a rectangular plate of finite width L in the x-coordinate direction and of width W in the y-coordinate direction. This is illustrated in Fig. 4.1.

Let the temperature of the edges of the plate at $x = 0$, $x = L$, and $y = 0$ be maintained at a constant temperature t_c and that of the edge at $y = W$ be

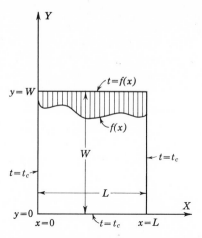

Fig. 4.1. The rectangular plate with a specified temperature distribution on one edge, the other edges at constant temperature.

maintained at values that vary along that edge. Let this variation be represented as $f(x)(0 \leq x \leq L)$, remembering that $f(x)$, although unspecified, is to be treated as known.

The differential equation and the boundary conditions to be satisfied are

$$\frac{\partial^2 t}{\partial x^2} + \frac{\partial^2 t}{\partial y^2} = 0.$$

$$At\ x = 0: \quad t = t_c.$$

$$At\ x = L: \quad t = t_c.$$

$$At\ y = 0: \quad t = t_c.$$

$$At\ y = W: \quad t = f(x).$$

The problem is, then, for the system of equations noted above: What is the temperature at any specified point within the plate? That is, find the function $t = t(x, y)$ that gives the distribution of temperature in the plate. This problem may be simplified somewhat by rendering the boundary conditions homogeneous through the introduction of the temperature difference variable

$$\theta = t - t_c.$$

Then the system to be solved is

$$\frac{\partial^2 \theta}{\partial x^2} + \frac{\partial^2 \theta}{\partial y^2} = 0.$$

At $x = 0$:	$\theta = 0.$	(1)
At $x = L$:	$\theta = 0.$	(2)
At $y = 0$:	$\theta = 0.$	(3)
At $y = W$:	$\theta = f(x) - t_c.$	(4)

(4.4)

The solution of the differential equation in Eq. (4.4) is of the same form as Eq. (4.3), which is the solution to Eq. (4.1). Thus,

$$\theta = (B_1 \sinh \lambda y + B_2 \cosh \lambda y)(B_3 \sin \lambda x + B_4 \cos \lambda x).$$

Application of condition (3) of Eq. (4.4) shows that $B_2 = 0$, so

$$\theta = B_1 \sinh \lambda y (B_3 \sin \lambda x + B_4 \cos \lambda x).$$

In order to satisfy condition (1) of Eq. (4.4), $B_4 = 0$. Thus,

$$\theta = \sinh \lambda y \cdot B \sin \lambda x, \qquad (4.5)$$

where B replaces the product $B_1 B_3$. Substitution of condition (2) gives

$$0 = \sinh \lambda y \cdot B \sin \lambda L.$$

The only way that this may be satisfied for *all* values of y is for

$$\sin \lambda L = 0.$$

This expression is satisfied for $\lambda = 0, \pi/L, 2\pi/L, \ldots$, or, in general,

$$\lambda_n = \frac{n\pi}{L}, \qquad n = 0, 1, 2, 3, \ldots. \qquad (4.6)$$

Each of the λ's of Eq. (4.6) gives rise to a separate solution of Eq. (4.5), and since the general solution will be the sum of these individual solutions, one has

$$\theta = \sum_{n=0}^{\infty} B_n (\sinh \lambda_n y)(\sin \lambda_n x).$$

The symbol B_n represents the constant B for each of the solutions. Since $\lambda_n = 0$ for $n = 0$, no contribution is made by the first term,

$$\theta = \sum_{n=1}^{\infty} B_n(\sinh \lambda_n y)(\sin \lambda_n x). \tag{4.7}$$

Applying, finally, condition (4) of Eq. (4.4) for $y = W$, one obtains

$$[f(x) - t_c] = \sum_{n=1}^{\infty} B_n \sinh \lambda_n W \sin \lambda_n x,$$

$$\lambda_n = \frac{n\pi}{L}; \qquad n = 1, 2, 3, \ldots; \qquad 0 \le x \le L. \tag{4.8}$$

Comparing this result with the facts concerning the orthogonal functions in Eqs. (D.9), (D.10), and (D.11) of Appendix D, and recognizing that $(B_n \sinh \lambda_n W)$ is a constant, one sees that the constants B_n of Eq. (4.8) may be expressed in terms of the constants C_n of Eq. (D.10):

$$B_n \sinh \lambda_n W = C_n = \frac{2}{L} \int_0^L [f(x) - t_c] \sin \lambda_n x.$$

Since $f(x)$, unspecified here as to form, is a *known* boundary condition, the above integration could be performed [perhaps numerically, depending on the form of $f(x)$]. Still being general in not specifying $f(x)$ yet, one finds that the final solution of Eq. (4.7) is

$$\theta = \frac{2}{L} \sum_{n=1}^{\infty} \frac{\sinh\left(\dfrac{n\pi y}{L}\right)}{\sinh\left(\dfrac{n\pi W}{L}\right)} \sin\left(\frac{n\pi x}{L}\right) \int_0^L [f(x) - t_c] \sin\left(\frac{n\pi x}{L}\right) dx. \tag{4.9}$$

One Edge at a Uniform Temperature. A special case in which the edge at $y = W$ is maintained at a constant temperature [i.e., $f(x) = t_0$, a constant] is illustrated in Fig. 4.2. For this case Eq. (4.9) becomes

$$\frac{t - t_c}{t_0 - t_c} = \frac{\theta}{t_0 - t_c} = 2 \sum_{n=1}^{\infty} \frac{1 - (-1)^n}{n\pi} \frac{\sinh\left(\dfrac{n\pi y}{L}\right)}{\sinh\left(\dfrac{n\pi W}{L}\right)} \sin\left(\frac{n\pi x}{L}\right). \tag{4.10}$$

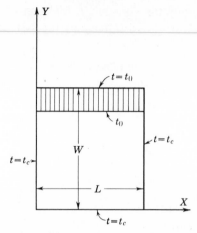

Fig. 4.2. The rectangular plate with one edge at a uniform temperature, all other edges at constant temperature.

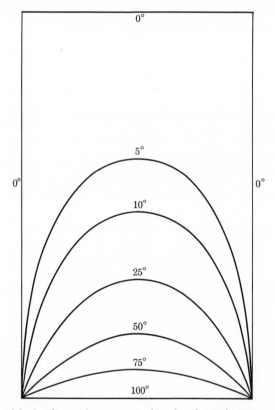

Fig. 4.3. Isotherms in a rectangular plate heated on one edge.

This relation results from the evaluation of

$$\int_0^L (t_0 - t_c) \sin\left(\frac{n\pi x}{L}\right) dx.$$

Equation (4.10) would then enable one to compute the temperature at any desired point in the plate. As an illustration, Fig. 4.3 shows a rectangular plate, 10 by 6, with one edge held at 100° and all others held at 0°. The isothermal lines within the plate were plotted by use of Eq. (4.10).

The Rectangular Plate with a Specified Temperature Distribution on More Than One Edge

A rectangular plate with specified temperature functions on more than one edge may be reduced to the above problem by a simple superposition. Consider, for example, the situation illustrated in Fig. 4.4, in which the plate has a distribution $t = f(x)$ specified at $y = W$, and $t = \varphi(x)$ is imposed at $y = 0$. Let the other edges be maintained at $t = t_c$. Then one must solve

$$\frac{\partial^2 \theta}{\partial x^2} + \frac{\partial^2 \theta}{\partial y^2} = 0,$$

$$\theta = f(x) - t_c \quad \text{at } y = W,$$
$$\theta = \varphi(x) - t_c \quad \text{at } y = 0, \qquad (4.11)$$
$$\theta = 0 \qquad \text{at } x = 0,$$
$$\theta = 0 \qquad \text{at } x = L,$$

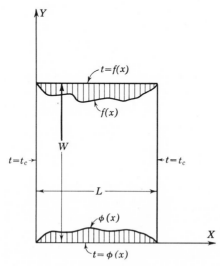

Fig. 4.4

where $\theta = t - t_c$ has again been introduced. Because the differential equation in Eq. (4.11) is linear, it may be reduced to two simpler systems by defining u and v such that

$$\theta = u + v. \tag{4.12}$$

The symbols u and v are used to denote the solutions to the following two systems:

$$\frac{\partial^2 u}{\partial x^2} + \frac{\partial^2 u}{\partial y^2} = 0,$$

$$u = f(x) - t_c \quad \text{at } y = W,$$

$$u = 0 \qquad\qquad \text{at } y = 0, \tag{4.13}$$

$$u = 0 \qquad\qquad \text{at } x = 0,$$

$$u = 0 \qquad\qquad \text{at } x = L.$$

$$\frac{\partial^2 v}{\partial x^2} + \frac{\partial^2 v}{\partial y^2} = 0,$$

$$v = 0 \qquad\qquad \text{at } y = W,$$

$$v = \varphi(x) - t_c \quad \text{at } y = 0, \tag{4.14}$$

$$v = 0 \qquad\qquad \text{at } x = 0,$$

$$v = 0 \qquad\qquad \text{at } x = L.$$

The solution to the system of Eq. (4.13) is the same as that expressed in Eq. (4.9) with u replacing θ. By a simple change of variable the solution given in Eq. (4.9) may be made to apply to the determination of v from Eq. (4.14):

$$v = \frac{2}{L} \sum_{n=1}^{\infty} \frac{\sinh\left(\dfrac{n\pi(W - y)}{L}\right)}{\sinh\left(\dfrac{n\pi W}{L}\right)} \sin\left(\frac{n\pi x}{L}\right) \int_0^L \varphi(x) \sin\left(\frac{n\pi x}{L}\right) dx. \tag{4.15}$$

Then, by Eq. (4.12), the solution to the system of Eq. (4.11) is the sum of Eqs. (4.9) and (4.15).

4.3   Steady Conduction in a Circular Cylinder of Finite Length

As a final example of steady conduction in two space dimensions consider, as depicted in Fig. 4.5, a solid circular cylinder of finite length. The symbol R denotes the outer radius of the cylinder and L denotes its length.

Cylindrical coordinates are the natural ones to use in this case because of the ease in specification of the boundary conditions. Of special interest are steady state conduction problems in such a cylindrical configuration but with the boundary conditions so chosen that *axial symmetry* exists. That is, if one considers problems (in the steady state) in which there is no dependence on the circumferential coordinate, the distribution of temperature in the cylinder depends only on the radial coordinate, r, and the axial coordinate, z. Thus, the general conduction equation in cylindrical coordinates, Eq. (1.10), reduces to the following:

$$\frac{\partial^2 \theta}{\partial r^2} + \frac{1}{r} \frac{\partial \theta}{\partial r} + \frac{\partial^2 \theta}{\partial z^2} = 0. \tag{4.16}$$

In Eq. (4.16) anticipation has been made of the eventual introduction of the temperature difference variable $\theta = t - t_c$. Then, again, a two-dimensional conduction problem is obtained. The imposed boundary conditions must, of course, be independent of the circumferential coordinate.

One obtains the general solution to Eq. (4.16) by use of the procedure that was employed in Sec. 4.2 to arrive at the solution of Eq. (4.1). That is, one seeks a solution in the form

$$\theta = \mathcal{R}(r) \cdot Z(z), \tag{4.17}$$

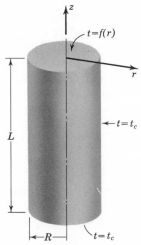

Fig. 4.5

in which $\mathscr{R}(r)$ and $Z(z)$ denote the functions of r and z only, respectively. Substitution of Eq. (4.17) into Eq. (4.16) and subsequent rearrangement leads to

$$\frac{1}{\mathscr{R}}\frac{d^2\mathscr{R}}{dr^2} + \frac{1}{r}\frac{1}{\mathscr{R}}\frac{d\mathscr{R}}{dr} = -\frac{1}{Z}\frac{d^2Z}{dz^2}.$$

In the equation, the variables have been separated—the right side being a function of z alone and the left side a function of r alone. Since the two are equal and since each is a function of a single independent variable, the two sides can equal only a constant. Calling the constant λ^2, one obtains two ordinary differential equations in place of the original partial differential equation:

$$\frac{d^2\mathscr{R}}{dr^2} + \frac{1}{r}\frac{d\mathscr{R}}{dr} + \lambda^2\mathscr{R} = 0, \tag{4.18}$$

$$\frac{d^2Z}{dz^2} - \lambda^2 Z = 0. \tag{4.19}$$

Equation (4.18) is recognized as Bessel's equation of zero order (Appendix C), whereas Eq. (4.19) leads to the hyperbolic functions. Thus, upon applying the definition of $\mathscr{R}$ and Z in Eq. (4.17), the solution of the differential equation in Eq. (4.16) may be expressed as

$$\theta = [B_1 J_0(\lambda r) + B_2 Y_0(\lambda r)](B_3 \sinh \lambda z + B_4 \cosh \lambda z). \tag{4.20}$$

Taking a particular problem, let the surface of the cylinder noted in Fig. 4.5 be maintained at a constant temperature t_c at every surface except for the circular end at $z = L$. On this surface let the temperature be specified as a known function of the radius r [i.e., $f(r)$]. Thus, the problem is now: Find an equation giving the distribution of the temperature throughout the cylinder [i.e., find $\theta = g(r, z)$] for the following conditions:

$$\text{At } z = 0: \quad t = t_c, \quad \text{or } \theta = 0. \tag{1}$$

$$\text{At } z = L: \quad t = f(r), \text{ or } \theta = f(r) - t_c. \tag{2} \qquad (4.21)$$

$$\text{At } r = R: \quad t = t_c, \quad \text{or } \theta = 0. \tag{3}$$

One additional condition exists: that the temperature must be finite at $r = 0$. Knowing that $Y_0(\lambda r) \to \infty$ as $\lambda r \to 0$ (see Table C.1), one obtains $B_2 = 0$. Condition (1) of Eq. (4.21) makes B_4 also vanish, so one has

$$\theta = B \sinh \lambda z \, J_0(\lambda r),$$

in which B replaces $B_1 B_3$. Application of condition (3) demands that

$$B \sinh \lambda z \, J_0(\lambda R) = 0.$$

The only way that this latter condition can be satisfied for all values of z between 0 and L is for

$$J_0(\lambda R) = 0.$$

Examination of the tables of $J_0(\lambda R)$ in Appendix C shows that J_0 has a succession of zeros that differ by an interval approaching π as $\lambda R \to \infty$. Hence there are a countably infinite number of λ's satisfying the defining relation:

$$J_0(\lambda_n R) = 0. \tag{4.22}$$

The first five are (Ref. 1) $\lambda_1 R = 2.4048$, $\lambda_2 R = 5.5201$, $\lambda_3 R = 8.6537$, $\lambda_4 R = 11.7915$, and $\lambda_5 R = 14.9309$.

Hence, the general solution is the sum of all the solutions corresponding to each of the λ_n's:

$$\theta = \sum_{n=1}^{\infty} (B_n \sinh \lambda_n z) J_0(\lambda_n r).$$

Finally, application of condition (2) of Eq. (4.21) determines the unknown B_n's since this gives

$$f(r) - t_c = \sum_{n=1}^{\infty} (B_n \sinh \lambda_n L) J_0(\lambda_n r).$$

Comparison of this last equation with Eq. (D.27) indicates that, since the λ's are defined by $J_0(\lambda_n R) = 0$, the constant coefficients $(B_n \sinh \lambda_n L)$ are given by

$$B_n \sinh \lambda_n L = \frac{\displaystyle\int_0^R r[f(r) - t_c] J_0(\lambda_n r) \, dr}{\dfrac{R^2}{2} J_1^2(\lambda_n R)}.$$

Finally, then, the solution to the problem giving the distribution of the temperature through the cylinder is

$$\theta = \frac{2}{R^2} \sum_{n=1}^{\infty} \frac{\sinh \lambda_n z}{\sinh \lambda_n L} \frac{J_0(\lambda_n r)}{J_1^2(\lambda_n R)} \int_0^R r[f(r) - t_c] J_0(\lambda_n r) \, dr. \tag{4.23}$$

One End at Uniform Temperature. For the special case in which $f(r) = t_0$ (a constant), i.e., if the cylinder is maintained at zero temperature on all surfaces except one end where the temperature is t_0, the solution reduces to

$$\frac{t - t_c}{t_0 - t_c} = \frac{\theta}{t_0 - t_c} = 2 \sum_{n=1}^{\infty} \frac{1}{\lambda_n R} \frac{\sin \lambda_n z}{\sinh \lambda_n L} \frac{J_0(\lambda_n r)}{J_1(\lambda_n R)}. \tag{4.24}$$

The relations of Appendix D have been used to show that

$$\int_0^R r J_0(\lambda_n r)\, dr = \frac{R}{\lambda_n} J_1(\lambda_n R).$$

The principle of superposition may be used to build up other cases. For example, the temperature in a cylinder heated to uniform temperatures at each end and zero temperature on its other surface may be found by adding two solutions of the form of Eq. (4.24).

Figure 4.6 plots the axial temperature distribution ($r = 0$) for the solid cylinder heated to constant temperature on one end. Also, the case of a cylinder heated on both ends is plotted. This is easily obtained by superposition from the first case.

EXAMPLE 4.1:   A solid cylinder 4 in. long, 8 in. in diameter is heated to 300°F on its cylindrical surface, 500°F on one end, and 200°F at its other end. What is the temperature at the center of the cylinder?

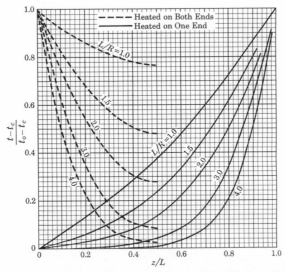

Fig. 4.6. Axial temperature distribution in solid cylinders of finite length.

Solution: By utilizing the principle of superposition, 300°F may be subtracted from all temperatures. Thus, the problem is equivalent to a cylinder with the cylindrical surface at 0°F and the ends at 200°F and -100°F, respectively. This, in turn, is equivalent to one cylinder at 0°F on all surfaces and 200°F on one end plus another cylinder at 0°F on all surfaces and -100°F on one end.

For the first of these, Fig. 4.6 gives (for $L/R = 1$, $z/L = 0.5$) the center temperature to be

$$t = 200 \times \frac{t - t_c}{t_0 - t_c} = 200 \times 0.385 = 77.0°F.$$

For the second

$$t = -100 \times \frac{t - t_c}{t_0 - t_c} = -100 \times 0.385 = -38.5°F.$$

Then the answer to the original problem, adding on the 300°F originally subtracted, is

$$t = 77 - 38.5 + 300 = 338.5°F \ (170.3°C).$$

4.4   Nonsteady Conduction in One Space Dimension

The problems of describing the temperature distribution and its variation with time for *nonsteady* heat conduction in only *one space dimension* has much similarity to the two-dimensional steady conduction problems discussed above in that both cases involve the determination of the temperature in terms of two independent variables. First, transient problems in which the boundary conditions are suddenly changed and maintained at new constant values will be discussed—in both rectangular and cylindrical coordinates. Instances in which the boundary temperature of a solid body is subjected to a sudden change will be considered as well as the more practical cases in which the temperature of a surrounding convecting fluid is suddenly changed from one value to another. In Sec. 4.8 a particular case in which the boundary conditions vary periodically with time will be investigated.

4.5   Transient Conduction in the Infinite Slab

In order that the conduction of heat will depend on only one *space* variable, it will be necessary to select geometrical configurations of special types. In the case in which rectangular coordinates are to be used, one considers conduction through a plane slab having a finite thickness in one direction but having an infinite extent in the other directions. That is, edge effects are to be neglected so that only the coordinate measured in the direction of the finite thickness is needed to describe positions. Take, as illustrated in Fig. 4.7, the x coordinate as the one-dimensional coordinate. Then, with time, τ, as a variable the general conduction equation, Eq. (1.8), reduces to the following with internal heat generation absent:

$$\frac{\partial t}{\partial \tau} = \alpha \frac{\partial^2 t}{\partial x^2}.$$

As in the cases considered in Sec. 4.4, it will prove desirable to express solutions in terms of a temperature difference variable, θ, to be defined specifically in each case which follows, so that the basic differential equation becomes

$$\frac{\partial \theta}{\partial \tau} = \alpha \frac{\partial^2 \theta}{\partial x^2} \qquad (4.25)$$

where x is the single space variable and α is the thermal diffusivity defined in Chapter 1.

The solution to this equation can be found in the same manner as used in Sec. 4.3. Seeking solutions of the separable form, one assumes that the solution

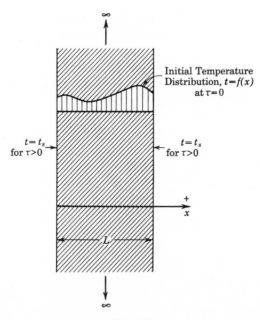

Fig. 4.7

is representable as

$$\theta = X(x) \cdot T(\tau), \tag{4.26}$$

in which $X(x)$ and $T(\tau)$ represent functions of x and τ only. Introducing this into Eq. (4.25), one obtains

$$X \frac{dT}{d\tau} = \alpha T \frac{d^2 X}{dx^2}.$$

This equation separates into

$$\frac{1}{\alpha T} \frac{dT}{d\tau} = \frac{d^2 X}{dx^2} \frac{1}{X} = -\lambda^2.$$

The separation parameter, λ^2, has been introduced in the above equality since each side is a function of only a single variable. The sign of this constant has been taken to be negative to ensure a negative exponential solution in time. This separation of variables leads to the following two ordinary differential equation with the indicated solutions:

$$\frac{dT}{d\tau} + \alpha \lambda^2 T = 0, \qquad T(\tau) = B_1' e^{-\alpha \lambda^2 \tau}$$

$$\frac{d^2 X}{dx^2} + \lambda^2 X = 0, \qquad X(x) = B_2' \sin \lambda x + B_3' \cos \lambda x.$$

The B's are arbitrary constants. From the definition in Eq. (4.26) the general solution to Eq. (4.25) is

$$\theta = e^{-\lambda^2 \alpha \tau}(B_1 \sin \lambda x + B_2 \cos \lambda x). \tag{4.27}$$

Certain particular problems in which the boundary conditions are used to evaluate the constants will now be discussed.

Sudden Changes in the Boundary Temperature of an Infinite Slab

Consider first the infinite slab which is initially heated to some known, but for the present unprescribed, temperature distribution. Call this initial distribution $f(x)$. Then at time $\tau = 0$ let the bounding surfaces have their temperatures reduced to and maintained at a constant temperature t_s. This is shown in Fig. 4.7.

If the temperature-difference variable is defined as $\theta = t - t_s$, the conditions on Eq. (4.27) are

$$\text{For } \tau = 0: \quad t = f(x), \text{ or } \theta = f(x) - t_s. \qquad (1)$$

$$\text{For } \tau \geq 0: \quad \text{at } x = 0, \quad t = 0 \text{ or } \theta = 0 \qquad (2) \qquad (4.28)$$

$$\text{at } x = L, \quad t = 0 \text{ or } \theta = 0. \qquad (3)$$

Application of conditions (2) and (3) to Eq. (4.27) yields

$$B_2 = 0 \quad \text{and} \quad \sin \lambda_n L = 0.$$

Thus, following the procedure used in the earlier discussions, one obtains the solution

$$\theta = \sum_{n=1}^{\infty} e^{-\lambda_n^2 \alpha \tau} B_n \sin \lambda_n x,$$

$$\lambda_n = \frac{n\pi}{L}; \quad n = 1, 2, 3, \cdots. \qquad (4.29)$$

Application of the initial condition at $\tau = 0$ gives

$$f(x) - t_s = \sum_{n=1}^{\infty} B_n \sin \lambda_n x.$$

Upon comparing these results with Eq. (D.9), Eq. (D.10) shows that

$$B_n = \frac{2}{L} \int_0^L [f(x) - t_s] \sin \lambda_n x \, dx.$$

Thus, the equation of the distribution of temperature throughout the slab is

$$\theta = \frac{2}{L} \sum_{n=1}^{\infty} e^{-(n\pi/L)^2 \alpha \tau} \sin \left(\frac{n\pi x}{L} \right) \int_0^L f(x) \sin \left(\frac{n\pi x}{L} \right) dx, \qquad (4.30)$$

where $f(x)$ is the known distribution of temperature in the slab at the time $\tau = 0$. This last expression is a general one in that it gives the temperature distribution in the slab as a function of time and as dependent on the specified initial temperature distribution.

Uniform Initial Temperature Distribution. For the special case of a uniform initial distribution [i.e., $f(x) = t_i$ (a constant)], one has the problem of a slab initially heated to a uniform temperature and which suddenly has its surface temperatures reduced to and maintained at constant temperature. Equation (4.30) for the temperature–time history of the slab becomes

$$\frac{t - t_s}{t_i - t_s} = \frac{\theta}{t_i - t_s} = 2 \sum_{n=1}^{\infty} e^{-(n\pi/L)^2 \alpha\tau} \frac{1 - (-1)^n}{n\pi} \sin\left(\frac{n\pi x}{L}\right).$$

This is most easily expressed graphically when written in a dimensionless form in terms of the following parameters:

Dimensionless position parameter: $\xi = \dfrac{x}{L}$.

Dimensionless time, the Fourier modulus: $N_{\text{FO}} = \dfrac{\alpha\tau}{L^2}$.

Then the dimensionless temperature–time history in the slab is given by

$$\frac{t - t_s}{t_i - t_s} = \frac{2}{\pi} \sum_{n=1}^{\infty} e^{-(n\pi)^2 N_{\text{FO}}} \frac{1 - (-1)^n}{n} \sin n\pi\xi. \tag{4.31}$$

The rate of heat flow out of the slab is (per unit area normal to the x direction)

$$\frac{q}{A} = \left[-k\left(\frac{\partial t}{\partial x}\right)_{x=0} \right] 2.$$

The factor 2 appears due to symmetry and the fact that heat flows out *two* faces of the slab, at $x = 0$ and $x = L$. The total heat flow up to time τ is then

$$\frac{Q}{A} = \int_0^\tau \frac{q}{A} \, d\tau.$$

Introduction of Eq. (4.31) gives, since $\alpha = k/\rho c_p$,

$$\frac{Q}{ALpc_p(t_i - t_s)} = 4 \sum_{n=1}^{\infty} \frac{1 - (-1)^n}{(n\pi)^2} [1 - e^{-(n\pi)^2 N_{\text{FO}}}].$$

The denominator is recognizable as the total heat stored in the slab initially—measured relative to the fixed boundary temperature. Call this $Q_i = ALpc_p(t_i - t_s)$, so the ratio of the total heat flow out of the slab up to time τ to that initially in the slab is

$$\frac{Q}{Q_i} = 4 \sum_{n=1}^{\infty} \frac{1 - (-1)^n}{(n\pi)^2} [1 - e^{-(n\pi)^2 N_{\text{FO}}}]. \tag{4.32}$$

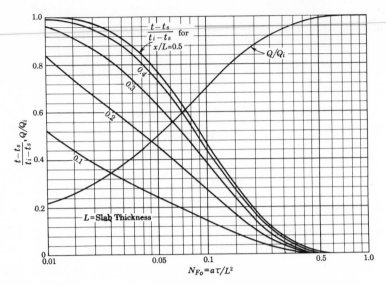

Fig. 4.8. The time variation of the temperature distribution and heat flow in an infinite slab initially at temperature t_i, which has its surface temperature suddenly changed to t_s.

Figure 4.8 shows plots of Eqs. (4.31) and (4.32) so that the temperature–time history and the total heat flow from the slab may be readily determined as functions of time.

EXAMPLE 4.2:   A 10-in. wall of common brick is initially at 180°F and suddenly has its surfaces reduced to 60°F. Find the temperature at a plane 4 in. from the surface after 2 hr have passed. How much heat has been conducted out of the wall during that time?

Solution: Table A.2 gives

$$\rho = 100 \text{ lb}_m/\text{ft}^3, \qquad c_p = 0.2 \text{ Btu/lb}_m\text{-}°\text{F},$$
$$\alpha = 0.02 \text{ ft}^2/\text{hr}, \qquad k = 0.4 \text{ Btu/hr-ft-}°\text{F}.$$

For the given conditions, the Fourier modulus is

$$N_{FO} = \frac{\alpha \tau}{L^2} = \frac{0.02 \times 2}{(10/12)^2} = 0.0576.$$

For $\xi = x/L = 4/10 = 0.4$, Fig. 4.8 gives

$$\frac{t - t_s}{t_i - t_s} = 0.685, \qquad \frac{Q}{Q_i} = 0.540.$$

Thus,

$$t = (0.685)(180 - 60) + 60$$

$$= 143°F\ (61.7°C).$$

Since

$$Q_i = AL\rho c_p(t_i - t_s),$$

$$\frac{Q_i}{A} = \frac{10}{12} \times 100 \times 0.2(180 - 60) = 2000\ \text{Btu/ft}^2,$$

then

$$\frac{Q}{A} = 2000 \times 0.540$$

$$= 1080\ \text{Btu/ft}^2\ (1.227 \times 10^7\ \text{J/m}^2).$$

Sudden Changes in the Temperature of the Fluid Surrounding an Infinite Slab

A situation having more practical significance than that discussed above is one in which the slab is surrounded by a convecting fluid. For example, the quenching of a heated solid in a liquid bath is a nonsteady conduction case in which the temperature of the surrounding fluid, not the surface of the solid, is suddenly changed to and maintained at some temperature different from the initial solid temperature.

Consider, then, an infinite slab, as shown in Fig. 4.9(a), of thickness $2L$ in which at time $\tau = 0$ a known temperature distribution, $\varphi(x)$, exists. The slab is surrounded on both sides by a convecting fluid so that a film coefficient, h, exists at the surface. Then let the temperature of this surrounding fluid be suddenly reduced to t_f (or any constant) and maintained at that temperature for all subsequent times. This is equivalent to plunging a heated bar into a quenching bath.

The condition to impose on the differential equation at the boundary is obtained from Newton's law of cooling as expressed in Eq. (1.15):

$$\frac{\partial t}{\partial x} = -\frac{h}{k}(t - t_f).$$

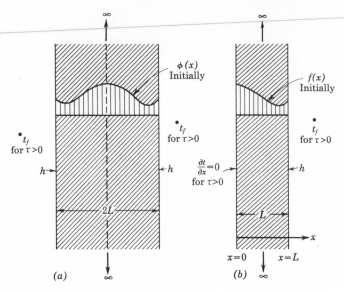

Fig. 4.9

Without much loss in generality as far as practical problems are concerned, let it be assumed that the initial distribution of the slab temperature, $\varphi(x)$, is symmetrical with respect to the midline of the slab. If this is so, there will be *no* heat flow across the midline and the temperature history in the slab will be the same on each side of the midline since the fluid on each side is reduced to the same temperature. In such an instance the problem may be reduced to that shown in Fig. 4.9(b), in which one considers an infinite slab of thickness L which is initially heated to a known distribution $f(x)$ and which has its face at $x = 0$ insulated so that no heat can cross that boundary. Then for all subsequent times, the fluid surrounding the slab at its $x = L$ face is reduced to and maintained at temperature t_f while convection takes place at $x = L$ through a film coefficient of h. The film coefficient will be treated as a constant.

The problem then is to evaluate the solution to Eq. (4.25), i.e.,

$$\theta = e^{-\lambda^2 \alpha \tau}(B_1 \sin \lambda x + B_2 \cos \lambda x).$$

(4.27)

In this case the temperature difference variable is taken to be

$$\theta = t - t_f.$$

The boundary conditions are

$$\text{For } \tau = 0: \quad t = f(x), \text{ or } \theta = f(x) - t_f. \tag{1}$$

$$\text{For } \tau \geq 0: \quad \text{at } x = 0, \frac{\partial t}{\partial x} = 0 \text{ or } \frac{\partial \theta}{\partial x} = 0. \tag{2}$$

$$\text{(4.33)}$$

$$\text{at } x = L, \frac{\partial t}{\partial x} = -\frac{h}{k}(t - t_f).$$

$$\text{or } \frac{\partial \theta}{\partial x} = -\frac{h}{k}\theta. \tag{3}$$

Application of condition (2) gives

$$\frac{\partial \theta}{\partial x}\bigg|_{x=0} = 0 = e^{-\lambda^2 \alpha \tau}\lambda(B_1 \cos \lambda x - B_2 \sin \lambda x)\bigg|_{x=0},$$

or

$$B_1 = 0.$$

So the solution reduces to

$$\theta = e^{-\lambda^2 \alpha \tau}B \cos \lambda x.$$

Then condition (3) gives

$$[e^{-\lambda^2 \alpha \tau}B\lambda(-\sin \lambda x)]_{x=L} = -\frac{h}{k}[e^{-\lambda^2 \alpha \tau}B \cos \lambda x]_{x=L},$$

or

$$\lambda \sin \lambda L = \frac{h}{k}\cos \lambda L.$$

This is equivalent to

$$\tan \lambda L = \frac{hL}{k}\frac{1}{\lambda L}. \tag{4.34}$$

For purposes of visualization, rewrite Eq. (4.34) as

$$\cot \lambda L = \frac{k}{hL}\lambda L. \tag{4.35}$$

Plotting each side of Eq. (4.35) as functions of λL (see Fig. 4.10), one observes that the equation is satisfied for an infinite succession of values of the parameter λL so that for a given L, Eq. (4.34) defines the value of λ. This succession of values of λ will be denoted by λ_n, which are seen to depend on the magnitude of the parameter hL/k. This latter parameter, a dimensionless parameter dependent only on the physical system, is termed *Biot's modulus* and is symbolized

$$N_{BI} = \frac{hL}{k}. \tag{4.36}$$

This parameter expresses the relative resistance of the interior of the solid to that of the film at the boundary. The values of λ_n have been evaluated and tabulated (Ref. 2). For example, for $N_{BI} = 10$, the first five roots of Eq. (4.38) are $\lambda_1 L = 1.4289$, $\lambda_2 L = 4.3058$, $\lambda_3 L = 7.2281$, $\lambda_4 L = 10.2003$, $\lambda_5 L = 13.2143,\ldots$.

Thus, one now has, as the solution,

$$\theta = \sum_{n=1}^{\infty} B_n \, e^{-\lambda_n^2 \alpha \tau} \cos \lambda_n x,$$

with λ_n being the nth root of

$$\lambda_n L \tan \lambda_n L - N_{BI} = 0. \tag{4.37}$$

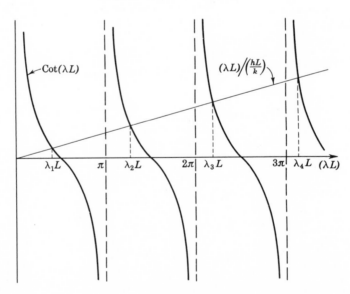

Fig. 4.10

Application of the initial condition (1) of Eq. (4.33) gives

$$f(x) - t_f = \sum_{n=1}^{\infty} B_n \cos \lambda_n x. \tag{4.38}$$

This implies that one wishes to represent $f(x)$ as a series of cosines when λ_n is defined as a root of Eq. (4.37). This is discussed in Appendix D, where comparison of the above Eqs. (4.37) and (4.38) with Eqs. (D.16), (D.19), and (D.20) shows that the coefficient, B_n, must be

$$B_n = \frac{\displaystyle\int_0^L [f(x) - t_f] \cos \lambda_n x \, dx}{\dfrac{L}{2} + \dfrac{\sin \lambda_n L \cos \lambda_n L}{2\lambda_n}}.$$

The final expression for the temperature in the slab, as a function of position and time, is then

$$\theta = 2 \sum_{n=1}^{\infty} \lambda_n e^{-\lambda_n^2 \alpha \tau} \frac{\cos \lambda_n x}{\lambda_n L + \sin \lambda_n L \cos \lambda_n L} \int_0^L [f(x) - t_f] \cos \lambda_n x \, dx. \tag{4.39}$$

Uniform Initial Temperature Distribution. Of particular practical interest is the case in which the initial distribution, $f(x)$, is a constant, say, t_i. Then Eq. (4.39) becomes

$$\frac{t - t_f}{t_i - t_f} = \frac{\theta}{t_i - t_f} = 2 \sum_{n=1}^{\infty} e^{-\lambda_n^2 \alpha \tau} \frac{\sin \lambda_n L}{\lambda_n L + \sin \lambda_n L \cos \lambda_n L} \cos \lambda_n x. \tag{4.40}$$

The total heat removed from the slab up to time τ is, as in the previous section,

$$Q = \int_0^\tau q \, d\tau$$

$$= -kA \int_0^\tau \left(\frac{\partial t}{\partial x} \right)_{x=L} d\tau.$$

Performing the indicated differentiation and integration on Eq. (4.40) and again introducing $Q_i = kAL(t_i - t_f)/\alpha = AL\rho c_p(t_i - t_f)$ as the heat initially stored in the *half-slab* (since, due to symmetry, only half of the full slab of width $2L$ is being treated), one obtains

$$\frac{Q}{Q_i} = 2 \sum_{n=1}^{\infty} \frac{1}{\lambda_n L} \frac{\sin^2 \lambda_n L}{\lambda_n L + \sin \lambda_n L \cos \lambda_n L} (1 - e^{-\lambda_n^2 \alpha \tau}). \qquad (4.41)$$

As above, these expressions, Eqs. (4.40) and (4.41), can be written a little more generally by introducing the dimensionless parameters:

Dimensionless position parameter: $\xi = \dfrac{x}{L}$.

Dimensionless time, Fourier modulus: $N_{FO} = \dfrac{\alpha \tau}{L^2}$.

Dimensionless relative resistance, Biot modulus: $N_{BI} = \dfrac{hL}{k}$.

One then has

$$\frac{t - t_f}{t_i - t_f} = 2 \sum_{n=1}^{\infty} e^{-\delta_n^2 N_{FO}} \frac{\sin \delta_n}{\delta_n + \sin \delta_n \cos \delta_n} \cos \delta_n \xi, \qquad (4.42)$$

$$\frac{Q}{Q_i} = 2 \sum_{n=1}^{\infty} \frac{1}{\delta_n} \frac{\sin^2 \delta_n}{\delta_n + \sin \delta_n \cos \delta_n} (1 - e^{-\delta_n^2 N_{FO}}), \qquad (4.43)$$

with δ_n being the nth root of

$$\delta_n \tan \delta_n - N_{BI} = 0. \qquad (4.44)$$

The first of these equations, for the temperature distribution, has been evaluated numerically by a number of investigators for ranges of the three parameters ξ (position), N_{FO} (time), and N_{BI} (relative resistance). The results of Reference 3 are presented in Figs 4.11 and 4.12. In these figures the dimensionless temperature distribution of Eq. (4.42) is plotted as a function of N_{FO} and N_{BI} for selected values of the position parameter ξ. It should be remembered in using these charts that L represents the slab *half-thickness*. Figure 4.13 presents the data of Gröber (Ref. 4) for the relative heat flow as expressed in Eq. (4.43) plotted as a function of N_{FO} and N_{BI}. These charts may be used to analyze practical problems of transient conduction in plates if the edge effects are negligible so that the one-dimensional assumption is reasonably well satisfied. The quantity Q/Q_i may be used for purposes other

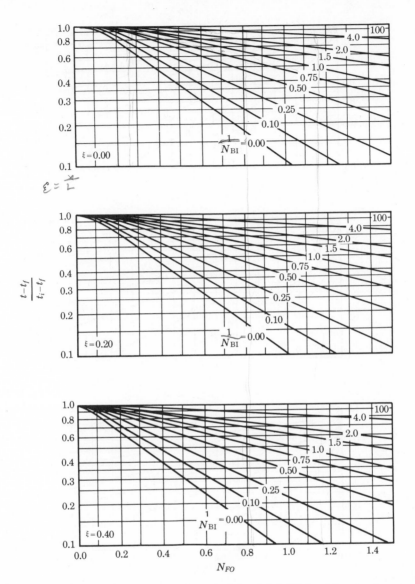

Fig. 4.11. Temperature–time history in an infinite slab initially at temperature t_i and placed in a medium at t_f, for $\xi = 0$, 0.2, and 0.4. (From *Heat Transfer Notes* by L. M. K. Boelter, V. H. Cherry, H. A. Johnson, and R. C. Martinelli. Copyright 1965. McGraw-Hill Book Company, New York. Used by permission.)

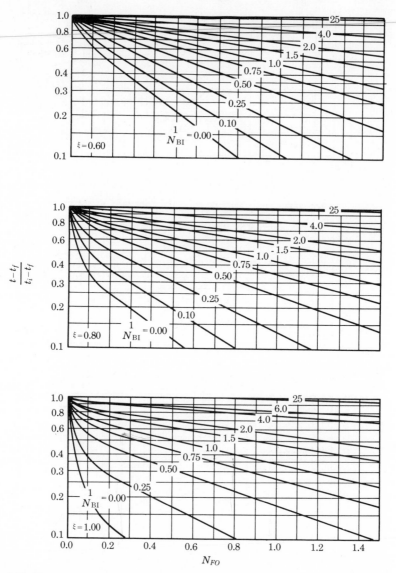

Fig. 4.12. Temperature–time history in an infinite slab initially at temperature t_i and placed in a medium at t_f, for $\xi = 0.6$, 0.8, and 1.0. (From *Heat Transfer Notes* by L. M. K. Boelter, V. H. Cherry, H. A. Johnson, and R. C. Martinelli. Copyright 1965. McGraw-Hill Book Company, New York. Used by permission.)

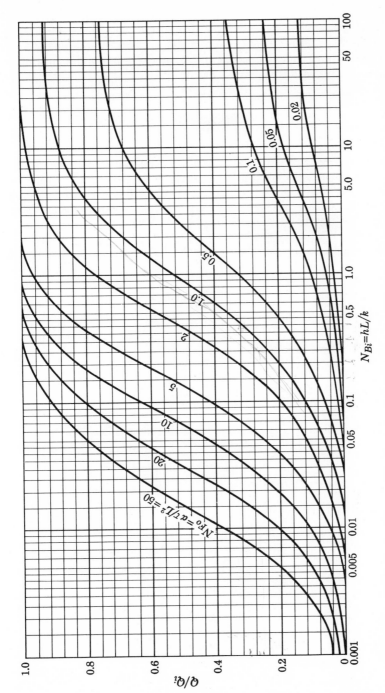

Fig. 4.13. The heat flow from an infinite slab as a function of time and thermal resistance. The slab is initially at a uniform temperature and is placed in a medium at constant temperature. (From H. Gröber, *Einführung in die Lehre der Wärmeübertragung*, Berlin, Springer, 1926. Used by permission.)

than the calculation of heat flow. Since Q/Q_i represents the relative heat flow out of the half slab, $(1 - Q/Q_i)$ represents the relative amount of heat left in the slab at any time. Consequently, $(1 - Q/Q_i)$ is the dimensionless average temperature within the slab. That is,

$$1 - \frac{Q}{Q_i} = \frac{t_{av} - t_f}{t_i - t_f}.$$

EXAMPLE 4.3: The brick wall of the previous example (10 in. thick, initially at 180°F) has a medium of 60°F suddenly placed in contact with it. For a film coefficient of $h = 4$ Btu/hr-ft²-°F, find the temperature at a 1-in. depth after 10 hr have passed. Find the heat that has flowed out of the wall during that time and the average temperature within the wall at the end of the 10 hr.

Solution: For the properties quoted in the previous example and taking $L = 10/2 = 5$ in.:

$$N_{FO} = \frac{\alpha\tau}{L^2} = \frac{0.02 \times 10}{(5/12)^2} = 1.15,$$

$$N_{BI} = \frac{hL}{k} = \frac{4 \times 5/12}{0.4} = 4.17,$$

$$\xi = \frac{x}{L} = \frac{4}{5} = 0.8.$$

Figures 4.12 and 4.13 give

$$\frac{t - t_f}{t_i - t_f} = 0.13,$$

$$\frac{Q}{Q_i} = 0.92.$$

The initially stored heat is, per unit area,

$$\frac{Q_i}{A} = \frac{10/2}{12} \times 100 \times 0.2 \times (180 - 60) = 1000 \text{ Btu/ft}^2.$$

Thus,

$$t = 0.13 \times (180 - 60) + 60 = 75.6°F \ (24.2°C),$$

$$\frac{Q_i}{A} = 1000 \times 0.92 = 920 \text{ Btu/ft}^2 \text{ for one side}$$

$$= 1840 \text{ Btu/ft}^2 \ (2.09 \times 10^7 \text{ J/m}^2) \text{ for both sides,}$$

$$t_{av} = (1 - 0.92) \times (180 - 60) + 60 = 69.6°F \ (20.9°C).$$

4.6   Transient Radial Conduction in a Long Solid Cylinder

Equally important as the slab configuration just discussed is the solid circular cylinder. In order to ensure one-dimensional conduction so that the transient problem will involve only two independent variables, it will be assumed that the cylinder is infinitely long and that axial symmetry exists. Thus, the heat conduction can be considered to be in a radial direction only. Then the heat conduction equation in cylindrical coordinates, Eq. (1.10), reduces to the following with internal heat generation absent:

$$\frac{\partial \theta}{\partial \tau} = \alpha \left(\frac{\partial^2 \theta}{\partial r^2} + \frac{1}{r} \frac{\partial \theta}{\partial r} \right). \tag{4.45}$$

Again seeking solution of the separable form, one finds that

$$\theta = \mathscr{R}(r) \cdot T(\tau), \tag{4.46}$$

where $\mathscr{R}(r)$ and $T(\tau)$ are functions of only the indicated variable. Then Eq. (4.45) separates into

$$\frac{1}{\alpha T} \frac{dT}{d\tau} = \frac{1}{\mathscr{R}} \left(\frac{d^2 \mathscr{R}}{dr^2} + \frac{1}{r} \frac{d\mathscr{R}}{dr} \right) = -\lambda^2.$$

In the above equation λ^2 is once more used to represent the separation constant created because of the fact that each portion of this equation is a function of only one of the variables. Again, the separation constant is taken as negative to obtain a negative exponential solution in τ.

The two resulting ordinary differential equations and their solutions are

$$\frac{dT}{d\tau} + \lambda^2 \alpha T = 0,$$

$$T = B'_1 e^{-\lambda^2 \alpha \tau},$$

$$\frac{d^2 \mathscr{R}}{dr^2} + \frac{1}{r} \frac{d\mathscr{R}}{dr} + \lambda^2 \mathscr{R} = 0,$$

$$\mathscr{R} = B'_2 J_0(\lambda r) + B'_3 Y_0(\lambda r).$$

Since the cylinder is solid and since no undefined solution can occur at $r = 0$ (the center of the cylinder), B'_3 must be zero because Y_0 is undefined for zero argument. So the solution [Eq. (4.46)] to Eq. (4.45) must be

$$\theta = B e^{-\lambda^2 \alpha \tau} J_0(\lambda r). \tag{4.47}$$

The constants B and λ are to be determined by the initial and boundary conditions.

Two particular cases of this configuration will now be discussed—that of transient conduction in a cylinder with specified boundary temperature and that of conduction in a cylinder of specified ambient fluid temperature.

Sudden Change in the Boundary Temperature of an Infinitely Long Cylinder

Consider now a solid cylinder of outer radius R and infinite in length. Let it be initially heated to some temperature distribution which is a function of the radial coordinate r only, $f(r)$, in keeping with the assumed axial symmetry. Then let the temperature of its outer surface be suddenly reduced to a constant, t_s, and maintained at that value for all subsequent times. The problem then is: What is the distribution of the temperature in the cylinder at any subsequent time?

The procedure and relations used in this section are the same as those used in Secs. 4.3 and 4.5. Hence, only the briefest description will be given.

The solution is

$$\theta = Be^{-\lambda^2\alpha\tau}J_0(\lambda r), \tag{4.47}$$

and when the temperature difference variable is defined as $\theta = t - t_s$, the boundary conditions are

$$
\begin{aligned}
&\text{For } \tau = 0: \quad t = f(r) \text{ or } \theta = f(r) - t_s &&(1)\\
&\text{For } \tau \geq 0: \quad t = t_s \text{ or } \theta = 0 \text{ at } r = R. &&(2)
\end{aligned}
\tag{4.48}
$$

Condition (2) of Eq. (4.48) determines λ_n as the roots of the equation

$$J_0(\lambda_n R) = 0, \qquad n = 1, 2, 3, \ldots, \tag{4.49}$$

as in Sec. 4.4. Thus,

$$\theta = \sum_{n=1}^{\infty} B_n e^{-\lambda_n^2\alpha\tau}J_0(\lambda_n r).$$

Then application of condition (1) leads to

$$f(r) - t_s = \sum_{n=1}^{\infty} B_n J_0(\lambda_n r). \tag{4.50}$$

Then, as in Sec. 4.3, Appendix D shows that under the conditions of Eqs. (4.49) and (4.50) the B_n's are

$$B_n = \frac{\int_0^R [f(r) - t_s] J_0(\lambda_n r)\, dr}{\frac{R^2}{2} J_1^2(\lambda_n R)}.$$

Thus, the final solution is

$$\theta = \frac{2}{R^2} \sum_{n=1}^{\infty} e^{-\lambda_n^2 \alpha \tau} \frac{J_0(\lambda_n r)}{J_1^2(\lambda_n R)} \int_0^R [f(r) - t_s] J_0(\lambda_n r)\, dr.$$

$$J_0(\lambda_n R) = 0.$$

Uniform Initial Temperature Distribution. For the special case of a uniform initial distribution, $f(r) = t_i$ (a constant),

$$\frac{t - t_s}{t_i - t_s} = \frac{\theta}{t_i - t_s} = 2 \sum_{n=1}^{\infty} e^{-\lambda_n^2 \alpha \tau} \frac{1}{\lambda_n R} \frac{J_0(\lambda_n r)}{J_1(\lambda_n R)}. \tag{4.51}$$

The total heat conducted out of a unit length of the cylinder up to a time τ is

$$\frac{Q}{L} = -k(2\pi R) \int_0^\tau \left(\frac{\partial t}{\partial r} \right)_{r=R} d\tau. \tag{4.52}$$

When Eq. (4.51) is introduced, and the total heat stored in the cylinder initially is defined as

$$\frac{Q_i}{L} = \pi R^2 \rho c_p (t_s - t_i) = \pi R^2 k \frac{(t_s - t_i)}{\alpha},$$

one obtains

$$\frac{Q}{Q_i} = 4 \sum_{n=1}^{\infty} \left(\frac{1}{\lambda_n R} \right)^2 (1 - e^{-\lambda_n^2 \alpha \tau}). \tag{4.53}$$

Sudden Change in the Temperature of the Fluid Surrounding an Infinitely Long Cylinder

Consider now the more practical case of the long solid cylinder (outer radius R) which is heated initially to a known, axially symmetric, distribution of temperature, $f(r)$, and which is suddenly placed in contact with a convecting fluid of constant temperature. This is illustrated in Fig. 4.14. The equation to be solved is then

$$\theta = Be^{-\lambda^2\alpha\tau}J_0(\lambda r) \tag{4.47}$$

with the temperature difference variable defined in this case as

$$\theta = t - t_f.$$

The boundary conditions are

For $\tau = 0$: $t = f(r)$ or $\theta = f(r) - t_f$.   (1)

For $\tau \geq 0$:   at $r = R$, $\dfrac{\partial t}{\partial r} = -\dfrac{h}{k}(t - t_f)$ or $\dfrac{\partial \theta}{\partial r} = -\dfrac{h}{k}\theta$.   (2)

$$(4.54)$$

Application of condition (2) of Eq. (4.54) leads to

$$Be^{-\lambda^2\alpha\tau}[-\lambda J_1(\lambda r)]_{r=R} = -\frac{h}{k}[Be^{-\lambda^2\alpha\tau}J_0(\lambda r)]_{r=R}.$$

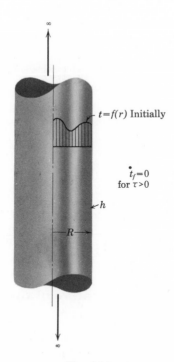

$t = f(r)$ Initially

$\dot{t}_f = 0$
for $\tau > 0$

h

R

Fig. 4.14

This gives the following defining relation for the λ_n's:

$$\lambda_n R \frac{J_1(\lambda_n R)}{J_0(\lambda_n R)} = \frac{hR}{k}.$$

Defining the Biot modulus for the cylinder to be

$$N_{\text{BI}} = \frac{hR}{k},$$

one finds that the λ_n's are the roots of

$$\lambda_n R \frac{J_1(\lambda_n R)}{J_0(\lambda_n R)} - N_{\text{BI}} = 0. \qquad (4.55)$$

Then the solution is

$$\theta = \sum_{n=1}^{\infty} B_n e^{-\lambda_n^2 \alpha t} J_0(\lambda_n r).$$

Thus, upon application of initial condition (1) of Eq. (4.54),

$$f(r) - t_f = \sum_{n=1}^{\infty} B_n J_0(\lambda_n r). \qquad (4.56)$$

Referring once more to Appendix D, Eqs. (D.31) and (D.32), one sees that the conditions of Eqs. (4.55) and (4.56) lead to the following relation for the B_n's:

$$B_n = \frac{\dfrac{2}{R^2} \displaystyle\int_0^R r[f(r) - t_f] J_0(\lambda_n r)\, dr}{J_0^2(\lambda_n R) + J_1^2(\lambda_n R)}.$$

Thus, one finally obtains, for the temperature distribution,

$$\theta = \frac{2}{R^2} \sum_{n=1}^{\infty} e^{-\lambda_n^2 \alpha t} \frac{J_0(\lambda_n r)}{J_0^2(\lambda_n R) + J_1^2(\lambda_n R)} \int_0^R r[f(r) - t_f] J_0(\lambda_n r)\, dr. \quad (4.57)$$

Uniform Initial Temperature Distribution. Just as in all the foregoing cases, the case of the circular bar initially heated to a uniform temperature is of particular practical value. So, with $f(r) = t_i$ (a constant), the above expression becomes

$$\frac{t - t_f}{t_i - t_f} = \frac{\theta}{t_i - t_f} = 2 \sum_{n=1}^{\infty} \frac{1}{\lambda_n R} e^{-\lambda_n^2 \alpha \tau} \frac{J_0(\lambda_n r) J_1(\lambda_n R)}{J_0^2(\lambda_n R) + J_1^2(\lambda_n R)}. \tag{4.58}$$

Applying Eq. (4.52) for the total heat flow from the cylinder up to time τ, one obtains the result

$$\frac{Q}{Q_i} = 4 \sum_{n=1}^{\infty} \frac{1}{(\lambda_n R)^2} \frac{J_1^2(\lambda_n R)}{J_0^2(\lambda_n R) + J_1^2(\lambda_n R)} (1 - e^{-\lambda_n^2 \alpha \tau}). \tag{4.59}$$

The three foregoing expressions are subject to the definition of the λ_n's:

$$\lambda_n R \frac{J_1(\lambda_n R)}{J_0(\lambda_n R)} - N_{\mathrm{BI}} = 0. \tag{4.55}$$

As in the case of the slab, it is best to write these last four expressions in terms of the nondimensional parameters:
Dimensionless position parameter: $\rho = r/R$.
Dimensionless time Fourier modulus: $N_{\mathrm{FO}} = \alpha \tau / R^2$.
Dimensionless relative film coefficient, Biot modulus: $N_{\mathrm{BI}} = hR/k$.
Then one has, in place of Eqs. (4.58) and (4.59):

$$\frac{t - t_f}{t_i - t_f} = 2 \sum_{n=1}^{\infty} \frac{1}{\delta_n} e^{-\delta_n^2 N_{\mathrm{FO}}} \frac{J_0(\delta_n \rho) J_1(\delta_n)}{J_0^2(\delta_n) + J_1^2(\delta_n)}. \tag{4.60}$$

$$\frac{Q}{Q_i} = 4 \sum_{n=1}^{\infty} \frac{1}{\delta_n^2} \frac{J_1^2(\delta_n)}{J_0^2(\delta_n) + J_1^2(\delta_n)} (1 - e^{-\delta_n^2 N_{\mathrm{FO}}}), \tag{4.61}$$

$$\delta_n \frac{J_1(\delta_n)}{J_0(\delta_n)} - N_{\mathrm{BI}} = 0. \tag{4.62}$$

Figures 4.15 and 4.16 give the dependence of the temperature on time (N_{FO}) the relative resistance (N_{BI}) and the relative position ρ. Figure 4.17 gives Gröber's results for the relative heat flow as a function of N_{BI} and N_{FO}. These charts are of great value in predicting the transient conduction in a real solid cylinder if its length is such that the radial heat flow assumed is closely approximated.

EXAMPLE 4.4: A cylindrical steel bar 8 in. diameter, heated to 1800°F, is quenched in an oil bath at 100°F in which the film coefficient h is 100 Btu/hr-ft²-°F. How

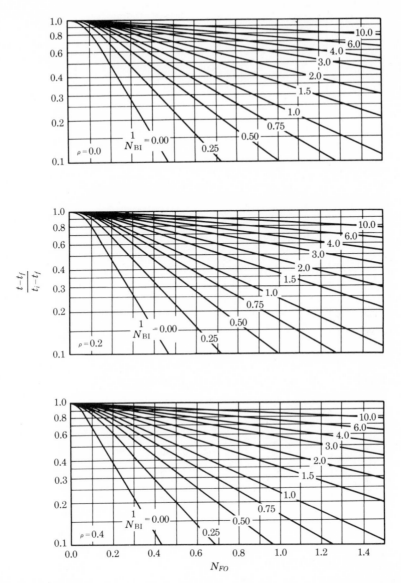

Fig. 4.15. Temperature–time history in a solid cylinder initially at temperature t_i and placed in a medium at t_f, for $\rho = 0.0$, 0.2, and 0.4. (From *Heat Transfer Notes* by L. M. K. Boelter, V. H. Cherry, H. A. Johnson, and R. C. Martinelli. Copyright 1965. McGraw-Hill Book Company, New York. Used by permission.)

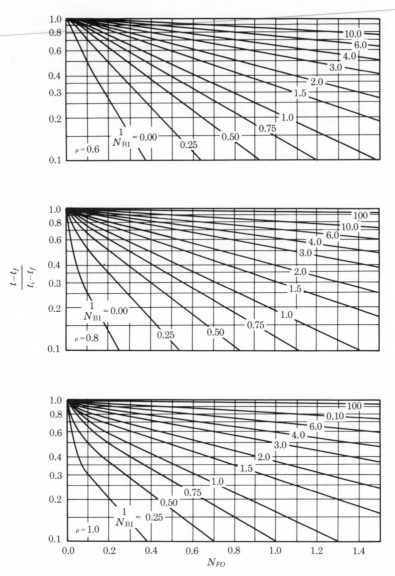

Fig. 4.16. Temperature–time history in a solid cylinder initially at temperature t_i and placed in a medium of t_f, for $\rho = 0.6$, 0.8, and 1.0. (From *Heat Transfer Notes* by L. M. K. Boelter, V. H. Cherry, H. A. Johnson, and R. C. Martinelli. Copyright 1965. McGraw-Hill Book Company, New York. Used by permission.)

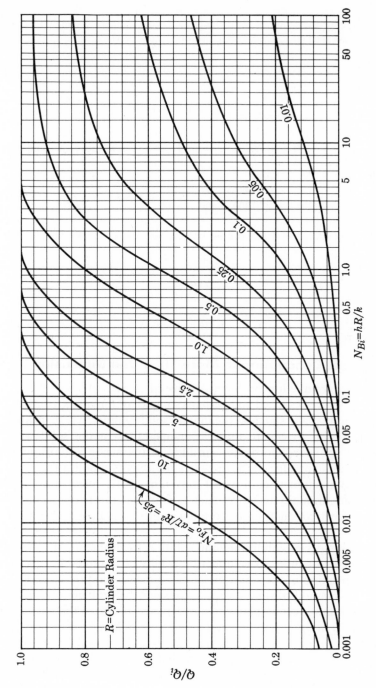

Fig. 4.17. The heat flow from an infinitely long cylinder as a function of time and thermal resistance. The cylinder is initially at a uniform temperature and is placed in a medium at constant temperature. (From H. Gröber, *Einführung in die Lehre der Wärmeübertragung,* Berlin, Springer, 1926. Used by permission.)

long will it take for the center of the cylinder to reach a temperature of 500°F? The steel is a stainless steel (18% Cr, 8% Ni).

Solution: Table A.1 gives

$$\rho = 488 \text{ lb/ft}^3, \qquad \alpha = 0.172 \text{ ft}^2/\text{hr},$$

$$c_p = 0.11 \text{ Btu/lb-°R}, \qquad k = 9.4 \text{ Btu/hr-ft-°F}.$$

Thus $N_{BI} = hR/k = (100 \times 4)/(9.4 \times 12) = 3.54$. Now, based above the fluid temperature of 100°F,

$$\frac{t - t_f}{t_i - t_f} = \frac{500 - 100}{1800 - 100} = \frac{400}{1700} = 0.235.$$

Figure 4.15 shows that at $\rho = 0$ N_{FO} must be 0.52. Since $N_{FO} = \alpha\tau/R^2$, then the time is

$$\tau = 0.52\frac{(4/12)^2}{0.172}$$

$$= 0.336 \text{ hr}.$$

4.7   Transient Conduction in More Than One Dimension

The transient conduction problems discussed in Secs. 4.5 and 4.6 were ones limited to the very special and perhaps somewhat unrealistic configurations of the infinite plane slab and the infinite solid cylinder. These idealized shapes were chosen to ensure that the temperature of the solid would depend on only one space coordinate—in addition to time. In certain applications, neglecting end or edge effects (to which the above simplification of one-dimensional conduction is equivalent) may not greatly affect the desired results. However, in many instances such a simplification may not be possible, and transient conduction in more than one space dimension must be considered.

Under certain special conditions, the solution to problems of transient conduction in two or three space dimensions may be obtained by a simple product superposition of the solutions of one-dimensional problems. As an example of this method of superposition, consider the problem of transient conduction in a long rectangular bar—as depicted in Fig. 4.18(a). The bar has a thickness D in the x direction and a thickness W in the y direction. It is infinite in extent in the z direction so that the conduction will occur only in the x and y directions. Thus, a two-dimensional, transient problem results if the temperatures are nonsteady. For instance, let the bar be heated so that, initially, the temperature distribution depends only on x and y. Let this initial

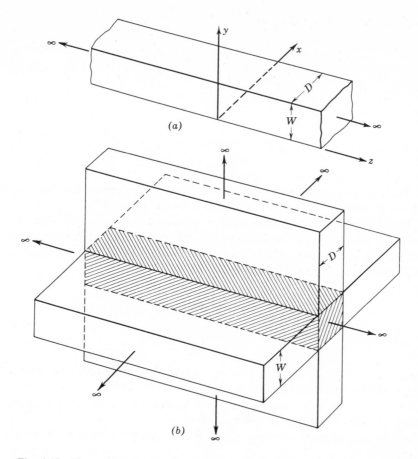

Fig. 4.18. The rectangular bar formed by the intersection of two infinite slabs.

distribution be denoted $f(x, y)$. Then at time $\tau = 0$ assume that the bar is placed in contact with a convecting fluid at a constant temperature t_f. The film coefficient, h, is assumed to be constant over the entire surface. If the temperature difference variable is defined as

$$\theta = t - t_f, \tag{4.63}$$

then the differential equation to be solved is

$$\frac{1}{\alpha}\frac{\partial\theta}{\partial\tau} = \frac{\partial^2\theta}{\partial x^2} + \frac{\partial^2\theta}{\partial y^2} \tag{4.64}$$

and the boundary conditions are

$$\text{For } \tau = 0: \quad \theta = f(x, y) - t_f = F(x, y).$$

$$\text{For } \tau \geq 0: \quad \text{at } x = 0 \text{ and } x = D, \frac{\partial \theta}{\partial x} = \pm\frac{h}{k}\theta, \qquad (4.65)$$

$$\text{at } y = 0 \text{ and } y = W, \frac{\partial \theta}{\partial y} = \pm\frac{h}{k}\theta.$$

[In the boundary conditions given in Eq. (4.65) the $+$ signs apply at $x = 0$ and $y = 0$, whereas the $-$ signs apply at $x = D$ and $y = W$.]

Now, if the initial temperature-*difference* distribution function, $F(x, y)$, is of such a form that it can be factored into a product of two other functions, each of which involves only one of the independent space variables, the initial condition may be replaced by

$$\text{At } \tau = 0: \quad \theta = F(x, y)$$
$$= F_1(x) \cdot F_2(y). \qquad (4.66)$$

Thus, $F_1(x)$ represents a function of only x and $F_2(y)$ represents a function of only y. Under these conditions, the solution to Eq. (4.64), subjected to the conditions of Eq. (4.65), may be expressed as the *product* of two one-dimensional, transient solutions, as will now be shown.

Let the sought-after solution, $\theta(x, y, \tau)$, be represented as the following product:

$$\theta = \theta_x(x, \tau) \cdot \theta_y(y, \tau). \qquad (4.67)$$

Here, $\theta_x(x, \tau)$ is a function of only x and time, τ, and $\theta_y(y, \tau)$ is a function of only y and τ. Substitution of Eq. (4.67) into Eq. (4.64) gives

$$\frac{1}{\alpha}\left(\theta_y\frac{\partial \theta_x}{\partial \tau} + \theta_x\frac{\partial \theta_y}{\partial \tau}\right) = \theta_y\frac{\partial^2 \theta_x}{\partial x^2} + \theta_x\frac{\partial^2 \theta_y}{\partial y^2}.$$

Upon rearrangement, this becomes

$$\theta_y\left(\frac{1}{\alpha}\frac{\partial \theta_x}{\partial \tau} - \frac{\partial^2 \theta_x}{\partial x^2}\right) + \theta_x\left(\frac{1}{\alpha}\frac{\partial \theta_y}{\partial \tau} - \frac{\partial^2 \theta_y}{\partial y^2}\right) = 0. \qquad (4.68)$$

The boundary conditions, Eq. (4.65), and the initial condition, Eq. (4.66), become

$$\text{At } \tau = 0: \quad \theta = \theta_x \cdot \theta_y = F_1(x) \cdot F_2(y).$$

$$\text{For } \tau \geq 0: \quad \text{at } x = 0 \text{ and } x = D, \theta_y \frac{\partial \theta_x}{\partial x} = \pm \frac{h}{k} \theta_x \theta_y,$$

$$\text{at } y = 0 \text{ and } y = W, \theta_x \frac{\partial \theta_y}{\partial y} = \pm \frac{h}{k} \theta_x \theta_y. \tag{4.69}$$

Examination of Eqs. (4.68) and (4.69) will quickly show that they are satisfied if $\theta_x(x, \tau)$ and $\theta_y(y, \tau)$ are each obtained as the solutions of the two following one-dimensional problems:

$$\frac{1}{\alpha} \frac{\partial \theta_x}{\partial \tau} = \frac{\partial^2 \theta_x}{\partial x^2}.$$

$$\text{At } \tau = 0: \quad \theta_x = F_1(x).$$

$$\text{For } \tau \geq 0: \quad \text{at } x = 0, \frac{\partial \theta_x}{\partial x} = \frac{h}{k} \theta_x,$$

$$\text{at } x = D, \frac{\partial \theta_x}{\partial x} = -\frac{h}{k} \theta_x. \tag{4.70}$$

$$\frac{1}{\alpha} \frac{\partial \theta_y}{\partial \tau} = \frac{\partial^2 \theta_y}{\partial y^2}.$$

$$\text{At } \tau = 0: \quad \theta_y = F_2(y).$$

$$\text{For } \tau \geq 0: \quad \text{at } y = 0, \frac{\partial \theta_y}{\partial y} = \frac{h}{k} \theta_y,$$

$$\text{at } y = W, \frac{\partial \theta_y}{\partial y} = -\frac{h}{k} \theta_y. \tag{4.71}$$

The above shows that the solution to the two-dimensional, transient conduction problem set forth in Eqs. (4.64) and (4.65) may be obtained by taking the product of the two simpler one-dimensional problems given in Eqs. (4.70) and (4.71) when the initial temperature distribution is expressible in the factored form shown in Eq. (4.66). The case of a constant initial temperature distribution, as discussed in the two preceding sections, is of this form.

The last two sets of equations, Eqs. (4.70) and (4.71), are seen to be identical in form to the equations which govern the transient conduction of heat in the infinite plane slab discussed in Sec. 4.5. Thus, the solution to the problem of transient heat conduction in the rectangular bar depicted in Fig. 4.18(a) is obtained by taking the product of the two solutions for the two infinite slabs whose intersection forms the bar in question. This is illustrated in Fig. 4.18(b). For the instance of the rectangular bar heated initially to a uniform temperature, the solutions (and graphical results) obtained in Sec. 4.5 for a plane slab initially at a uniform temperature may be utilized directly. It should be noted that the Biot and Fourier moduli for each of the two slabs making up the bar will be based on different nondimensionalizing factors unless the bar is square. It should also be remembered that the length L used in the formulas and charts of Sec. 4.5 is the half-width of the slab, so that L would have to be replaced by $D/2$ or $W/2$ before the results are applied to the problem set forth in this section.

The product superposition principle illustrated above for two-dimensional transient conduction in a rectangular bar can be extended to other configurations. Of particular practical value are the two configurations illustrated in Figs. 4.19 and 4.20. The solution for a parallelepiped of finite dimensions

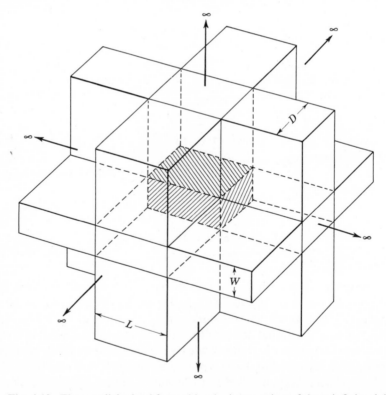

Fig. 4.19. The parallelepiped formed by the intersection of three infinite slabs.

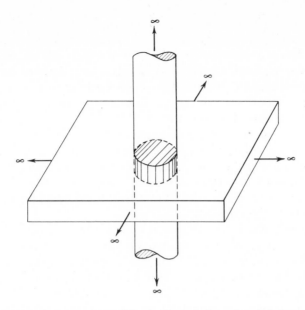

Fig. 4.20. The finite cylinder formed by the intersection of an infinite cylinder and an infinite slab.

may be obtained as the product of three infinite slabs. The solution for the finite solid circular cylinder may be obtained as the product of the solution for an infinite slab (Sec. 4.5) and the solution for a solid circular cylinder of infinite length (Sec. 4.6).

The product superposition principle described in this section is applicable only to cases in which the initial, dimensionless, temperature distribution can be factored into a form which is the product of functions, each involving only one of the independent space variables. The discussion of this chapter assumed that convection into a medium at constant temperature was taking place at the surface. The same superposition principle may be illustrated for cases in which the body surface temperature is suddenly changed to a constant value. In either instance the product superposition is valid only in the dimensionless form.

4.8   A Very Thick Wall Subjected to Periodic Surface Temperatures

As a final example of nonsteady heat conduction in one space dimension, a case in which the boundary conditions are dependent on time will be considered. To simplify the mathematics involved, a very special case of only one boundary will be considered.

Consider then a wall or slab of infinite thickness, shown in Fig. 4.21, which has its surface temperature varying in a sinusoidal, periodic manner. All the heat flow is assumed to be only in the direction normal to the wall—denoted by the x coordinate. Hence, the wall is considered to be infinite in extent in the y direction, or at least the edge effects are negligible. Also, the thickness of the wall in the x direction will be assumed to be so great that the temperature–time variation within the wall (at least in the region of interest) will depend only on the conditions imposed at the $x = 0$ surface. This is equivalent to the neglection of any effects of the other surface—or equivalent to the assumption that the wall is infinite in extent in the x direction. In such an instance, the solid is often referred to as a "semi-infinite solid." Later it will be shown under what conditions real walls of finite thickness may be adequately treated by this analysis.

Further, it will also be assumed that the cyclic temperature variation at the surface has been going for a sufficiently long time for the interior points to have reached an asymptotic state. That is, it is assumed that the temperatures of such points also vary periodically with time, repeating identical values each cycle. This is equivalent to the neglection of the initial heating-up period that would result if a uniformly cold solid was suddenly subjected to a periodic surface temperature variation. The type of conduction just described is sometimes referred to as "quasi-steady conduction." By this it is meant that the nonsteady temperature changes repeat themselves steadily. The problem will be, then: Given the physical properties of the solid and a surface temperature variation of known amplitude and period, find the period, amplitude, and time lag of the temperature variation on a plane at a distance x from the surface.

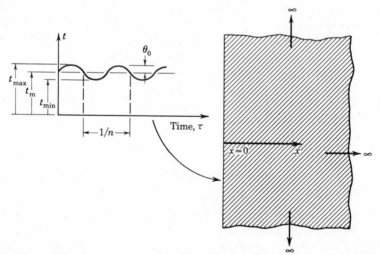

Fig. 4.21. Periodic temperature variation at the surface of a semi-infinite solid.

Now, then, the boundary condition at the surface of the solid, at $x = 0$, will be taken to be a sinusoidal variation in the temperature, as depicted in Fig. 4.21. The oscillation in the surface temperature will occur between a maximum and minimum value, t_{max} and t_{min}, respectively. Denoting the mean surface temperature by t_m, one finds that

$$t_m = \frac{t_{max} + t_{min}}{2}. \tag{4.72}$$

Defining a new temperature variable, θ, as

$$\theta = t - t_m, \tag{4.73}$$

one may express the boundary condition at the surface as

$$\theta_{x=0} = \theta_0 \sin 2\pi n\tau. \tag{4.74}$$

In the last equation, $\theta_0 = $ amplitude $ = (t_{max} - t_{min})/2$, and $n = $ frequency of oscillation. Now, the one-dimensional, nonsteady, conduction equation is

$$\frac{\partial t}{\partial \tau} = \alpha \frac{\partial^2 t}{\partial x^2}.$$

This becomes the following, in terms of the new variable θ, since t_m is a constant:

$$\frac{\partial \theta}{\partial \tau} = \alpha \frac{\partial^2 \theta}{\partial x^2}. \tag{4.75}$$

This last equation is identical with the expression given in Eq. (4.25). The solution found in Sec. 4.5 [Eq. (4.27)] is only one form of many possible forms of the solution. The choice of a particular form of solution depends on the nature of the boundary conditions which are imposed. The form of the solution chosen should be the one which enables the simplest application of the boundary conditions for evaluation of the integration constants. Hence, although the solution to the equation is unique, a different form of it will be sought here.

However, following the procedure of representing the solution of the equation above as the product of two different functions (each being a function of only one of the independent variables), one takes

$$\theta = T(\tau) \cdot X(x).$$

Upon substitution into Eq. (4.75) and upon subsequent rearrangement, this leads to

$$\frac{1}{\alpha T}\frac{dT}{d\tau} = \frac{1}{X}\frac{d^2 X}{dx^2} = \pm i\lambda^2.$$

In this instance the separation constant is taken to be an imaginary number (λ being considered real). The two resulting ordinary differential equations are

$$\frac{dT}{d\tau} - (\pm i\lambda^2 \alpha T) = 0,$$

$$\frac{d^2 X}{dx^2} - (\pm i\lambda^2 X) = 0.$$

$$(4.76)$$

Each of these has two solutions—one for each sign on λ. Taking first the positive sign, one has

$$T = B_1 e^{i\lambda^2 \alpha \tau},$$

$$X = B_2 e^{\lambda \sqrt{i}x} + B_3 e^{-\lambda \sqrt{i}x}.$$

Since $\theta = T \cdot X$, for the positive sign,

$$\theta_+ = B_1 e^{i\lambda^2 \alpha \tau}(B_2 e^{\lambda \sqrt{i}x} + B_3 e^{-\lambda \sqrt{i}x}).$$

Now, since $\sqrt{i} = (1 + i)/\sqrt{2}$, this may be written as

$$\theta_+ = B_1 e^{-\lambda x/\sqrt{2}}[B_2 e^{\sqrt{2}\lambda x} e^{i(\lambda^2 \alpha \tau + \lambda x/\sqrt{2})} + B_3 e^{i(\lambda^2 \alpha \tau - \lambda x/\sqrt{2})}].$$

Similarly, when the negative sign is taken one may write the solution as

$$\theta_- = B_1' e^{-\lambda x/\sqrt{2}}[B_2' e^{\sqrt{2}\lambda x} e^{-i(\lambda^2 \alpha \tau + \lambda x/\sqrt{2})} + B_3' e^{-i(\lambda^2 \alpha \tau - \lambda x/\sqrt{2})}].$$

Now, the sum of these two solutions is also a solution, i.e.,

$$\theta = \theta_+ + \theta_-.$$

But one notes that the constants $B_2 = B_2' = 0$, since otherwise θ would become infinite as $x \to \infty$. So, the solution is

$$\theta = \theta_+ + \theta_- = B_1 e^{-\lambda x/\sqrt{2}}[B_3 e^{i(\lambda^2 \alpha \tau - \lambda x/\sqrt{2})}] + B_1' e^{-\lambda x/\sqrt{2}}[B_3' e^{-i(\lambda^2 \alpha \tau - \lambda x/\sqrt{2})}].$$

From the identity,

$$e^{\pm i\beta} = \cos \beta \pm i \sin \beta,$$

the above solution can be written as

$$\theta = e^{-\lambda x/\sqrt{2}}[C \cos (\lambda^2 \alpha \tau - \lambda x/\sqrt{2}) + D \sin (\lambda^2 \alpha \tau - \lambda x/\sqrt{2})].$$

The symbols C and D denote constants composed of B_1, B_3, B_1', and B_3'. As noted in Eq. (4.74), the boundary condition at $x = 0$ is

$$\theta_{x=0} = \theta_0 \sin 2\pi n\tau, \tag{4.74}$$

where θ_0 and n are the amplitude and frequency of the imposed periodic temperature. By applying this to the above solution,

$$\theta_0 \sin 2\pi n\tau = C \cos \lambda^2 \alpha \tau + D \sin \lambda^2 \alpha \tau.$$

This shows that

$$C = 0,$$

$$D = \theta_0,$$

$$\lambda = \sqrt{\frac{2\pi n}{\alpha}}.$$

So the solution for the temperature variation with time at some plane within the wall is

$$\theta = \theta_0 e^{-x\sqrt{n\pi/\alpha}} \sin \left(2\pi n\tau - x\sqrt{\frac{\pi n}{\alpha}}\right). \tag{4.77}$$

Comparison of this with the imposed variation at the surface, $x = 0$, shows that at a given distance in from the surface the temperature variation is a periodic variation of the same period but of an amplitude that decreases exponentially with distance. That is, the amplitude at a distance x is

$$(\theta_m)_x = \theta_0 e^{-x\sqrt{\pi n/\alpha}}. \tag{4.78}$$

Also, the periodicity at a distance x from the surface is of the same period as that at the surface, but lags by a time increment equal to

$$\Delta \tau = \frac{x}{2}\sqrt{\frac{1}{\alpha \pi n}}. \tag{4.79}$$

These facts are illustrated in Fig. 4.22.

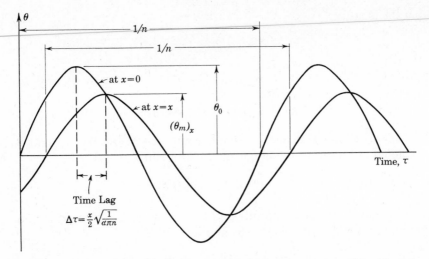

Fig. 4.22. Temperature distribution at two depths in a semi-infinite solid subjected to a periodic surface temperature.

The solution, Eq. (4.77), also shows that at any given instant, τ, the temperature distribution through the solid, is an exponentially damped sine wave. This is illustrated in Fig. 4.23, where the temperature distributions for two different times are shown.

Also of interest are the instantaneous heat flow rate and the cumulative heat flow at the surface—both as functions of time. At the surface the instantaneous heat flow rate is, per unit surface area,

$$\left(\frac{q}{A}\right)_{x=0} = -k\left(\frac{\partial t}{\partial x}\right)_{x=0} = -k\left(\frac{\partial \theta}{\partial x}\right)_{x=0}.$$

This becomes, upon introduction of the solution in Eq. (4.77),

$$\left(\frac{q}{A}\right)_{x=0} = -k\theta_0 e^{-x\sqrt{n\pi/\alpha}}\left(-\sqrt{\frac{n\pi}{\alpha}}\right)\left[\cos\left(2\pi n\tau - x\sqrt{\frac{n\pi}{\alpha}}\right)\right]_{x=0}$$

$$- k\theta_0\left(-\sqrt{\frac{n\pi}{\alpha}}\right)e^{-x\sqrt{n\pi/\alpha}}\left[\sin\left(2\pi n\tau - x\sqrt{\frac{n\pi}{\alpha}}\right)\right]_{x=0}$$

$$= k\theta_0\sqrt{\frac{n\pi}{\alpha}}\left[\sin\left(2\pi n\tau\right) + \cos\left(2\pi n\tau\right)\right]$$

$$= k\theta_0\sqrt{\frac{2n\pi}{\alpha}}\sin\left(2\pi n\tau + \frac{\pi}{4}\right). \qquad (4.80)$$

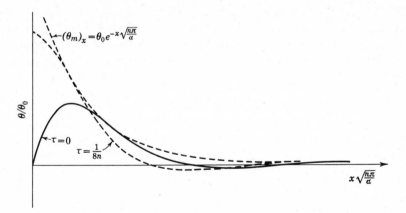

Fig. 4.23. Temperature distribution at two times in a semi-infinite solid subjected to a periodic surface temperature.

Thus, the surface heat flow rate is a periodic function of time, leading the temperature variation by a time increment of $\frac{1}{8}n$.

The cumulative heat flow then is alternately into and out of the wall:

$$\left(\frac{Q}{A}\right)_{x=0} = \int \left(\frac{q}{A}\right)_{x=0} d\tau = -k\theta_0 \sqrt{\frac{1}{2\pi n\alpha}} \cos\left(2\pi n\tau + \frac{\pi}{4}\right).$$

The latter functions, $(q/A)_{x=0}$ and $(Q/A)_{x=0}$, are shown in Fig. 4.24 together with the surface temperature variation for comparison purposes.

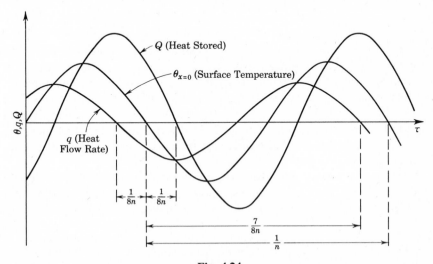

Fig. 4.24

The foregoing analyses were based on the assumption of infinitely thick slabs. A study of the diurnal temperature waves in the earth caused by periodic atmospheric changes may be accomplished with the aid of the equations above. For other important engineering applications in which periodic temperature variations are encountered (such as in the cylinder walls of combustion engines or in the walls of building structures subjected to periodic atmospheric and solar temperatures), the solids involved will, of course, be of finite thickness. However, the thermal properties of the solid may be such that the temperature wave is damped to a negligibly small amplitude in a relatively small distance into the wall. If δ represents the ratio of the amplitude of the temperature wave at a distance x to the amplitude of the surface, the distance x is, from Eq. (4.78),

$$x_\delta = -\frac{\ln \delta}{\sqrt{n\pi/\alpha}}. \tag{4.81}$$

The distance at which the wave is damped to a specified fraction, x_δ, is then easily calculated if the material diffusivity, α, and the frequency, n, are known.

For example, for a typical carbon steel (say 1 per cent carbon), Table A.1 gives $\alpha = 0.452 \, \text{ft}^2/\text{hr}$. For an engine operating at 3000 rpm on a four-stroke cycle, the temperature wave due to the variation of the cylinder temperature is damped to 1 per cent of its surface value at a depth of

$$x = -\frac{\ln(1/100)}{\sqrt{\dfrac{1500 \times 60 \times \pi}{0.452}}} \times 12 = 0.07 \, \text{in}.$$

Thus, for most applications the wall of an engine cylinder may be considered infinitely thick. The above calculation presumes that the surface film on the inside of the engine cylinder has a negligible thermal resistance. This is, of course, not true, but the results are at least of the right order of magnitude.

Most masonry materials and concrete have a diffusivity of about $\alpha = 0.02$ ft^2/hr. Thus, for a daily heating cycle with a frequency $n = \frac{1}{24}$, the temperature wave is damped to 5 per cent of its surface value in about 14 in. This indicates that the assumption of infinite thickness may not apply so well in such an application.

The above relations are also used to determine the diffusivity experimentally by measuring either the decrease in amplitude or the time lag in a sample of the material which has a known temperature wave imposed on it.

EXAMPLE 4.5: The earth is assumed to have an average thermal diffusivity of 0.015 ft^2/hr. If the 24-hr variation in the temperature at a certain locality is from 70°F to 95°F, find the amplitude of the temperature variation at a depth of 6 in.

What is the time lag of the temperature wave at this depth? Assume that a sinusoidal variation exists at the surface.

Solution: The amplitude of the surface variation is $(95 - 70)/2 = 12.5°F$, and the frequency is $n = \frac{1}{24} \, 1/\text{hr}$. Thus, at a depth of 6 in. $= 0.5$ ft, the amplitude is [from Eq. (4.78)]

$$(\theta_m)_x = 12.5 \, e^{-0.5\sqrt{\pi/(24 \times 0.015)}} = 2.85°F.$$

Equation (4.79) gives the time lag to be

$$\Delta\tau = \frac{0.5}{2} \sqrt{\frac{24}{0.015\pi}} = 5.64 \text{ hr} = 5 \text{ hr}, 38 \text{ min}.$$

4.9   The Transient Response of Bodies with Negligible Internal Resistance

The solutions of the cases of transient conduction treated in Secs. 4.5, 4.6, and 4.7 were expressed in forms which gave a dimensionless temperature distribution in terms of dimensionless spatial coordinates, a dimensionless time variable (the Fourier modulus, $N_{FO} = \alpha\tau/l^2$), and, in cases in which surface convection was present, a dimensionless surface conductance (the Biot modulus, $N_{BI} = hl/k$). As noted at the time of the introduction of the Biot modulus, in Eq. (4.36), one interpretation that may be made is that it is the ratio of the internal thermal resistance of the body, l/k, and the external, surface, resistance, $1/h$. (The symbol l is used to denote a characteristic dimension of the body.)

The instances in which specified boundary temperatures were considered may be envisioned as ones in which the surface resistance is negligible and $N_{BI} \rightarrow \infty$. The other cases considered were those in which both the surface and internal resistances are finite. This section will consider cases in which the *internal* resistance may be taken as negligible and $N_{BI} \rightarrow 0$. Such a case may arise if the body has a high-enough thermal conductivity, compared with the surface film coefficient, that the interior temperature of the body may be taken as uniform at any instant of time. Thus, the entire temperature–time history of the body is regulated by the surface resistance. Under these circumstances, a heat balance on the body yields

$$hA_s(t - t_f) = -\rho V c_p \frac{dt}{d\tau}. \tag{4.82}$$

In Eq. (4.82) A_s and V represent the exposed surface area and volume of the body, respectively. The density and specific heat of the body are denoted by ρ

and c_p, while h represents the surface film coefficient between the body and an ambient fluid at temperature t_f.

Equation (4.82) may be rewritten as

$$\frac{d(t - t_f)}{t - t_f} = -\frac{hA_s}{\rho c_p V} d\tau.$$

Let it be presumed that the body is initially at a temperature t_i before it is thrust into the fluid at temperature t_f. Integration of the equation above and application of the initial condition that $t = t_i$ at $\tau = 0$ yields the following expression for the subsequent temperature of the body t, as a function of time:

$$\frac{t - t_f}{t_i - t_f} = e^{-(hA_s/\rho c_p V)\tau} \tag{4.83}$$

$$= e^{-\tau/\varphi}.$$

The parameter

$$\varphi = \frac{\rho c_p V}{hA_s} \tag{4.84}$$

is termed the *time constant* of the body—the larger the value of φ, the slower the body is to respond to the change in temperature. For given thermal properties, the time constant φ is proportional to the ratio V/A_s; thus, the smaller the surface area of a body, compared with its volume, the slower it will respond.

The instantaneous rate of heat flow between the body and the ambient fluid is given by

$$q = hA_s(t - t_f),$$

or, in dimensionless form,

$$\frac{q}{hA_s(t_i - t_f)} = e^{-\tau/\varphi}. \tag{4.85}$$

The cumulative heat flow between time $\tau = 0$ and $\tau = \tau$ is

$$Q = \int_0^\tau q \, d\tau,$$

so Eq. (4.85) leads to

$$\frac{Q}{\rho V c_p(t_i - t_f)} = 1 - e^{-\tau/\varphi}. \tag{4.86}$$

Equations (4.83), (4.85), and (4.86) afford simple means for the rapid determination of the temperature response of a body as long as the basic assumption of a uniform interior temperature may be made.

EXAMPLE 4.6: A 1-in.-diameter pure copper sphere and a 1-in. pure copper cube are each heated to 1200°F. They are then placed in air at 200°F with a surface film coefficient of 12 Btu/hr-ft²-°F. Find the temperature of each body after a lapse of 5 min.

Solution: Table A.1 gives, for copper, $\rho = 559$ lb$_m$/ft³, $c_p = 0.0915$ Btu/lb$_m$-°F. For the sphere, $V/A_s = \frac{4}{3}\pi r^3/4\pi r^2 = r/3 = \frac{1}{3}$ in. For the cube, $V/A_s = d^3/6d^2 = d/6 = \frac{1}{6}$ in. Thus, the time constants are

$$\text{Sphere:} \quad \varphi = \frac{559 \times 0.0915}{12 \times 3 \times 12} = 0.118 \text{ hr.}$$

$$\text{Cube:} \quad \varphi = \frac{559 \times 0.0915}{12 \times 6 \times 12} = 0.059 \text{ hr.}$$

Thus, at $\tau = 5$ min $= \frac{1}{12}$ hr, for the sphere,

$$\tau/\varphi = \frac{\frac{1}{12}}{0.118} = 0.706$$

$$\frac{t - 200}{1200 - 200} = e^{-0.706} = 0.492$$

$$t = 692°\text{F.}$$

Similarly, for the cube,

$$\tau/\varphi = 1.412$$

$$\frac{t - 200}{1200 - 200} = 0.244$$

$$t = 444°\text{F.}$$

References

1. JAHNKE, E., and F. EMDE, *Tables of Functions*, 4th ed., New York, Dover, 1945.
2. CARSLAW, H. S., and J. C. JAEGER, *Conduction of Heat in Solids*, New York, Oxford U.P., 1957.
3. BOELTER, L. M. K., V. H. CHERRY, H. A. JOHNSON, and R. C. MARTINELLI, *Heat Transfer Notes*, New York, McGraw-Hill, 1965.
4. GRÖBER, H., *Einführung in die Lehre der Wärmeübertragung*, Berlin, Springer, 1926.

Problems

4.1. Write the solution for the steady state temperature distributions in a rectangular plate which is infinite in extent in the y direction, of width L in the x direction, has a temperature t_e imposed at $x = 0$ and $x = L$, and has a temperature distribution imposed on its edge at $y = 0$ as given by

$$t_{y=0} = (t_0 - t_e) \sin\left(\frac{\pi x}{L}\right), \qquad 0 \le x \le L.$$

4.2. A rectangular plate has the dimensions of 6 in. × 10 in. and has its edges maintained at $0°F$ except at one of the 6-in. edges. On this edge the temperature is maintained at

$$t = (100°F) \sin\left(\frac{\pi x}{6}\right), \qquad 0 \le x \le 6 \text{ in.}$$

Draw the isothermal lines within the plate for $100°F$, $90°F$, $80°F, \ldots, 10°F$.

4.3. A rectangular plate 6 in. × 8 in. has its two 8-in. sides maintained at $100°F$, one of its 6-in. sides maintained at $300°F$, and its other 6-in. side maintained at $500°F$. Find the temperature at the center of the plate.

4.4. A solid circular cylinder (outside radius R, finite length L) has its two circular ends maintained at $0°F$. On the cylindrical surface the temperature is maintained at a distribution that depends *only* on the axial coordinate z. Denote this distribution by $f(z)$. Show that the steady state solution for the temperature distribution is

$$t = \frac{2}{L} \sum_{n=1}^{\infty} \frac{I_0(\lambda_n r)}{I_0(\lambda_n R)} \sin \lambda_n z \int_0^L f(z) \sin \lambda_n z \, dz, \qquad \lambda_n = \frac{n\pi}{L}.$$

4.5. A slab, 10 in. thick in the x direction and infinite in extent in the y direction, is initially heated to a uniform temperature of $600°F$. It then has its surface temperature suddenly reduced to and maintained at $120°F$. If the material has a diffusivity $\alpha = 2.2 \text{ ft}^2/\text{hr}$, plot the variation of the centerline temperature and the heat removed (per square foot) as functions of time up to 1 hr for $k = 50$ Btu/hr-ft-°F.

4.6. A slab of 1.5 % carbon steel, 1 in. thick and infinite in extent in the other directions, is heated to 1800°F and then quenched in an oil bath at 200°F. If the convective film coefficient is 100 Btu/hr-ft²-°F, find the time required to reduce the center temperature to 800°F. What is the temperature at a depth of $\frac{1}{4}$ in. from the surface at this time? How much heat (per square foot of surface) has been removed up to this time?

4.7. Plot the time variation of the surface temperature of the plate in Prob. 4.6.

4.8. A steel manufacturer anneals large slabs of steel by placing them in an oven in which the ambient temperature is maintained at 1800°F. The slabs are at an initially uniform temperature of 300°F when placed in the oven. It is desired to raise the *average* temperature of the slab to 1440°F, but the surface temperature must not be allowed to rise above 1650°F. Find the maximum thickness of slabs that may be processed in this manner if the ambient heat transfer coefficient is 20 Btu/hr-ft²-°F. For the steel use $\rho = 500 \text{ lb}_m/\text{ft}^3$, $k = 10$ Btu/hr-ft-°F, $c_p = 0.1$ Btu/lb$_m$-°R.

4.9. A concrete wall 5 in. thick and insulated on the inside is initially at equilibrium with the surrounding ambient air at 80°F on the outside. The outside air temperature suddenly drops to 60°F. A heat transfer coefficient of 1.5 Btu/hr-ft²-°F exists at the outside surface. Find the temperature of the inside and outside surfaces after 2 hr and 6 hr have elapsed.

4.10. An infinite slab of thickness L in the x direction is initially heated to a known temperature distribution, $f(x)$. If both surfaces normal to the x direction are then insulated so that they are impervious to heat flow, write the equation for the distribution of the temperature in the slab as a function of time. If the initial distribution $f(x)$ is a constant t_i, to what does the solution just obtained reduce?

4.11. A cylindrical bar of stainless steel (18 % Cr, 8 % Ni) 4 in. in diameter is removed from an annealing furnace where it has been maintained at 1800°F. It is then placed in air at 90°F to cool. If the convection film coefficient at the surface is estimated to be 2 Btu/hr-ft²-°F, what is its center temperature after 2 hr have passed?

4.12. A cylindrical bar of Duralumin is chilled to $-150°F$ and then heated in an atmosphere at 100°F in which the film coefficient of 25 Btu/hr-ft²-°F can be expected. If the bar is 1 in. in diameter, when does the surface reach a temperature of 60°F?

4.13. A 5-in. diameter bar of brass is initially at 1200°F. It is placed to cool in air at 100°F through a film coefficient of 12 Btu/hr-ft²-°F. How long will it be until the surface cools to 650°F? What will the center temperature be at this time?

4.14. A solid cylinder of radius R and infinite length is initially heated to a known temperature distribution which depends on r (the radial coordinate) only. Find the distribution of temperature in the cylinder as a function of time if the surface is insulated.

4.15. Plot the temperature–time history for the geometric center of a steel bar (carbon steel) 3 in. in diameter and 3 in. long if it is initially heated to $1500°F$ and then allowed to cool in air at $100°F$. Take the film coefficient to be 5 Btu/hr-ft²-°F.

4.16. An ordinary house brick ($2\frac{1}{4}$ in. × $3\frac{1}{2}$ in. × 8 in.) is fired in a kiln at $2600°F$. It is then allowed to cool in air at $100°F$. If the film coefficient is 5 Btu/hr-ft²-°F, how long does it take for the surface temperature to reach $150°F$ in the center of the $3\frac{1}{2}$ × 8 side?

4.17. A 4-in. × 4-in. wood timber is initially at $75°F$. It is suddenly exposed to flames at $1300°F$ through a heat transfer coefficient of 3.0 Btu/hr-ft²-°F. If the wood ignites at $900°F$, how much time will elapse before the timber first starts burning? Use $\rho = 50 \text{ lb}_m/\text{ft}^3$, $c_p = 0.6 \text{ Btu/lb}_m\text{-}°F$, $k = 0.2 \text{ Btu/hr-ft-}°F$.

4.18. A furnace wall is composed of chrome brick and is considered to be infinitely thick. Its surface temperature varies periodically from $900°F$ to $1000°F$ in a 6-hr period. Assuming this variation to be a sine wave function, find
(a) the amplitude, maximum, and minimum temperature at a depth of 3 in.;
(b) the temperature at a depth of 3 in. when the surface is at its maximum temperature;
(c) the temperature at the surface when the temperature at a 3-in. depth is at its maximum;
(d) the heat flow, per unit wall area, past the surface during one-half of the cycle.

4.19. A thick concrete wall is subjected to a daily temperature variation due to solar radiation which varies between $50°F$ and $120°F$. At what depth is the temperature wave dampened to 5% of its surface amplitude?

4.20. A very long copper cylinder, 2 in. in diameter, is heated to $90°F$ and then plunged into a liquid bath at $30°F$. If, after 3 minutes, the cylinder temperature is $40°F$, determine the average film coefficient at the surface if the internal thermal resistance is negligible.

4.21. A fine wire is heated, initially, to $350°F$ and then exposed to an environment at $100°F$ through a surface film coefficient of 10 Btu/hr-ft²-°F. The wire has a diameter of 0.03 in. Find the wire temperature and the total heat loss to the surroundings after 30 sec if the wire is made of (a) copper, (b) aluminum.

4.22. A plate 1 in. thick is made of 10% nickel steel. It is heated to some initial temperature and then exposed to air at $100°F$ through a film coefficient of $h = 2$ Btu/hr-ft²-°F. What is the initial temperature of the plate if its temperature after 10 min is $1000°F$?

CHAPTER 5

Numerical and Analog Methods for Heat Conduction

5.1   Introductory Remarks

The problems of heat conduction treated in Chapters 3 and 4 have stressed the use of mathematical methods. This approach, which treats the conducting body as a continuum, yields much information of a general nature concerning the particular problem being treated. By the successful application of analysis one can ascertain the temperature at *any* point, at *any* time, within the given body. Since the results are given in closed, analytical form, it is possible to deduce much useful information—the effect of various parameters, the effect of altering the body geometry, etc.

Although a great number of problems (some of practical value, some not) have been solved analytically, only a limited number of relatively simple geometrical shapes (e.g., cylinders, spheres, infinite slabs, etc.) can be handled. Also, only those boundary conditions which can be easily expressed mathematically may usually be applied. There are many heat conduction problems of considerable practical value for which no analytical solution is feasible. These problems usually involve geometrical shapes of a mathematically inconvenient sort. The absence of an analytical solution does not remove the need of an answer, and some other approach must be sought.

Numerical and analog techniques exist which are able to handle almost any problem of any degree of complexity. The detail and accuracy of the answer obtained depends mainly upon the amount of effort one wishes to expend.

165

The various numerical methods all yield *numerical* values for temperatures at selected, *discrete*, points within the body being considered and only at *discrete* time intervals. Thus answers are obtained only for a given set of conditions, a given set of discrete points and discrete time intervals. One must give up the generality of the analytical solution in order to obtain an answer.

Several different techniques of numerical analysis of heat conduction problems exist. The references at the end of the chapter may be consulted in this regard. Some of the techniques developed have been based on the use of hand calculations or desk calculators. The availability of high speed digital computers has considerably reduced the value of these methods. The formulations presented and the techniques recommended are not the most efficient or concise from the standpoint of hand or desk calculator solution; they are presented in forms most suitable for computer programing.

Both steady state and transient methods are treated. Formulations of steady state numerical methods are not presented in the most concise fashion for desk calculation, but rather are formulated in a way that carries over most easily to the transient methods.

Attention will first be directed toward the development of finite difference approximations to the heat conduction equation (upon which all numerical methods are based) and the degree of accuracy of these approximations. On the basis of the finite difference formulation, an electrical network analogy will become apparent. Generalizations to complex problems will then be made on the basis of this analogy because of its greater physical appeal.

5.2   Finite Difference Approximations

The basic principle of the numerical approach to a heat conduction problem is the replacement of the differential equation for the continuous temperature distribution in a heat conducting solid by a finite difference equation which must be satisfied at only certain points in the solid. The relation of the finite difference expression to the differential equation can be understood best by deriving the latter from the former, via the use of a Taylor's expansion. Consider, then, a function of two independent variables:

$$f = f(\xi, \eta). \tag{5.1}$$

The general variables ξ and η have been used since the results will be applied to spatial variable (x, y, and z) as well as time (τ).

Let h_1 represent an increment in the variable ξ. A *forward* expansion of the function at $\xi = \xi + h_1, \eta = \eta$, in terms of its value at $\xi = \xi, \eta = \eta$, is

$$f(\xi + h_1, \eta) = f(\xi, \eta) + h_1\left(\frac{\partial f}{\partial \xi}\right)_{\xi, \eta} + \frac{h_1^2}{2}\left(\frac{\partial^2 f}{\partial \xi^2}\right)_{\xi, \eta} + \frac{h_1^3}{6}\left(\frac{\partial^3 f}{\partial \xi^3}\right)_{\xi, \eta} + \mathcal{O}(h_1^4). \tag{5.2}$$

The notation $\mathcal{O}(h_1^4)$ is used to indicate that subsequent terms are of the order of h_1^4, and *higher*. If terms of the order of h_1^2, and greater, are neglected, the following *forward* finite difference approximation to the first derivative is obtained:

$$\left[\left(\frac{\partial f}{\partial \xi}\right)_{\xi,\eta}\right]_{\text{fwd}} \approx \frac{1}{h_1}[f(\xi + h_1, \eta) - f(\xi, \eta)]. \tag{5.3}$$

Similarly, for a forward displacement of h_2 in η, the forward finite difference approximation for the first derivative is

$$\left[\left(\frac{\partial f}{\partial \eta}\right)_{\xi,\eta}\right]_{\text{fwd}} \approx \frac{1}{h_2}[f(\xi, \eta + h_2) - f(\xi, \eta)]. \tag{5.4}$$

In like fashion, *backward* finite difference expressions for the first derivatives may be obtained by writing Taylor's expansions for an increment in ξ of $-h_1$ or an increment in η of $-h_2$:

$$f(\xi - h_1, \eta) = f(\xi, \eta) - h_1\left(\frac{\partial f}{\partial \xi}\right)_{\xi,\eta} + \frac{h_1^2}{2}\left(\frac{\partial^2 f}{\partial \xi^2}\right)_{\xi,\eta} - \frac{h_1^3}{6}\left(\frac{\partial^3 f}{\partial \xi^3}\right)_{\xi,\eta} + \mathcal{O}(h_1^4). \tag{5.5}$$

Again neglecting terms of order h_1^2, and greater, one obtains

$$\left[\left(\frac{\partial f}{\partial \xi}\right)_{\xi,\eta}\right]_{\text{bkwd}} \approx \frac{1}{h_1}[f(\xi, \eta) - f(\xi - h_1, \eta)]. \tag{5.6}$$

Also,

$$\left[\left(\frac{\partial f}{\partial \eta}\right)_{\xi,\eta}\right]_{\text{bkwd}} \approx \frac{1}{h_2}[f(\xi, \eta) - f(\xi, \eta - h_2)]. \tag{5.7}$$

The *central* finite difference approximation to the second derivatives may be found by adding Eqs. (5.2) and (5.5):

$$f(\xi + h_1, \eta) + f(\xi - h_1, \eta) = 2f(\xi, \eta) + h_1^2\left(\frac{\partial^2 f}{\partial \xi^2}\right)_{\xi,\eta} + \mathcal{O}(h_1^4). \tag{5.8}$$

When terms of the order of h_1^4, and greater, are neglected,

$$\left[\left(\frac{\partial^2 f}{\partial \xi^2}\right)_{\xi,\eta}\right]_{\text{cent}} \approx \frac{1}{h_1^2}[f(\xi + h_1, \eta) - 2f(\xi, \eta) + f(\xi - h_1, \eta)]. \tag{5.9}$$

Also,

$$\left[\left(\frac{\partial^2 f}{\partial \eta^2}\right)_{\xi,\eta}\right]_{cent} \approx \frac{1}{h_2^2}[f(\xi, \eta + h_2) - 2f(\xi, \eta) + f(\xi, \eta - h_2)]. \quad (5.10)$$

It is worth noting that the first derivative finite difference approximations neglect terms of the order of h^2, while the second derivative approximations neglect terms of the order of h^4. It should be pointed out that other approximations to the second derivative may be written (e.g., forward, backward, etc.). However, the central difference given here is most commonly used for heat conduction analyses.

Section 5.3 applies these approximations to steady state heat conduction, and subsequent sections illustrate their solutions. Section 5.7 and subsequent sections apply the difference equations to the case of transient conduction.

5.3 Steady State Numerical Methods

From Chapter 1 the steady state conduction equation is known to be

$$k\left(\frac{\partial^2 t}{\partial x^2} + \frac{\partial^2 t}{\partial y^2} + \frac{\partial^2 t}{\partial z^2}\right) + q^* = 0. \quad (5.11)$$

The thermal conductivity, k, is left as a factor in Eq. (5.11) since this will be a useful form when extensions are made to nonsteady conduction. Also, reference to the derivation of the heat conduction equation in Sec. 1.4 reveals that the expression given on the left side of Eq. (5.11) represents the net rate, *per unit volume*, at which heat is stored at a point in the conducting body, and q^* represents the internal generated heat, *per unit volume*.

Consider, first, the one-dimensional case:

$$k\frac{d^2 t}{dx^2} + q^* = 0. \quad (5.12)$$

Figure 5.1 depicts a slender bar whose lateral faces are insulated. Thus, all heat conduction occurs in the x direction and Eq. (5.12) applies. If, however, the second derivative in Eq. (5.12) is replaced by the finite difference approximation in Eq. (5.9) (with the coordinate x replacing ξ, the space increment δx replacing h_1, and the temperature t replacing f), the following expression results, which relates the temperature at x, $x + \delta x$, and $x - \delta x$:

$$\frac{k}{(\delta x)^2}[t(x + \delta x) - 2t(x) + t(x - \delta x) + q_x^*] = 0.$$

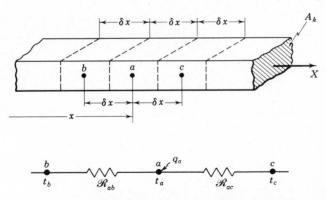

Fig. 5.1. Finite difference approximation and corresponding nodal network for one-dimensional conduction.

In keeping with the central difference approximation used, the internally generated heat is evaluated at the central point x.

If, as suggested in Fig. 5.1, the points at x, $x - \delta x$, and $x + \delta x$ are denoted by a, b, c, respectively, the equation above is expressed more simply as

$$\frac{k}{(\delta x)^2}(t_b - 2t_a + t_c)\, q_a = 0. \tag{5.13}$$

Equation (5.13) is the result of a Taylor's expansion of the heat conduction equation around the point a, and should thus be interpreted as a relation for the temperature t_a in terms of the temperatures at the neighboring points, t_b and t_c. A similar expression may be written for the point b—in terms of t_a and the temperature of the point δx to the left of b. Likewise, an expression may be written for point c, involving t_c, t_a, and the point δx to the right of c. Thus, if the body is subdivided into n distinct points, n such equations may be written. This set of n equations may then be solved for the temperatures at the n points presuming the q's are known. Consequently, one obtains the temperatures at these specific points without any consideration of the temperatures between them. The development of Eq. (5.9), upon which Eq. (5.13) is based, showed that the accuracy of the results obtained will increase with decreasing δx (i.e., by increasing the number of points), and, in fact, the error will be of the order of $(\delta x)^4$. Several methods of solving the set of equations resulting from the application of Eq. (5.13) at each body point will be discussed in Sec. 5.6.

The representation just developed for the one-dimensional case may be extended to the two-dimensional and three-dimensional cases. Each term in Eq. (5.11) may be approximated in identical form to Eq. (5.13). The two-dimensional case is illustrated in Fig. 5.2. A plate of unit thickness, which is

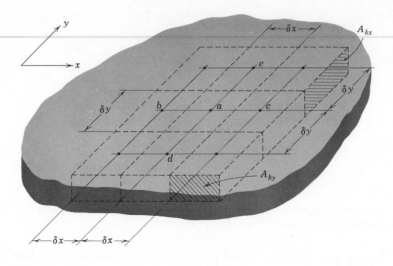

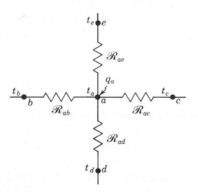

Fig. 5.2. Finite difference approximation and corresponding nodal network for two-dimensional conduction.

insulated on its plane faces, is subdivided in the x direction by increments of δx and in the y direction by increments of δy. The resulting finite difference approximation for the temperature at point a in terms of the temperatures at the neighboring points b, c, d, and e is

$$\frac{k}{(\delta x)^2}(t_b - 2t_a + t_c) + \frac{k}{(\delta y)^2}(t_d - 2t_a + t_e) + q_a^* = 0. \qquad (5.14)$$

As before, such an equation is to be satisfied at each point in the plate—in terms of its four neighboring points. A body divided into an $n \times m$ network will require the solution of $(n \times m)$ such equations. Quite apparently, considerable simplification will result if one chooses $\delta x = \delta y$. This could prove to be valuable if hand calculations were to be employed; however, it is of little consequence when digital computers are used.

The three-dimensional case is not illustrated, but it should be apparent that at each point in a three-dimensional body one must satisfy an equation of the form

$$\frac{k}{(\delta x)^2}(t_b - 2t_a + t_c) + \frac{k}{(\delta y)^2}(t_d - 2t_a + t_e) + \frac{k}{(\delta z)^2}(t_f - 2t_a + t_g) + q_a^* = 0.$$

$$(5.15)$$

The points f and g are the neighboring points in the z direction—at a spacing of δz.

5.4   Nodal Network Representations for the Steady State

The finite difference equations presented in the preceding section may be interpreted in a physical way that will permit rather convenient generalizations. Consider first the one-dimensional case. It will be recalled from Chapter 3 that for conduction through a slab of thickness Δx, a *unit* thermal resistance could be defined,

$$R = \frac{\Delta x}{k},$$

so the heat *flux* through the slab would be

$$\frac{q}{A} = \frac{t_1 - t_2}{R_{12}}.$$

A *total* thermal resistance, $\mathscr{R}$, may be defined which gives the total heat flow, q, rather than the heat flux:

$$\mathscr{R} = \frac{\Delta x}{kA_k},$$

$$(5.16)$$

$$q = \frac{t_1 - t_2}{\mathscr{R}_{12}}.$$

The subscript on the area symbol A_k is used to denote the fact that the area considered is the *conduction* area normal to the direction of heat flow.

Now, by reference to Fig. 5.1, imagine the one-dimensional bar to be divided into *lumps* of length δx so located that the points a, b, and c lie at the center of these lumps. The area A_k is the cross-sectional area normal to the one-dimensional coordinate x. Since the left side of Eq. (5.13) represents the heat flow *per unit volume,* multiply it by the volume of the lump surrounding point *a,* namely $(A_k \times \delta x)$:

$$\frac{kA_k}{\delta x}(t_b - 2t_a - t_c) + q_a = 0,$$

where

$$q_a = q_a^*(A_k\, \delta x)$$

is the total heat generated, per unit time, to the lump volume around point *a.* The term $\delta x/kA_k$ is seen to be the thermal resistance between point *a* and either *b* or *c,* so the above expression may be written

$$\frac{t_b - t_a}{\mathscr{R}} + \frac{t_c - t_a}{\mathscr{R}} = 0,$$

$$\mathscr{R} = \frac{\delta x}{kA_k}. \tag{5.17}$$

This latter expression suggests the electrical network analogy identified earlier and illustrated in Fig. 5.1. The lumps into which the conducting body is divided are presumed to be of uniform temperature—equal to the temperature of the node it surrounds. In the network representation, these isothermal lumps are symbolized by the nodal points *a, b,* and *c,* and the thermal resistances are symbolized by the electrical resistors shown connecting the nodes. The expression given in Eq. (5.17) is simply a statement of the total heat flow into the node *a* through the resistances connecting this node to all neighboring nodes. The lump temperatures applied to the nodes are analogous to electrical potentials in the network representation, and the heat balance of Eq. (5.17) is the sum of the currents flowing into a node if q_a is interpreted as a current flow into node *a* from some external potential source. In the steady state this summation must be zero.

In a similar fashion, the two-dimensional plate, of unit thickness, pictured in Fig. 5.2 may be subdivided into lumps $\delta x \times \delta y \times 1$ in volume, centered on the nodal points *a, b, ..., e.* The finite difference expression given in Eq. (5.14) reduces to

$$\frac{t_b - t_a}{\mathscr{R}_x} + \frac{t_c - t_a}{\mathscr{R}_x} + \frac{t_d - t_a}{\mathscr{R}_y} + \frac{t_e - t_a}{\mathscr{R}_y} + q_a = 0,$$

$$\mathscr{R}_x = \frac{\delta x}{kA_{kx}}, \qquad A_{kx} = \delta y \times 1, \tag{5.18}$$

$$\mathscr{R}_y = \frac{\delta y}{kA_{ky}}, \qquad A_{ky} = \delta x \times 1.$$

The equivalent nodal network is shown in Fig. 5.2. The three-dimensional case can be treated similarly, and the resultant expression should be apparent.

All the above results may be generalized into a single expression. For a nodal network representation of any steady state problem, if a thermal resistance is provided between a given node and each of its neighbors with which it conducts heat, each node i must satisfy the equations

$$\sum_j \frac{t_j - t_i}{\mathscr{R}_{ij}} + q_i = 0,$$

$$\mathscr{R}_{ij} = \frac{\delta_{ij}}{kA_{k_{ij}}}.$$

(5.19)

In Eq. (5.19) j denotes all neighboring nodes connected to node i, δ_{ij} denotes the conduction distance between node i and node j, $A_{k_{ij}}$ is the cross-sectional area for heat conduction normal to δ_{ij}, and q_i is the heat generated in or added to the volume lump at i. This representation permits the inclusion of unequal nodal spacings in any particular coordinate direction. Particular attention must then be paid to the evaluation of the various resistors, since a different conduction distance, δ_{ij}, would be involved in each case. Similarly, the volume of the lump and the conduction area associated with each node might be different. As will be seen later, the volume of the body lump associated with each node will become of considerable significance in the nonsteady case. In general, it is desirable to use equal net spacings, particularly if hand calculations are used, since all resistances become identical and can be eliminated from Eq. (5.19) altogether. In the instance of equal net spacings and zero heat addition, the two-dimensional case given by Eq. (5.18) reduces to

$$t_b + t_c + t_d + t_e - 4t_a = 0.$$

(5.20)

With the use of digital computers, the advantage of equal net spacings becomes of less value.

The existence of convective losses from body lumps bounded by a free surface exposed to an ambient fluid may be accounted for by adding to the network another node representing the fluid. The node representing the body lump is connected to the node representing the fluid by a resistor whose value is found from the heat transfer coefficient (a unit conductance) according to

$$\mathscr{R} = \frac{1}{hA_c}.$$

(5.21)

Here A_c denotes the surface area of the body lump exposed to the connecting ambient fluid. The expression which must be satisfied at each point then takes the form: At each node i,

$$\sum_j \frac{t_j - t_i}{\mathscr{R}_{ij}} + q_i = 0,$$

$$\mathscr{R}_{ij} = \begin{cases} \dfrac{\delta_{ij}}{kA_{k_{ij}}} & \text{for conduction,} \\[2ex] \dfrac{1}{h_{ij}A_{c_{ij}}} & \text{for convection.} \end{cases} \qquad (5.22)$$

In Eq. (5.22) q_i represents the heat added at a node by means other than surface convection. In cases in which internal heat generation is present, the q_i's are known. In the event that the node under consideration is one subjected to a boundary condition of specified temperature, t_i is then known and q_i becomes an unknown—namely the heat flux necessary to maintain the node at the desired temperature. In the remaining discussions of this chapter no internal heat generation will be considered, and thus the latter instances of specified boundary temperature will be the only cases in which the q_i term must be included.

The convective resistance was included in order to write the general form of the equation as it is shown in Eq. (5.22). Actually, this convective condition is just one of several boundary conditions that could be encountered. Before illustrating the solution of the set of equations resulting from the application of Eq. (5.22) to actual networks, these boundary conditions must be discussed. A sample nodal network for a complex system is shown in Fig. 5.3 as an illustration of the application of the concepts discussed here.

5.5   Boundary Conditions

The discussion so far has been restricted mainly to points interior to the body in which the heat conduction is taking place. In order to apply the heat balances represented by Eqs. (5.19) or (5.22), the imposed boundary conditions must also be satisfied. One such boundary condition, that of convection, has already been mentioned. Other typical conditions will be discussed in connection with two-dimensional conduction; however, the special form of these conditions for the one-dimensional case and their generalization to the three-dimensional case are sufficiently apparent to be left as exercises for the reader. Attention will first be directed to the instance in which regular network spacings (i.e., equal spacings in any one direction) are used, and the boundary coincides with a nodal point in this equal spacing.

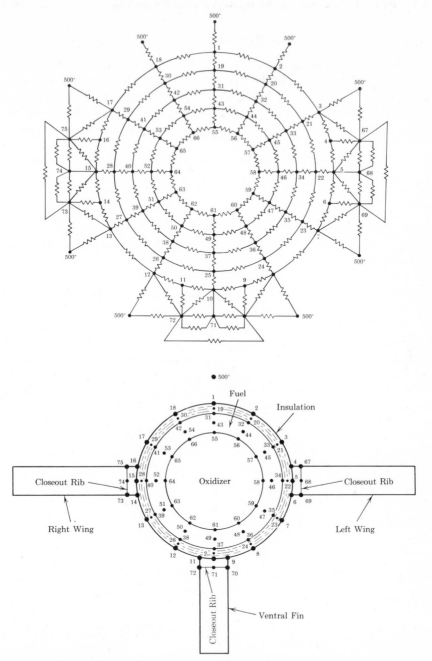

Fig. 5.3. The complex nodal representation for the thermal analysis of a missile fuel tank. (From "Temperature Control Systems for Space Vehicles," *ASD Rept. TRD-62-493, Part II*, Air Force Flight Dynamics Laboratory, Wright-Patterson Air Force Base, Ohio, 1963.)

175

Regular Spacing, Plane Boundary

Figure 5.4 illustrates a two-dimensional case in which nodal spacings of δx in the x direction and δy in the y direction are used. The conducting solid is presumed to have a unit thickness in the plane of the paper. A plane boundary parallel to the y direction coincides with one column of nodal points. The limits of the volume lumps surrounding each node are shown. It is apparent that the lump volumes associated with the boundary nodes are one-half of the volume of a regular, interior node. Evaluation of the heat balances on node a will be considered.

$$\mathscr{R}_{ab} = \frac{\delta x}{k\,(\delta y \times 1)}$$

$$\mathscr{R}_{ae} = \mathscr{R}_{ad} = \frac{\delta y}{k\,(\tfrac{1}{2}\delta x \times 1)}$$

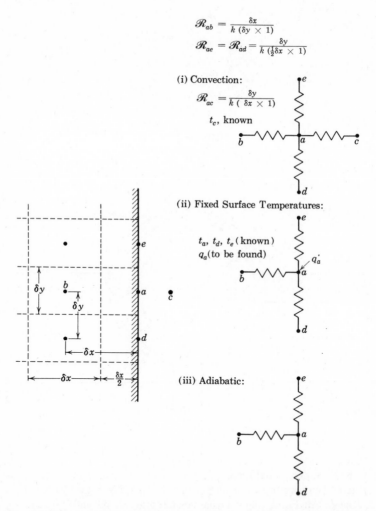

(i) Convection:

$$\mathscr{R}_{ac} = \frac{\delta y}{k\,(\,\delta x \times 1)}$$

t_c, known

(ii) Fixed Surface Temperatures:

$t_a,\ t_d,\ t_e$ (known)
q_a (to be found)

$\dot{q}_a$

(iii) Adiabatic:

Fig. 5.4. The boundary conditions at a plane, two-dimensional surface.

The value of the conduction resistor connecting a and b is the same as for any other resistor connecting two interior points spaced in the x direction:

$$\mathscr{R}_{ab} = \frac{\delta_{ij}}{kA_{k_{ij}}} = \frac{\delta x}{k(\delta y \times 1)}. \tag{5.23}$$

For resistors $\mathscr{R}_{ad}$ and $\mathscr{R}_{ae}$, the conduction distance is δy, as for an interior resistance in the same direction; however, the conduction area between e and a or d and a is one-half of what it would be otherwise:

$$\mathscr{R}_{ad} = \mathscr{R}_{ae} = \frac{\delta y}{k(\frac{1}{2}\delta x \times 1)}. \tag{5.24}$$

The three resistances just discussed will remain unchanged regardless of the boundary condition to be imposed. The resistance $\mathscr{R}_{ac}$ (if any), the heat input at node a, etc., *will* depend on the imposed condition. Some conditions are listed below.

Convection to a Fluid of Known Temperature. If a fluid of known temperature is in contact with the body surface, and if the film coefficient, h, for connective heat transfer between the solid surface and the fluid is known, the fluid may be represented by a node—call it c—maintained at the known fluid temperature. Then the resistance connecting a and c is

$$\mathscr{R}_{ac} = \frac{1}{hA_c} = \frac{1}{h(\delta y \times 1)}. \tag{5.25}$$

The application of the heat balance in Eq. (5.22) is, at node a:

$$\frac{t_b - t_a}{\mathscr{R}_{ab}} + \frac{t_d - t_a}{\mathscr{R}_{ad}} + \frac{t_e - t_a}{\mathscr{R}_{ae}} + \frac{t_c - t_a}{\mathscr{R}_{ac}} = 0. \tag{5.26}$$

The $\mathscr{R}$'s are given in Eqs. (5.23) through (5.25). The applicable network is shown in Fig. 5.4. For a square network ($\delta x = \delta y = \delta$) this equation reduces to

$$t_b + \tfrac{1}{2}(t_d + t_e) + \frac{h\delta}{k}t_c - \left(2 + \frac{h\delta}{k}\right)t_a = 0. \tag{5.27}$$

No heat balance is necessary at node c, since its temperature is known.

Specified Surface Temperature. A temperature distribution may be specified and maintained at the boundary. If so, the three temperatures t_a, t_d, and t_e are known. The resistance between these nodes was given above. In order that the temperature at node a remain fixed, an unknown amount of heat must flow between the node and its surroundings. The network representation is shown and Eq. (5.22) yields, at node a:

$$\frac{t_b - t_a}{\mathscr{R}_{ab}} + \frac{t_d - t_a}{\mathscr{R}_{ad}} + \frac{t_e - t_a}{\mathscr{R}_{ae}} + q_a = 0. \qquad (5.28)$$

Since t_a is known, the unknown quantities in the above equation are q_a and t_b. For a square network, Eq. (5.28) becomes

$$t_b + \tfrac{1}{2}(t_d + t_e) - 2t_a + \frac{q_a}{k} = 0. \qquad (5.29)$$

If it is not desired to find the heat flux at a, this nodal equation can be eliminated completely from consideration.

If the surface is *isothermal*, $t_a = t_d = t_e$, so the nodal balance reduces to

$$\frac{t_b - t_a}{\mathscr{R}_{ab}} + q_a = 0, \qquad (5.30)$$

or

$$t_b - t_a + \frac{q_a}{k} = 0 \qquad (5.31)$$

for square networks.

An Adiabatic Boundary. If the boundary is insulated, no interaction with the surroundings occurs, and node c is unnecessary. Thus,

$$\frac{t_b - t_a}{\mathscr{R}_{ab}} + \frac{t_d - t_a}{\mathscr{R}_{ad}} + \frac{t_e - t_a}{\mathscr{R}_{ae}} = 0, \qquad (5.32)$$

where the $\mathscr{R}$'s are given above. For the square network

$$t_b + \tfrac{1}{2}(t_d + t_e) - 2t_a = 0. \qquad (5.33)$$

Regular Spacing, Exterior Corner

Figure 5.5 illustrates a two-dimensional case in which the boundary of the solid is an exterior corner, and the regular nodal grid results in a node being located at the corner. In this instance the lump volume associated with nodes b and d is one-half the regular volume and that of node a is one-fourth that of an interior point. Without detailed discussion, the results are

$$\mathscr{R}_{ab} = \frac{\delta x}{k(\frac{1}{2}\delta y \times 1)},$$

$$\mathscr{R}_{ad} = \frac{\delta y}{k(\frac{1}{2}\delta x \times 1)}.$$

(5.34)

$$\mathscr{R}_{ab} = \frac{\delta x}{k\,(\frac{1}{2}\delta y \times 1)}$$

$$\mathscr{R}_{ad} = \frac{\delta y}{k\,(\frac{1}{2}\delta x \times 1)}$$

(i) Convection:

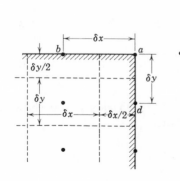

$$\mathscr{R}_{ae} = \frac{1}{h\,(\frac{1}{2}\delta x \times 1)} \qquad \mathscr{R}_{ac} = \frac{1}{h\,(\frac{1}{2}\delta y \times 1)}$$

$$t_c = t_e, \text{ known}$$

(ii) Fixed Surface Temperatures:

t_a, t_b, t_d (known)
q_a (to be found)

(iii) Adiabatic:

Fig. 5.5. The boundary conditions at a two-dimensional exterior corner.

Convection. Using nodes c and e to represent the fluid (with $t_e = t_c$), one finds that

$$\mathscr{R}_{ae} = \frac{1}{h(\frac{1}{2}\delta x \times 1)},$$

$$\mathscr{R}_{ac} = \frac{1}{h(\frac{1}{2}\delta y \times 1)}.$$

(5.35)

Equation (5.26) still applies, with the $\mathscr{R}$'s now given by Eqs. (5.34) and (5.35), and $t_e = t_c$. For the square network,

$$t_b + t_d + 2\frac{h\delta}{k}t_c - 2\left(1 + \frac{h\delta}{k}\right)t_a = 0.$$

(5.36)

Specified Surface Temperature. With t_a, t_b, and t_c known, Eq. (5.22) gives

$$\frac{t_b - t_a}{\mathscr{R}_{ab}} + \frac{t_d - t_a}{\mathscr{R}_{ad}} + q_a = 0,$$

(5.37)

and for the square network,

$$t_b + t_d - 2t_a + \frac{2q_a}{k} = 0.$$

(5.38)

If the surface is isothermal, either Eq. (5.37) or (5.38) yields $q_a = 0$, showing that no heat balance is necessary in this instance.

An Adiabatic Boundary. One has, simply,

$$\frac{t_b - t_a}{\mathscr{R}_{ab}} + \frac{t_d - t_a}{\mathscr{R}_{ad}} = 0$$

(5.39)

which is, for the square network,

$$t_b + t_d - 2t_a = 0.$$

(5.40)

Irregular Boundary Points

In many practical applications of the numerical methods discussed here, the boundaries of the solid are often shaped such that it is either inconvenient or impossible to arrange net spacings so that the boundary points coincide with regular points of the net. In such cases special relations must be developed for the resistances connecting the boundary points to interior points. The case of a two-dimensional network is shown in Fig. 5.6. For simplicity of discussion, a square network, of spacing δ, is considered. A curved boundary is shown which passes between nodes in the square network. Points b and e represent boundary points on the network connected to the interior point a. The points b' and e' represent points spaced the regular distance, δ, from point a, and lying outside the boundary of the solid. The symbol δ' will denote the distance from node a to node b, the real boundary point, while δ'' denotes the spacing from a to e. Apparently, both δ' and δ'' are less than δ.

In order to apply the heat balance of Eq. (5.22) to nodes a, b, c, d, and e, the resistances $\mathscr{R}_{ab}, \mathscr{R}_{ac}, \mathscr{R}_{ad}$, and $\mathscr{R}_{ae}$ must be evaluated. The method described here is developed in full in Ref. 1 wherein it is shown that the proposed representations neglect terms of the order of δ^3, and greater. It will be recalled from the discussion of Secs. 5.2 and 5.3 that the representation given by Eq. (5.22) for interior points neglects terms of the order of δ^4, and greater.

The recommended procedure consists of defining the limits of the solid lump surrounding node a as that rectangle which cuts the network lengths

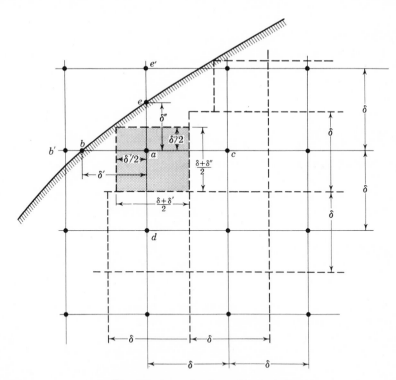

Fig. 5.6. Irregular boundary points.

a-b, *a-c*, *a-d*, and *a-e* into equal parts. The results may be observed in Fig. 5.6—the lump volume associated with node *a* being shaded. The nodal point *a* is not located at the geometric center of the lump. With this representation, the conduction areas between node *a* and its neighbors are seen to be

$$A_{k_{ab}} = A_{k_{ac}} = \frac{\delta + \delta''}{2},$$

$$A_{k_{ad}} = A_{k_{ae}} = \frac{\delta + \delta'}{2}.$$

Thus, the resistances needed become

$$\mathscr{R}_{ab} = \frac{\delta'}{k(\delta + \delta'')/2},$$

$$\mathscr{R}_{ac} = \frac{\delta}{k(\delta + \delta'')/2},$$

$$\mathscr{R}_{ad} = \frac{\delta}{k(\delta + \delta')/2}, \qquad (5.41)$$

$$\mathscr{R}_{ae} = \frac{\delta''}{k(\delta + \delta')/2}.$$

Application of these relations is made in the examples of following sections.

5.6 Numerical Solution of Steady State Network Equations

The foregoing sections have been devoted to the development of finite difference approximations for the heat conduction equation, network representations, boundary conditions, etc. The result of applying these concepts to a given conduction situation is that the *differential* equation for heat conduction in a body is replaced by an *algebraic* expression at each nodal subdivision of the body at which the temperature (or the heat flux in the case of an isothermal node) is desired. The expression for the temperature at a node involves the temperatures of the neighboring nodes. Thus, for a case in which N nodes of unknown temperature are involved, N algebraic equations are obtained which involve these unknowns. It remains, then, to be shown how such a system of N equations may be solved.

Several methods are available for the solution of a system of simultaneous algebraic equations (e.g., Refs. 3, 4, and 5). Certain methods particularly suited to the form of the equations involved here (usually linear) have been developed to a high degree and may be found discussed in detail in Refs. 1,

2, and 3. Those methods particularly adaptable to digital computer use will be described here. These methods will be briefly discussed and then illustrated by example. The examples will be chosen to illustrate as many of the points developed in the foregoing sections as possible.

Solution by Iteration

The iteration technique is a method of successive approximations and follows a fixed sequence of operations. As such, no opportunity exists for modification of the sequence of the operations at the discretion of the user. Hence, the method is particularly adaptable to digital computer use.

Equation (5.22), the nodal heat balance, may be written for each node i:

$$t_i = \frac{q_i + \sum_j \dfrac{t_j}{\mathcal{R}_{ij}}}{\sum_j 1/\mathcal{R}_{ij}}. \tag{5.42}$$

The symbol j, it is recalled, represents all neighboring nodes with which node i exchanges heat. In general, the j nodes will number from one to, perhaps, six, depending on the dimensionality of the problem and the boundary conditions. The iteration method (properly referred to as the Gauss–Seidel method) consists of assuming an initial set of values for the nodal temperatures (i.e., $[t_i]_1$). Then equations of the form of Eq. (5.42) are used to calculate new values of the t_i's (i.e., $[t_i]_2$) at each point. With the new set of temperatures thus generated, a third set of t_i's are calculated, using the same equations. This process is continued until there is a sufficiently small difference between two successive values of each temperature. That is, the operation

$$[t_i]_{n+1} = \frac{q_i + \sum_j \dfrac{[t_i]_n}{\mathcal{R}_{ij}}}{\sum_j 1/\mathcal{R}_{ij}}$$

is repeated at each nodal point until

$$|[t_i]_{n+1} - [t_i]_n| \leq \epsilon$$

for every i. The quantity ϵ is a preset convergence criterion and n represents the number of iterations executed. The above discussion is directed toward the instances in which the heat fluxes, q_i, are known and the temperatures are unknown. If a node is one of specified temperature, t_i is known and q_i becomes the sought-for unknown.

The above process will converge since Eq. (5.42) represents a weighted averaging process. Somewhere in the network, known, fixed, temperatures (or q's) are specified, and the iteration procedure continually averages these fixed values with the incorrect unknowns. Slowness of convergence should not be confused with accuracy of convergence. The accuracy of the final results depends upon the network spacing parameters and the convergence criterion ϵ. Improved accuracy may be obtained by using finer nets or smaller values of ϵ at the expense of increasing the number of computations and the required computing time, and no definite rule can be formulated for the choice of these parameters.

Quite apparently, the closeness of the initial assumptions for the t_i's to the correct solution has a profound effect upon the computing time. In some instances it may be advisable to make an initial calculation using a rather coarse net and then to use the results obtained to interpolate initial values for a much finer net.

The example which follows illustrates the iteration method.

EXAMPLE 5.1:   Solve, by numerical means, the problem stated in Example 3.8 for the case of the iron rod. It is desired to find the temperature distribution and the total heat flux from a $\frac{1}{2}$-in.-diameter iron rod ($k = 33$ Btu/hr-ft-°F) which is 12 in. long, heated to 250°F on one end, exposed to a convecting fluid at 70°F through a film coefficient of $h = 1.6$ Btu/hr-ft²-°F, and insulated on its free end.

Solution: Figure 5.7 illustrates the situation at hand. In keeping with the one-dimensional treatment employed in Chapter 3 for extended surfaces, it will be assumed that the conduction in the rod occurs in the longitudinal direction only. The rod is,

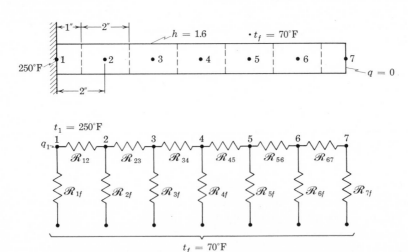

Fig. 5.7. Nodal representation of a one-dimensional spine.

thus, divided into five, equal, 2-in. lumps with 1-in. half-lumps at each end. This division results in seven equally spaced nodal points—with one at each end of the rod. Node 1 is an isothermal node maintained at $t_1 = 250°F$, and node 7 is a boundary node on an adiabatic surface. All other nodes are regular interior points. Each lump exchanges heat by convection with the ambient fluid at $t_f = 70°F$. The resultant nodal network is also shown in Fig. 5.7.

The conduction area for the conduction resistors $\mathscr{R}_{12} = \mathscr{R}_{23} = \cdots \mathscr{R}_{67}$ is the same in all cases:

$$A_k = \pi(1/2)^2/(4 \times 144) = 0.00136 \text{ ft}^2.$$

Thus,

$$\mathscr{R}_{12} = \mathscr{R}_{23} = \cdots = \mathscr{R}_{67} = \frac{\delta x}{kA_k} = \frac{\frac{2}{12}}{33 \times 0.00136}$$

$$= 3.704 \frac{\text{hr-°F}}{\text{Btu}}.$$

The convection area for the five interior nodes is $A_c = \pi(\frac{1}{12})(2)/144 = 0.0218 \text{ ft}^2$, so

$$\mathscr{R}_{2f} = \mathscr{R}_{3f} = \cdots = \mathscr{R}_{6f} = \frac{1}{hA_c} = \frac{1}{1.6 \times 0.0218}$$

$$= 28.65 \frac{\text{hr-°F}}{\text{Btu}}.$$

Since the convection area for nodes 1 and 7 is half that of the other nodes, the resistance is doubled:

$$\mathscr{R}_{1f} = \mathscr{R}_{7f} = 57.3 \frac{\text{hr-°F}}{\text{Btu}}.$$

Recognizing the fact that an unknown heat flux, q_1, exists at node 1 and that no other node has a heat flux associated with it, one finds that the application of Eq. (5.42) at each node gives

$$q_1 = 70.64 - 0.2700t_2,$$

$$t_2 = 0.4696t_3 + 121.66,$$

$$t_3 = 0.4696(t_2 + t_4) + 4.250,$$

$$t_4 = 0.4696(t_3 + t_5) + 4.250,$$

$$t_5 = 0.4696(t_4 + t_6) + 4.250,$$

$$t_6 = 0.4696(t_5 + t_7) + 4.250,$$

$$t_7 = 0.9393t_6 + 4.250.$$

Table 5.1 illustrates the results of the application of the iteration technique outlined above, starting with an assumed linear temperature distribution and using a convergence criterion of $\epsilon = 0.1°F$. Forty-two iterations were necessary to achieve this degree of convergence. The final results for an $\epsilon = 0.01°F$ and $1.0°F$ are also shown and the effect of the convergence factor and the required iterations is apparent. The results of the analytic methods of Example 3.8 are given in Table 5.1 for comparison purposes.

Table 5.1

Number of Iterations	q_1	t_2	t_3	t_4	t_5	t_6	t_7
0	$(t_1 = 250)$	220.00	190.00	160.00	130.00	100.00	70.00
1	11.241	210.89	182.71	154.54	126.36	98.18	98.18
2	13.700	207.47	175.90	149.40	122.93	109.70	96.47
3	14.624	204.26	171.85	144.58	125.94	107.29	107.29
4	15.492	202.37	168.08	144.10	122.54	113.78	105.02
5	16.001	200.60	166.97	140.74	125.36	111.12	111.12
6	16.480	200.07	164.55	141.54	122.53	115.31	108.63
7	16.620	198.94	164.69	139.08	124.88	112.81	112.56
8	16.927	199.00	163.00	140.24	122.55	115.76	110.21
9	16.910	198.21	163.57	138.35	124.48	113.56	112.98
10	17.124	198.48	162.31	139.53	122.56	115.77	110.92
11	17.051	197.89	163.00	138.04	124.15	113.90	112.99
12	17.210	198.21	162.02	139.11	122.57	115.62	111.24
13	17.124	197.75	162.67	137.90	123.88	114.06	112.85
14	17.248	198.05	161.89	138.82	122.58	115.43	111.38
15	17.166	197.69	162.46	137.85	123.66	114.13	112.67
⋮	⋮	⋮	⋮	⋮	⋮	⋮	⋮
42 ($\epsilon = 0.1$)	17.275	197.67	161.80	137.89	122.62	114.30	111.53
$\epsilon = 0.01$ 66 iterations	17.275	197.65	161.80	137.83	122.62	114.23	111.53
$\epsilon = 1.0$ 19 iterations	17.21	197.7	162.2	137.8	123.3	114.2	112.3
Analytic solution	17.01	197.98	161.83	137.69	122.40	113.95	111.26

Solution by Matrix Inversion

As was noted in the introductory discussion of this section, the problem at hand is that of solving N algebraic (usually linear) equations in N unknowns, where N is the number of nodal points at which either an unknown temperature or an unknown heat flux must be found. The application of Eq. (5.22) at each node generates a set of N equations of the form

$$a_{11}t_1 + a_{12}t_2 + a_{13}t_3 + \cdots + a_{1N}t_N = C_1$$
$$a_{21}t_1 + a_{22}t_2 + \cdots \qquad\qquad = C_2$$
$$a_{31}t_1 + \cdots \qquad\qquad\qquad = C_3 \qquad\qquad (5.43)$$
$$\vdots \qquad\qquad\qquad\qquad \vdots$$
$$a_{N1}t_1 + a_{N2}t_2 + \cdots \qquad + a_{NN}t_N = C_N.$$

The coefficients, $a_{11}, a_{12}, \cdots a_{NN}$ involve the resistances between the nodes and will be different from zero only when the nodes indicated by the subscripts are in thermal communication. The C's are constant terms resulting from specified boundary conditions or heat fluxes.

If one defines the following matrix representations:

$$[A] = \begin{bmatrix} a_{11} & a_{12} & \cdots & a_{1N} \\ a_{21} & a_{22} & \cdots & \\ a_{31} & & & \\ \vdots & & & \\ a_{N1} & a_{N2} & \cdots & a_{NN} \end{bmatrix}, \quad [C] = \begin{bmatrix} C_1 \\ C_2 \\ \vdots \\ C_N \end{bmatrix}, \quad [t] = \begin{bmatrix} t_1 \\ t_2 \\ \vdots \\ t_N \end{bmatrix},$$

the above set of equations may be written

$$[A][t] = [C].$$

Reference 5 may be consulted for information on the rules of matrix algebra. If $[A]^{-1}$ represents the *inverse* of $[A]$, the solution for the temperatures is given by

$$[t] = [A]^{-1} \cdot [C].$$

If $[A]^{-1}$ has the elements

$$[A]^{-1} = \begin{bmatrix} b_{11} & b_{12} & \cdots & b_{1N} \\ b_{21} & \cdots & & \\ \vdots & & & \\ b_{N1} & b_{N2} & \cdots & b_{NN} \end{bmatrix},$$

the required t's are

$$t_1 = b_{11}C_1 + b_{12}C_2 + b_{13}C_3 + \cdots + b_{1N}C_N,$$
$$t_2 = b_{21}C_1 + \cdots$$
$$\vdots$$
$$t_N = b_{N1}C_1 + b_{N2}C_2 + \cdots \qquad\qquad + b_{NN}C_N. \tag{5.44}$$

The problem, then, reduces to the inversion of a matrix. The process of matrix inversion (Ref. 5) is usually a laborious one when done by hand, and this method is not recommended in that instance, although the matrix $[A]$ usually contains a large number of zero elements. Most digital computer installations, however, have matrix inversion and matrix multiplication routines available as parts of the system library. In such cases the above operations may be carried out rather simply. Even so, if the number of nodes, and hence the rank of the matrix $[A]$, is large, the inversion process may become lengthy. Generally speaking, the matrix inversion method will usually be speedier for a modest number of nodes, but as the number of nodes increases, the iteration technique may prove to be faster.

EXAMPLE 5.2:   Find the temperature distribution for Example 5.1 by matrix inversion.

Solution: Since in this case the heat flux is not desired, eliminate the heat balance at node 1 and solve for only the 6 temperatures t_2 through t_7. The number of equations is, then, correspondingly reduced to six. The coefficient matrix and the matrix of the nonhomogeneous terms are

$$[A] = \begin{bmatrix} 1 & -0.4696 & 0 & 0 & 0 & 0 \\ -0.4696 & 1 & -0.4696 & 0 & 0 & 0 \\ 0 & -0.4696 & 1 & -0.4696 & 0 & 0 \\ 0 & 0 & -0.4696 & 1 & -0.4696 & 0 \\ 0 & 0 & 0 & -0.4696 & 1 & -0.4696 \\ 0 & 0 & 0 & 0 & -0.9393 & 1 \end{bmatrix},$$

$$[C] = \begin{bmatrix} 121.66 \\ 4.250 \\ 4.250 \\ 4.250 \\ 4.250 \\ 4.250 \end{bmatrix}.$$

The inverse of A is

$$[A]^{-1} = \begin{bmatrix} 1.5084 & 1.0830 & 0.7988 & 0.6188 & 0.5195 & 0.2439 \\ 1.0830 & 2.3071 & 1.7018 & 1.3183 & 1.1068 & 0.5195 \\ 0.7988 & 1.7018 & 2.8267 & 2.1898 & 1.8384 & 0.8629 \\ 0.6188 & 1.3183 & 2.1898 & 3.3467 & 2.8097 & 1.3189 \\ 0.5195 & 1.1068 & 1.8384 & 2.8097 & 4.1475 & 1.9468 \\ 0.4880 & 1.0396 & 1.7268 & 2.6391 & 3.8956 & 2.8286 \end{bmatrix}.$$

Since

$$\begin{bmatrix} t_2 \\ t_3 \\ t_4 \\ t_5 \\ t_6 \\ t_7 \end{bmatrix} = [A]^{-1}[C],$$

$$t_2 = 197.38, \qquad t_5 = 121.97,$$
$$t_3 = 161.31, \qquad t_6 = 113.57,$$
$$t_4 = 137.22, \qquad t_7 = 110.92.$$

These results compare favorably with those found by iteration.

Solution by Relaxation

The relaxation method of Southwell (Ref. 3) is particularly suited to the solution of the equations involved here—especially when hand calculations are contemplated. The nodal heat balance equations may be written

$$a_{11}t_1 + a_{12}t_2 + a_{13}t_3 + \cdots + a_{1N}t_N - C_1 = r_1$$
$$a_{21}t_1 + a_{22}t_2 + \cdots \qquad\qquad -C_2 = r_2$$
$$a_{31}t_1 + a_{32}t_2 + \cdots \qquad\qquad\qquad\qquad (5.45)$$
$$\vdots$$
$$a_{N1}t_1 + a_{N2}t_2 + \cdots \qquad\qquad + a_{NN}t_N - C_N = r_N.$$

The quantities $r_1, r_2, \ldots, r_N$ are referred to as the *residuals* and should all be zero when correct values of $t_1, t_2, \ldots, t_N$ are substituted into Eq. (5.45). If *estimated* values of the t's are introduced, the residuals will be, generally, different from zero, and the problem is to modify, intelligently, the estimated values of the t's until the residuals *are* zero.

Examination of Eq. (5.45) shows that if an adjustment is made in, say, the temperature t_1 by a unit amount ($\Delta t_1 = 1$), the residual r_1 is altered by an amount equal to a_{11}, r_2 by a_{21}, r_3 by a_{31}, etc. A *relaxation pattern* may be established as shown in Table 5.2. This table shows the alterations in the various residuals for unit changes in each of the unknowns. The coefficients in the relaxation table are seen to form an array which is the transpose of the matrix $[A]$ used in the matrix inversion method. The last line shows the changes which occur in the residuals when unit changes are made simultaneously in *all* the temperatures.

Table 5.2   A Relaxation Table

	Δr_1	Δr_2	Δr_3	$\cdots$	Δr_N
$\Delta t_1 = +1$	a_{11}	a_{21}	a_{31}	$\cdots$	a_{N1}
$\Delta t_2 = +1$	a_{12}	a_{22}	a_{32}		$\vdots$
$\Delta t_3 = +1$	a_{13}				
$\vdots$	$\vdots$				
$\Delta t_N = +1$	a_{1N}	a_{2N}	a_{3N}	$\cdots$	a_{NN}
All Δt's $= +1$	$\sum_N a_{1N}$	$\sum_N a_{2N}$	$\sum_N a_{3N}$		$\sum_N a_{NN}$

The relaxation procedure is performed as follows:

(a) Make initial assumptions for all the nodal temperatures and compute the initial values of the residuals which result according to Eq. (5.45).

(b) Find the node with the largest residual and "relax" it to zero (or conveniently close to zero) by making a suitable change in the temperature of that node.

(c) Compute, according to the relaxation table of Table 5.2, the changes in the residuals of all nodes influenced by the temperature change noted in (b) and record new values of the residuals.

(d) Selecting again the largest residual, repeat steps (b) and (c) and continue the process until all residuals are reduced sufficiently close to zero as the desired accuracy requires.

(e) Experience will show that considerable increase of the progression to convergence can be obtained by "overrelaxing" or "underrelaxing" in an intelligent way. That is, a residual may be overrelaxed by applying a sufficiently large Δt that it changes sign—with the knowledge that subsequent relaxation of an adjacent node that must be relaxed will counteract the overrelaxation. Underrelaxing denotes a similar concept.

The primary advantage of the relaxation procedure may be seen to be the introduction of choice and discretion on the part of the individual performing the calculation. That is, an intelligent use of overrelaxation, underrelaxation, and block relaxation (i.e., simultaneous relaxation of more than one node) may hasten convergence. This is not possible in the iterative technique described earlier. In that instance no discretionary choice of the user may be introduced, and one must continue, step by step, through the entire iterative procedure to obtain an answer. Also, in the relaxation method, one may always stop at any time and recalculate the residuals, based on current values of the temperatures, in order to check the accuracy. In the iterative technique numerical errors are not so easily identified and tend to increase the time required to reach convergence.

EXAMPLE 5.3:   A long circular shaft of 8 in. O.D. has a 2 in. × 2 in. square hole located on its axis. The shaft is so long that the conduction down its length may be neglected. The surface of the square hole is maintained at 100°F, and the external cylindrical surface is maintained at 0°F. Since the thermal conductivity will not affect the temperature distribution, choose it to be $k = 1$ Btu/hr-ft-°F and consider the two-dimensional conduction in a length of the shaft of 1 ft. Determine the temperature distribution.

Solution: As may be noted in Fig. 5.8, symmetry considerations show that only one-eighth of the cross section need be considered. A $\delta x = \delta y = 1$-in.-square nodal network was chosen as shown in the figure. The symmetry requirement along radial lines 0-a and 0-b is handled by including in the network symmetrical points for all nodes which are thermally connected to nodes located on the lines of symmetry. Specifically, image points for nodes 4 (twice), 7, and 8 are included. The resulting resistor network is also shown. No resistors are shown for connections along the isothermal surfaces (i.e., points 1-2 and points 6-11-12-13) since no conduction will take place in these directions. Similarly, no heat balance expression will be written for these nodes since their temperatures are known, and there is no desire to find the associated heat fluxes.

The resulting equivalent network is also shown. Heat balance expressions need to be written only for the six interior nodes: 3, 4, 5, 7, 8, 9. Seventeen resistances must be found; however, many are identical for the regular interior nodes. By use of Eq. (5.23),

$$\mathcal{R}_{14} = \mathcal{R}_{25} = \mathcal{R}_{45} = \mathcal{R}_{59} = \mathcal{R}_{9\text{-}13} = \mathcal{R}_{34} = 1.0 \ \frac{\text{hr-°F}}{\text{Btu}}.$$

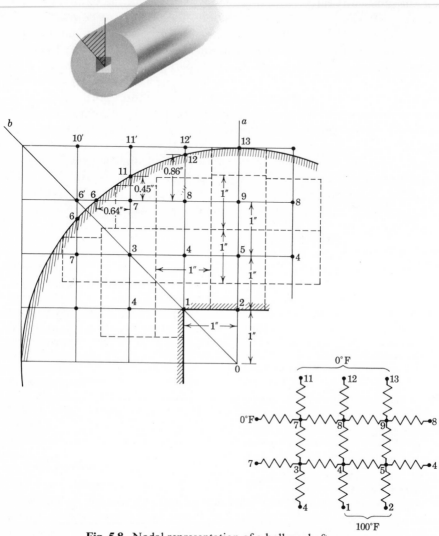

Fig. 5.8. Nodal representation of a hollow shaft.

Nodes 7 and 8 have irregularly spaced nodes connected with them (nodes 6, 11, 12) and the lengths of the unequal spacings are shown in the figure. By application of the relations given in Eq. (5.41):

$$\mathscr{R}_{67} = 0.882, \quad \mathscr{R}_{7\text{-}11} = 0.548, \quad \mathscr{R}_{78} = 1.38, \quad \mathscr{R}_{37} = 1.22,$$

$$\mathscr{R}_{89} = 1.075, \quad \mathscr{R}_{8\text{-}12} = 0.86, \quad \mathscr{R}_{48} = 1.0 \ \frac{\text{hr-}°\text{F}}{\text{Btu}}.$$

By applying Eq. (5.22) and using the facts that $t_1 = t_2 = 100, t_6 = t_{11} = t_{12} = t_{13} = 0$, one obtains:

$$\text{node 3:} \quad r_3 = -2.22t_3 + 1.22t_4 + t_7,$$

$$\text{node 4:} \quad r_4 = t_3 - 4t_4 + t_5 + t_8 + 100,$$

$$\text{node 5:} \quad r_5 = 2t_4 - 4t_5 + t_9 + 100,$$

$$\text{node 7:} \quad r_7 = 1.13t_3 - 6.32t_7 + t_8,$$

$$\text{node 8:} \quad r_8 = 1.38t_4 + t_7 - 5.26t_8 + 1.28t_9,$$

$$\text{node 9:} \quad r_9 = t_5 + 1.86t_8 - 3.86t_9.$$

These expressions for the residuals generate the relaxation table shown in Table 5.3. Table 5.4 illustrates the results of a relaxation solution for the required temperatures, using an initial assumption of $t_3 = t_4 = t_5 = 50°F, t_7 = 20°F, t_8 = t_9 = 30°F$.

Table 5.3   A Relaxation Table

	Δr_3	Δr_4	Δr_5	Δr_7	Δr_8	Δr_9
$\Delta t_3 = 1$	−2.22	1.0	—	1.13	—	—
$\Delta t_4 = 1$	1.22	−4.0	2.0	—	1.38	—
$\Delta t_5 = 1$	—	1.0	−4.0	—	—	1.0
$\Delta t_7 = 1$	1.0	—	—	−6.32	1.0	—
$\Delta t_8 = 1$	—	1.0	—	1.0	−5.26	1.86
$\Delta t_9 = 1$	—	—	1.0	—	1.28	−3.86
All $\Delta t's = 1$	0	−1.0	−1.0	−4.19	−1.60	−1.0

5.7   Nonsteady Numerical Methods

The material presented thus far has been limited to the case of steady state conduction. Of considerable practical importance is the application of numerical techniques to the analysis of nonsteady conduction problems. The finite difference representation of the second derivatives in the space variables, the nodal resistance network representations, the boundary conditions, etc., developed in Secs. 5.3 through 5.5 for the steady state will be equally applicable for nonsteady conduction analysis. The significant difference between the steady and nonsteady cases lies in finite difference representations which may be written for the partial derivative of temperature with respect to time. As will be recalled from Sec. 5.1, two possible finite difference expressions may be written for a first derivative, the forward difference and the backward

Table 5.4 The Relaxation Table for Example 5.3

	Node 3			Node 4			Node 5			Node 7			Node 8			Node 9		
	t_3	r_3	Δr_3	t_4	r_4	Δr_4	t_5	r_5	Δr_5	t_7	r_7	Δr_7	t_8	r_8	Δr_8	t_9	r_9	Δr_9
Initial values	50.00	−30.00		50.00	30.00		50.00	30.00		20.00	−39.90		30.00	−30.40		30.00	−10.00	
	−20.00	−40.00	−10.00		30.00	—		30.00	—	−10.00	23.30	63.20		−40.40	−10.00		−10.00	—
		4.40	44.40		10.00	−20.00		30.00	—		0.70	22.60	−10.00	−40.40	52.60		−10.00	—
		4.40	—		0.00	−10.00	7.00	30.00	—		−9.30	−10.00		12.20	—	−5.00	−28.60	−18.60
		4.40	—		7.00	7.00		2.00	−28.00		−9.30	—		12.20	−6.40		−21.60	7.00
	3.00	4.40	−6.66	3.00	7.00	—		−3.00	−5.00		−9.30	3.39	2.00	5.80	—		−2.30	19.30
		−2.26	3.66		10.00	3.00		−3.00	—		−5.91	—		5.80	4.14		−2.30	—
		1.40	—		−2.00	−12.00		3.00	6.00	−0.50	−5.91	2.00		9.94	−10.52		−2.30	3.72
		1.40	−0.50		0.00	2.00		3.00	—		−3.91	3.16		−0.58	−0.50		1.42	—
		0.90	—		0.00	—	1.00	3.00	—		−0.75	—		−1.08	—	0.70	1.42	1.00
		0.90	—		1.00	1.00		−1.00	−4.00		−0.75	—		−1.08	0.90		2.42	−2.70
	0.70	0.90	0.61	0.50	1.00	−2.00		−0.30	0.70		−0.75	0.79	2.00	−0.18	0.69		−0.28	—
		1.51	−1.55		−1.00	0.70		0.70	1.00		−0.75	—		0.51	—		−0.28	0.20
		−0.04	—		−0.30	0.20		0.70	—		0.04	0.10	0.10	0.51	−0.53		−0.28	0.19
		−0.04	0.03		−0.10	0.10	0.20	−0.10	−0.80	0.03	0.04	−0.19		0.51	0.03	0.05	−0.08	—
		−0.04	—		0.00	—		−0.10	—		0.14	—		−0.02	0.06		0.11	−0.19
		−0.01	—		0.00	—		−0.05	0.05		−0.05	0.01	0.01	0.01	−0.05		0.11	0.02
		−0.01	—		0.00	0.01		−0.05	—		−0.05	—		0.07	—		−0.08	0.01
		−0.01	—		0.01	−0.01	−0.01	−0.01	0.04		−0.04	—		0.02	−0.03	−0.02	−0.06	0.08
		−0.01	—		0.00	—		−0.03	−0.02		−0.04	—		0.02	—		−0.07	
		−0.01			0.00						−0.04			−0.01			0.01	
Final values	33.70	−0.01		53.50	0.00		58.19	−0.03		9.53	−0.04		22.11	−0.01		25.73	0.01	

difference. These representations lead to two different numerical methods for treatment of nonsteady problems.

In either instance, it is important to note that in addition to calculating temperatures at points spaced discrete intervals apart in space, in the nonsteady case the temperatures at these points are calculated at discrete intervals of time. That is, after the temperatures are known at all the spatial points, a finite increment of time is selected and all the temperatures are recalculated at the end of this time. Time is then progressively incremented, the spatial temperatures being calculated at each increment.

The nonsteady methods to be described here may be applied to the solution of steady state problems. With time-independent boundary conditions, an assumed temperature distribution may be regarded as an initial distribution of a nonsteady problem. As the nonsteady solution is carried foward in time, the desired steady state solution is eventually approached to any desired degree of accuracy.

Difference Equations

The nonsteady conduction equation is, from Eq. (1.8),

$$k\left(\frac{\partial^2 t}{\partial x^2} + \frac{\partial^2 t}{\partial y^2} + \frac{\partial^2 t}{\partial z^2}\right) = \rho c_p \frac{\partial t}{\partial \tau}. \tag{5.46}$$

Written in this form, each side of the conduction equation represents the time rate of heat storage, per unit volume, at a point. Finite difference approximations will now be written for this expression, but as noted above, two possible representations are possible. For conciseness of expression, the representations for the one-dimensional case will be written first. In this instance,

$$k\frac{\partial^2 t}{\partial x^2} = \rho c_p \frac{\partial t}{\partial \tau}. \tag{5.47}$$

Explicit Formulation. The explicit formulation is obtained by using the forward difference expression for the first derivative in place of the time derivative on the right side of Eq. (5.47). The central difference for the second derivative is used for the left side. In other words, to expand Eq. (5.47) about $x = x$ and $\tau = \tau$, use Eqs. (5.4) and (5.9) with t replacing f, τ replacing η, x replacing ξ, $\delta\tau$ replacing h_2, and δx replacing h_1. The result is

$$\frac{k}{(\delta x)^2}[t(x + \delta x, \tau) - 2t(x, \tau) + t(x - \delta x, \tau)] = \frac{\rho c_p}{\delta \tau}[t(x, \tau + \delta \tau) - t(x, \tau)]. $$

$$\tag{5.48}$$

If, as used in the steady state case and as suggested in Fig. 5.9, the subscripts a, b, and c are used to denote nodal locations at x, $x - \delta x$, and $x + \delta x$, respectively, and if t' is used to denote a temperature at time $\tau + \delta\tau$ while t is used simply to denote a temperature at time τ, Eq. (5.48) is more concisely stated as

$$\frac{k}{(\delta x)^2}(t_b - 2t_a + t_c) = \frac{\rho c_p}{\delta\tau}(t'_a - t_a). \tag{5.49}$$

This expression gives the *future* temperature at node a, t'_a, in terms of the *current* temperatures at node a and its surrounding nodes. This forward difference approximation for the nonsteady conduction equation is also termed the "explicit" formulation.

The equivalent two-dimensional, explicit, formulation is easily shown to be (see Fig. 5.9)

$$\frac{k}{(\delta x)^2}(t_b - 2t_a + t_c) + \frac{k}{(\delta y)^2}(t_d - 2t_a + t_e) = \frac{\rho c_p}{\delta\tau}(t'_a - t_a). \tag{5.50}$$

Implicit Formulation. The implicit representation is obtained by expanding Eq. (5.47) about $x = x$ and $\tau = \tau + \delta\tau$. This is done by use of Eqs. (5.7) and (5.9) with t replacing f, $\tau + \delta\tau$ replacing η, x replacing ξ, $\delta\tau$ replacing h_2, and δx replacing h_1. The result is

$$\frac{k}{(\delta x)^2}[t(x + \delta x, \tau + \delta\tau) - 2t(x, \tau + \delta\tau) + t(x - \delta x, \tau + \delta\tau)]$$

$$= \frac{\rho c_p}{\delta\tau}[t(x, \tau + \delta\tau) - t(x, \tau)].$$

Using the same abbreviated notation as employed above, one obtains

$$\frac{k}{(\delta x)^2}(t'_b - 2t'_a + t'_c) = \frac{\rho c_p}{\delta\tau}(t'_a - t_a). \tag{5.51}$$

This implicit formulation gives the *future* temperature of point a, t'_a, in terms of the current temperature at a, t_a, and the *future* temperatures of the neighboring points. The two-dimensional case reduces to

$$\frac{k}{(\delta x)^2}(t'_b - 2t'_a + t'_c) + \frac{k}{(\delta y)^2}(t'_d - 2t'_a + t'_e) = \frac{\rho c_p}{\delta\tau}(t'_a - t_a). \tag{5.52}$$

Explicit versus Implicit Formulation. A rather obvious advantage of the explicit representation over the implicit is the fact the forward difference equation gives the *future* temperature of a single node in terms of *current* temperatures of that node and its neighbors. Thus, if at the end of a certain time period, all the nodal temperatures are known, then each of the nodal temperatures at the end of the next moment, $\delta\tau$, may be explicitly found, node by node. The implicit equation, however, expresses a *future* nodal temperature in terms of its current value and the *future* values of its neighbors' temperatures. Thus, to progress from one time step to the next, a system of equations of the form of Eqs. (5.51) or (5.52) must be solved.

It would appear, then, that the explicit representation would be preferred to the implicit representation. However, as will be seen later, in the explicit case a serious restriction must be placed upon the magnitude of the time step $\delta\tau$ in relation to the spatial increment δx. Thus, although more direct, the explicit formulation may actually involve more calculation time than the less-direct implicit method.

Nodal Networks in the Nonsteady State

Directing attention for the moment to the one-dimensional case shown in Fig. 5.9, one may divide the conducting solid into lumps centered on the nodal points, as in the steady state cases discussed earlier. The total heat flow at a node may be had by multiplying Eq. (5.49), or (5.51), by the lump volume, $V_a = A_k \, \delta x$. Thus, in the explicit case:

$$\frac{t_b - t_a}{\dfrac{\delta x}{kA_k}} + \frac{t_c - t_a}{\dfrac{\delta x}{kA_k}} = \frac{V\rho c_p}{\delta\tau}(t'_a - t_a).$$

The resistances $\mathscr{R}_{ab} = \mathscr{R}_{ac} = \delta x/kA_k$ are defined as before. The term

$$C_a = V_a \rho c_p$$

is recognized as the thermal capacity of the lump volume surrounding the node a. Thus, the above expression becomes

$$\frac{t_b - t_a}{\mathscr{R}_{ab}} + \frac{t_c - t_a}{\mathscr{R}_{ac}} = C_a \frac{t'_a - t_a}{\delta\tau}. \tag{5.53}$$

Similarly, for the implicit case

$$\frac{t'_b - t'_a}{\mathscr{R}_{ab}} + \frac{t'_c - t'_a}{\mathscr{R}_{ac}} = C_a \frac{t'_a - t_a}{\delta\tau}. \tag{5.54}$$

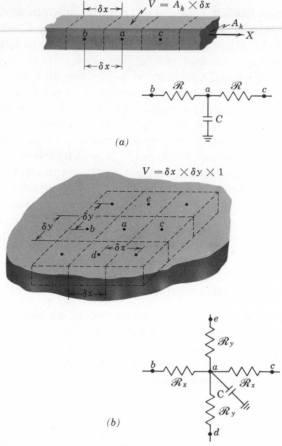

Fig. 5.9. Nodal representations of nonsteady conduction: (a) one-dimensional; (b) two-dimensional.

Either Eq. (5.53) or (5.54) suggests the analogous electrical network illustrated in Fig. 5.9, wherein a capacitance is associated with each node—representing the thermal capacitance of the associated lump.

Generalizing upon the concepts just discussed, and drawing upon the concepts developed in Sec. 5.4, one may conclude that if a complex geometry (i.e., two or more dimensions) is subdivided into a nodal network of, perhaps unequal spacings, the following expression must be satisfied at each nodal point i:

$$\text{Explicit:} \quad \sum_j \frac{t_j - t_i}{\mathcal{R}_{ij}} + q_i = \frac{C_i}{\delta\tau}(t_i' - t_i). \tag{5.55}$$

$$\text{Implicit:} \quad \sum_j \frac{t_j' - t_i'}{\mathcal{R}_{ij}} + q_i = \frac{C_i}{\delta\tau}(t_i' - t_i). \tag{5.56}$$

The term q_i again denotes the rate of heat addition (external or internal) to the node. Thus, the future temperature of node i is determined by

$$\text{Explicit}: \quad t_i' = t_i + \delta\tau\left(\sum_j \frac{t_j - t_i}{\mathcal{R}_{ij}C_i} + \frac{q_i}{C_i}\right). \tag{5.57}$$

$$\text{Implicit}: \quad t_i' = t_i + \delta\tau\left(\sum_j \frac{t_j' - t_i'}{\mathcal{R}_{ij}C_i} + \frac{q_i}{C_i}\right), \tag{5.58}$$

where

$$\mathcal{R}_{ij} = \frac{\delta_{ij}}{kA_{k_{ij}}}, \qquad \text{for conduction,}$$

$$\mathcal{R}_{ij} = \frac{1}{h_{ij}A_{c_{ij}}}, \qquad \text{for convection,} \tag{5.59}$$

$$C_i = V_i\rho c_p.$$

These equations just given express the temperatures at a given node in terms of discrete spatial steps (i.e., δ_{ij}) and discrete time steps (i.e., $\delta\tau$). The method of solution depends on which formulation (explicit or implicit) is used, but the important point to be noted is that the continuous solution of the original differential equation is replaced by the numerical solution of these difference equations at particular points in space and at particular intervals of time.

Boundary Conditions in the Nonsteady State

Equations of the form of Eq. (5.57) or (5.58) must be satisfied at each nodal point into which a conducting body is subdivided. As in the steady state case, certain special precautions must be taken at boundary points. In general, this amounts to calculating the resistances (conduction and convection) convected to the boundary nodes by use of special relations rather than those given in Eq. (5.59). The boundary conditions given in Sec. 5.5 for the steady case are equally applicable to the nonsteady case. Thus, no repetition will be made here for those conditions.

The case of time-dependent boundary temperatures enters as a possibility in the nonsteady case, however. In such cases, it is usual practice to treat such a case as that of a fixed temperature boundary condition [see Eqs. (5.28), (5.29), and (5.30)]. During any time increment, the fixed boundary temperatures are chosen as the mean of the temperatures occurring at the boundary points at the beginning and the end of that time increment. As the calculations proceed in time, these boundary point temperatures must be continually revised according to the prescribed temperature function imposed at the surface. One of the examples given later will illustrate this procedure.

Stability of the Solution

As has been pointed out previously, the numerical methods presented in this chapter represent approximate solutions to the original differential equations since derivatives are replaced by finite differences. Terms of the order of the fourth power of the spatial step size are neglected and, in the transient case, terms of the order of the square of the time increment are neglected. Errors introduced by these approximations are termed *truncation* errors, and the degree to which the approximate solution approaches the exact solution is termed the *convergence* of the finite difference representation. *Numerical* errors are also introduced into a solution by virtue of the inability, or impracticality, of performing the numerical computations with a sufficient number of significant figures.

In nonsteady numerical problems, the *stability* of the difference equations must also be considered. This matter is related to the way in which numerical and truncation errors introduced at one point in time either damp out or propagate and amplify in succeeding time steps. Detailed analyses of the stability properties of the equations used in this chapter are quite complex (see Refs. 1 and 6), but simple stability criteria may be developed on an elementary, intuitive, basis.

Rewrite Eqs. (5.57) and (5.58) in the following forms—expressing the sought-for future temperature at node i, t'_i, in terms of the other quantities:

$$\text{Explicit:}\quad t'_i = t_i\left(1 - \sum_j \frac{\delta\tau}{\mathcal{R}_{ij}C_i}\right) + \sum_j \frac{t_j\delta\tau}{\mathcal{R}_{ij}C_i}. \tag{5.60}$$

$$\text{Implicit:}\quad t'_i = \frac{t_i + \sum_j t'_j \dfrac{\delta\tau}{\mathcal{R}_{ij}C_i}}{1 + \sum_j \dfrac{\delta\tau}{\mathcal{R}_{ij}C_i}}. \tag{5.61}$$

For simplicity of discussion, any internal heat generation, q_i, has been considered zero.

It may be noted that in the case of the explicit formulation, the coefficient of the t_i term might become negative—particularly if the time increment, $\delta\tau$, is chosen large enough. If the coefficient of t_i *does* become negative, t'_i could conceivably be less than t_i. This implies that for certain values of $\sum_j \delta\tau/\mathcal{R}_{ij}C_i$, the greater is the temperature t_i at time τ, the *smaller* it will be at time $\tau + \delta\tau$. This fact does not make sense, thermodynamically, and it can be seen as a trend which will cause the temperature t_i to oscillate wildly from one time period to the next. Although this is not a precise analysis of stability, one would be certain of a stable procedure so long as

$$\sum_j \frac{\delta\tau}{\mathscr{R}_{ij}C_i} \leq 1. \tag{5.62}$$

More sophisticated analyses (Refs. 1 and 6) lead to less stringent stability criteria; however, the above limiting relation is generally used in practical cases.

The implication of Eq. (5.62) is that the choice of the time step $\delta\tau$ is intimately connected with the choice of the spatial increment, δ_{ij}, which is involved in the resistance $\mathscr{R}_{ij}$. As smaller spatial increments are chosen to reduce the associated truncation error, smaller time increments must also be used in order to satisfy Eq. (5.62). Thus, increased accuracy in the spatial network is obtained at the cost of smaller time increments—perhaps leading to prohibitive computation times. This is the principal disadvantage of the explicit method, and it will be examined in more detail in Sec. 5.9.

Examination of Eq. (5.61) for the implicit formulation reveals that no stability limitation exists as a result of possible negative coefficients. Thus, the time increment is not restricted by the choice of spatial increments. The only limitation on $\delta\tau$ in this instance is that imposed by the minimization of truncation errors in time.

5.8   Solution of Network Equations for the Implicit Case

The nodal heat balance in the implicit case is given by

$$t'_i = \frac{t_i + \sum_j t'_j \dfrac{\delta\tau}{\mathscr{R}_{ij}C_i}}{1 + \sum_j \dfrac{\delta\tau}{\mathscr{R}_{ij}C_i}}. \tag{5.61}$$

If all the resistances of a network are known, and if the capacitances of all the associated lumps are known, then with a specified *initial* temperature distribution, equations of the type of Eq. (5.61) may be written for each nodal point—care being taken to satisfy the established boundary conditions. The net result is that a set of equations are obtained for the *future* temperatures, t'_i, of the nodal points—in terms of the *future* temperatures of neighboring points—once a time step, $\delta\tau$, is chosen. There will be as many equations as there are unknown future temperatures. Once this set of equations is solved, the resulting temperatures become the initial temperatures for the next time step calculation.

Thus, the implicit technique reduces to the solution of a set of simultaneous algebraic equations at each time increment. Consequently, the methods discussed in Secs. 5.3 through 5.6 for the solution of the steady state equations are applicable here—iteration, matrix inversion, or relaxation. The first two are recommended for digital computer usage. The implicit formulation is seen, then, to reduce to a series of steady state calculations at each time step. It has the advantage of not having a restricted time step—in fact, the time step may be varied during the progression of the calculation. It has the disadvantage of requiring a *set* of calculations (i.e., iteration or matrix inversion) at each step in time, leading to increased calculation time and storage requirements as the number of nodes becomes large. Since the calculations do not differ, in principle, from those already shown for the steady state, no examples will be given.

5.9 Solution of Network Equations for the Explicit Case

As discussed earlier, the explicit formulation avoids the need of iterative or matrix inversion techniques, since each future nodal temperature can be individually calculated for a time increment $\delta\tau$ from only the current nodal temperatures. Thus, from an equation of the form

$$t'_i = t_i\left(1 - \sum_j \frac{\delta\tau}{\mathscr{R}_{ij}C_i}\right) + \sum_j t_j \frac{\delta\tau}{\mathscr{R}_{ij}C_i}, \tag{5.60}$$

new temperatures are successively calculated at each node, starting with the given initial temperature distribution in a network for a given $\delta\tau$. Time is then incremented and the calculations are repeated. No iterations or matrix inversions are required. Only the stability requirement stated in Eq. (5.62) need be satisfied:

$$\sum_j \frac{\delta\tau}{\mathscr{R}_{ij}C_i} \le 1. \tag{5.62}$$

Special Stability Criteria

Examination of the stability requirement for special instances is appropriate. The relations to be developed emphasize the *upper* limit of the allowable time increment. In practice it is wise to use a time increment smaller than the maximum. The reason for this practice is a combination of the desires to improve accuracy by reducing truncation error in time and to avoid instabilities by staying safely below the upper limit of $\delta\tau$.

Irregular Networks. For the most general instance in which the network contains irregular net spacings, convective resistors as well as conductive resistors, etc., the criterion given in Eq. (5.62) must be evaluated at each nodal point. Since it is necessary at any one time step to carry *each* node forward in time by the same $\delta\tau$, the time increment for the entire calculation is controlled by the node for which stability criterion yields the smallest time increment. That is,

$$\delta\tau \leq \left(\frac{1}{\sum_j \frac{1}{\mathcal{R}_{ij}C_i}} \right)_{min} \tag{5.63}$$

In general, nets that include lumps which are significantly smaller than the others should be avoided. Such small lumps usually have small capacitances and consequently control the time increment for the entire network.

Equally Spaced Networks. In the event that a net is one of equal spacings, special forms result for Eq. (5.63). Figure 5.10 illustrates one- and two-dimensional cases in which regular spacings are used. In the case of the two-dimensional network, regular spacing implies the use of a square grid. In either the one- or two-dimensional cases a half-lump must be provided at the boundary.

Consider first the one-dimensional case. The body dimensions in the two directions, other than the one in which conduction takes place, are taken as l_1 and l_2, respectively. (Actually the body may be of infinite extent in these directions, in which case one would carry out the calculation on a unit flow area basis and simply take $l_1 = l_2 = 1$.) For the case in which there is conduction only (i.e., no convection at the boundary) the equal spacing of δx results in identical resistors—two connected to each interior node and one at the surface node. The resistors all have the value

$$\mathcal{R} = \frac{\delta x}{k(l_1 l_2)}.$$

The capacitors of the interior nodes are all equal to

$$C = \rho V c_p = \rho l_1 l_2 \, \delta x c_p,$$

while the surface node has a capacitance of $C/2$. Since the nodes with capacitance C have two resistors, $\mathcal{R}$, connected to them and that with capacitance $C/2$ has only one resistor connected to it, the stability criterion of Eq. (5.62)

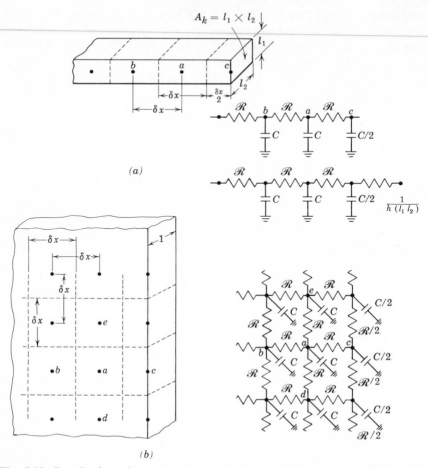

Fig. 5.10. Regular boundary points in nonsteady conduction: (a) one-dimensional; (b) two-dimensional.

yields the same result for *all* nodes:

$$2\frac{\delta\tau}{\mathcal{R}C} \leq 1.$$

Rewritten, the above expression becomes

$$\delta\tau \leq \tfrac{1}{2}\mathcal{R}C$$

$$\leq \frac{1}{2}\frac{\rho c_p}{k}(\delta x)^2$$

$$\leq \frac{1}{2}\frac{(\delta x)^2}{\alpha}. \tag{5.64}$$

The latter form shows more clearly the restriction placed on the time increment by the choice of spatial increment.

For the two-dimensional case shown in Fig. 5.10, and for the three-dimensional case with an equally spaced net, arguments similar to that just given yield

$$\text{Two dimensions:} \quad \delta\tau \le \tfrac{1}{4}\mathcal{R}C$$

$$\le \frac{1}{4}\frac{(\delta x)^2}{\alpha}. \tag{5.65}$$

$$\text{Three dimensions:} \quad \delta\tau \le \tfrac{1}{6}\mathcal{R}C$$

$$\le \frac{1}{6}\frac{(\delta x)^2}{\alpha}. \tag{5.66}$$

If the upper limit of the time increment is used, the nodal heat balance, Eq. (5.60), takes on particularly simple forms:

$$\text{One dimension:} \quad t_a' = \frac{t_b + t_c}{2}. \tag{5.67}$$

$$\text{Two dimensions:} \quad t_a' = \frac{t_b + t_c + t_d + t_e}{4}. \tag{5.68}$$

The future temperature at a node is seen to be a simple mean of the surrounding nodes. A similar relation results for the three-dimensional case. These relations prove to be most useful for hand calculations and also form the basis of certain graphical methods (Ref. 1).

In the event that convection occurs at the boundary, the above stability criteria must be altered. In the one-dimensional case, also pictured in Fig. 5.10, the stability criterion yields the same result as given in Eq. (5.64) when applied at the interior nodes. At the surface node, however, there is a conduction resistance and a convection resistance:

$$\mathcal{R}_{\text{cond}} = \frac{\delta x}{k(l_1 l_2)},$$

$$\mathcal{R}_{\text{conv}} = \frac{1}{h(l_1 l_2)},$$

$$C = \tfrac{1}{2}\rho l_1 l_2\, \delta x c_p.$$

Thus, Eq. (5.62) yields

$$\frac{\delta\tau}{\tfrac{1}{2}\rho l_1 l_2\, \delta x c_p}\left[\frac{k l_1 l_2}{\delta x} + h l_1 l_2\right] \le 1,$$

or,

$$\delta\tau \leq \frac{1}{2}\frac{(\delta x)^2}{\alpha}\left[\frac{1}{1 + (h\,\delta x/k)}\right]. \tag{5.69}$$

Since the term in brackets in Eq. (5.69) is always less than 1, it is apparent that the stability criterion applied at the surface node yields a smaller time increment than that resulting from consideration of the interior nodes. Thus, the surface node becomes the controlling factor as far as the selection of the time increment is concerned.

A similar analysis when applied to the two-dimensional case shown in Fig. 5.10 and in the three-dimensional case gives

$$\text{Two-dimensional:}\quad \delta\tau \leq \frac{1}{4}\frac{(\delta x)^2}{\alpha}\left[\frac{1}{1 + \frac{1}{2}(h\,\delta x/k)}\right].$$

$$\text{Three-dimensional:}\quad \delta\tau \leq \frac{1}{6}\frac{(\delta x)^2}{\alpha}\left[\frac{1}{1 + \frac{1}{3}(h\,\delta x/k)}\right]. \tag{5.70}$$

Relations similar to those given in Eq. (5.70) may be developed for other special configurations, such as an external corner, etc. These may be found in the references at the end of the chapter. When the surface film coefficient becomes quite large the permissible time increment allowed by the boundary nodes may become considerably less than that allowed by the interior nodes. In such an instance, the computation of the temperature history may become prohibitively long. Certain schemes have been developed whereby such severe limitations may be circumvented (Refs. 4, 5, and 6.).

The implicit and explicit techniques just described are only two, rather direct, of many possible numerical methods available for solution of the heat conduction equation. For certain problems, accelerating techniques are available. These techniques, although more complex to carry out, are hybrid methods which attempt to combine the inherent stability of the implicit approach with the matrix-inversion-free advantages of the explicit approach. Reference 5 may be consulted for information on such techniques.

EXAMPLE 5.4: An infinite slab 10 in. thick is initially at a uniform temperature of 100°F. The temperature at both surfaces is suddenly raised to 500°F and maintained at that value. The slab is composed of a material with the following properties: $c_p = 0.2$ Btu/lb$_m$-°F, $\rho = 144$ lb$_m$/ft^3, $k = 1.00$ Btu/hr-ft-°F, $\alpha = 0.0347$ ft^2/hr. Determine, numerically, the temperature history of the slab during the first hour after the surface temperature is increased. Use several values of the time increment.

Solution: Because of symmetry considerations only one-half of the slab need be considered. Six nodal points, spaced 1 in. apart, are used, with the associated lumps

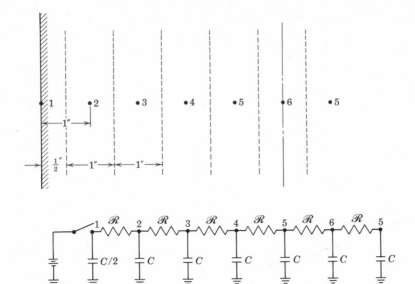

Fig. 5.11. The nonsteady circuit for a plane slab.

as shown in Fig. 5.11. The network subdivision satisfies the requirements for Eq. (5.64), so the sum $\sum_j 1/\mathcal{R}_{ij}C_i$ is the same for all nodes. Specifically,

$$\sum_j \frac{1}{\mathcal{R}_{ij}C_i} = \frac{2}{\mathcal{R}C},$$

$$\mathcal{R}C = \frac{(\delta x)^2}{\alpha} = \frac{(1/12)^2}{0.0347} = 0.2 \, \text{hr}.$$

According to Eq. (5.62) or (5.64), the maximum permissible time increment is

$$\delta\tau \le \left(\sum_j \frac{1}{\mathcal{R}_{ij}C_i}\right)^{-1} = \frac{\mathcal{R}C}{2}$$

$$\le 0.1 \, \text{hr}.$$

For this time increment the nodal equations are all of the form of Eq. (5.67). Table 5.5 shows the results of the calculations.

For node 1, the temperatures are specified. In keeping with the principle mentioned in association with the boundary conditions, the temperature of node 1 during the first time increment is taken as the mean of its value at the beginning ($100°F$) and the end ($500°F$) of the increment, namely $300°F$. At $\tau = 0$, the initial temperatures of nodes 2 through 6 are known to be $100°F$. Then, at $\tau = 0.1$, the new temperatures at nodes 2 through 6 are found by use of Eq. (5.67), or, equivalently, Eq. (5.60), upon the completion of each row of the table, the next row is constructed in a similar manner.

Figure 5.12 compares the results of the numerical calculations at selected time intervals with the results of the exact solution available from Sec. 4.5 and Fig. 4.8.

Table 5.5 Numerical Calculations for Infinite Slab

Time, hr	Node 1	2	3	4	5	6
0.0	300	100.0	100.0	100.0	100.0	100.0
0.1	500	200.0	100.0	100.0	100.0	100.0
0.2	500	300.0	150.0	100.0	100.0	100.0
0.3	500	325.0	200.0	125.0	100.0	100.0
0.4	500	350.0	225.0	150.0	112.5	100.0
0.5	500	362.5	250.0	168.8	125.0	112.5
0.6	500	375.0	265.7	188.5	140.7	125.0
0.7	500	382.9	281.8	203.2	156.8	140.7
0.8	500	390.0	293.1	219.3	172.0	156.8
0.9	500	396.6	305.1	232.6	188.1	172.0
1.0	500	402.6	314.6	246.6	202.3	188.1

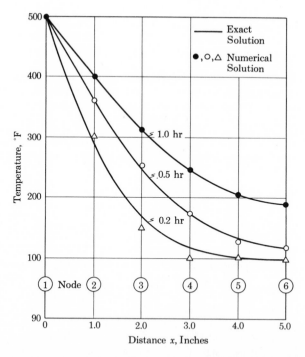

Fig. 5.12. Comparison of analytical and numerical results for a plane slab, showing the temperature distribution at various times when using the maximum time increment.

Also compared in Fig. 5.13 are the temperature–time histories at the slab center (node 6) which result from using the maximum time increment, as above, $\delta\tau = (\delta\tau)_{max} = \frac{1}{2}(\delta x)^2/\alpha$, and a time increment one-half of that, $\delta\tau = \frac{1}{2}(\delta\tau)_{max} = \frac{1}{4}(\delta x)^2/\alpha$. The improvement in accuracy with decreased time increment is apparent. Also shown is the unstable character of the results which are obtained when one uses a time increment which exceeds the maximum: $\delta\tau = 1.2(\delta\tau)_{max} = 0.60(\delta x)^2/\alpha$.

EXAMPLE 5.5:   The rod described in Example 5.1 is initially at equilibrium with the ambient fluid at 70°F, the base of the rod being maintained at that temperature. For time $\tau > 0$, the base temperature of the rod is increased, linearly, to 250°F over a time period of 15 min. Find the temperature time history for the rod for a total time span of 1 hr.

Solution: Use the same nodal subdivision for the rod as used in the example pictured in Fig. 5.7. The resulting network will be identical to that shown in Fig. 5.7, except that a capacitance will be associated with each node. Using $\rho = 490\ \mathrm{lb_m/ft^3}$ and

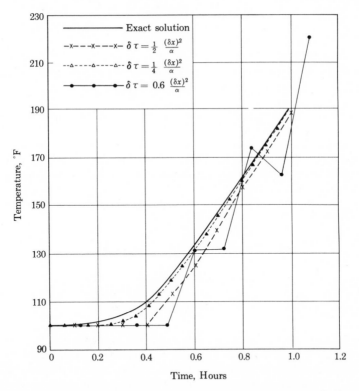

Fig. 5.13. Comparison of analytical and numerical results for a plane slab, showing the temperature–time history of the centerline for different time increments.

$c_p = 0.11$ Btu/lb$_m$-°F for iron (the other data remain the same as in Example 5.1), the nodal capacitances become

$$C_2 = C_3 = \cdots = C_6 = \frac{\pi(1/2)^2 \times 2 \times 490 \times 0.11}{4 \times 1728}$$

$$= 0.0122 \text{ Btu/°F},$$

$$C_1 = C_7 = 0.0061 \text{ Btu/°F}.$$

The resistances connected to each node were calculated in the previous examples, resulting in

Nodes 1 and 7: $\sum_j \dfrac{1}{\mathscr{R}_{ij}C_i} = \dfrac{1}{0.0061}\left(\dfrac{1}{57.2} + \dfrac{1}{3.71}\right) = 46.93\dfrac{1}{\text{hr}}$,

Nodes 2 through 6: $\sum_j \dfrac{1}{\mathscr{R}_{ij}C_i} = \dfrac{1}{0.0122}\left(\dfrac{1}{28.6} + \dfrac{2}{3.71}\right) = 46.93\dfrac{1}{\text{hr}}$.

Thus, the time restriction imposed by the stability criterion is the same for all nodes, namely

$$\delta\tau \sum_j \frac{1}{\mathscr{R}_{ij}C_i} \le 1,$$

$$\delta\tau \le 0.0213 \text{ hr.}$$

For convenience of computation and to ensure stability, use a time increment $\delta\tau = 0.02$ hr. The temperature of node 1 is specified, rising 180°F in 0.25 hr or 14.4°F per time increment. The initial temperature imposed at node 1 will be the mean between 70°F and the value after one increment, 84.4°F. Thus, at $\tau = 0$, $t_1 = 77.2$°F, while $t_2 = t_3 = \cdots = t_7 = 70$°F. For each subsequent time interval t_1 is incremented by 14.4°F until it reaches 250°F at the twelfth increment. Thereafter t_1 is maintained at 250°F. The temperatures at nodes 2 through 7 are determined at each time step by Eq. (5.60) with $\Sigma \, \delta\tau/\mathscr{R}_{ij}C_i = 0.02/0.0213 = 0.9386$:

$$t'_n = 0.0614t_n + 0.4408(t_{n-1} + t_{n+1} + 9.0511).$$

The results of the calculations, carried out for 50 time steps (i.e., 1 hour), are shown in Fig. 5.14 at selected time intervals. The approach to the distribution as $\tau \to \infty$ (i.e., the steady state solution in Example 5.1) is apparent.

EXAMPLE 5.6:   In the case of the Example 5.3, the circular shaft with a central hole, analyzed for the steady state, determine the maximum allowable time step that stability limitations would permit if the same geometry were to be analyzed for the nonsteady case. Assume that the boundary conditions would not involve convection. The material of the shaft has the properties $k = 1$ Btu/hr-ft-°F, $\rho = 20$ lb$_m$/ft^3, $c_p = 0.2$ Btu/lb$_m$-°F, $\alpha = 0.25$ ft^2/hr.

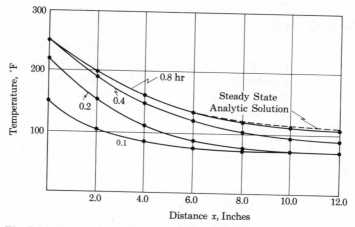

Fig. 5.14. Comparison of analytical and numerical results for a spine.

Solution: As in the steady state example pictured in Fig. 5.8, use a length of the shaft of 1 ft. In that instance, the resistances found in the steady state example are applicable here if the same network is used. Using the dimensions shown in Fig. 5.8, one may calculate the nodal capacitances shown below. The values of $\Sigma\, 1/\mathcal{R}_{ij}$ are those determined from the previously determined $\mathcal{R}$'s:

Node 3: $\dfrac{1}{C_3} = 36\dfrac{°F}{Btu}$; $\sum_j \dfrac{1}{\mathcal{R}_{3j}} = 3.64\dfrac{Btu}{hr\text{-}°F}$; $\sum_j \dfrac{1}{\mathcal{R}_{3j}C_3} = 131\dfrac{1}{hr}$.

Node 4: $\dfrac{1}{C_4} = 36$; $\sum_j \dfrac{1}{\mathcal{R}_{4j}} = 4.0$; $\sum_j \dfrac{1}{\mathcal{R}_{4j}C_4} = 144$.

Node 5: $\dfrac{1}{C_5} = 36$; $\sum_j \dfrac{1}{\mathcal{R}_{5j}} = 4.0$; $\sum_j \dfrac{1}{\mathcal{R}_{5j}C_5} = 144$.

Node 7: $\dfrac{1}{C_7} = 60.5$; $\sum_j \dfrac{1}{\mathcal{R}_{7j}} = 4.41$; $\sum_j \dfrac{1}{\mathcal{R}_{7j}C_7} = 267$.

Node 8; $\dfrac{1}{C_8} = 38.7$; $\sum_j \dfrac{1}{\mathcal{R}_{8j}} = 3.83$; $\sum_j \dfrac{1}{\mathcal{R}_{8j}C_8} = 148$.

Node 9: $\dfrac{1}{C_9} = 36$; $\sum_j \dfrac{1}{\mathcal{R}_{9j}} = 4.162$; $\sum_j \dfrac{1}{\mathcal{R}_{9j}C_9} = 150$.

From the largest value of $\Sigma_j\, 1/\mathcal{R}_{ij}C_i = 267$, the associated maximum time increment is found to be

$$\delta\tau = 1/267 = 0.00374 \text{ hr}.$$

All other nodes will yield larger maxima increments, so the above value must be used—or any smaller value.

5.10   Analog Methods

Appropriate to a discussion of the use of numerical methods in the analysis of heat conduction problems is a brief mention of analog techniques. When two distinct physical phenomena are described mathematically, in terms of their individual independent variables, by equations of identical form, a mathematical *analogy* is said to exist. The heat conduction equation is an equation which occurs frequently in physics for the description of a number of other phenomena. Hence, a variety of analogies exist by which the conduction problem may be replaced in favor of another physical system in which direct measurements may be more easily made.

In the case of steady state heat conduction, the equation for the temperature distribution reduces to Laplace's equation—for which many analogous situations exist. Formerly, the most popular system for analogous solution of Laplace's equation in two dimensions was the *membrane analogy*. It can be shown (Ref. 1) that the deflection of a uniformly loaded thin membrane (e.g., a soap film) attached to a specified contour must satisfy Laplace's equation, and measurements of this deflection may be interpreted as temperatures in a two-dimensional conducting body of the same contour as the membrane. However, the use of the membrane analogy requires considerable experimental skill, and simpler analogies exist for the solution of two-dimensional steady conduction problems. Commercially available devices may be obtained in which the conduction of electricity in a sheet of paper of low electrical conductivity is used to simulate two-dimensional temperature distributions in similarly shaped conducting slabs. Sheets of the conducting paper are cut into the desired two-dimensional contour, and imposed surface temperatures are simulated by impressed electrical voltages. (Adiabatic surfaces are simulated by electrical isolation.) By the aid of a probe and a null-indicating voltmeter, lines of constant potential, analogous to isotherms, may be drawn on the paper.

If one chooses to represent one-, two-, or three-dimensional conduction by discrete lumps, Eq. (5.22) suggests that an equivalent electrical network may be constructed. Specified temperatures may then be imposed electrically and sought-for nodal temperatures measured with a voltmeter. Alternating current may be employed if care is taken to eliminate capacitance or inductance in the network—otherwise direct current should be employed. The discussion of the numerical techniques in this chapter has relied heavily upon this analogy.

Nonsteady conduction problems may also be treated by analogy. Electrical network analogs are used, almost exclusively. If, in Eqs. (5.55) and (5.56), the finite difference representation of the time-dependent lump temperature in a nodal network, $C_i(t_i' - t_i)/\delta\tau$, is replaced by $C_i \, dt_i/d\tau$, the resulting expression is analogous to the time response of voltages in a resistance–

capacitance, dc, electrical network. This analogy was used extensively in the discussion of the nonsteady numerical methods. Based upon this analogy, *ℛ-C network analyzers* have been used extensively for the analogous solution of nonsteady conduction problems (Refs. 1, 7, and 8). Two approaches are possible. One may construct an individual *ℛ-C* network analogy from individual electrical components. Large-scale network analog facilities are commercially available which have a large variety of permanently installed resistances and capacitors (variable through certain ranges) which may be interconnected to represent the particular problem at hand. By careful choice of scale factors, and with a sufficient quantity of components available, a complex conduction problem (involving, perhaps, many hours in real time) may be reduced to the measurement of the time response of electrical voltages in a matter of a few minutes. Generally, such *analog computers* also incorporate integrating circuits and function generators in order to simulate complex boundary conditions and to evaluate integrated heat flows, etc.

It should be noted that a significant difference exists between the network analog and the numerical solution of transient conduction problems. The numerical approach yields a stepwise solution in time while the circuit analog yields a continuous solution in time.

References

1. SCHNEIDER, P. J., *Conduction Heat Transfer*, Reading, Mass., Addison-Wesley, 1955.
2. DUSINBERRE, G. M., *Heat-Transfer Calculations by Finite Differences*, Scranton, Pa., International Textbook, 1961.
3. SOUTHWELL, R. V., *Relaxation Methods in Theoretical Physics*, New York, Oxford U.P., 1946.
4. HAWKINS, G. A., and J. T. AGNEW, "The Solution of Transient Heat-Conduction Problems by Finite Differences," *Purdue Univ. Eng. Expt. Sta. Res. Ser. 98*, Lafayette, Ind., 1947.
5. HILDEBRAND, F. B., *Methods of Applied Mathematics*, Englewood Cliffs, N.J., Prentice-Hall, 1952.
6. FOWLER, C. M., "Analysis of Numerical Solutions of Transient Heat Flow Problems," *Quart. Appl. Math.*, Vol. 3, 1945, p. 361.
7. McCANN, G. D., and C. H. WILTS, "Application of Electric-Analog Computers to Heat-Transfer and Fluid Flow Problems," *J. Appl. Mech.*, Vol. 16, 1949, p. 247.
8. LIEBMANN, G., "A New Electrical Analog Method for the Solution of Transient Heat-Conduction Problems," *Trans. ASME*, Vol. 78, 1956, p. 655.

Problems

5.1. A $\frac{1}{4}$-in.-diameter pure copper rod is 18 in. long and is heated to 200°F at *each* end. The rod surface is exposed to an ambient fluid at 80°F through a surface film coefficient of 4.5 Btu/hr-ft²-°F. By numerical means, find the temperature distribution in the rod using a nodal spacing of 1.5 in. How much heat is dissipated by the rod? How much heat is dissipated by the first 4.5 in. of the rod? Compare the results with an analytic solution.

5.2. A straight fin of uniform thickness is attached to a surface at 300°F. The fin is 2 in. thick and 6 in. long and is composed of 20% nickel steel. The surface and the end of the fin are exposed to an ambient fluid at 50°F through a film coefficient of 5.0 Btu/hr-ft²-°F. Find, numerically, the temperature distribution in the fin, recognizing that the thickness–length ratio is too large to treat the fin as one-dimensional.

5.3. The accompanying figure depicts a section of a chimney made of common brick. The inside surface is maintained at 350°F while the outside surface is maintained at 100°F. Assuming that the chimney is quite tall so that the heat conduction through it may be considered as two-dimensional, find by numerical means the temperature distribution in the chimney. Use a network spacing of 4 in.

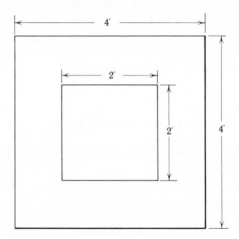

Problem 5.3

5.4. Repeat Prob. 5.3 if a gas at 400°F flows inside the chimney with a film coefficient of 10 Btu/hr-ft²-°F at the surface. Air at 70°F surrounds the outside of the chimney and the film coefficient there is 2 Btu/hr-ft²-°F. Use a 6-in. net spacing.

5.5. A concrete block, 12 in. square, has a 6-in.-O.D. steam pipe buried concentrically inside. The steam pipe dissipates 200 Btu/hr per foot of length. The exterior surfaces of the concrete are maintained at 50°F. Find, numerically, the temperature distribution in the concrete. Presume that the block is long enough to treat the heat flow as two-dimensional.

5.6. Imagine that the chimney in Prob. 5.3 is initially at 350°F throughout and that the outside surface temperature is subsequently lowered suddenly to 100°F. Determine, numerically, the temperature–time history in the chimney for a time span of 10 hr.

5.7. A large plate of pure aluminum is 1.5 in. thick. It is in contact with convecting fluids on each side. On one side the film coefficient is 5 Btu/hr-ft²-°F, and on the other side it is 12 Btu/hr-ft²-°F. Initially both fluids are at 70°F, and the plate is in equilibrium. Suddenly the fluid temperatures are raised to 200°F. Determine, numerically, how long it takes the center of the plate to reach 175°F. Compare the result with analytical predictions based on the material of Chapter 4.

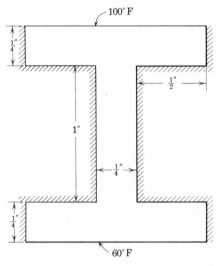

Problem 5.8

5.8. The accompanying figure depicts the cross section of a very long Duralumin *I* beam. All surfaces, other than the upper and lower faces, are insulated. Determine, numerically, the steady state rate of heat flow (per foot of beam length) which results when the upper surface is maintained at 100°F and the lower surface at 60°F.

5.9. For the network chosen in Prob. 5.8, determine the maximum time increment that could be used to numerically analyze the transient response of the *I* beam to a sudden change in one of the surface temperatures.

5.10. Derive relations analogous to Eqs. (5.34) through (5.40) for the applicable boundary conditions at an *internal* corner.

5.11. Derive the facts given in Eqs. (5.65), (5.66), and (5.70).

5.12. Verify, numerically, the results given in Fig. 5.13 for the temperature distribution in the rod at 6 min from the start of the transient process.

CHAPTER 6

The Fundamental Principles of Viscous Fluid Motion and Boundary Layer Motion

6.1   The Fluid Mechanical Aspects of Convection

In the foregoing chapters, primary attention has been directed to the problems of heat conduction in solids. The mode of heat transfer known as convection has been considered only as one type of boundary condition to be applied at the surface of a conducting solid. This boundary condition has been treated in terms of a gross parameter, the surface heat transfer coefficient, h, defined by Newton's law of cooling in Eq. (1.14). In these applications h has been presumed known. The purpose of the next few chapters will be to focus attention on the process of heat convection in a fluid and to develop methods of predicting the value of the film coefficient that will likely result under a given set of conditions.

As discussed earlier in Secs. 1.3 and 1.5, convection is the term applied to the energy transfer process which is observed to occur in fluids mainly because of the transport of the energy by means of the motion of the fluid itself. The process of conduction of energy by molecular interchange is, of course, still present, but high energy (or hot) portions of the fluid are brought into contact with the lower energy regions (cooler regions) by virtue of the fact that there are gross displacements of the fluid particles. When the fluid motion is caused by the imposition of external forces in the form of pressure differences, this mechanism is called *forced convection*. The pumping of a fluid through or past solid surfaces of a temperature different from the fluid is an example of forced convection. When no external forces are applied to a fluid it may be set into motion by differences in density that would be caused by an immersed solid body whose temperature is different from that of the fluid. Such a heat exchange situation is termed *free convection* and may be observed in a heated pan of boiling water, in the air surrounding a room radiator, etc.

In either case, an analytical approach to the determination of the film co-efficient would involve the finding of the temperature distribution in the fluid surrounding the body. If, as is usually the case, the fluid motion in the region immediately adjacent to the surface is laminar, then the heat flux from the surface may be evaluated in terms of the fluid temperature gradient at the surface. Then the definition of the film coefficient as the ratio of the heat flux to the difference between the surface temperature and the fluid temperature enables one to write

$$h = \frac{-k_f(dt/dn)_s}{t_s - t_f},$$

where

$\quad k_f$ = thermal conductivity of the fluid,

$\quad t_s$ = temperature of surface,

$\quad t_f$ = bulk temperature of fluid far removed from surface,

$\left(\dfrac{dt}{dn}\right)_s$ = fluid temperature gradient, measured at the surface in a direction normal to the surface.

Figure 6.1 illustrates these concepts.

The actual execution of an analytical determination of h by the procedure just outlined is of such a highly complex nature that it has been carried out in only a relatively few cases of practical importance. Even so, it is instructive to discuss the fundamental equations involved in such an analytical approach in order to justify some of the rather far-reaching simplifications that will have to be made in order to obtain solutions by other, approximate, methods.

From this brief discussion of the convective mechanism, it is apparent that any theoretical analysis must involve the use of the fundamental equations governing the motion of a viscous fluid. These equations must be solved in conjunction with the application of the principle of conservation of energy and conservation of mass in order to relate the heat conducted through the fluid

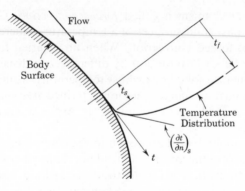

Fig. 6.1

to the energy stored in the fluid and the work done by the fluid as it moves against the forces of pressure and viscosity.

For the general case of the motion of a convecting fluid in three dimensions, a complete description of the fluid motion would require the determination of the three velocity components, the fluid temperature, the fluid pressure, and the fluid density—all as functions of position and time. Thus, six equations are required to determine these six unknowns. Application of Newton's law of motion in each of the coordinate directions yields three partial differential equations, and two additional partial differential equations are obtained from the principles of conservation of mass and conservation of energy. The equation of state of the fluid furnishes a sixth, algebraic equation. A general solution would then require the simultaneous solution of these six equations. Such a solution is beyond the present limits of mathematical analysis, and rather severe limitations must be put on the system to make an analytical solution possible.

For the sake of simplicity this chapter will consider fluid motion in only *two space dimensions*. This restriction eliminates one unknown and one independent variable. Also, it will be presumed that the fluid may be treated as *incompressible*—i.e., the density will be taken as constant. This assumption eliminates another unknown, making it necessary to have only four equations to describe the flow. In some instances it will be assumed that *steady flow* exists. By this it is meant that at a given point in the fluid domain there is no dependence of the flow parameters (velocity, pressure, etc.) on time.

These fundamental equations will be discussed in the following sections. These discussions are not meant to be rigorous derivations of the equations, but, rather, are meant to illustrate the physical significance of the various terms and to justify the accepted standard forms of these relations. Complete discussions of these equations may be found in Refs. 1 and 2.

6.2   The Continuity Equation—The Conservation of Mass

Since motion in only two space dimensions will be considered, it is convenient to think of the fluid flowing in the x-y plane (as illustrated in Fig. 6.2) with unit depth in the z direction (normal to the plane of the figure). All flow properties would then be constant over this unit depth.

Generally, then, the x-y plane represents a field in which the fluid velocity varies throughout. Representing the velocity vector, $\vec{v}$, by the x and y components, v_x and v_y, one has, in general, a variation of these components throughout the x-y plane, each component being a function of x and y.

Selecting arbitrarily a volume element of space having dimensions δx, δy, and unit depth, one sees that the rate of mass flow into the element must equal the rate of outflow if the principle of conservation of mass is to hold for the case in which no sources or sinks of mass exist within the fluid domain. This statement is true whether the flow is steady or not since no mass may be stored in the element if the fluid is incompressible. Locating one corner of the element at (x, y) as shown in Fig. 6.2, one can evaluate the rate of mass flow across the boundaries. Considering first flow in the x direction only, one finds that fluid enters or leaves the element across only the faces at x and $x + \delta x$. If δy is sufficiently small, it is possible to say that the fluid velocity across the y face at x is uniform and equal to v_x. Since the velocity components vary throughout the space, the flow velocity across the y face at $x + \delta x$ differs from v_x but may be expressed in terms of v_x by a Taylor's series:

At x, velocity $= v_x$.

$$\text{At } x + \delta x, \text{ velocity} = v_x + \frac{\partial v_x}{\partial x}\delta x + \frac{\partial^2 v_x}{\partial x^2}\frac{(\delta x)^2}{2!} + \cdots .$$

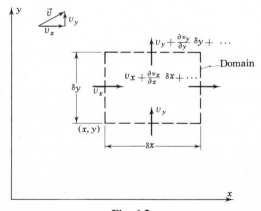

Fig. 6.2

The density, ρ, is constant; thus, the excess of flow out of the element over that into the element due to x-direction motion is (since the mass flow rate is the product of the velocity, the density, and the cross-sectional flow area):

$$\text{Excess} = \rho\left[\left(v_x + \frac{\partial v_x}{\partial x}\delta x + \frac{\partial^2 v_x}{\partial x^2}\frac{(\delta x)^2}{2!} + \cdots\right)\delta y \times 1 - v_x\,\delta y \times 1\right]$$

$$= \rho\,\delta x\,\delta y\left[\frac{\partial v_x}{\partial x} + \frac{\partial^2 v_x}{\partial x^2}\frac{\delta x}{2!} + \cdots\right].$$

In a similar fashion, the excess of outflow over inflow due to y-direction motion across the x faces may be obtained so that the *total* excess of outflow for motion in *both* directions is obtained as

$$\rho\,\delta x\,\delta y\left(\frac{\partial v_x}{\partial x} + \frac{\partial^2 v_x}{\partial x^2}\frac{\delta x}{2!} + \cdots + \frac{\partial v_y}{\partial y} + \frac{\partial^2 v_y}{\partial y^2}\frac{\delta y}{2!} + \cdots\right).$$

Conservation of mass demands that the above expression be zero. This means that

$$\frac{\partial v_x}{\partial x} + \frac{\partial^2 v_x}{\partial x^2}\frac{\delta x}{2!} + \cdots + \frac{\partial v_y}{\partial y} + \frac{\partial^2 v_y}{\partial y^2}\frac{\delta y}{2!} + \cdots = 0.$$

This equation must be satisfied for all elements in the domain, so letting $\delta x \to \delta y \to 0$, one obtains the *continuity equation* for incompressible flow:

$$\frac{\partial v_x}{\partial x} + \frac{\partial v_y}{\partial y} = 0. \tag{6.1}$$

As noted earlier, this relation is valid for steady or nonsteady flow.

6.3   Viscous Resistance for Plane Laminar Fluid Motion

Before discussing the equations of motion of a viscous fluid it will be necessary to extend the definition of viscosity made in Chapter 2. The present discussion and the resulting equations of motion will be restricted to *laminar* flow conditions. The concept of turbulence and the interpretation of the equations of motion for turbulent flow will be reserved for discussion in later sections.

In Sec. 2.7 the coefficient of dynamic viscosity, μ, was defined for laminar motion in one direction. That is, for motion all in one direction, the dynamic viscosity was defined as the proportionality factor between the shear stress in the fluid and the velocity gradient produced by that shearing resistance. Figure 6.3(a) shows an infinitesimal element of a fluid moving in only one direction. In this case, viscous shear acts only in that direction—the x direction in Fig. 6.3—and deforms the element as shown. The initially rectangular element deforms into a parallelogram if it is assumed infinitesimal in size and if only first order effects are considered.

Now, μ is defined by Newton's relation [Eq. (2.8)],

$$\tau_l = \mu \frac{\partial v_x}{\partial y}. \tag{6.2}$$

The symbol τ_l is used to denote the *laminar* shear stress. The subscript l is used to emphasize the fact that the definition applies only to the laminar case and to offer a distinction from the symbol τ, which is used to denote time. This use of the same symbol to denote shear stress and time may appear to be confusing; however, the context should make it clear which quantity is meant, and rarely will the two be involved in the same equation.

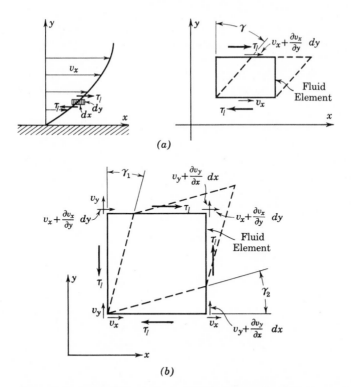

Fig. 6.3. Angular deformation of fluid elements by shear stresses.

Let the velocity of the side of the element at y be v_x. Taking a linear velocity distribution in the y direction—i.e., only two terms in a Taylor's series—one finds the velocity at the side of the fluid element at $y + dy$ to be $v_x + (\partial v_x/\partial y)\,dy$. Thus, the *relative* sliding rate between the two sides of the fluid element at y and $y + dy$ is $(\partial v_x/\partial y)\,dy$. If γ denotes (as shown) the amount by which an originally right angle is deformed, the time rate of change of this angular deformation is

$$\frac{d\gamma}{d\tau} = \frac{\dfrac{\partial v_x}{\partial y}\,dy}{dy} = \frac{\partial v_x}{\partial y}. \tag{6.3}$$

Thus, a combination of Eqs. (6.2) and (6.3) shows that the dynamic viscosity may be defined as the proportionality factor between the time rate of angular deformation and the shearing stress producing that deformation:

$$\tau_l = \mu \frac{d\gamma}{d\tau}.$$

If, now, one is concerned with motion in two directions, one type of deformation that may occur is that illustrated in Fig. 6.3(b), wherein a fluid element (assumed initially square for simplicity) is acted on by shear stresses in two directions. In the absence of any rotation of the fluid element, the shear stress must be equal in both directions and oriented as shown. Using τ_l to denote this shear stress, one finds that the element will be deformed as shown. If γ_1 and γ_2 represent the relative angular deformation of two sides of the rectangle, initially at right angles, the *total rate* of angular deformation of the particle is

$$\frac{d\gamma_1}{d\tau} + \frac{d\gamma_2}{d\tau} = \frac{\left(v_x + \dfrac{\partial v_x}{\partial y}\,dy\right) - v_x}{dy} + \frac{\left(v_y + \dfrac{\partial v_y}{\partial x}\,dx\right) - v_y}{dx}$$

$$= \frac{\partial v_x}{\partial y} + \frac{\partial v_y}{\partial x}.$$

Extending the above-mentioned concept of dynamic viscosity to this case, define μ such that

$$\tau_l = \mu\left(\frac{d\gamma_1}{d\tau} + \frac{d\gamma_2}{d\tau}\right)$$

$$= \mu\left(\frac{\partial v_x}{\partial y} + \frac{\partial v_y}{\partial x}\right). \tag{6.4}$$

This extended definition of μ includes the earlier one as a special case.

The deformations considered above have been only angular deformations in the absence of rotation. It is also possible to deform a fluid element by normal stresses (tensile or compressive) producing a linear deformation of the element. Consider as an oversimplified example that illustrated in Fig. 6.4(a) wherein a fluid element, considered initially square, has been deformed by application of a normal stress, σ_x, acting in the x direction. Such a state could be imagined to occur if, all other motions being considered uniform for simplicity, there is a relative motion in the x direction between the two sides of the element designated as $\overline{AD}$ and $\overline{BC}$.

To a first order approximation the *rate of linear strain* of the element in the x direction is given by $\partial v_x / \partial x$. This strain (which moves $\overline{BC}$ to $\overline{B'C'}$) causes a contraction in the y direction, moving $\overline{CD}$ to $\overline{C'D'}$ and $\overline{AB}$ to $\overline{A'B'}$. Since the fluid is taken to be incompressible the area of the square $(ABCD)$ must equal that of the rectangle $(A'B'C'D')$. From the geometry of the figure, this means that, for small strains, the side $\overline{DC}$ moves one-half the distance moved by $\overline{BC}$. That is, since the center of the element moves one-half the distance that $\overline{BC}$ does,

$$\overline{DD'} = \overline{EE'}. \tag{6.5}$$

The diagonals of the element are deformed and rotated (i.e., $\overline{DB}$ moves to $\overline{D'B'}$ and $\overline{AC}$ moves to $\overline{A'C'}$). If one denotes the rotation of each diagonal by $\gamma/2$, then γ denotes the *total* angular deformation of the angle AED which was originally 90°. This deformation produces the shear stresses τ on the diagonals

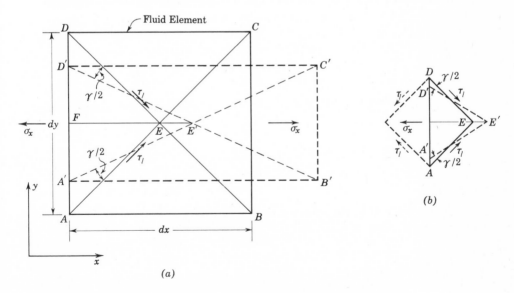

(a)

(b)

Fig. 6.4. Deformation of a fluid element by normal stress.

shown acting on the triangular element (ADE) depicted in Fig. 6.4(b). From the above definition of μ as the proportionality between shear stresses and angular deformation,

$$\tau_l = \mu \frac{d\gamma}{d\tau}. \tag{6.6}$$

Examination of the subelement (ADE) in Fig. 6.4(b) shows, further, that a balance of forces gives (for unit depth)

$$[\sigma_x(\overline{AD})]^2 = [\tau(\overline{ED})]^2 + [\tau(\overline{AE})]^2$$

which, since $(\overline{AD})^2 = (\overline{ED})^2 + (\overline{AE})^2$, gives

$$\sigma_x = \tau_l. \tag{6.7}$$

The angular displacements of the diagonals can be expressed in terms of the linear strain since, for small displacements,

$$\frac{\gamma}{2} = \frac{(\overline{DD'})/\sqrt{2}}{\overline{DE}/2}.$$

Upon applying Eq. (6.5) and the fact that $\overline{DE} = \sqrt{2(EF)}$, this equation becomes

$$\frac{\gamma}{2} = \frac{\overline{EE'}}{\overline{EF}}.$$

This latter quantity is the linear strain of the element in the x direction, and hence the strain *rate* is

$$\frac{d}{d\tau}\frac{\overline{EE'}}{\overline{EF}} = \frac{1}{2}\frac{d\gamma}{d\tau}.$$

As mentioned above, the strain rate is given by the gradient $\partial v_x/\partial x$, and Eq. (6.6) related the angular deformation $d\gamma/d\tau$ to the shear τ. So this gives

$$\frac{\partial v_x}{\partial x} = \frac{1}{2}\frac{\tau_l}{\mu}.$$

Applying, finally, the relation between the shear and normal stress deduced above and noted in Eq. (6.7), one finds that the normal stress is related to the normal velocity gradient:

$$\sigma_x = 2\mu \frac{\partial v_x}{\partial x}. \tag{6.8}$$

Motion in the y direction will give rise to a normal stress in that direction, $\sigma_y = 2\mu(\partial v_y/\partial y)$, in exactly the same fashion.

Generally speaking, fluid motion in two dimensions will give rise to stresses acting on a fluid element which are a combination of the above normal and shear stresses. The shear stresses are proportional to the cross derivatives of the velocity field, $\partial v_x/\partial y$ and $\partial v_y/\partial x$, whereas the normal stresses are proportional to the normal derivatives $\partial v_x/\partial x$ and $\partial v_y/\partial y$. Summarizing, the stresses to be accounted for in the two-dimensional laminar motion of an incompressible fluid are

$$\text{Shear: } \tau_l = \mu\left(\frac{\partial v_x}{\partial y} + \frac{\partial v_y}{\partial x}\right).$$

$$\text{Normal: } \sigma_x = 2\mu \frac{\partial v_x}{\partial x}, \tag{6.9}$$

$$\sigma_y = 2\mu \frac{\partial v_y}{\partial y}.$$

With the stresses acting on a moving fluid element thus related to the velocity changes in the fluid field, Newton's second law of motion may be applied to deduce the basic relation governing the motion of a fluid. This equation is discussed in Sec. 6.5. However, before Newton's law is applied it is necessary to obtain a representation of the acceleration of a fluid element. This will be done in the next section.

6.4   The Substantial Derivative

The fundamental laws of physics (e.g., Newton's law of motion, the laws of conservation of energy, etc.) are usually expressed as applying to a constant collection of matter. In the analysis of fluid motion, then, it is necessary to apply these principles to a given fluid element or "particle" which always consists of the same mass of fluid. The fluid is moving with respect to a selected, fixed coordinate system; and it is desirable, for practical reasons, to obtain expressions of these fundamental laws in terms of this coordinate system.

Since the fluid is moving relative to the coordinate system, a given fluid element will experience changes in the flow properties because of the fact that the element has moved from one point in the fluid domain to another point where the flow properties (pressure, temperature, velocity, etc.) have different values. Since it takes a certain length of time for the fluid element to move from one point to another, the above-described changes in the flow properties may be interpreted as changes with respect to time. These changes may also be accompanied by changes caused by a local dependence of the flow properties on time. In the case of steady flow this latter time dependence is absent, but the fluid element will still undergo time changes in its properties due to the mechanism described above. For this reason, care must be taken to distinguish between partial derivatives and total derivatives taken with respect to time in a moving fluid field.

To present this argument in a more formal way, consider the fluid element shown in Fig. 6.5. If φ is used to symbolize any scalar flow parameter (e.g., pressure, temperature, velocity component, etc.), the element experiences a change $d\varphi$ as it moves from one point, (x, y), to another point, $(x + dx, y + dy)$. This change, $d\varphi$, is composed of the two changes described above. In general, since φ is dependent on time and the coordinate location of the fluid element, one may write

$$\varphi = \varphi(x, y, \tau),$$

where τ represents time. Then, the change in φ as the element moves from (x, y) to $(x + dx, y + dy)$ is given by the total differential of φ:

$$d\varphi = \frac{\partial \varphi}{\partial \tau} d\tau + \frac{\partial \varphi}{\partial x} dx + \frac{\partial \varphi}{\partial y} dy. \tag{6.10}$$

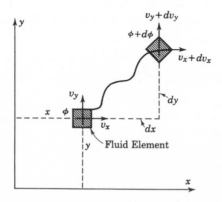

Fig. 6.5

If $d\varphi$ is the change, then dx and dy represent the coordinate displacements of the fluid element, and $d\tau$ represents the lapse of time during this displacement. Then the total time rate of change of φ is

$$\frac{d\varphi}{d\tau} = \frac{\partial \varphi}{\partial \tau} + \frac{\partial \varphi}{\partial x}\frac{dx}{d\tau} + \frac{\partial \varphi}{\partial y}\frac{dy}{d\tau}.$$

Since dx and dy are the x and y displacements of the fluid element, $dx/d\tau$ and $dy/d\tau$ are identified as the x and y components of the velocity of the fluid element—i.e., $v_x = dx/d\tau$ and $v_y = dy/d\tau$. Thus,

$$\frac{d\varphi}{d\tau} = \frac{\partial \varphi}{\partial \tau} + v_x\frac{\partial \varphi}{\partial x} + v_y\frac{\partial \varphi}{\partial y}. \tag{6.11}$$

This last expression is usually referred to as the *substantial derivative* of φ with respect to time. It expresses the total time rate of change of φ for a fluid element moving in a time-dependent velocity field. The first term on the right side of Eq. (6.11), $\partial\varphi/\partial\tau$, is called the *local derivative* of φ with respect to time and is the rate of change of φ due to local variations. The second term, $v_x(\partial\varphi/\partial x) + v_y(\partial\varphi/\partial y)$, is called the *convective derivative* of φ and is due to the fact that the element is moving in a field in which φ varies from point to point.

The term *steady flow* is used to denote the flow condition in which there is no *local* dependence on time of any of the flow properties. That is, "steady flow" means that $\partial\varphi/\partial\tau = 0$, *but* $d\varphi/d\tau \neq 0$, necessarily. In the case of steady flow the substantial derivative consists solely of the convective part:

$$\frac{d\varphi}{d\tau} = v_x\frac{\partial \varphi}{\partial x} + v_y\frac{\partial \varphi}{\partial y}. \tag{6.12}$$

Thus, a time rate of change of a flow parameter may occur in steady flow when attention is directed to a particular fluid element.

Since acceleration is the time rate of change of velocity, Eq. (6.11) may be applied to find the components of the acceleration of a fluid element. If one lets a_x and a_y denote the x and y components of the acceleration, Eq. (6.11) gives (interpreting φ as v_x or v_y) the following expressions:

$$a_x = \frac{dv_x}{d\tau} = \frac{\partial v_x}{\partial \tau} + v_x\frac{\partial v_x}{\partial x} + v_y\frac{\partial v_x}{\partial y},$$

$$a_y = \frac{dv_y}{d\tau} = \frac{\partial v_y}{\partial \tau} + v_x\frac{\partial v_y}{\partial x} + v_y\frac{\partial v_y}{\partial y}. \tag{6.13}$$

In the case of steady flow, the fluid element still undergoes an acceleration as it moves because its velocity changes. In this case

$$a_x = v_x \frac{\partial v_x}{\partial x} + v_y \frac{\partial v_x}{\partial y},$$

$$a_y = v_x \frac{\partial v_y}{\partial x} + v_y \frac{\partial v_y}{\partial y}.$$

(6.14)

Since v_x and v_y are, in general, functions of position, then a_x and a_y are also functions of position.

6.5   The Equation of Motion

In Sec. 6.3 equations were developed which expressed the viscous forces acting on a fluid element in terms of the velocity gradients of the fluid field. The acceleration of the fluid element was related to the velocity field in Sec. 6.4 [Eq. (6.13)]. These two expressions may be combined to yield a differential equation in the fluid velocity by application of Newton's second law. This equation will also involve other forces which may act on the fluid element— namely forces due to gradients in the fluid pressure and "body forces" which act throughout the fluid domain. The forces of gravity (or magnetic forces in the case of liquid metals) are examples of body forces.

Newton's law states, for a particle of mass m and acceleration a, which is acted on by a set of forces ΣF, that

$$ma = \Sigma F.$$

The acceleration is directed in the sense of the resultant of the forces. Since Newton's law thus stated applies to a system of constant mass, its application to the analysis of fluid motion must consider a particular element of the fluid —as discussed in Sec. 6.4.

Figure 6.6. considers, in detail, the fluid element (of dimensions δx, δy, unit depth). All the forces acting on the particle that contribute to its acceleration in the x direction are shown. A similar set of forces act in the y direction but are not shown for the sake of clarity. Acting in the x direction, then, are

Static pressure forces on the δy faces $= P$,

Normal stresses on the δy faces $= \sigma_x$,

Shear stresses on the δx faces $= \tau_t$,

A body force (per unit mass) $= X$.

The symbol X will be used to denote the body force component *per unit mass* in the x direction. If the body force is gravity, and if the coordinates are chosen so that the gravity acts parallel to the y coordinate, X may be zero. However, it is included here for generality.

Noting that the static pressure, P, and the normal stress, σ_x, act on areas (per unit depth) equal to $\delta y \cdot 1$ whereas the shear stresses act on areas equal to $\delta x \cdot 1$, one finds that Newton's law applied to the motion in the x direction gives

$$a_x(\delta x \cdot \delta y)\rho = \left[P - \left(P + \frac{\partial P}{\partial x}\delta x + \frac{\partial^2 P}{\partial x^2}\frac{(\delta x)^2}{2!} + \cdots \right) \right]\delta y$$

$$- \left[\sigma_x - \left(\sigma_x + \frac{\partial \sigma_x}{\partial x}\delta x + \frac{\partial^2 \sigma_x}{\partial x^2}\frac{(\delta x)^2}{2!} + \cdots \right) \right]\delta y$$

$$- \left[\tau_l - \left(\tau_l + \frac{\partial \tau_l}{\partial y}\delta y + \frac{\partial^2 \tau_l}{\partial y^2}\frac{(\delta y)^2}{2!} + \cdots \right) \right]\delta x$$

$$+ [\delta x\, \delta y]\rho X.$$

In the above expression ρ denotes the fluid density and a_x is the acceleration of the element in the x direction. Dividing by the particle mass $(\delta x \cdot \delta y)\rho$ and allowing the particles to shrink to a point $(\delta x \to 0, \delta y \to 0)$, one obtains

$$a_x = X - \frac{1}{\rho}\frac{\partial P}{\partial x} + \frac{1}{\rho}\left(\frac{\partial \sigma_x}{\partial x} + \frac{\partial \tau_l}{\partial y} \right).$$

Upon introducing the definition of σ_x and τ from Eq. (6.9),

$$a_x = X - \frac{1}{\rho}\frac{\partial P}{\partial x} + \frac{\mu}{\rho}\left(2\frac{\partial^2 v_x}{\partial x^2} + \frac{\partial^2 v_x}{\partial y^2} + \frac{\partial^2 v_y}{\partial x\, \partial y} \right).$$

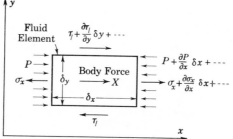

Fig. 6.6

The last term may be altered in the following way by application of the continuity equation, Eq. (6.1):

$$\frac{\partial^2 v_y}{\partial x \, \partial y} = \frac{\partial}{\partial x} \frac{\partial v_y}{\partial y}$$

$$= \frac{\partial}{\partial x}\left(-\frac{\partial v_x}{\partial x}\right)$$

$$= -\frac{\partial^2 v_x}{\partial x^2}.$$

Thus,

$$a_x = X - \frac{1}{\rho}\frac{\partial P}{\partial x} + \frac{\mu}{\rho}\left(\frac{\partial^2 v_x}{\partial x^2} + \frac{\partial^2 v_x}{\partial y^2}\right).$$

Now, in Sec. 6.4 the acceleration, a_x, was related to the velocity field in Eq. (6.13). Introducing this relation gives

$$\frac{\partial v_x}{\partial \tau} + v_x \frac{\partial v_x}{\partial x} + v_y \frac{\partial v_x}{\partial y} = X - \frac{1}{\rho}\frac{\partial P}{\partial x} + \nu\left(\frac{\partial^2 v_x}{\partial x^2} + \frac{\partial^2 v_x}{\partial y^2}\right). \qquad (6.15a)$$

Similarly, motion in the y direction must satisfy

$$\frac{\partial v_y}{\partial \tau} + v_x \frac{\partial v_y}{\partial x} + v_y \frac{\partial v_y}{\partial y} = Y - \frac{1}{\rho}\frac{\partial P}{\partial y} + \nu\left(\frac{\partial^2 v_y}{\partial x^2} + \frac{\partial^2 v_y}{\partial y^2}\right). \qquad (6.15b)$$

The last two equations govern the laminar, nonsteady motion of an incompressible fluid in two dimensions and are known as the *Navier–Stokes equations of motion*. It should be noted that in Eqs. (6.15) the ratio of the dynamic viscosity to the density has been replaced by the kinematic viscosity, ν, defined as μ/ρ in Sec. 2.7. In certain fluids and under the appropriate conditions, the effect of the viscous forces may be neglected in comparison to the inertia forces in the above equations. If such is the case, setting $\nu = 0$ in the Navier–Stokes equations yields the simpler *Euler equations* of motion of an inviscid fluid:

$$\frac{\partial v_x}{\partial \tau} + v_x \frac{\partial v_x}{\partial x} + v_y \frac{\partial v_x}{\partial y} = X - \frac{1}{\rho}\frac{\partial P}{\partial x},$$

$$\frac{\partial v_y}{\partial \tau} + v_x \frac{\partial v_y}{\partial x} + v_y \frac{\partial v_y}{\partial y} = Y - \frac{1}{\rho}\frac{\partial P}{\partial y}. \qquad (6.16)$$

For steady motion $\partial v_x/\partial \tau = \partial v_y/\partial \tau = 0$, and the Navier–Stokes equations become

$$v_x \frac{\partial v_x}{\partial x} + v_y \frac{\partial v_x}{\partial y} = X - \frac{1}{\rho} \frac{\partial P}{\partial x} + v\left(\frac{\partial^2 v_x}{\partial x^2} + \frac{\partial^2 v_x}{\partial y^2}\right),$$

$$v_x \frac{\partial v_y}{\partial x} + v_y \frac{\partial v_y}{\partial y} = Y - \frac{1}{\rho} \frac{\partial P}{\partial y} + v\left(\frac{\partial^2 v_y}{\partial x^2} + \frac{\partial^2 v_y}{\partial y^2}\right).$$

$$(6.17)$$

6.6 The Energy Equation—The First Law of Thermodynamics

If only the hydrodynamic picture of a given flow condition is desired, the quantities required to describe completely the flow (for the two-dimensional, incompressible case) are the velocity components, v_x and v_y, and the fluid pressure P. Thus, for a given set of boundary conditions one needs to determine how these three quantities depend on time and on the coordinates x and y. The continuity equation, Eq. (6.1), and the two Navier–Stokes equations, Eqs. (6.15), are the three equations needed to find the three unknowns.

If heat convection is being considered, it is also necessary to determine the fluid temperature, t, at each point in the field of flow. This fourth dependent variable requires the use of a fourth fundamental equation. This additional relation, known as the *energy equation*, is obtained from the first law of thermodynamics.

Since, as in the derivation of the Navier–Stokes equations, a particular fluid element (always consisting of the same mass of fluid) is being considered, it is thermodynamically a "closed system." For a closed system, the first law of thermodynamics may be written in the following way:

$$\frac{dQ}{d\tau} = \frac{dU}{d\tau} + \frac{dW}{d\tau}.$$

Here $dQ/d\tau$ is used to symbolize the rate at which heat is crossing the boundary of the fluid element (presumably by conduction), $dU/d\tau$ is the rate of change of the internal energy of the element, and $dW/d\tau$ is the rate at which work is done on the element. Since only incompressible fluids are being considered, no work may be done in the form of compression or expansion of the element. Thus, the only form in which work will be present is that of dissipation through the action of the viscous forces. If $dW_d/d\tau$ is used to represent this rate of dissipatively done work,

$$\frac{dQ}{d\tau} = \frac{dU}{d\tau} + \frac{dW_d}{d\tau}.$$

If, now, $dQ'/d\tau$ and $dW_d'/d\tau$ are used to denote the rates of heat addition and viscous dissipative work *per unit volume* and if u denotes the *specific* internal energy,

$$\frac{dQ'}{d\tau} = \rho\frac{du}{d\tau} + \frac{dW_d'}{d\tau}. \tag{6.18}$$

By assuming that the thermal internal energy, u, is dependent on temperature only, this equation may be written

$$\frac{dQ'}{d\tau} = \rho c\frac{dt}{d\tau} + \frac{dW_d'}{d\tau},$$

where c is the specific heat of the fluid. The specific heat which is customarily used is that at constant pressure, although this actually involves the making of additional assumptions.* Thus, for the incompressible fluid,

$$\frac{dQ'}{d\tau} = \rho c_p\frac{dt}{d\tau} + \frac{dW_d'}{d\tau}.$$

---

  * A general thermodynamic relation (see, for instance, Ref. 4) between the specific heats at constant pressure and constant volume is

$$c_p - c_v = T\frac{\beta^2 B}{\rho}.$$

In this expression β is the coefficient of volume expansion (see Sec. 2.6), B is the isothermal bulk modulus, T is the absolute temperature, and ρ is the density. Since the bulk modulus is defined as $B = \rho(\partial p/\partial\rho)_T$, then

$$c_p - c_v = \frac{T\beta^2}{(\partial\rho/\partial p)_T}.$$

If by "incompressible" it is meant that the density is truly a constant, this means that ρ does not change with *pressure or temperature*; i.e., no thermal expansion is allowed. If this is taken as the definition of an incompressible fluid, then $c_p = c_v$ since β^2 goes to zero faster than $(\partial\rho/\partial p)_T$. Thus, the use of c_p in the energy equation, Eq. (6.19), is justified. In some cases, however, this equation is used even though $\beta \neq 0$, although the fluid may be treated as incompressible as far as continuity principles are concerned. In such cases (e.g., low-speed air flow) the use of Eq. (6.19) is only an approximation.

The same result is often obtained a second way by introducing the enthalpy, h, into Eq. (6.18). Since $h = u + p/\rho$, Eq. (6.18) is

$$\frac{dQ'}{d\tau} = \rho\frac{dh}{d\tau} + \frac{dW_d'}{d\tau} - \rho\frac{d}{d\tau}\left(\frac{p}{\rho}\right).$$

Since ρ is constant $d(p/\rho)/d\tau = 1/\rho(dp/d\tau)$. Then, by using the fact that $dh = c_p\,dt$, Eq. (6.19) is obtained *if* it is assumed that the substantial derivative of p is negligible—i.e., if $dp/d\tau = \partial p/\partial\tau + v_x(\partial p/\partial x) + v_y(\partial p/\partial y) = 0$. Neglecting $dp/d\tau$ is equivalent to assuming that the fluid is incompressible and $\beta = 0$.

Now, $dt/d\tau$ is the substantial derivative of the fluid element temperature so that Eq. (6.11) leads to

$$\frac{dQ'}{d\tau} = \rho c_p \left(\frac{\partial t}{\partial \tau} + v_x \frac{\partial t}{\partial x} + v_y \frac{\partial t}{\partial y} \right) + \frac{dW'_d}{d\tau}. \tag{6.19}$$

The heat that is added to the fluid element from its surroundings is added by the mechanism of *conduction* arising from the temperature difference between the element and the surrounding fluid. The derivation of the heat conduction equation in Sec. 1.4 showed that the heat conducted per *unit time* per *unit volume* into an element is (for two dimensions)

$$\frac{dQ'}{d\tau} = k \left(\frac{\partial^2 t}{\partial x^2} + \frac{\partial^2 t}{\partial y^2} \right). \tag{6.20}$$

The expression for $dW'_d/d\tau$, the rate at which dissipative work is done on the fluid element by the viscous forces per unit volume is difficult to obtain in a concise fashion. Hence, only the result is given here—the reader being referred to texts on fluid dynamics, such as Ref. 1, for a derivation of the fact that for two-dimensional, incompressible flow

$$\frac{dW'_d}{d\tau} = - \left[\sigma_x \frac{\partial v_x}{\partial x} + \sigma_y \frac{\partial v_y}{\partial y} + \tau_l \left(\frac{\partial v_x}{\partial y} + \frac{\partial v_y}{\partial x} \right) \right].$$

Introducing the definitions of the viscous stress given by Eq. (6.9), one obtains

$$\frac{dW'_d}{d\tau} = -\mu \left[2 \left(\frac{\partial v_x}{\partial x} \right)^2 + 2 \left(\frac{\partial v_y}{\partial y} \right)^2 + \left(\frac{\partial v_x}{\partial y} + \frac{\partial v_y}{\partial x} \right)^2 \right]. \tag{6.21}$$

This term represents the rate at which mechanical energy is being dissipated into thermal energy by the action of the viscous forces. It must, because of the requirements of the second law of thermodynamics, be negative; i.e., there must be a loss of mechanical energy and a corresponding gain in thermal energy.

Substitution of Eqs. (6.20) and (6.21) into Eq. (6.19) gives, finally, the accepted form of the *energy equation* for two-dimensional, incompressible, laminar flow of a viscous fluid:

$$k \left(\frac{\partial^2 t}{\partial x^2} + \frac{\partial^2 t}{\partial y^2} \right) + \mu \left[2 \left(\frac{\partial v_x}{\partial x} \right)^2 + 2 \left(\frac{\partial v_y}{\partial y} \right)^2 + \left(\frac{\partial v_x}{\partial y} + \frac{\partial v_y}{\partial x} \right)^2 \right]$$

$$= c_p \rho \left(v_x \frac{\partial t}{\partial x} + v_y \frac{\partial t}{\partial y} + \frac{\partial t}{\partial \tau} \right), \tag{6.22}$$

or

$$\alpha\left(\frac{\partial^2 t}{\partial x^2} + \frac{\partial^2 t}{\partial y^2}\right) + \frac{\mu}{\rho c_p}\left[2\left(\frac{\partial v_x}{\partial x}\right)^2 + 2\left(\frac{\partial v_y}{\partial y}\right)^2 + \left(\frac{\partial v_x}{\partial y} + \frac{\partial v_y}{\partial x}\right)^2\right]$$

$$= \left(v_x\frac{\partial t}{\partial x} + v_y\frac{\partial t}{\partial y} + \frac{\partial t}{\partial \tau}\right).$$

In the latter form the thermal diffusivity defined in Eq. (1.9) has been introduced.

In problems of heat convection in a two-dimensional incompressible
laminar fluid motion this equation, Eq. (6.22), must be solved simultaneously
with the two Navier–Stokes equations, Eqs. (6.15), and the continuity equation, Eq. (6.1), to find the unknown quantities v_x, v_y, P, and t for a given set of
boundary conditions (such as the shape and temperature of the heated body
about which the fluid flows, the fluid velocity and temperature far removed
from the body, etc.). Once the temperature distribution of the fluid is known,
the convective film coefficient could be determined at a particular point on the
body by the method outlined in the introductory paragraphs of this chapter.

Obviously, the actual execution of such an analytical procedure would be
an extremely complex and laborious, if not impossible, undertaking—even
for the simplest of geometrical bodies and boundary conditions.

The main reason that these basic equations governing the convection
mechanism are presented here is not to impress the student with a complex
system of differential equations merely for the reason of mathematical ele-
gance. The complexity of the equations and the inability to obtain solutions
quickly to practical convection problems do not eliminate the need of the
engineer to have answers to his *real* problems. These basic equations are presented to justify making some rather far-reaching, simplifying assumptions
for the very practical reason that there is no other way to obtain a useful
engineering answer. These basic laws will also provide a rational basis for such
simplifications. They also aid the engineer in being logical in his approach
when an empirical attack on a problem is necessary. Thus, in this respect the
study of heat transfer involves the effective combination of powerful analytical methods and the empirical results of experiment.

6.7   The Reynolds Number and Its Significance

The Navier–Stokes equations derived above for the laminar motions of a
viscous, incompressible fluid are among the most complex equations of applied mathematics. No general methods are available for the solution of
these equations. Solutions for very special cases of simple geometry are

known, but even these are few in number. Hence, most of the work done in the solution of these equations has been limited to cases in which the viscosity is either very large or very small. In these cases some terms become negligible in comparison to others, enabling one to reduce the equations to forms which can be solved.

Even procedures such as these just mentioned, are often quite difficult, mathematically, so that a great deal of experimental work has been (and still must be) done in order to achieve usable engineering results. Thus, many results are obtained by experimental *model studies*, and these results are then extended to other full-scale situations if the proper conditions for similarity exist between the model and the prototype.

Certain conditions for such similarity may be deduced from the differential equations themselves without knowing their solutions. Two different flow situations are considered to by dynamically similar if the bounding solids past which the fluids flow are geometrically similar and if the velocity and pressure fields about the bodies are also similar.

The question then is: Given flow around or through bodies having *geometric* similarity [such as illustrated in Fig. 6.7(a), (b), (c)], under what conditions will the resultant fluid velocity and pressure fields be similar? Any of

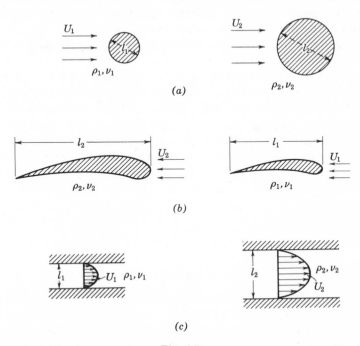

Fig. 6.7

the flow fields must satisfy the Navier–Stokes equations, repeated here for convenience:

$$v_x \frac{\partial v_x}{\partial x} + v_y \frac{\partial v_x}{\partial y} = -\frac{1}{\rho} \frac{\partial P}{\partial x} + v \left(\frac{\partial^2 v_x}{\partial x^2} + \frac{\partial^2 v_x}{\partial y^2} \right),$$

$$v_x \frac{\partial v_y}{\partial x} + v_y \frac{\partial v_y}{\partial y} = -\frac{1}{\rho} \frac{\partial P}{\partial y} + v \left(\frac{\partial^2 v_y}{\partial x^2} + \frac{\partial^2 v_y}{\partial y^2} \right).$$

(6.17)

For simplicity, the steady state has been assumed to exist. If one assumes that the viscosity is independent of temperature and that the fluid is incompressible, these two equations and the continuity equation are sufficient to determine the velocity field of the fluid since temperature does not enter. On this basis, these equations are adequate for an analysis of the dynamics of the flow.

For purposes of this discussion, only flows without a free surface will be considered. If, then, the pressure, P, is interpreted as the difference between the pressure at a point when the fluid is moving and when it is stationary, the body-force terms, X and Y, may be omitted as above.

To analyze similarity of flow around two geometrically similar bodies of different sizes, all distances must be measured in reference to the body size. That is, two points in the two different flow fields are located in "similar" positions if they have the same coordinates relative to the body size. Hence, all distances must be nondimensionalized with respect to the body dimension.

Let l then denote a characteristic length dimension of the bodies being considered. The choice of this length is purely arbitrary as long as it is chosen the same in all geometrically similar bodies. Figure 6.7(a), (b), (c) indicates some possibilities. In the case of flow about a cylinder, it is customary to pick the diameter as the characteristic length, although one could choose the circumference if such a choice would be rigidly adhered to in the future. Thus, the choice of the characteristic length becomes merely a gentlemen's agreement always to use a certain dimension to characterize geometrically similar shapes. Now, define a new coordinate system measured relative to l; i.e., define ξ and η:

$$\xi = \frac{x}{l},$$

$$\eta = \frac{y}{l}.$$

(6.23)

In an analogous fashion, one chooses a characteristic velocity to which all velocity measurements are referred. Call this characteristic velocity U. Again its choice is arbitrary as long as consistency is used. Customarily, for flow around bodies, the "free stream" velocity of the flow measured at some point far removed from the influence of the body is used. In cases of flow through pipes the maximum velocity or, perhaps, the average velocity is used. Now the velocity components measured relative to U will be defined:

$$v_\xi = \frac{v_x}{U},$$

$$v_\eta = \frac{v_y}{U}. \tag{6.24}$$

Since the fluid is incompressible, no characteristic density needs to be chosen since ρ is a constant. Since a characteristic velocity, U, has already been picked, the pressure may be nondimensionalized by the total dynamic pressure at the characteristic velocity, $\frac{1}{2}\rho U^2$. So, define a dimensionless pressure:

$$\Pi = \frac{P}{\frac{1}{2}\rho U^2}. \tag{6.25}$$

Substitution of the definitions of ξ, η, v_ξ, v_η, and Π given by Eqs. (6.23) through (6.25) into the Navier–Stokes equations results in the following dimensionless presentation of these equations after a bit of manipulation:

$$v_\xi \frac{\partial v_\xi}{\partial \xi} + v_\eta \frac{\partial v_\xi}{\partial \eta} = -\frac{1}{2}\frac{\partial \Pi}{\partial \xi} + \frac{\nu}{Ul}\left(\frac{\partial^2 v_\xi}{\partial \xi^2} + \frac{\partial^2 v_\xi}{\partial \eta^2}\right),$$

$$v_\xi \frac{\partial v_\eta}{\partial \xi} + v_\eta \frac{\partial v_\eta}{\partial \eta} = -\frac{1}{2}\frac{\partial \Pi}{\partial \eta} + \frac{\nu}{Ul}\left(\frac{\partial^2 v_\eta}{\partial \xi^2} + \frac{\partial^2 v_\eta}{\partial \eta^2}\right).$$

One may now observe that two different flow situations about geometrically similar bodies will have similar flow fields at geometrically similar points (i.e., at $\xi_1 = \xi_2$ and $\eta_1 = \eta_2$, 1 and 2 denoting the two flows) if $v_{\xi_1} = v_{\xi_2}$ and $v_{\eta_1} = v_{\eta_2}$. The above dimensionless equations show that similarity of flow will occur if the combination ν/Ul is the same for both cases, i.e., if

$$\frac{\nu_1}{U_1 l_1} = \frac{\nu_2}{U_2 l_2}.$$

If one defines the *Reynolds number* (named for Osborne Reynolds, who first recognized its significance) as

$$N_{RE} = \frac{Ul}{\nu} = \frac{Ul\rho}{\mu},$$ (6.26)

then flow similarity exists if the Reynolds numbers of the two flows are the same:

$$N_{RE_1} = N_{RE_2}.$$

The experimental implications of this are great. The measured results for the flow about a particular model may be extended to prototype cases when the same Reynolds number exists. Also, in order to analyze the dependence of a certain phenomenon on the conditions of flow, fluid properties, etc., one does not need to make measurements in cases covering all possible flow velocities or all possible fluid viscosities, densities, etc. One needs only to vary, experimentally, the Reynolds number by variation of any of the quantities involved in it.

The Reynolds number is thus seen to be a flow parameter of the greatest significance. Other useful similarity parameters having significance in convective heat transfer will be deduced in the following section.

6.8   Similarity Parameters in Heat Transfer

The preceding section showed how the Reynolds number serves to evaluate the existence of dynamic similarity between two different fluid flows about geometrically similar bodies. In the event one is also concerned with the heat transfer between the bodies and the fluids, results obtained for the heat transfer coefficient in one instance may be applied to the other if one is assured that *energetic* similarity also exists.

The insistence on energetic similarity will introduce similarity parameters in addition to the Reynolds number already obtained. These additional similarity parameters may be generated by examining the entire set of the fundamental equations—motion, continuity, and energy. These expressions are given in Eqs. (6.1), (6.15), and (6.22). Consider, for simplicity, the case of steady flow of a fluid at temperature t_f flowing about the exterior of a heated body of surface temperature t_s. In such a case, a body force due to buoyancy effects may exist. By selecting the x axis in the direction of the gravity vector, the body forces in the Navier–Stokes equations become $Y = 0, X = g\beta(t - t_f)$, where β is the coefficient of expansion and t is the total fluid temperature. (This expression for the buoyant forces involves some assumptions which are discussed in more detail in Chapter 9.) Under these conditions, Eqs. (6.1), (6.15), and (6.22) become

$$\frac{\partial v_x}{\partial x} + \frac{\partial v_y}{\partial y} = 0,$$

$$v_x \frac{\partial v_x}{\partial x} + v_y \frac{\partial v_x}{\partial y} = -\frac{1}{\rho}\frac{\partial P}{\partial x} + v\left(\frac{\partial^2 v_x}{\partial x^2} + \frac{\partial^2 v_x}{\partial y^2}\right) + g\beta(t - t_f),$$

$$v_x \frac{\partial v_y}{\partial x} + v_y \frac{\partial v_y}{\partial y} = -\frac{1}{\rho}\frac{\partial P}{\partial y} + v\left(\frac{\partial^2 v_y}{\partial x^2} + \frac{\partial^2 v_y}{\partial y^2}\right),$$

$$k\left(\frac{\partial^2 t}{\partial x^2} + \frac{\partial^2 t}{\partial y^2}\right) + \mu\left[2\left(\frac{\partial v_x}{\partial x}\right)^2 + 2\left(\frac{\partial v_y}{\partial y}\right)^2 + \left(\frac{\partial v_x}{\partial y} + \frac{\partial v_y}{\partial x}\right)^2\right]$$

$$= \rho c_p\left(v_x \frac{\partial t}{\partial x} + v_y \frac{\partial t}{\partial y}\right).$$

As in Sec. 6.7, the existence of similarity parameters may be ascertained by nondimensionalizing the above equations. To this end, select a characteristic length l, a characteristic velocity U, and a characteristic temperature difference $\Delta t = t_s - t_f$. Then the dimensionless variables may be defined:

Coordinates: $\xi = x/l, \eta = y/l$.

Velocity: $v_\xi = v_x/U, v_\eta = v_y/U$.

Temperature: $\varphi = (t - t_f)/(t_s - t_f) = (t - t_f)/\Delta t$.

Pressure: $\Pi = P/\frac{1}{2}\rho U^2$.

When these variable changes are made in the above equations, the following results are obtained:

$$\frac{\partial v_\xi}{\partial \xi} + \frac{\partial v_\eta}{\partial \eta} = 0,$$

$$v_\xi \frac{\partial v_\xi}{\partial \xi} + v_\eta \frac{\partial v_\xi}{\partial \eta} = -\frac{1}{2}\frac{\partial \Pi}{\partial \xi} + \frac{v}{Ul}\left(\frac{\partial^2 v_\xi}{\partial \xi^2} + \frac{\partial^2 v_\xi}{\partial \eta^2}\right) + \frac{g\beta l \, \Delta t}{U^2}\varphi,$$

$$v_\xi \frac{\partial v_\eta}{\partial \xi} + v_\eta \frac{\partial v_\eta}{\partial \eta} = -\frac{1}{2}\frac{\partial \Pi}{\partial \eta} + \frac{v}{Ul}\left(\frac{\partial^2 v_\eta}{\partial \xi^2} + \frac{\partial^2 v_\eta}{\partial \eta^2}\right),$$

$$\left(\frac{\partial^2 \varphi}{\partial \xi^2} + \frac{\partial^2 \varphi}{\partial \eta^2}\right) + \frac{\mu U^2}{k \, \Delta t}\left[2\left(\frac{\partial v_\xi}{\partial \xi}\right)^2 + 2\left(\frac{\partial v_\eta}{\partial \eta}\right)^2 + \left(\frac{\partial v_\xi}{\partial \eta} + \frac{\partial v_\eta}{\partial \xi}\right)^2\right]$$

$$= \frac{\rho c_p Ul}{k}\left(v_\xi \frac{\partial \varphi}{\partial \xi} + v_\eta \frac{\partial \varphi}{\partial \eta}\right).$$

The main object here of analyzing the fluid motion is that of determining the temperature distribution in the fluid for subsequent evaluation of the heat transfer coefficient. Imagine, then, that the four equations above are solved for the dimensionless temperature, φ. This solution may be represented by the functional relation

$$\varphi = f\left(\xi, \eta, \frac{v}{Ul}, \frac{g\beta l\, \Delta t}{U^2}, \frac{\mu U^2}{k\, \Delta t}, \frac{\rho c_p Ul}{k}\right).$$

In addition to the dimensionless space variables, the solution is seen to depend upon four dimensionless groupings of the physical parameters of the problem. These four groupings could be used as similarity parameters as they stand; however, it is customary to rearrange them in the following way:

$$\varphi = f\left(\xi, \eta, \frac{v}{Ul}, \frac{gl^3\beta\, \Delta t}{v^2}, \frac{v^2}{U^2 l^2}, \frac{U^2}{c_p\, \Delta t}, \frac{\mu c_p}{k}, \frac{Ul}{v}\frac{\mu c_p}{k}\right).$$

Certain groups of properties are seen to appear more than once, and since only a functional dependence is being examined, one may rewrite the above as

$$\varphi = f\left(\xi, \eta, \frac{Ul}{v}, \frac{gl^3\beta\, \Delta t}{v^2}, \frac{U^2}{c_p\, \Delta t}, \frac{\mu c_p}{k}\right). \tag{6.27}$$

The four groups of physical parameters in Eq. (6.27) are, thus, the sought-for similarity parameters for the general case of a viscous fluid flowing past a heated surface. The first of these, Ul/v, is recognized as the Reynolds number developed in Sec. 6.7. The others bear special names:

$$N_{RE} = \frac{Ul}{v} = \text{the Reynolds number,}$$

$$N_{GR} = \frac{gl^3\beta\, \Delta t}{v^2} = \text{the Grashof number,}$$

$$N_{EC} = \frac{U^2}{c_p\, \Delta t} = \text{the Eckert number,}$$

$$N_{PR} = \frac{\mu c_p}{k} = \frac{v}{\alpha} = \text{the Prandtl number.}$$

$$(6.28)$$

The Prandtl number, $N_{PR} = v/\alpha$, is a quantity previously discussed in Sec. 2.7. Unlike the others, it is composed only of physical properties of the fluid and represents one of the most significant heat transfer parameters—being a measure of the relative magnitudes of momentum and thermal diffusion in the fluid. The Grashof number results from the inclusion of buoyancy forces in the equations of motion and would be absent in cases in which buoyancy is absent. Similarly, the Eckert number arises from the term representing the rate of dissipation of energy in the energy equation.

If an analytical expression of the form of Eq. (6.27) were to be actually obtained it would then be used to determine the heat transfer coefficient h:

$$h = -\frac{k\left(\dfrac{\partial t}{\partial n}\right)_s}{t_s - t_f},$$

where $(\partial t / \partial n)_s$ represents the fluid temperature gradient, at the body surface, taken in the direction normal to the surface n. When the dimensionless variables $\varphi = (t - t_f)/(t_s - t_f)$, $\bar{n} = n/l$ are introduced, this expression for h becomes

$$N_{\mathrm{NU}} = \frac{hl}{k} = -\left(\frac{\partial \varphi}{\partial \bar{n}}\right)_s. \tag{6.29}$$

Thus, a dimensionless heat transfer coefficient has been defined:

$$N_{\mathrm{NU}} = \frac{hl}{k} = \text{the Nusselt number.} \tag{6.30}$$

By use of this dimensionless representation of the heat transfer coefficient, Eqs. (6.28) and (6.29) yield

$$N_{\mathrm{NU}} = f(\xi, \eta, N_{\mathrm{RE}}, N_{\mathrm{GR}}, N_{\mathrm{EC}}, N_{\mathrm{PR}}). \tag{6.31}$$

If an average heat transfer coefficient is evaluated by integration over the body surface, the spatial variables are eliminated, and the resultant average Nusselt number is related in the following way:

$$N_{\mathrm{NU}} = f(N_{\mathrm{RE}}, N_{\mathrm{GR}}, N_{\mathrm{EC}}, N_{\mathrm{PR}}). \tag{6.32}$$

The implications of Eqs. (6.31) and (6.32) are twofold. First, in the event that an experimental investigation is being conducted for the solution of a heat convection problem, the data would be expected to correlate when analyzed in terms of the above parameters. Thus, a considerable saving of effort in the taking of data, and the subsequent analysis of this data, may be obtained. Second, the similarity analysis just performed gives advance notice of the form of expression that any analytic solution would be expected to take. Both these aspects of the similarity parameters will be used in subsequent chapters.

Certain special forms of the above results will also be encountered. In *forced convection* applications, one often neglects any effects of buoyant forces, so that

$$N_{\mathrm{NU}} = f(N_{\mathrm{RE}}, N_{\mathrm{EC}}, N_{\mathrm{PR}}) \qquad (6.33)$$

when viscous dissipation is included, and

$$N_{\mathrm{NU}} = f(N_{\mathrm{RE}}, N_{\mathrm{PR}}) \qquad (6.34)$$

when viscous dissipation is neglected. In the case of free convection, the main fluid stream is absent, so that N_{RE} becomes no longer significant. In addition, the flow is generally so slow that viscous dissipation does not enter. In free convection applications, then, one should expect

$$N_{\mathrm{NU}} = f(N_{\mathrm{GR}}, N_{\mathrm{PR}}). \qquad (6.35)$$

6.9   Turbulent Flow

All the foregoing statements have been limited to the assumption of laminar fluid motion. In laminar flow the fluid streamlines are observed to run in a well-ordered manner with adjacent layers of fluid sliding relative to one another and without any motion or exchange taking place normal to the streamlines of the main flow.

Under certain conditions, however, close examination of a flow which appears to be a steady flow shows that although the main flow is predominantly steady and proceeds in a specific direction, there is superimposed upon this main flow a time-dependent fluctuating motion. These fluctuations are observed to occur in directions which are both parallel and transverse to the main flow. Such a motion is actually dependent on time, although the fluctuations in the flow velocity and direction occur about mean values so that, at a given point, the time-average velocity and direction are constant with time. Such fluid motion is termed *turbulent* flow. If $\bar{v}$ denotes the *time-average* velocity at a point and v' denotes the time-dependent velocity fluctuation about that average value, the true velocity may be written:

$$v_x = \bar{v}_x + v'_x,$$
$$v_y = \bar{v}_y + v'_y.$$

Figure 6.8 illustrates such a motion.

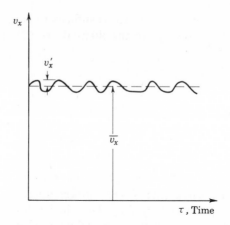

Fig. 6.8. Turbulent velocity fluctuations about a time average.

Thus the true velocity is a nonsteady one, and the paths of the individual fluid particles are found to be an interlacing network. This also means that a certain amount of mixing and energy exchange occurs in a direction transverse to that of the main flow. This transverse mixing is a characteristic that is missing in laminar flow and is of great importance in heat transfer problems, since such motions should tend to increase the rate of heat exchange within the fluid because of the associated movement of large-scale fluid "lumps." Indeed, it is a fact of experience that the rate of heat transfer is generally much greater for convection in turbulent flows than in laminar flows.

To make the above discussion a bit more formal, the velocity components of the flow are nonsteady functions oscillating about mean values:

$$v_x = \bar{v}_x + v'_x,$$
$$v_y = \bar{v}_y + v'_y,$$

(6.36)

in which $\bar{v}_x$ and $\bar{v}_y$ denote the time-average values of v_x and v_y (and are constant with time) and v'_x and v'_y are the fluctuations with time of v_x and v_y. Then for *steady turbulent* motion,

$$\bar{v}_x = \frac{1}{\tau_2 - \tau_1} \int_{\tau_1}^{\tau_2} v_x \, d\tau, \text{ independent of time,}$$

$$\bar{v}_y = \frac{1}{\tau_2 - \tau_1} \int_{\tau_1}^{\tau_2} v_y \, d\tau, \text{ independent of time.}$$

The time interval $(\tau_2 - \tau_1)$ is taken to be sufficiently large so that $\bar{v}_x$ and $\bar{v}_y$ are constant. The time average of the fluctuations (denoted by $\overline{v'_x}, \overline{v'_y}$) are then zero:

$$\overline{v'_x} = \frac{1}{\tau_2 - \tau_1} \int_{\tau_1}^{\tau_2} v'_x \, d\tau = 0,$$

$$\overline{v'_y} = \frac{1}{\tau_2 - \tau_1} \int_{\tau_1}^{\tau_2} v'_y \, d\tau = 0. \tag{6.37}$$

The mean motion $(\bar{v}_x, \bar{v}_y)$ is only an apparent motion. The actual particle velocities, which must satisfy the Navier–Stokes equations, are given by the time-dependent components v_x and v_y. The Navier–Stokes equations, Eqs. (6.15), for nonsteady motion in two dimensions are repeated here for convenience:

$$\frac{\partial v_x}{\partial \tau} + v_x \frac{\partial v_x}{\partial x} + v_y \frac{\partial v_x}{\partial y} = X - \frac{1}{\rho} \frac{\partial P}{\partial x} + v \left(\frac{\partial^2 v_x}{\partial x^2} + \frac{\partial^2 v_x}{\partial y^2} \right), \tag{6.15a}$$

$$\frac{\partial v_y}{\partial \tau} + v_x \frac{\partial v_y}{\partial x} + v_y \frac{\partial v_y}{\partial y} = Y - \frac{1}{\rho} \frac{\partial P}{\partial y} + v \left(\frac{\partial^2 v_y}{\partial x^2} + \frac{\partial^2 v_y}{\partial y^2} \right). \tag{6.15b}$$

Now, the velocity components for the turbulent motion given in Eq. (6.36) must satisfy these expressions along with the continuity equation:

$$\frac{\partial v_x}{\partial x} + \frac{\partial v_y}{\partial y} = 0. \tag{6.1}$$

If the expressions for v_x and v_y in Eq. (6.36) are substituted into the Navier–Stokes equations and the continuity equation, and if the time-average value of each term is then taken, considerable algebraic manipulation will give (see Ref. 1)

$$\frac{\partial \bar{v}_x}{\partial x} + \frac{\partial \bar{v}_y}{\partial y} = 0,$$

$$\bar{v}_x \frac{\partial \bar{v}_x}{\partial x} + \bar{v}_y \frac{\partial \bar{v}_x}{\partial y} = X - \frac{1}{\rho} \frac{\partial \bar{P}}{\partial x} + v \left(\frac{\partial^2 \bar{v}_x}{\partial x^2} + \frac{\partial^2 \bar{v}_x}{\partial y^2} \right) - \left[\frac{\partial \overline{(v'^2_x)}}{\partial x} + \frac{\partial \overline{(v'_x v'_y)}}{\partial y} \right],$$

$$\bar{v}_x \frac{\partial \bar{v}_y}{\partial x} + \bar{v}_y \frac{\partial \bar{v}_y}{\partial y} = Y - \frac{1}{\rho} \frac{\partial \bar{P}}{\partial y} + v \left(\frac{\partial^2 \bar{v}_y}{\partial x^2} + \frac{\partial^2 \bar{v}_y}{\partial y^2} \right) - \left[\frac{\partial \overline{(v'_x v'_y)}}{\partial x} + \frac{\partial \overline{(v'^2_y)}}{\partial y} \right]. \tag{6.38}$$

Examination of Eqs. (6.38) indicates that the usual steady state equations of motion, Eqs. (6.17), may be applied to the *mean flow* if certain additional terms are included. These additional terms are seen to involve the time averages of the square of the velocity fluctuation $(\overline{v_x'^2}, \overline{v_y'^2})$ and the cross product of the velocity fluctuations $(\overline{v_x'v_y'})$. Although the time averages of the velocity fluctuations $(\overline{v_x'}, \overline{v_y'})$ are defined as zero, these products are not necessarily zero. Formally, these additional terms appearing on the right side of the Navier–Stokes equation may be interpreted as being due to additional *apparent* stresses resulting from the turbulent fluctuations.

A physical interpretation of these apparent stresses may be illustrated in the following way. Imagine, for purposes of discussion, a turbulent flow parallel to a plane solid wall. That is, the flow is two-dimensional in the x-y plane, as shown in Fig. 6.9. However, the flow illustrated is a special type of two-dimensional flow called *parallel flow*, in that it will be assumed that the mean flow has a component of velocity only in the direction parallel to the wall— the x axis. In other words, it is assumed that the mean flow velocity components are

$$\bar{v}_x = \text{function of } y \text{ only},$$

$$\bar{v}_y = 0.$$

Superimposed on this mean parallel flow are the turbulent fluctuations v_x' and v_y', so the time-dependent turbulent velocity has both x and y components:

$$v_x = \bar{v}_x + v_x',$$

$$v_y = 0 + v_y'.$$

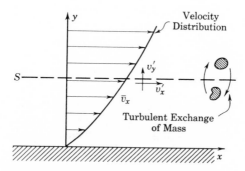

Fig. 6.9. Mass transport due to transverse turbulent fluctuations.

Consider now a plane, S, some distance away from the wall but parallel to it. At this plane the mean velocity component is $\bar{v}_x$. The total apparent shear stress, τ, on the plane S consists of a laminar and a turbulent shear stress. Using τ_l to denote the laminar stress, and τ_t to denote the turbulent stress, one obtains

$$\tau = \tau_l + \tau_t. \tag{6.39}$$

The laminar shear stress is given by the viscosity law of Eq. (6.2) and is

$$\tau_l = \mu \frac{\partial \bar{v}_x}{\partial y}. \tag{6.40}$$

An expression for τ_t may be deduced in the manner to be described next.

As discussed earlier in Sec. 2.7, a physical interpretation of laminar viscosity may be obtained from the molecular theory of gases and liquids. An exchange of molecules in the fluid layers on either side of the plane S will produce a change in the x-direction momentum of the fluid because of the existence of the gradient in the x-direction velocity. This momentum change produces a shearing force in the fluid which is directed in the x direction—the shear τ_l noted above. If turbulent velocity fluctuations occur both in the x and y directions about these mean values of $\bar{v}_x$ and $\bar{v}_y$, fluid masses which are large in comparison to molecular sizes may be transported across S due to the y-direction fluctuation, v'_y. For a *unit area* of S, the instantaneous time rate of transport of mass across S is $\rho v'_y$ (ρ being the density of the fluid). The momentum change in the x direction experienced by this mass will be dependent on the change in v_x which it experiences. If one considers particles which travel upward in Fig. 6.9 (i.e., v'_y positive), such motion will be associated with a negative v'_x since they are moving from a region where, on the whole, a smaller mean velocity v_x prevails. This will tend to produce a negative v'_x in the layer above the surface S. The momentum change experienced by such upward-moving particles will be, then, $(\rho v'_y)(-v'_x)$. By a similar argument, it may be seen that a negative v'_y will, on the whole, be most often associated with a positive v'_x, so that the momentum change in the x direction will be $(-\rho v'_y)(v'_x)$. The time-average value of the x-direction momentum thus exchanged across a unit area of S, $-\rho \overline{v'_x v'_y}$, will then be different from zero although the individual time averages, $\overline{v'_x}$ and $\overline{v'_y}$, are zero. This exchange of momentum will then give rise to a "turbulent" shear stress in the x direction:

$$\tau_t = -\rho \overline{v'_x v'_y}. \tag{6.41}$$

Combination of Eqs. (6.39), (6.40), and (6.41) gives the total apparent shear stress acting on the plane S as

$$\tau = \tau_l + \tau_t$$

$$= \mu \frac{\partial \bar{v}_x}{\partial y} - \rho \overline{v'_x v'_y}. \tag{6.42}$$

If a general two-dimensional turbulent motion exists rather than the parallel flow discussed above, turbulent normal stresses may also be deduced:

$$\sigma_{x_t} = -\rho \overline{v'^2_x},$$

$$\sigma_{y_t} = -\rho \overline{v'^2_y}.$$

Thus, the total stresses are given by

$$\tau = \mu \left(\frac{\partial \bar{v}_x}{\partial y} + \frac{\partial \bar{v}_y}{\partial x} \right) - \rho \overline{v'_x v'_y},$$

$$\sigma_x = 2\mu \frac{\partial \bar{v}_x}{\partial x} - \rho \overline{v'^2_x}, \tag{6.43}$$

$$\sigma_y = 2\mu \frac{\partial \bar{v}_y}{\partial y} - \rho \overline{v'^2_y}.$$

Inclusion of these definitions in the derivation of the Navier–Stokes equations based on the steady mean flow would then give rise to the additional terms found to exist in the Navier–Stokes equations noted in Eq. (6.38).

The above discussion and the associated equations are quite valid and help to illustrate the physical concepts which form the basis of turbulent flow theory. However, the equations cannot be used because the detailed nature of the fluctuations v'_x and v'_y are not known. In fact, the determination of the dependence of these quantities on the fluid motion, boundary conditions, etc., constitutes the central problem of the study of turbulent motion. Certain simplified representations of the turbulent shear stresses are available; however, these will not be discussed until after the notion of the boundary layer has been introduced.

A modification of the energy equation for turbulent flow may also be made—with results analogous to Eq. (6.38). This, too, will be deferred until after the following discussion of the concept of the boundary layer.

6.10   The Concept of the Boundary Layer

The Navier–Stokes equations of motion of a viscous fluid along with the energy equations are, mathematically speaking, very complex. Exact solutions to these equations are known for only a few cases, most of which have very specialized and often impractical boundary conditions. The complicated nature of these equations does not, however, eliminate the *need* for answers to problems of practical importance. For this reason an engineer must content himself with the compromise of accepting either an approximate solution to these fundamental equations, an exact solution to simplified or approximate versions of the equations, or, sometimes, an approximate solution of approximate equations. This does not destroy the value of the exact fundamental equations, for they are necessary in order to understand the physical implications of making any approximation or simplification.

Since the Navier–Stokes equations express a balance among inertia forces, viscous forces, pressure forces, and body forces, one such simplification may come from neglecting certain of the forces as being small in comparison to others when the conditions of the flow and the relative magnitude of the terms will permit. A very important case of this is found in instances in which one decides that *all* the viscous forces are quite negligible with respect to the inertia and pressure forces. If this is so, one can assume that the fluid viscosity is negligibly small and the Navier–Stokes equations reduce to the simpler Euler equations [as stated in Eq. (6.16)]. This simplification has many applications, the general theories of aerodynamics being an example.

However, when one considers the flow of a fluid past a solid surface the assumption of a vanishingly small viscosity may lead to results that are not verified in experiment. If the basic assumption is made that the fluid particles adjacent to the surface adhere to it and have zero velocity, there must exist velocity gradients in the fluid motion since the velocity must change from zero at the surface to some finite value at points removed from the surface. Immediately in the vicinity of the surface the velocity gradient may be so large that even if the fluid viscosity is small, the product of the velocity gradient and the viscosity (i.e., the viscous stress) may not be negligible. The extent of the region near the wall in which the viscous stresses may not be negligible, even in fluids of small viscosity, will depend on the properties of the fluid, the shape of the wall, the velocity of the main stream of the fluid, etc.

Even in such a region it may be possible that not all the viscous stresses are of a sizable nature, nor may all the components of the inertia forces be significant. It is on this basis that Ludwig Prandtl, in 1904, proposed his boundary layer theory. Prandtl's fundamental concept was that the motion of a fluid of small viscosity about a solid surface could be divided into two regions. One region, termed the *boundary layer*, near the body surface is defined as a region in which the velocity gradients are large enough so that the influence of viscosity cannot be neglected. The other region, termed the *potential region* (or *potential core* if flow inside a body is considered), is defined as the region in which the influence of the presence of the body has died out enough so that the velocity gradients are so small that the fluid viscosity can be ignored.

In actual flows the influence of the body extends to all regions of the fluid, but this influence and the associated changes in the fluid velocity decrease rapidly with the distance from the body—particularly if the viscosity is small and the main fluid velocity is large (i.e., at large Reynolds numbers). Since this decrease is continuous, it is obviously not possible to define a precise locus of the limit of the boundary layer and the beginning of the potential region. In practice, the limit of the boundary layer, the *boundary layer thickness*, is taken to be the distance away from the surface at which the flow velocity has achieved some arbitrary percentage of the undisturbed, free stream flow—say 90 or 99 per cent.

6.11 The Equation of Motion and the Energy Equation of the Laminar Boundary Layer

The application of Prandtl's boundary layer theory to the solution of real problems can be carried out in two ways. The first is based on differential equations of motion and energy that are obtained by simplifying the Navier–Stokes equations given above. These simplifications arise out of Prandtl's hypothesis that the boundary layer will be thin compared to the dimensions of the body about which it flows. This section will consider this approach.

The second method is based on integral equations of momentum and energy. This method attempts to describe, approximately, the over-all behavior of the boundary layer. Although the results obtained by this integral approach are not as complete and detailed as the results that may be obtained by the application of the differential equations, a greater variety of problems may be handled by this method. The integral equations of the boundary layer will be derived in Sec. 6.12.

Chapter 7 illustrates the solution of problems by these two methods.

The Equation of Motion

Visualize now the fluid motion depicted in Fig. 6.10(a), in which the fluid is moving past a solid wall, shown to be plane for simplicity. The x coordinate is measured parallel to the surface, and the y coordinate is measured normal to it. At any station, x, along the surface the x-component velocity distribution in the y direction varies somewhat as shown—from a zero value at the wall to a uniform value away from the wall.

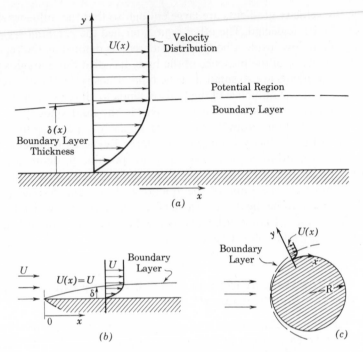

Fig. 6.10. Boundary layer flow past solid surfaces.

Defining the *boundary layer thickness* as the distance from the wall beyond which the velocity gradient is small enough that the viscous stresses may be neglected, one finds that this thickness increases in some manner with x as more and more of the fluid is influenced by the retarding effect of the wall—i.e., as the flow proceeds along the surface, more and more momentum must be fed into the boundary layer to overcome the viscous stresses in the region of the high velocity gradients. Thus, the boundary layer thickness is a function of the coordinate x and will be denoted by $\delta(x)$.

The velocity of the fluid in the region external to the boundary layer—the potential region—is indicated in Fig. 6.10(a) to be a function of x, namely $U(x)$. This is done for the sake of generality since there could be other bodies or surfaces present in the flow to cause a change in the nonviscous region. For example, another wall placed above the one shown (parallel or at an angle to the one drawn) would cause an acceleration or deceleration in the potential core. If the flow depicted is that resulting from a uniform stream approaching a flat plate placed parallel to the approaching flow [such as in Fig. 6.10(b)], then $U(x)$ would be a constant. If, however, the plate were oriented at even a slight angle with the oncoming flow, the potential region would have an x-direction velocity component which would not be constant. Flow about a curved surface, such as in Fig. 6.10(c), is another case in which a nonconstant $U(x)$ would result.

This analysis will consider only the steady flow case. If one assumes that the body forces, X and Y, are negligible compared to the other terms, the Navier–Stokes equations to be applied *in* the boundary layer are, from Eq. (6.17),

$$v_x \frac{\partial v_x}{\partial x} + v_y \frac{\partial v_x}{\partial y} = -\frac{1}{\rho} \frac{\partial P}{\partial x} + v \left(\frac{\partial^2 v_x}{\partial x^2} + \frac{\partial^2 v_x}{\partial y^2} \right),$$

$$v_x \frac{\partial v_y}{\partial x} + v_y \frac{\partial v_y}{\partial y} = -\frac{1}{\rho} \frac{\partial P}{\partial y} + v \left(\frac{\partial^2 v_y}{\partial x^2} + \frac{\partial^2 v_y}{\partial y^2} \right).$$

(6.44)

Now, from the definition of the boundary layer one must account for the viscous stresses only when $y < \delta(x)$. Prandtl's fundamental hypothesis for the boundary layer was that the thickness must be quite small in comparison to the linear dimensions of the body [i.e., $\delta(x) \ll x$]. With this fact, an examination of the above two equations may be made to determine the relative order of magnitude of the various terms. The order-of-magnitude analysis is complex and laborious, so only the results will be quoted here, although certain general justifications of the results may be made. For a detailed discussion of the analysis the reader is referred to Refs. 1 and 3.

Consider first the equation of motion for the x direction. If the layer is quite thin, the v_x component of the velocity must change very rapidly from 0 at $y = 0$ to $U(x)$ at $y = \delta$. Thus, the gradient $\partial v_x/\partial y$ may be expected to be of significant magnitude as will $\partial^2 v_x/\partial y^2$. If one takes the inertia force in the x direction, $[v_x(\partial v_x/\partial x) + v_y(\partial v_x/\partial y)]$, to be the standard order of magnitude, it is not surprising that the end results show that the viscous force $v(\partial^2 v_x/\partial y^2)$ may be of the same order, whereas the viscous force in the x direction due to $v(\partial^2 v_x/\partial x^2)$ will be of a considerably lower order of magnitude. Upon comparing the terms of the second equation of motion to the same standard order of magnitude, all the terms are found to be of a much smaller magnitude. This is due principally to the fact that the v_y component will be small if the boundary layer thickness is likewise small. If these terms of smaller orders of magnitude are neglected, the equations reduce to

$$v_x \frac{\partial v_x}{\partial x} + v_y \frac{\partial v_x}{\partial y} = -\frac{1}{\rho} \frac{\partial P}{\partial x} + v \frac{\partial^2 v_x}{\partial y^2},$$

$$\frac{\partial P}{\partial y} = 0.$$

(6.45)

An important fact to be observed here is, under the assumptions noted above, that the pressure gradient normal to the surface is zero throughout the boundary layer! This result implies the very useful fact that the longitudinal pressure gradient, $\partial P/\partial x$, in the boundary layer, taken in a direction parallel to the wall, is equal to the pressure gradient in the external potential flow region taken in the same direction. Thus, a basic knowledge of the laws of the flow of an inviscid fluid, as governed by the Euler equations [(6.16)], is necessary for a complete understanding of boundary layer theory.

The boundary layer equations of motion then reduce to the following single equation with a total derivative replacing the partial derivative of pressure in the x direction*:

$$v_x \frac{\partial v_x}{\partial x} + v_y \frac{\partial v_x}{\partial y} = -\frac{1}{\rho}\frac{dP}{dx} + v\frac{\partial^2 v_x}{\partial y^2}. \tag{6.46}$$

The important point to note is that the basic assumption leading to the simplified boundary layer equations shown above is that the layer is quite thin compared to the other linear dimensions. This assumption implies a fluid of low viscosity or a high external flow velocity—or both. Whether or not this condition is met in practice depends on the particular situation at hand, and the accuracy obtained by replacing the Navier–Stokes equations with these simpler expressions depends on how well this assumption is satisfied in reality.

In cases in which the fluid is flowing past nonplane surfaces (but still in two space dimensions), e.g., in the case of flow about airfoils or, say, cylinders as shown in Fig. 6.10(c), the same equations may be applied to the boundary layer formed along the curved surface if certain additional conditions are met.

Let the x coordinate be interpreted as the arc length measured along the curved surface, and let y be the coordinate measured normal to the local x direction. Thus the x-y coordinates are an intrinsic system attached to the flow. If, in this system of coordinates, the boundary layer thickness is presumed to be small in comparison with the radius of curvature ($\delta \ll R$), the same boundary layer equations are obtained if the following additional (but not unreasonable) conditions are met: (a) There must be no *sudden* changes in the radius of curvature, and (b) the radius of curvature must be large compared to the linear displacement along the surface. In such cases of nonplane surfaces the external potential flow velocity $U(x)$ will be a function of the arc coordinate x.

---

* In the case of free convection where the fluid is caused to move by density differences, the body force is the driving force and hence cannot be neglected as was done here. In such a case the body force must be related to the fluid temperature and the cofficient of thermal expansion. This is discussed in Sec. 6.8 and Chapter 9.

The Energy Equation

Application of the same order-of-magnitude considerations to the energy equations discussed in Sec. 6.6 and shown in Eq. (6.22) will also lead to a simplified version of that equation. Equation (6.22) is repeated here for the steady state:

$$k\left(\frac{\partial^2 t}{\partial x^2} + \frac{\partial^2 t}{\partial y^2}\right) + \mu\left[2\left(\frac{\partial v_x}{\partial x}\right)^2 + 2\left(\frac{\partial v_y}{\partial y}\right)^2 + \left(\frac{\partial v_x}{\partial y} + \frac{\partial v_y}{\partial x}\right)^2\right]$$

$$= c_p\rho\left(v_x\frac{\partial t}{\partial x} + v_y\frac{\partial t}{\partial y}\right).$$

If a difference exists between the temperature at the surface and the temperature of the main fluid stream, one presumes the presence of a *thermal boundary layer* of thickness δ_t. Rather than being related to the exchange of momentum as in the case of the velocity boundary layer, the thermal boundary layer results from an exchange of heat energy between the surface and the potential region. This thermal boundary layer thickness is also assumed to be quite small with respect to the linear dimensions of the surface, so that it is of the same order of magnitude as (but not necessarily equal to) the velocity boundary layer thickness defined above. Thus, the distribution of the temperature in the thermal boundary layer would be similar to that of velocity in the velocity boundary layer, i.e., a rapid change from some value at the surface to the temperature of the potential core, when moving in a direction normal to the surface, but with a relatively gradual gradient in the direction parallel to the surface if the rate of dissipation of mechanical energy is not too large. By taking the internal energy term (the right side of the above equation) as the standard order of magnitude, then, owing to the large possible gradient in the y direction, the first part of the first term of the left side of the equation may be expected to be negligible. In keeping with the assumptions made for the equations of motion, the viscous dissipation terms (i.e., those terms preceded by μ in the above equation) should again be negligible, excepting the term involving the velocity gradient normal to the wall, $\mu(\partial v_x/\partial y)^2$. The resulting simplified, approximate form of the energy equation would then be

$$v_x\frac{\partial t}{\partial x} + v_y\frac{\partial t}{\partial y} = \frac{k}{\rho c_p}\frac{\partial^2 t}{\partial y^2} + \frac{\mu}{\rho c_p}\left(\frac{\partial v_x}{\partial y}\right)^2. \tag{6.47}$$

The equations of the laminar, two-dimensional boundary layer just discussed are naturally simpler than the full Navier–Stokes and energy equations, but the analytical solution is still difficult because of the nonlinearity of the equations. A brief discussion of a so-called "exact" solution to these equations for flow parallel to a flat plate will be presented in Chapter 7. Before turning to that example, it is appropriate to discuss another approach to the problem of boundary layer flow, developed by Theodore von Kármán. Von Kármán's method consists of an integral approach to the gross behavior of the boundary layer without examining in detail the equations governing the motion of each fluid particle in the layer. This approach is considered in the next section.

6.12   The Integral Equations of the Laminar Boundary Layer

The *integral methods* of analyzing boundary layers consist of fixing one's attention on the over-all behavior of the boundary layer (i.e., as far as the conservation of energy and momentum principles are concerned) rather than on the detailed motion of an individual fluid particle. With this approach rather crude approximations may be made to the distribution of velocity and temperature in the boundary layer without appreciable error in the resultant over-all behavior. The necessary governing equations for the integral treatment of laminar incompressible boundary layers can be obtained directly by integration of the approximate equation of motion and energy equations [Eqs. (6.46) and (6.47)]. However, the equations may also be obtained by a direct derivation that will more clearly illustrate the physical process taking place.

The Momentum Integral Equation

Turning first to the velocity boundary layer, consider the flow picture shown in Fig. 6.11. Here boundary layer flow along a surface is shown. (The surface is assumed to be plane, although x may, in reality, represent the arclength displacement along a curved surface as discussed earlier.) It is assumed, in this analysis, that the boundary layer has a definite thickness, $\delta(x)$, which is a function of the coordinate x. It is this thickness and the way in which it varies with x, among other things, that one wishes to determine in order to analyze the convection of heat away from the wall.

At some station, say $x = x$, the distribution of the x component of the velocity is somewhat as shown in Fig. 6.11—a variation from 0, at $y = 0$, to a uniform value for all values of $y > \delta(x)$. At some station farther downstream, say $x = x + dx$, the variation is of the same nature, except that the boundary layer thickness may be greater, and the uniform potential value reached at the edge of the boundary layer may also be different. If a reference value of y, say y_r, is chosen so that it lies outside the boundary layer between x and

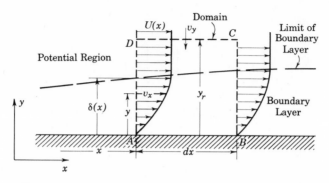

Fig. 6.11. Laminar boundary layer flow past a solid surface.

$x + dx$, the domain enclosed between x and $x + dx$ and between $y = 0$ and y_r is defined, and the conservation laws may be applied thereto. This domain is denoted as the rectangle $ABCD$ in Fig. 6.11, and unit depth in the plane of the paper is assumed.

The mass flow rates, denoted by $\dot{m}$, across each of the sides of the rectangle may be deduced. The results, for constant density, are summarized below:

$$\dot{m}_{AD} = \left[\rho \int_0^{y_r} v_x \, dy \right]_x,$$

$$\dot{m}_{BC} = \left[\rho \int_0^{y_r} v_x \, dy \right]_{x+dx} = \left[\rho \int_0^{y_r} v_x \, dy \right]_x + \frac{d}{dx}\left[\rho \int_0^{y_r} v_x \, dy \right]_x dx,$$

$$\dot{m}_{AB} = 0 \,(\text{solid wall}),$$ (6.48)

$$\dot{m}_{DC} = \dot{m}_{BC} - \dot{m}_{AD} - \dot{m}_{AB} = \frac{d}{dx}\left[\rho \int_0^{y_r} v_x \, dy \right]_x dx.$$

A Taylor expansion has been written above, to express the mass flow at $x + dx$ in terms of the mass flow at x, with only the first two terms considered since dx will subsequently be allowed to approach 0. The necessity for the existence of a flow velocity in the y direction across the line DC is shown by the existence of $\dot{m}_{DC}$, which must be present in order to account for the difference in the flows across AD and BC.

Denoting by $\dot{M}$ the momentum flux in the x *direction* across each of the sides of the rectangle, one may write the following expressions noting that the y direction flow carries x direction momentum into the domain due to the existence of $U(x)$:

$$\dot{M}_{AD} = \left[\int_0^{y_r} (\rho v_x) v_x \, dy \right]_x = \left[\rho \int_0^{y_r} v_x^2 \, dy \right]_x,$$

$$\dot{M}_{BC} = \left[\rho \int_0^{y_r} v_x^2 \, dy \right]_x + \frac{d}{dx} \left[\rho \int_0^{y_r} v_x^2 \, dy \right]_x dx,$$

$$\dot{M}_{AB} = 0, \tag{6.49}$$

$$\dot{M}_{DC} = \dot{m}_{DC} U(x) = U(x) \frac{d}{dx} \left[\rho \int_0^{y_r} v_x \, dy \right]_x dx.$$

The forces acting on the domain will consist of a shear force at $y = 0$ (denote the shear stress there by τ_0), and pressure and normal stresses on the AD and BC faces. Since the dimension y_r is chosen so that DC lies outside the boundary layer, there will be no viscous shear on that face. If the dominant velocity gradient is that of v_x in the y direction, the normal stresses on AD and BC may be ignored, and the following equations may be written for the x-direction forces, F, acting on each face:

$$F_{AD} = [P]_x y_r,$$

$$F_{BC} = \left[P + \frac{dP}{dx} dx \right]_x y_r,$$

$$F_{AB} = \tau_0 \, dx = \mu \left(\frac{\partial v_x}{\partial y} \right)_{y=0} dx, \tag{6.50}$$

$$F_{DC} = 0.$$

The momentum principle states that for steady flow the net flux of momentum over the influx of momentum (in a selected direction) from a domain must equal the algebraic sum of the forces acting on the domain in that direction. This principle may be applied to the x-direction motion for the domain $ABCD$ of Fig. 6.11, resulting in the following expression after substitution of Eqs. (6.49) and (6.50):

$$\left[\rho \int_0^{y_r} v_x^2 \, dy + \frac{d}{dx} \left(\rho \int_0^{y_r} v_x^2 \, dy \right) dx \right]_x - \left[\rho \int_0^{y_r} v_x^2 \, dy \right]_x$$

$$- \left[U(x) \frac{d}{dx} \left(\rho \int_0^{y_r} v_x \, dy \right) \right]_x dx = [P \cdot y_r]_x - \left[P + \frac{dP}{dx} dx \right]_x y_r - \tau_0 \, dx.$$

Canceling some terms and noting that ρ is constant, one finds that

$$\frac{d}{dx}\int_0^{y_r} v_x^2\, dy - U(x)\frac{d}{dx}\int_0^{y_r} v_x\, dy = -\frac{1}{\rho}\frac{dP}{dx}y_r - \frac{\tau_0}{\rho}.$$

If more of the Taylor expansion had been included, then division by dx and subsequent progression to the limit, $dx \to 0$, would, more rigorously, have achieved the same result. With dx thus interpreted as an infinitesimal, y_r may be set equal to the boundary layer thickness, $\delta(x)$, since the two integrals on the left of the above equation would yield identical results for $y > \delta$ [i.e., $v_x = U(x)$]. Thus, the following equation is obtained:

$$\frac{d}{dx}\int_0^{\delta} v_x^2\, dy - U(x)\frac{d}{dx}\int_0^{\delta} v_x\, dy = -\frac{\delta}{\rho}\frac{dP}{dx} - \frac{\tau_0}{\rho}. \tag{6.51}$$

This equation constitutes the *integral momentum equation* of the steady laminar incompressible boundary layer. For laminar flow the wall shear, τ_0, may be replaced with $\tau_0 = \mu(\partial v_x/\partial y)y = 0$, so

$$\frac{d}{dx}\int_0^{\delta} v_x^2\, dy - U(x)\frac{d}{dx}\int_0^{\delta} v_x\, dy = -\frac{\delta}{\rho}\frac{dp}{dx} - v\left(\frac{\partial v_x}{\partial y}\right)_0. \tag{6.52}$$

In the case of steady turbulent boundary layers, the same expression as Eq. (6.51) would result with the time-average velocity $\bar{v}_x$ replacing v_x; however, the simplification to the forms of Eq. (6.52) is not possible since the laminar shear law is not applicable. Some expression to represent the turbulent shear must be introduced. This is discussed in Sec. 6.13.

If the velocity distribution $v_x(y)$ is known, the integrands of the two integrals are known. The resulting expression may then be interpreted as a differential equation for δ, the boundary layer thickness, as a function of x. This approach will be investigated fully in Chapter 7.

The Energy Integral Equation

The very same approach as applied above may be used to arrive at an integral energy relation for the thermal boundary layer. As mentioned above, the thermal boundary layer thickness is defined as the distance away from the surface at which the fluid temperature approaches the uniform temperature of the free stream potential flow. This thickness, $\delta_t(x)$, will be of the same order of magnitude as, but not necessarily equal to, the velocity boundary layer thickness $\delta(x)$. As in the case of the momentum integral analysis, it will be assumed that the thermal layer thickness is a definite quantity, although in reality it would be difficult to define such a precise limit.

Figure 6.12 shows a fluid flowing past a surface heated to a constant temperature t_s. The velocity boundary layer, δ, is shown along with the thermal boundary layer, δ_t, which is assumed to be thinner than the velocity layer. The implication of this assumption will be discussed later. The symbol t_f denotes the uniform temperature of the potential flow and, for purposes of discussion, it is assumed that the fluid is hotter than the surface ($t_f > t_s$), although the reverse might be true. Once more a control domain, of length dx along the surface, is chosen with the ordinate y_r taken to be outside the velocity layer. The energy transported across the boundaries of the domain may be expressed in terms of the flux of enthalpy across those boundaries since kinetic energy effects will be ignored. Again using a linear approximation in the incremental distance dx, the excess of the enthalpy flux out the face BC (for unit depth) over the enthalpy flux in face AD is

$$\frac{d}{dx}\left[\int_0^{y_r} c_p(\rho v_x)t\,dy\right]dx.$$

The symbols t and v_x denote the velocity and temperature of the fluid at an ordinate y. As shown in Eqs. (6.48), the mass crossing face DC is

$$\dot{m}_{DC} = \frac{d}{dx}\left[\rho\int_0^{y_r} v_x\,dy\right]dx.$$

This mass carries into the domain an amount of energy equal to

$$c_p t_f \frac{d}{dx}\left[\rho\int_0^{y_r} v_x\,dy\right]dx.$$

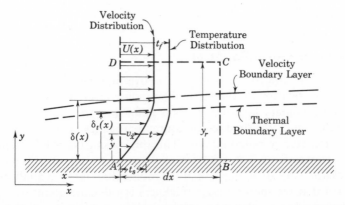

Fig. 6.12. Laminar velocity and thermal boundary layers on a heated surface.

Finally, the heat entering or leaving the domain by conduction from the surface may be expressed in terms of the wall heat flux:

$$\left(\frac{q}{A}\right)_0 dx.$$

The principle of conservation of energy gives then, neglecting the energy dissipated by friction and neglecting kinetic energy effects,

$$\rho c_p \frac{d}{dx} \int_0^{y_r} t_f v_x \, dy - \rho c_p \frac{d}{dx} \int_0^{y_r} t v_x \, dy = -\left(\frac{q}{A}\right)_0.$$

For values of $y > \delta_t$ the two integrals give identical results since t will be the same as t_f, so the integration needs to be carried only to $y = \delta_t(x)$, the thermal boundary layer thickness. The result is the *integral energy equation* of the laminar, incompressible boundary layer:

$$\frac{d}{dx} \int_0^{\delta_t} (t_f - t) v_x \, dy = -\frac{(q/A)_0}{\rho c_p}. \tag{6.53}$$

For laminar flow the wall heat flux may be written in terms of the wall temperature gradient:

$$\frac{d}{dx} \int_0^{\delta_t} (t_f - t) v_x \, dy = \alpha \left(\frac{\partial t}{\partial y}\right)_0. \tag{6.54}$$

As in the case of the integral momentum equation, the integral energy equation in the form of Eq. (6.53) may be applied in turbulent cases if an appropriate rate law for $(q/A)_0$ can be obtained.

6.13 Turbulent Boundary Layers

The boundary layers and their governing equations discussed thus far have been restricted to steady laminar motion. However, the viscous flow region of the boundary layer need not be laminar. Indeed, in the majority of cases arising in engineering applications, turbulent layers are encountered.

The unique features of turbulent boundary layers and their governing relations will be discussed in this section. This can best be done by reconsidering the laminar equations [(6.46) and (6.47)] and comparing them with the turbulent counterparts:

$$v_x \frac{\partial v_x}{\partial x} + v_y \frac{\partial v_x}{\partial y} = -\frac{1}{\rho} \frac{dP}{dx} + \frac{\mu}{\rho} \frac{\partial^2 v_x}{\partial y^2}$$

$$= -\frac{1}{\rho} \frac{dP}{dx} + \frac{1}{\rho} \frac{\partial}{\partial y} \tau_l, \tag{6.55a}$$

$$v_x \frac{\partial t}{\partial x} + v_y \frac{\partial t}{\partial y} = \frac{\mu}{\rho c_p} \left(\frac{\partial v_x}{\partial y} \right)^2 + \frac{k}{\rho c_p} \frac{\partial^2 t}{\partial y^2}$$

$$= \frac{1}{\rho c_p} \left(\tau_l \frac{\partial v_x}{\partial y} \right) + \frac{1}{\rho c_p} \frac{\partial}{\partial y} \left(\frac{-q}{A} \right)_l. \tag{6.55b}$$

In these equations the laminar expressions for shear and heat flux

$$\tau_l = \mu \frac{\partial v_x}{\partial y},$$

$$\left(\frac{q}{A} \right)_l = -k \frac{\partial t}{\partial y}, \tag{6.56}$$

have been introduced to emphasize the origin of the terms involved. The shear stress τ_l and heat flux $(q/A)_l$ represent the flux of momentum and energy in the y direction (normal to the boundary layer) due to molecular scale activity.

For the case of turbulent flow past a solid surface, the turbulent equations of motion, Eqs. (6.38), may be subjected to the same order-of-magnitude analysis described in Sec. 6.11 for laminar layers. These equations will not be repeated here because of their complexity. If one examines the relative order of magnitude of the various terms compared with the x-direction inertia term, it should not be surprising that all terms in the y-direction equation are negligible and only those viscous terms in the x-direction equation resulting from the strong y variations are retained. Thus, as for the laminar layer, the equations of motion for a turbulent boundary layer reduce to the single expression:

$$\bar{v}_x \frac{\partial \bar{v}_x}{\partial x} + \bar{v}_y \frac{\partial \bar{v}_x}{\partial y} = -\frac{1}{\rho} \frac{d\bar{P}}{dx} + \frac{\mu}{\rho} \frac{\partial^2 \bar{v}_x}{\partial y^2} - \frac{\partial}{\partial y} \overline{(v_x' v_y')}$$

$$= -\frac{1}{\rho} \frac{d\bar{P}}{dx} + \frac{1}{\rho} \frac{\partial}{\partial y} (\tau_l + \tau_t). \tag{6.57}$$

In Eq. (6.57) the concept of the "turbulent shear stress"

$$\tau_t = -\rho \overline{v'_x v'_y} \tag{6.58}$$

discussed in Sec. 6.9 [Eq. (6.41)] has been introduced.

A corresponding turbulent boundary layer energy equation may be developed by, first, introducing into Eq. (6.22) the turbulent representations of Eqs. (6.36),

$$
\begin{aligned}
v_x &= \bar{v}_x + v'_x, \\
v_y &= \bar{v}_y + v'_y,
\end{aligned} \tag{6.59}
$$

and a corresponding temperature representation,

$$t = \bar{t} + t'. \tag{6.60}$$

In Eq. (6.60) $\bar{t}$ is the time-average temperature and t' is the temporal variation around $\bar{t}$. Following this substitution the resulting terms are time-averaged. If, subsequently, one performs the boundary layer order-of-magnitude simplification, the following final result is obtained (Ref. 1):

$$\bar{v}_x \frac{\partial \bar{t}}{\partial x} + \bar{v}_y \frac{\partial \bar{t}}{\partial y} = \frac{\mu}{\rho c_p} \left(\frac{\partial \bar{v}_x}{\partial y} \right)^2 - \frac{1}{c_p} \overline{(v'_x v'_y)} \frac{\partial \bar{v}_x}{\partial y} + \frac{k}{\rho c_p} \frac{\partial^2 \bar{t}}{\partial y^2} - \frac{\partial}{\partial y} \overline{(v'_y t')}. \tag{6.61}$$

The last term in this equation represents the transport of energy in the y direction resulting from the larger-than-molecular-scale fluctuations, in direct analogy to the term $(\partial/\partial y)\overline{(v'_x v'_y)}$ representing the transport of momentum due to these fluctuations. Thus, the molecular flux of energy

$$\left(\frac{q}{A} \right)_l = -k \left(\frac{\partial t}{\partial y} \right)$$

is enhanced by the large-scale mixing due to the turbulence. As in the case of the momentum flux being interpreted as a *turbulent shear stress*, Eq. (6.58), a *turbulent heat flux* may be defined:

$$\left(\frac{q}{A} \right)_t = \rho c_p \overline{(v'_y t')}. \tag{6.62}$$

Thus, the turbulent boundary layer energy equation may be written as

$$\bar{v}_x \frac{\partial \bar{t}}{\partial x} + \bar{v}_y \frac{\partial \bar{t}}{\partial y} = \frac{1}{\rho c_p} (\tau_l + \tau_t) \frac{\partial \bar{v}_x}{\partial y} + \frac{1}{\rho c_p} \frac{\partial}{\partial y} \left[-\left(\frac{q}{A} \right)_l - \left(\frac{q}{A} \right)_t \right]. \tag{6.63}$$

Equations (6.57) and (6.63) are the equations of motion and energy of the turbulent boundary layer. Their correspondence to the laminar expressions, Eqs. (6.55a) and (6.55b), is obvious, particularly if one introduces the concept of an *apparent shear stress* and an *apparent heat flux*, which may be defined as the sum of the laminar and turbulent contributions,

$$\tau_{app} = \tau_l + \tau_t$$

$$= \mu\left(\frac{\partial \bar{v}_x}{\partial y}\right) - \rho\overline{v'_x v'_y}, \tag{6.64}$$

$$\left(\frac{q}{A}\right)_{app} = \left(\frac{q}{A}\right)_l + \left(\frac{q}{A}\right)_t$$

$$= -k\left(\frac{\partial t}{\partial y}\right) + \rho c_p\overline{v'_y t'}. \tag{6.65}$$

Although convenient conceptually, these "apparent" transport quantities do not eliminate the basic complexity of turbulent flow. The turbulent contributions, $\overline{v'_x v'_y}$ and $\overline{v'_y t'}$, are unknown quantities whose values depend on the state of the fluid motion.

Often the turbulent boundary layer equations are made to look, formally, like their laminar counterparts by defining a turbulent kinematic viscosity or *eddy viscosity*,

$$\epsilon = -\frac{\overline{v'_x v'_y}}{(\partial \bar{v}_x/\partial y)},$$

so that

$$\tau_t = \rho\epsilon\left(\frac{\partial \bar{v}_x}{\partial y}\right), \tag{6.66}$$

and

$$\tau_{app} = \rho(v + \epsilon)\frac{\partial \bar{v}_x}{\partial y}. \tag{6.67}$$

Likewise, an *eddy diffusivity* is defined as

$$\epsilon_H = -\frac{\overline{v'_y t'}}{(\partial t/\partial y)}, \tag{6.68}$$

so that

$$\left(\frac{q}{A}\right)_t = -\rho c_p \epsilon_H \left(\frac{\partial \bar{t}}{\partial y}\right) \tag{6.69}$$

and

$$\left(\frac{q}{A}\right)_{\text{app}} = -\rho c_p (\alpha + \epsilon_H)\frac{\partial \bar{t}}{\partial y}. \tag{6.70}$$

When the representations of Eqs. (6.67) and (6.70) are introduced into Eqs. (6.57) and (6.63), one has

$$\bar{v}_x \frac{\partial \bar{v}_x}{\partial x} + \bar{v}_y \frac{\partial \bar{v}_x}{\partial y} = -\frac{1}{\rho}\frac{dP}{dx} + (v + \epsilon)\frac{\partial^2 \bar{v}_x}{\partial y^2}, \tag{6.71}$$

$$\bar{v}_x \frac{\partial \bar{t}}{\partial x} + \bar{v}_y \frac{\partial \bar{t}}{\partial y} = \frac{(v + \epsilon)}{c_p}\left(\frac{\partial \bar{v}_x}{\partial y}\right)^2 + (\alpha + \epsilon_H)\frac{\partial \bar{t}}{\partial y}. \tag{6.72}$$

These two equations are seen to be formally the same as their laminar counterparts, Eqs. (6.55a) and (6.55b)—with $(v + \epsilon)$ replacing v; $(\alpha + \epsilon_H)$ replacing α; and $\bar{v}_x$ and $\bar{t}$ replacing v_x and t.

The eddy coefficients ϵ and ϵ_H, and the corresponding transport laws

$$\tau_{\text{app}} = \rho(v + \epsilon)\frac{\partial \bar{v}_x}{\partial y},$$

$$\left(\frac{q}{A}\right)_{\text{app}} = -\rho c_p(\alpha + \epsilon_H)\frac{\partial \bar{t}}{\partial y}, \tag{6.73}$$

are convenient formal representations, but they do not actually provide any insight or understanding of the basic complexities of turbulent fluid motion. The parameters ϵ and ϵ_H are *not* fluid properties as are the molecular quantities v and α. The eddy coefficients are dependent upon the nature of the fluctuating quantities v'_x and t' and are known to vary from point to point in the flow field itself since, for example, they must vanish at or near a solid surface where transverse fluctuations must disappear. At any rate, one may remember that while v and α are related to *molecular* scale transport of momentum and energy, ϵ and ϵ_H are functions of the larger-scale fluctuations. Many complex theories of turbulent flow exist and a comprehensive treatment of the subject requires much more space than is available here.

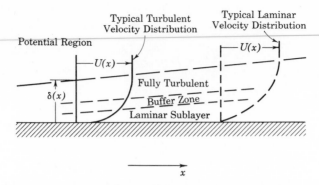

Fig. 6.13. A turbulent boundary layer on a flat surface.

Hence, certain simplified empirical laws representing ϵ and ϵ_H will be introduced in Chapter 7.

The general characteristics of turbulent boundary layers are similar to those of laminar layers—a thin region exists near the surface of a body where the time-average velocity varies rapidly from a uniform value in the potential core to zero at the surface. As a result of transverse turbulent fluctuations there exists a more uniform velocity distribution near the outer edge of a turbulent layer than in the corresponding laminar case. This is indicated in Fig. 6.13.

Often the turbulent layer is further divided into subregions as indicated in Fig. 6.13. Near the wall one may imagine a region in which the transverse fluctuations must be suppressed by the wall. In this region one would expect that the eddy viscosity and diffusivity are dominated by the laminar ones, or $\epsilon \ll v$ and $\epsilon_H \ll \alpha$. Thus, the apparent shear stress and heat flux given in Eq. (6.73) follow the laminar laws and the region is sometimes called the *laminar sublayer*. Near the outer edge of the turbulent layer a region is observed to exist in which the opposite is true: $\epsilon \gg v$, $\epsilon_H \gg \alpha$. Such a region is referred to as a "fully turbulent" zone, in which the eddy quantities dominate. There may be a region between the laminar sublayer and the fully turbulent zone in which $\epsilon \approx v$, $\epsilon_H \approx \alpha$, and it is called the *buffer zone*. The three regions just discussed (laminar sublayer, the buffer zone, and the fully turbulent region) are not distinctly identifiable, of course, and actually there must be a continuous variation from one to the other.

One final point concerning turbulent boundary layers may be made in conjunction with the integral equations developed in Sec. 6.12. Those equations in the forms given in Eqs. (6.51) and (6.53) may be applied to turbulent layers as long as the time-averaged velocity and temperature are used. The extension to the forms given in Eqs. (6.52) and (6.54) may be made only if v is replaced by $(v + \epsilon)$ and α by $(\alpha + \epsilon_H)$.

References

1. SCHLICHTING, H., *Boundary Layer Theory*, 6th ed., New York, McGraw-Hill, 1968.
2. PRANDTL, L., and O. G. TIETJENS, *Fundamentals of Hydro- and Aeromechanics*, New York, McGraw-Hill, 1934.
3. GOLDSTEIN, S. (ed.), *Modern Development in Fluid Dynamics*, Vols. I and II, New York, Oxford U.P., 1938.
4. ZEMANSKY, M. W., *Heat and Thermodynamics*, 4th ed., New York, McGraw-Hill, 1957.

Examples of the Application of Boundary Layer Theory to Problems of Forced Convection

7.1 Introductory Remarks

The application of the principles discussed in Chapter 6 to the analytical prediction of problems in forced convection—in particular the determination of values of the heat transfer coefficient, h—can become quite complex and can involve great quantities of computational labor. It is not the purpose of this chapter to equip the reader with the ability to apply the boundary layer theory technique to the solution of all subsequent heat convection problems that he may encounter. Rather it is meant to illustrate, by consideration of the simplest of examples, the basic methods and concepts behind such techniques and to develop an appreciation of such methods. Also, the knowledge gained from the consideration of these examples will familiarize the reader with the important physical parameters of forced convection. This will aid in the building of an adequate background to give a certain amount of logic to some of the empirical techniques to be introduced later.

Perhaps the simplest example of the application of boundary layer theory to the solution of forced convection heat transfer is that of the flow of a fluid of initial uniform velocity and temperature past a flat plate of constant temperature placed parallel to the incident flow. The reason for the simplicity of this example is that, in the absence of any other surface, the flow outside the boundary layer (the limit of the influence of the surface) must still be uniform.

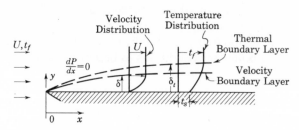

Fig. 7.1. Velocity and thermal boundary layers for laminar flow past a flat plate.

Hence, the potential velocity $U(x)$ defined in Chapter 6 must be constant, and the longitudinal pressure gradient (in either the potential region or the boundary layer) must be zero. Such a flow situation is depicted in Fig. 7.1.

Examination of the equations of motion and energy quoted above—either the differential boundary layer equations [Eqs. (6.46) and (6.47)] or the integral equations [Eqs. (6.52) and (6.54)]—shows that the solution for the velocity boundary layer (i.e., its velocity distribution, variation of the boundary layer thickness, etc.) may be obtained for the given boundary conditions of zero velocity at the wall and uniform velocity outside the layer without knowledge of the solution of the temperature boundary layer as long as the fluid properties (μ and ρ) are taken as independent of temperature. On the other hand, in order to apply the energy equation to the solution for the temperature distribution in the thermal boundary layer, one must first have the solution for the velocity distribution. Thus, it is necessary first to determine the solution for the velocity layer and then to apply the results for the determination of the temperature distribution and, eventually, the convection film coefficient.

Two approaches will be illustrated, one which uses the differential equations of energy and motion of the boundary layer, and another which uses the integral energy and momentum equations. Both the cases in which viscous dissipation in the boundary layer is neglected and accounted for will be treated.

7.2 The Solution of Laminar Forced Convection on a Flat Plate by Use of the Differential Equations of Motion and Energy of the Boundary Layer, Dissipation Neglected

This section is devoted to the determination of the distribution of the fluid velocity and temperature throughout the boundary layer in the case of laminar flow past a plane flat surface. This is done by the application of the differential equations of motion and energy developed in Sec. 6.11. These equations must be integrated to satisfy the conditions imposed at the plate surface and in the free stream of undisturbed fluid. If, as shown in Fig. 7.1, the fluid stream approaches the plate with a uniform velocity denoted by U and a temperature denoted by t_f and if the plate is maintained at a constant temperature t_s, the boundary conditions to be satisfied are

$$\text{At } y = 0: \quad t = t_s, v_x = v_y = 0.$$

$$\text{As } y \to \infty: \quad t \to t_f, v_x \to U.$$

The coordinates x and y are taken parallel and normal to the plate surface, respectively, the origin being taken at the leading edge. As noted above, the solution for the velocity boundary layer may be obtained independently of the solution for the thermal boundary layer. This will be done first.

The Velocity Boundary Layer and the Skin Friction Coefficient

For the determination of the velocity in the boundary layer of an incompressible fluid flowing past a flat plate, the equations to be satisfied are the continuity equation, Eq. (6.1), and the boundary layer equation of motion, Eq. (6.46), with $dP/dx = 0$:

$$\frac{\partial v_x}{\partial x} + \frac{\partial v_y}{\partial y} = 0, \tag{7.1}$$

$$v_x \frac{\partial v_x}{\partial x} + v_y \frac{\partial v_x}{\partial y} = v \frac{\partial^2 v_x}{\partial y^2}. \tag{7.2}$$

The boundary conditions to be satisfied are

$$\text{At } y = 0: \quad v_x = v_y = 0.$$

$$\text{As } y \to \infty: \quad v_x = U \text{ (constant)}. \tag{7.3}$$

An initial, rather crude analysis, based on order of magnitude alone, will give considerable insight into the problem. The terms on the left of Eq. (7.2) represent the inertia forces, and the first term, $v_x(\partial v_x/\partial x)$, may be taken as typical. Now, v_x will be of the order of magnitude of U (the free stream velocity), so this inertia term, at some downstream station x, will be of the order of $U(U/x)$. Considering now the viscous term on the right, one sees that the v_x velocity goes from 0 at the wall to U at $y = \delta$, δ being the unknown boundary layer thickness. Assuming a linear velocity distribution for order-of-magnitude considerations, one can expect the velocity gradient $\partial v_x/\partial y$ to be of the order of U/δ. Hence, $\partial^2 v_x/\partial y^2 = (\partial/\partial y)(\partial v_x/\partial y)$ will be of the order of U/δ^2—making the viscous term of the order of $\nu(U/\delta^2)$. The fundamental assumption leading to the simplified form of the boundary layer equation was that the inertia and viscous forces were of the same order of magnitude in the boundary layer. Thus, one may expect, *approximately*, that

$$\frac{U^2}{x} \approx \frac{\nu U}{\delta^2}, \quad \text{or} \quad \delta \approx \sqrt{\frac{\nu x}{U}}. \tag{7.4}$$

Thus, the boundary layer may be expected to grow as $\sqrt{x}$ as it moves down the plate. More generally, Eq. (7.4) shows that the ratio of the boundary layer thickness at a certain point to the distance of that point from the leading edge of the plate is

$$\frac{\delta}{x} \approx \sqrt{\frac{\nu}{Ux}} = \frac{1}{\sqrt{N_{RE_x}}}. \tag{7.5}$$

Equation (7.5) shows that the ratio δ/x is inversely proportional to the square root of the Reynolds number of the flow based on the distance from the leading edge as the reference length (see Sec. 6.7). The subscript x on the Reynolds number is used to note that the characteristic length upon which N_{RE} is based is the distance x measured from the leading edge of the plate.

The proportionality expressed by Eq. (7.4) may be used to simplify greatly the problem as stated in Eqs. (7.1), (7.2), and (7.3). Referring once again to Fig. 7.1, one sees that the absence of any characteristic length (i.e., the plate originates at $x = 0$ and extends indefinitely downstream) and the fact that the external stream has a constant velocity U lead one to expect that the shape of the curve of the v_x velocity distribution across the boundary layer would be *similar* at all stations in the boundary layer. That is, if one were to plot, at a given x, the dimensionless velocity distribution v_x/U against a dimensionless y based on the layer thickness at x (i.e., y/δ) the result would be identical for all values of x. In other words, one would expect that, relative to the local layer thickness, the velocity would vary from 0 to U in the same way. If this is true, the above partial differential equation may be reduced to an ordinary one by defining a dimensionless coordinate parameter:

$$\eta \sim \frac{y}{\delta}. \tag{7.6}$$

Now, since $\delta \approx \sqrt{\nu x/U}$ as noted in Eq. (7.4), let η be defined as

$$\eta = y\sqrt{\frac{U}{\nu x}}. \tag{7.7}$$

The assumption of similarity of velocity distribution is now expressed as

$$\frac{v_x}{U} = g(\eta), \tag{7.8}$$

where the $g(\eta)$ is some function of $\eta = y\sqrt{U/vx}$ and is now the quantity to be found. Actually, the resulting expressions become simpler when written in term of a second function of η, namely the integral of $g(\eta)$. That is, let

$$f(\eta) = \int g(\eta)\, d\eta$$

so that

$$\frac{v_x}{U} = \frac{df}{d\eta} = f'(\eta). \tag{7.9}$$

The continuity equation, Eq. (7.1), requires that

$$\frac{\partial(v_x/U)}{\partial x} + \frac{\partial(v_y/U)}{\partial y} = 0.$$

Since v_y/U is a function of x and y or, equivalently, η and y, this expression may be written as

$$\frac{d(v_x/U)}{d\eta}\frac{\partial \eta}{\partial x} + \frac{\partial(v_y/U)}{\partial \eta}\frac{\partial \eta}{\partial y} = 0.$$

$$\frac{d(v_x/U)}{d\eta}\left(-\frac{1}{2}\frac{y}{x}\sqrt{\frac{U}{vx}}\right) + \frac{\partial(v_y/U)}{\partial \eta}\sqrt{\frac{U}{vx}} = 0.$$

After substituting Eq. (7.9) for v_x/U, this equation can be solved for v_y/U, giving

$$\frac{v_y}{U} = \frac{1}{2}\sqrt{\frac{v}{Ux}}(\eta f' - f). \tag{7.10}$$

Substitution of Eqs. (7.9) and (7.10) for v_x and v_y into the equation of motion, Eq. (7.2), yields the following *ordinary* differential equation for $f(\eta)$:

$$ff'' + 2f''' = 0. \tag{7.11}$$

In this last equation, the primes denote differentiation with respect to η. Equation (7.11) and its solution were first obtained by Blasius (Ref. 9) in 1908. In terms of the function $f(\eta)$, the boundary conditions of Eq. (7.3) become

$$\text{At } \eta = 0: \quad f = f' = 0.$$

$$\text{As } \eta \to \infty: \quad f' \to 1.$$

The differential equation (7.11) and its boundary conditions do not yield to the usual methods of solution, so numerical methods have been applied. (See, again, references on boundary layer theory such as Ref. 1.) The results of this numerical integration are tabulated in Table 7.1. In this table the solution to Eq. (7.11), $f = f(\eta)$, and its first two derivatives, f' and f'', are tabulated. The velocity functions $v_x/U = f'(\eta)$ and v_y/U given by Eq. (7.10) are also plotted in Figs. 7.2 and 7.3. Also included in Fig. 7.2 are some experimental values of the velocity distribution in a laminar boundary layer for comparison purposes. The good agreement between the theoretical

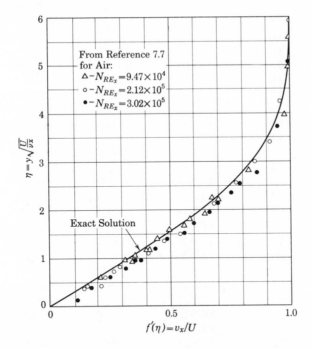

Fig. 7.2. The dimensionless results for the velocity distribution in the laminar boundary layer on a flat plate as given by the exact solution of the boundary layer equations. Experimentally determined values are shown for comparison.

Table 7.1 The Blasius Function for the Laminar Boundary Layer on a Flat Plate

η	f	f'	f''
0	0	0	0.33206
0.2	0.00664	0.06641	0.33199
0.4	0.02656	0.13277	0.33147
0.6	0.05974	0.19894	0.33008
0.8	0.10611	0.26471	0.32739
1.0	0.16557	0.32979	0.32301
1.2	0.23795	0.39378	0.31659
1.4	0.32298	0.45627	0.30787
1.6	0.42032	0.51676	0.29917
1.8	0.52952	0.57477	0.28293
2.0	0.65003	0.62977	0.26675
2.2	0.78120	0.68132	0.24835
2.4	0.92230	0.72899	0.22809
2.6	1.07252	0.77246	0.20646
2.8	1.23099	0.81152	0.18401
3.0	1.39682	0.84605	0.16136
3.2	1.56911	0.87609	0.13913
3.4	1.74696	0.09177	0.11788
3.6	1.92954	0.92333	0.09809
3.8	2.11605	0.94112	0.08013
4.0	2.30576	0.95552	0.06424
4.2	2.49806	0.96696	0.05052
4.4	2.69238	0.97587	0.03897
4.6	2.88826	0.98269	0.02948
4.8	3.08534	0.98779	0.02187
5.0	3.28329	0.99155	0.01591
5.2	3.48189	0.99425	0.01134
5.4	3.68094	0.99616	0.00793
5.6	3.88031	0.99748	0.00543
5.8	4.07990	0.99838	0.00365
6.0	4.27964	0.99898	0.00240
6.2	4.47948	0.99937	0.00155
6.4	4.67938	0.99961	0.00098
6.6	4.87931	0.99977	0.00061
6.8	5.07928	0.99987	0.00037
7.0	5.27926	0.99992	0.00022
7.2	5.47923	0.99996	0.00013
7.4	5.67924	0.99998	0.00007
7.6	5.87924	0.99999	0.00004
7.8	6.07923	1.00000	0.00002
8.0	6.27923	1.00000	0.00001
8.2	6.47923	1.00000	0.00001
8.4	6.67923	1.00000	0
8.6	6.87923	1.00000	0
8.8	7.07923	1.00000	0

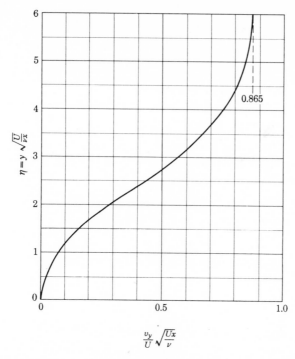

Fig. 7.3. The dimensionless results for the normal velocity component in the laminar boundary layer on a flat plate as given by the exact solution.

and experimental results is apparent and justifies the simplifying assumptions made in the development of the analytical results.

As can be noted in Table 7.1, the v_x velocity component rapidly approaches U, the uniform free stream value, as η increases, but according to the boundary conditions this occurs only as $y \to \infty$ ($\eta \to \infty$). However, for practical purposes it is desirable to know the region in which the viscous effects are most dominant. For this reason it is often customary to designate the boundary layer thickness, δ, as the point at which v_x is 99 per cent of U (i.e., $v_x/U = 0.99$). This magnitude appears to occur at a value of η very nearly equal to 5.0. Hence, it is customary to say that the boundary layer thickness for laminar flow on a flat plate is given by

$$\eta_\delta = 5.0,$$

$$\delta = 5.0\sqrt{\frac{vx}{U}}. \qquad (7.12)$$

Thus, δ does grow as $\sqrt{x}$, as predicted in Eq. (7.4), or the ratio of δ to the distance from the leading edge (in terms of the local Reynolds number based on x) is

$$\frac{\delta}{x} = \frac{5.0}{\sqrt{N_{RE_x}}}. \tag{7.13}$$

Since the velocity distribution through the boundary layer is now known, it is possible to determine the hydrodynamic "drag" on the surface due to the viscosity. This viscous drag may be expressed in terms of the shear stress at the surface:

$$\tau_0 = \mu \left(\frac{\partial v_x}{\partial y} \right)_{y=0}.$$

In terms of the dimensionless parameter $\eta = y\sqrt{U/vx}$ and the function $f(\eta)$, this expression becomes

$$\tau_0 = \mu U \left[\frac{d}{d\eta} \left(\frac{v_x}{U} \right) \right]_{\eta=0} \frac{\partial \eta}{\partial y}$$

$$= \mu U f''(0) \cdot \sqrt{\frac{U}{vx}}.$$

The numerical value of $f''(0)$ is thus seen to be of considerable significance. Table 7.1 shows this value to be

$$f''(0) = 0.33206. \tag{7.14}$$

It is customary to express the wall shear stress in terms of the dimensionless skin friction coefficient, C_f, defined as

$$C_f = \frac{\tau_0}{\frac{1}{2}\rho U^2}. \tag{7.15}$$

The above expression for τ_0 gives

$$C_f = 2f''(0)\sqrt{\frac{v}{Ux}}.$$

Thus, using Eq. (7.14), one obtains

$$C_f = 0.664\sqrt{\frac{v}{Ux}} = \frac{0.664}{\sqrt{N_{RE_x}}}. \tag{7.16}$$

If the plate has a total length L and a width b, the total drag force on one side of the plate is

$$D = b\int_0^L \tau_0\, dx.$$

The *drag coefficient* C_D is defined as

$$C_D = \frac{D}{(\frac{1}{2}\rho U^2)(bL)} = \frac{1}{L}\int_0^L \frac{\tau_0}{\frac{1}{2}\rho U^2}\, dx = \frac{1}{L}\int_0^L C_f\, dx. \tag{7.17}$$

Thus, the drag coefficient is the average value of C_f over the plate length and can be expressed as a function of only the Reynolds number based on the total plate length. When Eq. (7.16) is introduced into Eq. (7.17),

$$C_D = \frac{1}{L}0.664\int_0^L \sqrt{\frac{v}{Ux}}\, dx$$

$$= \frac{1.328}{\sqrt{N_{RE_L}}}. \tag{7.18}$$

In the last equation $N_{RE_L} = UL/v$ is used to denote the *length Reynolds number* based on the total plate length as the characteristic dimension.

The relations just derived will be used later to deduce information concerning the heat transfer from the plate by means of a concept known as *Reynolds' analogy*. This will be discussed in Sec. 7.4.

The Thermal Boundary Layer

If the thermal effects are now considered, it is necessary to employ the energy equation of the boundary layer and to prescribe the boundary conditions for the temperature. If the temperature of the undisturbed fluid is, as shown in Fig. 7.1, equal to t_f and the surface is maintained at t_s, a constant, one must solve the energy equation, Eq. (6.47):

$$v_x\frac{\partial t}{\partial x} + v_y\frac{\partial t}{\partial y} = \frac{k}{\rho c_p}\frac{\partial^2 t}{\partial y^2} + \frac{\mu}{\rho c_p}\left(\frac{\partial v_x}{\partial y}\right)^2$$

along with the conditions

$$\text{At } y = 0: \quad t = t_s.$$

$$\text{At } y = \infty: \quad t = t_f.$$

The above expression may be more conveniently written in terms of a temperature difference variable, θ, defined as

$$\theta = t - t_s. \tag{7.19}$$

If, in addition, the term expressing the heat produced by frictional dissipation is neglected, the system of equations to be solved is

$$v_x \frac{\partial \theta}{\partial x} + v_y \frac{\partial \theta}{\partial y} = \alpha \frac{\partial^2 \theta}{\partial y^2}.$$

$$\text{At } y = 0: \quad \theta = 0. \tag{7.20}$$

$$\text{As } y \to \infty: \quad \theta \to \theta_f = t_f - t_s.$$

The thermal diffusivity has replaced $k/\rho c_p$ and θ_f symbolizes the difference between the plate temperature and the fluid temperature.

At this point it is useful to note that if $\alpha = \nu$, Eq. (7.20) will be identical with those for the velocity boundary layer [see Eqs. (7.2) and (7.3)] if θ is replaced by v_x. This is an extremely useful fact since it implies that the temperature distribution and the velocity distribution are similar and that the velocity and thermal boundary layers are equal in thickness. The relation between the thermal diffusivity α and the kinematic viscosity ν is usually expressed by the Prandtl number, a dimensionless property defined in Sec. 2.7:

$$\text{Prandtl number} = N_{\text{PR}} = \frac{\nu}{\alpha} = \frac{\mu c_p}{k}. \tag{7.21}$$

This number, like the Reynolds number, is an important similarity parameter when one is dealing with problems involving heat convection. As may be noted in Appendix A, many gases have values of N_{PR} not too different from 1 so that the above noted similarity may be assumed to hold in many cases. In this way one may deduce relations for heat transfer from previously determined relations for momentum transport.

Return now to the case in which $N_{\text{PR}} \neq 1$. Let it be assumed, as in the case of the velocity boundary layer, that the temperature distributions at different values of x will be similar with respect to the thermal thickness, δ_t. Then the same similarity parameter may be used, namely $\eta = y\sqrt{U/vx}$. The assumption of similarity of the temperature distribution through the boundary layer at different locations along the plate means that

$$\frac{\theta}{\theta_f} = \varphi(\eta). \tag{7.22}$$

In Eq. (7.22) $\varphi(\eta)$ is a function of η only and must be determined. The foregoing solutions for v_x and v_y are applicable here. In terms of the function $f(\eta)$, now presumed known, Eqs. (7.9) and (7.10) give

$$v_x = Uf',$$

$$v_y = \frac{1}{2}\sqrt{\frac{Uv}{x}}(\eta f' - f).$$

Substitution of these facts and Eq. (7.22) for θ into Eq. (7.20), the differential equation of the boundary layer, gives

$$\frac{d^2\varphi}{d\eta^2} + \frac{1}{2}N_{PR}f(\eta)\frac{d\varphi}{d\eta} = 0.$$

At $\eta = 0$: $\varphi = 0$.   (7.23)

As $\eta \to \infty$: $\varphi \to 1$.

The solution to this equation, due to Pohlhausen (Ref. 2), may be obtained by noting that it is a linear, first order differential equation in $d\varphi/d\eta$. The solution is

$$\frac{t - t_s}{t_f - t_s} = \frac{\theta}{\theta_f} = \varphi(\eta) = 1 - \frac{\int_\eta^\infty (f'')^{N_{PR}}\,d\eta}{\int_0^\infty (f'')^{N_{PR}}\,d\eta}. \tag{7.24}$$

With $f(\eta)$ known (at least in numerical form) from the solution for the velocity boundary layer, Eq. (7.24) may be solved (numerically again) for the distribution of the temperature. These results are, of course, dependent on the Prandtl number N_{PR}. These results are shown graphically in Fig. 7.4 for various values of N_{PR}. The curve for $N_{PR} = 0.7$ is typical for air (and many other gases).

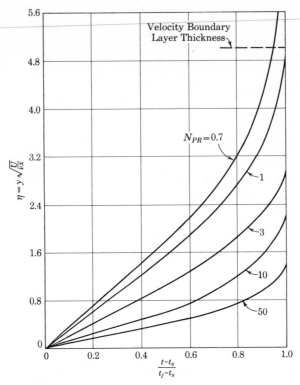

Fig. 7.4. Temperature distributions in the laminar boundary layer on a heated flat plate (dissipated energy neglected).

For the special case in which $N_{PR} = 1$ the above solution for $\varphi(\eta)$ becomes

$$\varphi(\eta) = 1 - \frac{[f']_\eta^\infty}{[f']_0^\infty}$$

$$= 1 - \frac{[v_x/U]_\eta^\infty}{[v_x/U]_0^\infty}$$

$$= 1 - \frac{1 - v_x/U}{1 - 0}$$

$$= \frac{t - t_s}{t_f - t_s} = \frac{v_x}{U}. \tag{7.25}$$

This calculation shows that, as predicted, for $N_{PR} = 1$ the thermal layer and velocity layer have equal thicknesses. Also, the dimensionless velocity distribution v_x/U is identical with the dimensionless temperature distribution

$(t - t_s)/(t_f - t_s)$. Figure 7.4 shows that for $N_{PR} > 1$ the thermal layer is thinner than the velocity layer, whereas for $N_{PR} < 1$ the reverse is true.

The Film Coefficient and the Nusselt Number

Of even greater use than the complete distribution of the temperature through the boundary layer is the gradient of this temperature distribution at the wall. This is true since this gradient will lead to an expression for the surface film coefficient, h. From Eq. (7.24), this gradient is

$$\varphi'(0) = \frac{\left(\dfrac{\partial t}{\partial y}\right)_{y=0}}{t_f - t_s}\sqrt{\frac{vx}{U}} = \frac{[f''(0)]^{N_{PR}}}{\displaystyle\int_0^\infty [f''(\eta)]^{N_{PR}}\,d\eta}$$

$$= \varphi_0'(N_{PR}). \tag{7.26}$$

The parameter $f''(0)$ noted in Eq. (7.14) is thus seen to be of significance again. The notation $\varphi_0'(N_{PR})$ has been introduced to emphasize the fact that the wall temperature gradient, $\varphi'(0)$, is a function of only the fluid Prandtl number. Table 7.2 shows numerical values of $\varphi_0'(N_{PR})$ for various values of N_{PR}.

By definition, the surface film coefficient is

$$h_x = -\frac{k\left(\dfrac{\partial t}{\partial y}\right)_{y=0}}{t_s - t_f} = k\varphi_0'(N_{PR})\sqrt{\frac{U}{vx}}. \tag{7.27}$$

Table 7.2   Laminar Forced Convection Functions*

N_{PR}	$\varphi_0'(N_{PR})$	$0.332(N_{PR})^{1/3}$	$\Psi_0(N_{PR})$	$N_{PR}^{1/2}$
0.6	0.276	0.280	0.770	0.775
0.7	0.293	0.295	0.835	0.837
0.8	0.307	0.308	0.895	0.894
0.9	0.320	0.321	0.950	0.949
1.0	0.332	0.332	1.000	1.000
1.1	0.344	0.343	1.050	1.049
7.0	0.645	0.635	2.515	2.646
10.0	0.730	0.715	2.865	3.162
15.0	0.835	0.819	3.535	3.873

* From H. SCHLICHTING, *Boundary Layer Theory*, 6th ed., New York, McGraw-Hill, 1968.

Equation (7.27) shows that the film coefficient varies with distance along the plate. Thus, it becomes necessary to identify the coefficient in Eq. (7.27) as a *local* film coefficient—hence the notation h_x. In many applications it is useful to employ an average heat transfer coefficient h over the whole plate length L:

$$h = \frac{1}{L} \int_0^L h_x \, dx$$

$$= \frac{k}{L} 2\varphi_0'(N_{PR}) \sqrt{\frac{UL}{v}}. \tag{7.28}$$

It is convenient, and more general, to nondimensionalize Eqs. (7.27) and (7.28), for the local and average film coefficients, by introduction of the *Nusselt number*:

$$N_{NU} = \frac{hl}{k}. \tag{7.29}$$

In Eq. (7.29) l is a characteristic geometric dimension of the system under consideration. The Nusselt number was discussed previously in Sec. 6.8.

By using N_{NU_x} to denote the *local* Nusselt number based on the local film coefficient h_x and the local x coordinate, and using N_{NU_L} to denote the *average* Nusselt number based on the average h and the total plate length, Eqs. (7.27) and (7.28) are expressed:

$$N_{NU_x} = \frac{h_x x}{k} = \varphi_0'(N_{PR}) \sqrt{\frac{Ux}{v}},$$

$$N_{NU_L} = \frac{hL}{k} = 2\varphi_0'(N_{PR}) \sqrt{\frac{UL}{v}}. \tag{7.30}$$

N_{RE_x} denotes the local Reynolds number based on the coordinate x, and N_{RE_L} denotes the length Reynolds number based on the total plate length.

Equations (7.30) may be used, as they stand, for determination of the film coefficients; however, the functional relation for $\varphi_0'(N_{PR})$ must be available—such as given in Table 7.2. A rather accurate approximation to this relation is

$$\varphi_0'(N_{PR}) \approx 0.332(N_{PR})^{1/3}. \tag{7.31}$$

Data given in Table 7.2 show numerical values of Eq. (7.31) compared with $\varphi_0'(N_{PR})$. Thus, for engineering applications, Eqs. (7.30) may be written as

$$N_{NU_x} = 0.332(N_{PR})^{1/3}(N_{RE_x})^{1/2},$$

$$N_{NU_L} = 0.664(N_{PR})^{1/3}(N_{RE_L})^{1/2}. \tag{7.32}$$

Equations (7.32) express the fact predicted in Sec. 6.8—that when viscous dissipation is neglected in forced convection, it may be expected that the result will take the form $N_{NU} = f(N_{PR}, N_{RE})$.

7.3 The Solution of Laminar Forced Convection on a Flat Plate by Use of the Integral Momentum and Energy Equations of the Boundary Layer

An alternative approach to the problem just solved is by means of the integral momentum and energy equations derived in Sec. 6.12.

For flow past a flat plate in which $U(x) = U$ (a constant) and $dP/dx = 0$, the integral momentum and energy equations [Eqs. (6.52) and (6.54)] become

$$\frac{d}{dx}\int_0^\delta v_x(v_x - U)\,dy = -\frac{\tau_0}{\rho} = -\nu\left(\frac{\partial v_x}{\partial y}\right)_{y=0},$$

$$\frac{d}{dx}\int_0^{\delta_t} v_x(t - t_f)\,dy = -\alpha\left(\frac{\partial t}{\partial y}\right)_{y=0}.$$

(7.33)

These equations will now be applied to illustrate this alternative approach to the determination of the velocity and temperature distribution (and the resultant heat transfer coefficient) in the laminar boundary layer on a flat plate. The results can then be compared with those obtained above by the first approach. It should be remembered that the results obtained in Sec. 7.2 represent an exact solution to approximate equations, whereas the method described here will yield an approximate solution to approximate equations.

The Velocity Boundary Layer

The method of application of the momentum integral equation involves, usually, the assumption that the velocity profiles are similar at different stations, x. That is, as in the above analysis, a similarity parameter, η, is defined as

$$\eta = \frac{y}{\delta}.$$ (7.34)

It should be noted that the η defined in Eq. (7.34) is *not* the same similarity parameter that was used in the analysis of Sec. 7.2. The boundary layer thickness, δ, is presumed to be a finite quantity—although it is a function of x, the distance along the plate. Also it is assumed that the dimensionless velocity profile, v_x/U, is a function of η only:

$$\frac{v_x}{U} = g(\eta). \tag{7.35}$$

Introducing this definition of $g(\eta)$ into the first equation of Eq. (7.33), the integral momentum equation, one obtains

$$\frac{d}{dx} \int_0^\delta \frac{v_x}{U}\left(\frac{v_x}{U} - 1\right) dy = -\frac{\nu}{U}\left[\frac{\partial(v_x/U)}{\partial y}\right]_{y=0},$$

$$\frac{d}{dx}\left\{\delta \int_0^1 g(\eta)[g(\eta) - 1]\, d\eta\right\} = -\frac{\nu}{U\delta}\left(\frac{dg}{d\eta}\right)_{\eta=0}. \tag{7.36}$$

If the shape of the velocity distribution profile, $g(\eta)$, is known, Eq. (7.36) results in a differential equation for δ as a function of x. The main advantage of this integral approach is that integration tends to suppress errors. That is, even if a velocity profile shape, $g(\eta)$, is *assumed* which differs from the correct result, the resultant integral will not differ as much from the correct integral. Experience shows that even rather crude guesses for $g(\eta)$ may yield very reasonable and usable results for δ and the associated film coefficient h. Customarily one represents the unknown velocity distribution as a polynomial in η,

$$\frac{v_x}{U} = g(\eta) = a + b\eta + c\eta^2 + d\eta^3 + \cdots. \tag{7.37}$$

As many terms may be taken as one desires—depending on what accuracy is wanted and how much algebraic labor can be tolerated. If, for instance, a third-degree polynomial is used, one has four constants to be determined. These may be fixed by application of conditions at the wall ($\eta = 0$) and at the outer limit of the boundary layer ($\eta = 1$).

At the outer limit of the boundary layer, $\eta = 1$, it is known that $v_x = U$, or $g(\eta) = 1$. Also for a smooth transition from the boundary layer to the potential region, all the derivatives of v_x with respect to y must vanish. This means that all the derivatives of $g(\eta)$ with respect to η are zero at $\eta = 1$ [i.e., $g'(1) = 0$, $g''(1) = 0, \ldots$].

At the solid surface $v_x = 0$, so $g(0) = 0$. The first derivative of v_x with respect to y at $y = 0$ determines the viscous shear stress there. Thus, $g'(0)$ may not be specified. Now, the boundary layer equation of motion, for zero pressure gradient, is [Eq. (7.2)]

$$v_x \frac{\partial v_x}{\partial x} + v_y \frac{\partial v_x}{\partial y} = \nu \frac{\partial^2 v_x}{\partial y^2}.$$

At the surface $v_x = v_y = 0$, and thus

$$\left(\frac{\partial^2 v_x}{\partial y^2}\right)_{y=0} = 0, \qquad \text{or} \qquad g''(0) = 0.$$

Differentiation of the equation of motion and evaluation at $y = 0$ will show that all the higher derivatives of $g(\eta)$ are also zero for $\eta = 0$.

These conditions at $\eta = 0$ and $\eta = 1$ may be employed to evaluate the constants of the polynomial for $g(\eta)$. As many will be employed as are required by the degree of the polynomial. The use of a polynomial as an *approximation* for $g(\eta)$ means, of course, that *all* the boundary conditions cannot be satisfied. It is best to satisfy as many conditions at the plate surface as at the outer limit of the boundary layer—alternating between one and the other as higher degree polynomials are used. Summarizing, the available conditions are

$$
\left.
\begin{array}{l}
g(1) = 1 \\
g(0) = 0
\end{array}
\right\} \text{first degree}
\left.
\begin{array}{l}
 \\
 \\
g'(1) = 0 \\
g''(0) = 0
\end{array}
\right\} \text{second degree}
\left.
\begin{array}{l}
 \\
 \\
 \\
 \\
g''(1) = 0 \\
g'''(0) = 0 \\
\vdots
\end{array}
\right\} \text{third degree.}
\tag{7.38}
$$

Considering, as an example, a third-degree polynomial

$$\frac{v_x}{U} = a + b\eta + c\eta^2 + d\eta^3,$$

one finds that the application of the first four conditions of Eq. (7.38) gives

$$1 = a + b + c + d,$$

$$0 = a,$$

$$0 = b + 2c + 3d,$$

$$0 = 2c.$$

Thus, $a = 0$, $b = \frac{3}{2}$, $c = 0$, and $d = -\frac{1}{2}$, so

$$\frac{v_x}{U} = g(\eta) = \frac{3}{2}\eta - \frac{1}{2}\eta^3. \tag{7.39}$$

This equation is, then, the assumed form of the velocity distribution through the boundary layer. It is, of course, approximate, but comparison with the exact solution of Blasius (Sec. 7.2) will show a close correspondence. Figure 7.5 plots the velocity distribution given by the exact solution (Table 7.1) with that given by Eq. (7.39).

Substitution of Eq. (7.39) into the momentum integral equation, Eq. (7.36), gives the following differential equation in δ:

$$\frac{d}{dx}\left[\delta \int_0^1 \left(\frac{1}{4}\eta^6 - \frac{6}{4}\eta^4 + \frac{1}{2}\eta^3 + \frac{9}{4}\eta^2 - \frac{3}{2}\eta\right) d\eta\right] = -\frac{v}{U\delta}\frac{3}{2},$$

$$\frac{d}{dx}\left(\frac{39}{280}\delta\right) = \frac{v}{U\delta}\frac{3}{2},$$

$$\delta\, d\delta = \frac{140}{13}\frac{v}{U}\, dx.$$

Since the boundary layer has zero thickness at the leading edge of the flat plate—where $x = 0$—the integration of the above equation gives

$$\delta^2 = \frac{280}{13}\frac{vx}{U}.$$

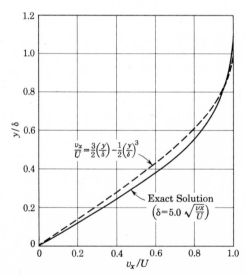

Fig. 7.5. Comparison of the velocity distribution for a laminar boundary layer on a flat plate as given by the exact solution and the momentum integral method.

Thus, the boundary layer thickness is given as the following function of x:

$$\delta = \sqrt{\frac{280}{13}}\sqrt{\frac{vx}{U}} = 4.64\sqrt{\frac{vx}{U}},$$

$$\frac{\delta}{x} = \sqrt{\frac{280}{13}}\sqrt{\frac{v}{Ux}} = \frac{4.64}{\sqrt{N_{RE_x}}}. \tag{7.40}$$

Comparison of these results with Eqs. (7.12) and (7.13)—the "exact" solution of the problem—again shows a close correspondence between the two methods.

The Thermal Boundary Layer and the Heat Transfer Coefficient

Turning now, as before, to the thermal boundary layer, one can apply the results just obtained to the integral energy equation to determine the temperature distribution in the fluid and, eventually, the heat transfer coefficient. Introducing, as before, the temperature difference parameter, one obtains

$$\theta = t - t_s,$$

and upon assuming similarity of the temperature difference distribution through the boundary layer, the problem reduces to the finding of a function of only a single variable. In this case η_t is the single variable, and it is defined as

$$\eta_t = \frac{y}{\delta_t}. \tag{7.41}$$

Here δ_t is the thermal layer thickness. Similarity of the temperature distributions implies the existence of a function φ:

$$\varphi(\eta_t) = \frac{\theta}{\theta_f} = \frac{t - t_s}{t_f - t_s}.$$

The energy equation [the second equation of Eqs. (7.33)] may now be rewritten in terms of these new variables, η_t and φ:

$$\frac{d}{dx}\left[\delta_t \int_0^1 g(\eta)[1 - \varphi(\eta_t)]\, d\eta_t\right] = \frac{\alpha}{U\delta_t}\left(\frac{d\varphi}{d\eta_t}\right)_{\eta_t = 0}. \tag{7.42}$$

In Eq. (7.42) the velocity distribution $v_x/U = g(\eta)$, from the above analysis of the velocity boundary layer, has been introduced. It will be remembered that in Sec. 6.12, in which the intergal energy equation was derived, it was assumed that the thermal boundary layer was thinner than the velocity boundary layer (i.e., $\delta_t < \delta$). This implies that $N_{PR} > 1$, if the plate is heated to t_s over its entire length from the leading edge.

It is reasonable to approximate the temperature distribution in the boundary layer with a polynomial of the same degree as was used for the velocity layer. In this case this is a third-degree polynomial. Hence, let

$$\frac{\theta}{\theta_f} = \varphi(\eta_t) = a + b\eta_t + c\eta_t^2 + d\eta_t^3.$$

By the same reasoning used for the velocity boundary layer, the following conditions may be applied to determine the constants $a, b, c,$ and d:

At $y = \delta_t, t = t_f,$ or at $\eta_t = 1, \varphi = 1.$

At $y = 0, \ t = t_s,$ or at $\eta_t = 0, \varphi = 0.$

At $y = \delta_t, \dfrac{\partial t}{\partial y} = 0,$ or at $\eta_t = 1, \dfrac{d\varphi}{d\eta} = 0.$

At $y = 0, \dfrac{\partial^2 t}{\partial y^2} = 0,$ or at $\eta_t = 0, \dfrac{d^2\varphi}{d\eta^2} = 0.$

These conditions determine the constants to be $a = 0, b = \frac{3}{2}, c = 0, d = -\frac{1}{2},$ so

$$\varphi(\eta_t) = \frac{3}{2}\eta_t - \frac{1}{2}\eta_t^3. \tag{7.43}$$

This expression for $\varphi(\eta_t)$ and the previously found expression for $g(\eta)$ in the velocity boundary layer may now be introduced into Eq. (7.42) to give

$$\frac{d}{dx}\left[\delta_t \int_0^1 \left(\frac{3}{2}\eta - \frac{1}{2}\eta^3\right)\left(1 - \frac{3}{2}\eta_t + \frac{1}{2}\eta_t^3\right) d\eta_t\right] = \frac{\alpha}{U\delta_t}\frac{3}{2}.$$

It is convenient to define a variable (a function of x only) ζ as the ratio of the thermal layer thickness to the velocity layer thickness:

$$\zeta = \frac{\delta_t}{\delta} = \frac{\eta}{\eta_t}. \tag{7.44}$$

Introducing this fact into the equation above, one obtains

$$\frac{d}{dx}\left[\delta_t\zeta\int_0^1\left(\frac{3}{2}\eta_t - \frac{1}{2}\zeta^2\eta_t^3\right)\left(1 - \frac{3}{2}\eta_t + \frac{1}{2}\eta_t^3\right)d\eta_t\right] = \frac{\alpha}{U\delta_t}\frac{3}{2}.$$

Upon evaluation of the integral this expression becomes

$$\frac{d}{dx}\left[\delta_t\zeta\left(\frac{3}{20} - \frac{3}{280}\zeta^2\right)\right] = \frac{\alpha}{U\delta_t}\frac{3}{2}.$$

A rather far-reaching simplification can be made at this point if one accepts the fact that ζ, the ratio of the boundary layer thicknesses, will be near 1. This will be true, as seen in the previous sections, if the Prandtl number, N_{PR}, is close to 1—a fact which is substantially met for a great number of fluids. If this is assumed, then $\frac{3}{280}$ can be neglected when compared with $\frac{3}{20}$, so the equation above becomes, approximately,

$$\frac{d}{dx}\left(\delta_t\zeta\frac{3}{20}\right) = \frac{3}{2}\frac{\alpha}{U\delta_t}.$$

Since $\delta_t = \delta\zeta$, this equation may be written as

$$\frac{d}{dx}\delta\zeta^2 = 10\frac{\alpha}{U\zeta\delta}.$$

Now, δ is a function of only x, and this functional relation has already been determined in the analysis of the velocity boundary layer, Eq. (7.40), as

$$\delta = \sqrt{\frac{280}{13}}\sqrt{\frac{vx}{U}}.$$

Thus, the above differential equation for δ is expressible as

$$\frac{d}{dx}(x^{1/2}\zeta^2) = \frac{13}{28}\frac{\alpha}{v}\frac{1}{\zeta x^{1/2}},$$

or

$$\zeta^3 + \frac{4}{3}x\frac{d}{dx}(\zeta^3) = \frac{13}{14}\frac{\alpha}{v}.$$

This latter expression is a first order, linear equation in ζ^3 which has the solution

$$\zeta^3 = \frac{13}{14}\frac{\alpha}{\nu} + Cx^{-3/4}.$$

The constant C in this equation must be zero to avoid an indeterminate solution at the leading edge, $x = 0$, so

$$\zeta = \left(\frac{13}{14}\frac{\alpha}{\nu}\right)^{1/3} \approx \left(\frac{\alpha}{\nu}\right)^{1/3}.$$

This is recognized as involving the Prandtl number, $N_{PR} = \nu/\alpha$, so

$$\zeta = \frac{1}{(N_{PR})^{1/3}}. \tag{7.45}$$

If the Prandtl number is close to 1, then $\zeta \approx 1$, as was assumed above when $\frac{3}{280}\zeta^2$ was neglected in comparison with $\frac{3}{20}$.

It is now possible to evaluate the heat transfer coefficient since

$$\left(\frac{\partial t}{\partial y}\right)_{y=0} = (t_f - t_s)\left(\frac{d\varphi}{d\eta}\right)_{\eta_t = 0}\frac{1}{\delta_t}$$

$$= (t_f - t_s)\frac{3}{2}\frac{1}{\delta_t}.$$

By using the same notation as before for the local heat transfer coefficient, h_x, Eq. (7.27) gives

$$h_x = -\frac{k\left(\dfrac{\partial t}{\partial y}\right)_{y=0}}{t_s - t_f}$$

$$= \frac{3}{2}\frac{k}{\delta_t}$$

$$= \frac{3}{2}\frac{k}{\delta\zeta}.$$

But δ and ζ are known from Eqs. (7.40) and (7.45), so

$$h_x = \frac{3}{2}k\sqrt{\frac{13}{280}}\sqrt{\frac{U}{vx}}(N_{PR})^{1/3}$$

$$h_x = 0.331k\sqrt{\frac{U}{vx}}(N_{PR})^{1/3} \tag{7.46}$$

$$N_{NU_x} = 0.331(N_{RE_x})^{1/2}(N_{PR})^{1/3}.$$

The average film coefficient, h, and Nusselt number for a plate of length L are, then,

$$h = 0.662k\sqrt{\frac{U}{vL}}(N_{PR})^{1/3},$$

$$N_{NU_L} = 0.662(N_{RE_L})^{1/2}(N_{PR})^{1/3}. \tag{7.47}$$

The close correspondence between these relations and those obtained in Sec. 7.2 using the differential boundary layer equations of motion [see Eqs. (7.27) through (7.32)] is rather striking. Figure 7.6 shows a plot of

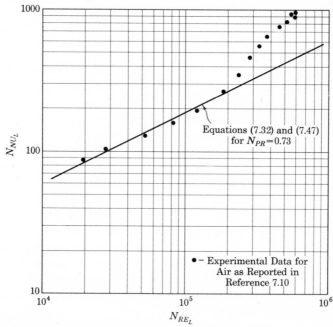

Fig. 7.6. Comparison of theory and experiment for heat transfer in laminar flow of air past a flat plate.

Eq. (7.32) or (7.47) for the Nusselt number. This plot is compared with the experimentally measured values of Ref. 10 for air. The close agreement between theory and experiment is apparent.

The discussions of this last section were presented mainly to illustrate the basic thoughts behind the methods of determining suitable descriptions of the behavior of boundary layer flow and the associated heat convection mechanism. Each of the two methods has various advantages over the other—the first approach attempts to satisfy exactly the boundary layer governing equations, whereas the second approach attempts to give an over-all picture of the processes and to satisfy conditions only at the limits of the boundary layer. The integral method may be made more exact by using polynomials of higher degree, but the labor of computation increases therewith. For cases other than flow past a flat plate the "exact" approach becomes very complex—and often impossible—whereas the integral method will usually yield a usable result. For certain very complex geometrical configurations neither method may be of value, and some other attack must be made to obtain suitable engineering answers. This is usually done by experimentation—using the dimensionless parameters developed in Sec. 6.8.

7.4 Reynolds' Analogy for Laminar Flow

It was noted in Sec. 7.2 that Eq. (7.2) for the velocity in the laminar boundary layer on a flat surface was identical in form to Eq. (7.20) for the temperature in the boundary layer if the kinematic viscosity, v, and the thermal diffusivity, α, of the fluid were the same (i.e., $N_{PR} = v/\alpha = 1$). Under this condition the boundary conditions were also identical in form. The implied similarity between the exchange of momentum and the exchange of heat in laminar motion was first recognized by Reynolds, and for that reason it has been termed "Reynolds' analogy." It is the purpose of this section to make a formal presentation of Reynolds' analogy.

Consider the laminar motion of a boundary layer past a flat surface as shown in Fig. 7.7. As usual, U and t_f denote the velocity and temperature of the undisturbed fluid stream, whereas t_s denotes the temperature of the plate—assumed to be constant. For a Prandtl number $N_{PR} = 1$, the equations of motion and energy of the boundary layer show that the dimensionless distribution of the velocity is identical to the dimensionless distribution of the temperature. This was shown to be the case in the discussion following Eq. (7.24). If $y = y_r$ denotes a reference plane in the boundary layer that is parallel to the plate, then the velocity and temperature distribution through the layer may be nondimensionalized in terms of the values at that plane —say v_r and t_r. Reynolds' analogy then implies that the distribution $(v_x - v_r)/(U - v_r)$ versus y is identical to the distribution of $(t - t_r)/(t_f - t_r)$.

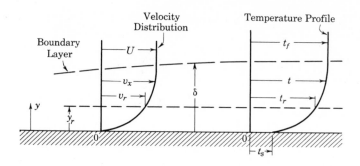

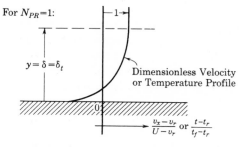

Fig. 7.7. Reynolds' analogy for laminar flow with $N_{PR} = 1$.

This means that

$$\frac{d}{dy}\left(\frac{v_x - v_r}{U - v_r}\right) = \frac{d}{dy}\left(\frac{t - t_r}{t_f - t_r}\right).$$

Thus,

$$\frac{1}{U - v_r}\frac{dv_x}{dy} = \frac{1}{t_f - t_r}\frac{dt}{dy}. \tag{7.48}$$

Now, in laminar flow the shear stress at the plane denoted by $y = y_r$ is

$$\tau_r = \mu\frac{dv_x}{dy} = \rho v\frac{dv_x}{dy},$$

and the heat flux across $y = y_r$ is

$$\frac{q_r}{A} = -k\frac{dt}{dy} = -\rho \alpha c_p\frac{dt}{dy}.$$

If the last two expressions are introduced into Eq. (7.48), one has

$$\frac{1}{U - v_r} \frac{\tau_r}{v} = \frac{q_r/A}{\alpha c_p} \frac{1}{t_r - t_f}.$$

Since $N_{PR} = v/\alpha = 1$ for the assumed analogy between momentum and heat exchange, the last equation becomes

$$\frac{\tau_r c_p}{U - v_r} = \frac{q_r/A}{t_r - t_f}. \tag{7.49}$$

This equation may be taken as a statement of Reynolds' analogy for laminar flow of a fluid with $N_{PR} = 1$. It expresses an equivalence between the viscous shear and the heat flux across a plane ($y = y_r$) in the laminar boundary layer—in terms of the velocity and temperature of the free stream.

Since the heat transfer coefficient is based on the heat flux evaluated at the surface of the plate, let $y_r = 0$ so that $t_r = t_s$ (plate surface temperature) and $v_r = 0$. Then Eq. (7.49) gives

$$\frac{\tau_0 c_p}{U} = \frac{q_0/A}{t_s - t_f}.$$

The right side of the above equation is recognized as the heat transfer coefficient h, so

$$h = \frac{\tau_0 c_p}{U}. \tag{7.50}$$

Equation (7.50) implies that for fluids of unity Prandtl number, the heat transfer coefficient may be simply determined from the shear stress at the plate surface. Thus, one could deduce probable values of h from relatively simple drag force measurements.

If Eq. (7.50) is written in a nondimensional form, the local Nusselt number is given by

$$N_{NU_x} = \frac{h_x x}{k} = \frac{\tau_0 x c_p}{kU}.$$

Since $N_{PR} = 1$, then $v = \alpha = k/\rho c_p$. Thus,

$$N_{NU_x} = \frac{\tau_0}{\rho U^2} \frac{Ux}{v}.$$

In terms of the local skin friction coefficient, C_f, defined in Eq. (7.15), N_{NU_x} becomes

$$N_{NU_x} = \tfrac{1}{2}C_f N_{RE_x}. \tag{7.51}$$

Equation (7.51) is a dimensionless statement of Reynolds' analogy for laminar flow.

In Sec. 7.2 it was shown that [see Eq. (7.16)] for laminar flow on a flat plate,

$$C_f = 0.664(N_{RE_x})^{-1/2}.$$

For this case Reynolds' analogy, Eq. (7.51), gives

$$N_{NU_x} = 0.332(N_{RE_x})^{1/2}.$$

The result obtained from the exact solution of the boundary layer equations is given in Eq. (7.32):

$$N_{NU_x} = 0.332(N_{RE_x})^{1/2}(N_{PR})^{1/3}.$$

The result obtained by Reynolds' analogy is clearly a special case of this for $N_{PR} = 1$. It appears that the effect of the Prandtl number differing from unity is expressed by a weighting factor equal to $(N_{PR})^{1/3}$. The latter fact is sometimes arbitrarily applied in other cases when an exact solution to the thermal boundary layer cannot be obtained, and experimental skin friction measurements are used to predict heat transfer coefficients.

7.5   Aerodynamic Heating—Laminar Forced Convection, with Viscous Dissipation, on a Flat Surface

The examples treated in Secs. 7.2, 7.3, and 7.4 have all been instances in which it was presumed that the rate of viscous dissipation was small enough, when compared with the other terms of the energy equation, that it could be neglected. Such a situation could be expected to occur at relatively low values of the free stream velocity U. However, situations occur in which the velocity U is large enough that the dissipation term must be included even though the flow is still laminar. This section will consider such a case—exactly the same problem as that posed in Sec. 7.2, except that the dissipation term will be retained in the energy equation. The solution to this problem is best described and understood if a simpler, subproblem, is considered first.

Laminar Flow Past an Adiabatic Surface, the Recovery Temperature

First, rather than treat the case in which the surface temperature is fixed and the surface temperature gradient is sought, consider instead the case in which the surface temperature gradient is fixed as zero and the resultant surface temperature is sought.

Consider, then, a stream at velocity U, temperature t_f, flowing past a flat plate which is otherwise thermally isolated. After sufficient time has elapsed, the plate will reach an equilibrium temperature, and at the surface $(\partial t/\partial y)_{y=0}$ will be zero, since no heat will flow into or out of the plate when such equilibrium is reached. This equilibrium temperature, called the "recovery temperature," t_r, is to be found. Under these conditions, the energy equation, given in Eq. (6.47), must be solved under the conditions given below:

$$v_x \frac{\partial t}{\partial x} + v_y \frac{\partial t}{\partial y} = \alpha \frac{\partial^2 t}{\partial y^2} + \frac{\mu}{\rho c_p} \left(\frac{\partial v_x}{\partial y} \right)^2.$$

$$\text{At } y = 0: \quad \frac{\partial t}{\partial y} = 0. \tag{7.52}$$

$$\text{As } y \to \infty: \quad t \to t_f.$$

As before, introduce the similarity parameter

$$\eta = y \sqrt{\frac{U}{vx}},$$

so that

$$\frac{v_x}{U} = f'(\eta),$$

$$\frac{v_y}{U} = \frac{1}{2} \sqrt{\frac{v}{Ux}} (\eta f' - f),$$

as in the Blasius solution of Sec. 7.2. If, in addition, a dimensionless temperature parameter, Ψ, is defined

$$\Psi = \frac{t - t_f}{U^2/2c_p}, \tag{7.53}$$

Eq. (7.52) may be transformed into the following form if Ψ is assumed to depend upon η only:

$$\frac{d^2\Psi}{d\eta} + \frac{N_{\mathrm{PR}}}{2} f(\eta) \frac{d\Psi}{d\eta} = -2N_{\mathrm{PR}}[f''(\eta)]^2,$$

$$\Psi(\infty) = 0, \tag{7.54}$$

$$\Psi'(0) = 0.$$

The solution to the system given in Eq. (7.54) may be shown to be

$$\Psi(\eta) = 2N_{\mathrm{PR}} \int_{\eta}^{\infty} (f'')^{N_{\mathrm{PR}}} \left\{ \int_{0}^{\eta} [f''(\zeta)]^{2-N_{\mathrm{PR}}} \, d\zeta \right\} d\eta, \tag{7.55}$$

where ζ is a dummy integration variable. Since it is the equilibrium surface temperature, t_r, which is sought,

$$\frac{t_r - t_f}{U^2/2c_p} = \Psi(0) = 2N_{\mathrm{PR}} \int_{0}^{\infty} (f'')^{N_{\mathrm{PR}}} \left[\int_{0}^{\eta} [f''(\zeta)]^{2-N_{\mathrm{PR}}} \, d\zeta \right] d\eta$$

$$= \Psi_0(N_{\mathrm{PR}}). \tag{7.56}$$

The notation $\Psi_0(N_{\mathrm{PR}})$ is introduced to point out the fact that $\Psi(0)$ is a function of *only* the Prandtl number. $\Psi_0(N_{\mathrm{PR}})$ is a constant for a given fluid, and values of it are tabulated for various N_{PR} in Table 7.2.

Since $\Psi_0(N_{\mathrm{PR}}) \geq 0$, the recovery temperature is always greater than the ambient free stream. The difference $(t_r - t_f)$ is a measure of the rate of viscous dissipation in the boundary layer. The rise of t_r above t_f depends upon only the fluid and its velocity. For air, $N_{\mathrm{PR}} \cong 0.7$ and $c_p \cong 0.24$ Btu/lb$_{\mathrm{m}}$-°R. Thus, Table 7.2 gives $\Psi_0(N_{\mathrm{PR}}) = 0.835$, so that the values of the temperature rise, $t_r - t_f$, shown in Table 7.3, result. The dissipative effect of viscosity at high velocities is readily apparent.

The quantity $\Psi_0(N_{\mathrm{PR}})$ is often referred to as the *recovery factor*. As an approximate relation, it may be observed that the dependence of Ψ_0 on N_{PR} is rather accurately given by

$$\Psi_0(N_{\mathrm{PR}}) \approx N_{\mathrm{PR}}^{1/2}. \tag{7.57}$$

Table 7.3   Recovery Temperature in
Air Due to Viscous Dissipation

Air velocity, U, ft/sec	$t_r - t_f$, °F
50	0.2
100	0.7
200	2.8
400	11.1
500	17.4
600	25.0
1000	69.5
2000	278

Laminar, Viscous, Flow Past a Heated or Cooled Flat Plate

Return now to the original problem of this section—what would be the expected heat transfer (i.e., temperature gradient) at the surface of a flat plate when viscous dissipation is accounted for and when the plate is maintained at a temperature, t_s? The system of equations to be solved is

$$v_x \frac{\partial t}{\partial x} + v_y \frac{\partial t}{\partial y} = \alpha \frac{\partial^2 t}{\partial y^2} + \frac{\mu}{\rho c_p} \left(\frac{\partial v_x}{\partial y} \right)^2.$$

$$\text{At } y = 0: \quad t = t_s. \tag{7.58}$$

$$\text{As } y \to \infty: \quad t \to t_f.$$

Once again, the Blasius parameter $\eta = y\sqrt{U/vx}$ may be introduced so that v_x and v_y are known functions—according to Eqs. (7.9) and (7.10). If the dimensionless temperature ratio $(t - t_f)/(t_s - t_f)$ is assumed to depend upon η, only, the solution of Eq. (7.58) may be shown (Ref. 1) to be

$$\frac{t - t_f}{t_s - t_f} = [1 - \varphi(\eta)] \left[1 - \frac{\Psi_0(N_{PR})}{2} \frac{U^2}{c_p(t_s - t_f)} \right] + \frac{1}{2} \Psi(\eta) \frac{U^2}{c_p(t_s - t_f)}. \tag{7.59}$$

In Eq. (7.59) the functions $\varphi(\eta)$ and $\Psi(\eta)$ are the functions given in Eqs. (7.24) and (7.55). Thus, the temperature distribution may be calculated for any given set of conditions. However, for heat transfer calculations, the temperature gradient at the surface is desired.

Equation (7.59) yields

$$\left(\frac{dt}{d\eta}\right) \Big/ (t_s - t_f) = -\varphi'(\eta)\left[1 - \frac{1}{2}\Psi_0(N_{PR})\frac{U^2}{c_p(t_s - t_f)}\right] + \frac{\Psi'(\eta)}{2}\frac{U^2}{c_p(t_s - t_f)}.$$

Since at $\eta = 0$, $\varphi'(0) = \varphi'_0(N_{PR})$ and $\Psi'(0) = 0$,

$$\left(\frac{dt}{d\eta}\right)_0 = -(t_s - t_f)\varphi'_0(N_{PR})\left[1 - \frac{1}{2}\Psi_0(N_{PR})\frac{U^2}{c_p(t_s - t_f)}\right]. \tag{7.60}$$

Equation (7.60) shows an interesting fact. When the plate is cooler than the ambient fluid, $t_s < t_f$, $(dt/d\eta)_0$ is always positive and heat flows *into* the surface. However, when the plate is hotter than the fluid, $t_s > t_f$, $(dt/d\eta)_0$ may be positive or negative depending on the relative sizes of $\Psi_0(N_{PR})$ and $U^2/c_p(t_s - t_f)$. Thus, even though the surface is hotter than the fluid, heat may still flow *into* it!

These facts are better illustrated in terms of the recovery temperature, t_r, defined in Eq. (7.56):

$$\frac{t_r - t_f}{U^2/2c_p} = \Psi_0(N_{PR}).$$

In terms of t_r, Eq. (7.60) becomes

$$\left(\frac{dt}{d\eta}\right)_0 = -(t_s - t_f)\varphi'_0(N_{PR})\left(1 - \frac{t_r - t_f}{t_s - t_f}\right)$$

$$= -(t_s - t_r)\varphi'_0(N_{PR}). \tag{7.61}$$

In this formulation, the sign of $(dt/d\eta)_0$, and the direction of the heat flow, are determined by the temperature difference $(t_s - t_r)$ rather than the customary difference $(t_s - t_f)$. The recovery temperature is that temperature which the surface would achieve if it were not maintained at t_s but allowed to come to equilibrium with the fluid stream flowing past. Thus, the heat flux is dependent upon the difference between this equilibrium temperature and the imposed t_s. Figure 7.8 illustrates the possible situations. For $t_s > t_r$, heat flows *from* the wall; for $t_s = t_r$, no heat flows; for $t_s < t_r$ heat flows into the wall.

Since $t_r > t_f$, the case of heat flow *from* the wall is always associated with $t_s > t_f$. However, for heat flow into the wall with $t_s < t_r$, the surface may be hotter or cooler than the free stream temperature—depending on the relative sizes of $(t_r - t_s)$ and $(t_r - t_f)$. A case in which $t_s > t_f$ but in which

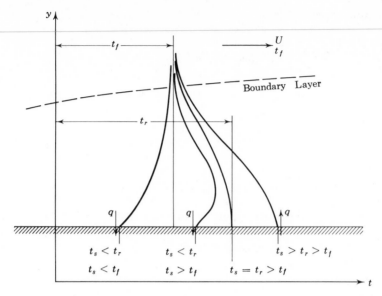

Fig. 7.8. Temperature distributions in a laminar boundary layer when viscous dissipation is present.

heat flows *into* the wall is shown in Fig. 7.8. An inversion of the temperature profile is apparent, indicating that such a high rate of viscous dissipation occurs in the boundary layer that heat flows *both* into the fluid stream *and* the wall.

If the conventional definition of the film coefficient is used, based on $(t_s - t_f)$ as in Eq. (7.27), the above discussion indicates that a negative h could result even though $t_s > t_f$. A more rational basis for the definition of the film coefficient is the temperature difference $t_s - t_r$:

$$h_x = \frac{(q/A)_0}{t_s - t_r} = -\frac{k\left(\frac{\partial t}{\partial y}\right)_0}{t_s - t_r}. \tag{7.62}$$

Thus, Eq. (7.61) gives

$$h_x = k\varphi_0'(N_{\mathrm{PR}})\sqrt{\frac{U}{\nu x}}.$$

This latter expression is identical to that obtained in Eq. (7.27) of Sec. 7.2. Thus, in terms of the Nusselt numbers, the local and average film coefficients are given by Eqs. (7.30) and (7.32):

$$N_{NU_x} = \frac{h_x x}{k} = \varphi'_0(N_{PR})N_{RE_x}^{1/2}$$

$$\approx 0.332 N_{PR}^{1/3} N_{RE_x}^{1/2},$$

(7.63)

$$N_{NU_L} = \frac{hL}{k} = 2\varphi'_0(N_{PR})N_{RE_L}^{1/2}$$

$$\approx 0.664 N_{PR}^{1/3} N_{RE_L}^{1/2}.$$

In order to use these equations with Eq. (7.62) for the heat flux, the recovery temperature must be found from the known values of t_s, t_f, and U:

$$\frac{t_r - t_f}{U^2/2c_p} = \Psi_0(N_{PR}),$$

or

$$\frac{t_r - t_f}{t_s - t_f} = \frac{1}{2}\Psi_0(N_{PR}) \cdot \frac{U^2}{c_p(t_s - t_f)}.$$

The last factor in this equation is the *Eckert number* identified in Eq. (6.28), so that

$$\frac{t_r - t_f}{t_s - t_f} = \frac{1}{2}\Psi_0(N_{PR})N_{EC}$$

$$\approx \frac{1}{2}N_{PR}^{1/2}N_{EC}.$$

(7.64)

Equations (7.62), (7.63), and (7.64), when used together to find the heat flux when viscous dissipation is present, verify the predictions of Sec. 6.8, where it was found that the Nusselt, Reynolds, Prandtl, and Eckert numbers would be involved in any such analysis.

Since $N_{EC} \to 0$ as $U \to 0$, $t_r \to t_f$ under such conditions, and the definition of h given in Eq. (7.62) reduces to the definition used in the cases in which dissipation was neglected.

7.6   Turbulent Boundary Layers on Flat Surfaces and the Transition from Laminar Flow

The examples set forth in the preceding sections have all been restricted to laminar motion. Even though rather simple examples were considered, it should be apparent that in general an analytical treatment is quite complex. In the case of turbulent motion, the additional complexities introduced by the turbulent fluctuations make an analytical approach even more involved. This is due, primarily, to the fact that the eddy viscosity and eddy diffusivity defined in Sec. 6.13 are not simply fluid properties but depend upon the motion of the fluid itself. Nevertheless, in spite of these complexities the need still exists for the prediction of the characteristics of turbulent velocity and thermal boundary layers, and a certain amount of empiricism must be introduced.

Although a vast amount of data and theories exist for the description of turbulent boundary layers, only the simplest will be discussed here since they yield sufficiently accurate results for the majority of engineering heat transfer calculations. Basically two kinds of empirical information are required: skin friction (or wall shear stress) and velocity distribution.

The Turbulent Skin Friction Coefficient

On the basis of extensive measurements in smooth tubes, Blasius (Ref. 11) deduced the following relation for the skin friction coefficient in turbulent flow on a smooth flat plate:

$$C_f = \frac{\tau_0}{\frac{1}{2}\rho U^2} = 0.0456\left(\frac{\nu}{U\delta}\right)^{1/4}, \qquad 5 \times 10^5 < N_{RE_x} < 10^7. \qquad (7.65)$$

In this expression U is the velocity of the main fluid stream well outside the boundary layer and δ is the local turbulent layer thickness at the station x from the leading edge of the plate. The term in the parentheses on the right side of Eq. (7.65) is recognized as a Reynolds number based on the boundary layer thickness and is sometimes called the *thickness Reynolds number*, N_{RE_δ}.

Power Law Velocity Distribution

Velocity profiles in turbulent boundary layers have been obtained by a number of investigators. As one would expect, the layer would be expected to behave much in a laminar fashion very near the surface (the laminar "sublayer" discussed in Sec. 6.13) with an eventual transition to a fully turbulent region farther away from the wall. A detailed description of this variation is complicated, and one such representation is described in the next section. However, for certain applications the gross behavior of a turbulent boundary layer may be represented adequately by the following power law*:

$$\frac{v_x}{U} = \left(\frac{y}{\delta}\right)^{1/7}. \tag{7.66}$$

Figure 7.9 shows a plot of this one-seventh power law along with measured data for air. The agreement between the empirical law and experiment is

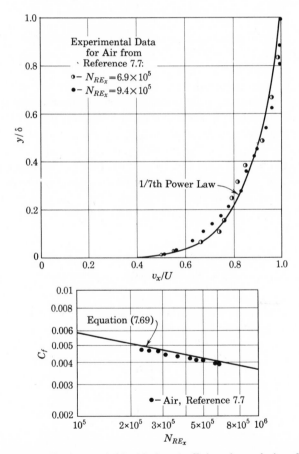

Fig. 7.9. Velocity distribution and skin friction coefficient for turbulent flow of air past a flat plate.

---

* In this equation, and all others in this section relating to turbulent motion, the bar notation for the time-average turbulent velocity will be dropped—with the understanding that all turbulent velocities referred to are time-average velocities.

apparent. This power law represents well the observed behavior of a turbulent layer for local length Reynolds numbers in the range $5 \times 10^5 < N_{RE_x} < 10^7$.

Although Eq. (7.66) describes the velocity distribution in the turbulent layer rather well for most of the layer thickness, it cannot be applied at the surface. Equation (6.66) relating the turbulent shear stress to the velocity gradient is

$$\tau_t = \rho\epsilon\frac{\partial v_x}{\partial x},$$

so

$$\tau_t = \rho\epsilon U\frac{1}{7}\frac{1}{\delta^{1/7}}\frac{1}{y^{6/7}}.$$

This relation leads to an infinite value of the shear at the wall. Since this is not physically reasonable, some other representation must be applied near the surface, as discussed in the next section. Before doing this, however, one may use the above to gain an insight into the growth of turbulent layers.

As discussed in Sec. 6.13, the momentum integral equation derived for laminar flow will apply equally well to turbulent flow as long as the velocities used are time-average velocities and as long as wall shear term, τ_0, is adequately represented for turbulent flow [i.e., Eq. (7.65)]. For flow past a flat plate ($dP/dx = 0$, $U = $ constant) Eq. (6.51) becomes, under these conditions, for turbulent flow:

$$\frac{d}{dx}\int_0^\delta v_x(v_x - U)\,dy = -\frac{\tau_0}{\rho}. \tag{7.67}$$

If one introduces now the empirical expressions for v_x/U and τ_0 given in Eqs. (7.65) and (7.66), the momentum equation gives

$$\frac{d}{dx}\int_0^\delta \left(\frac{y}{\delta}\right)^{1/7}\left[\left(\frac{y}{\delta}\right)^{1/7} - 1\right]dy = -\frac{0.0456}{2}\left(\frac{v}{U\delta}\right)^{1/4},$$

$$\frac{d}{dx}\left[\delta\int_0^1 \eta^{1/7}(\eta^{1/7} - 1)\,d\eta\right] = -0.0228\left(\frac{v}{U\delta}\right)^{1/4}.$$

After integration and rearrangement one has the following differential equation for δ, the boundary layer thickness:

$$\delta^{1/4}\,d\delta = 0.0228\frac{72}{7}\left(\frac{v}{U}\right)^{1/4}dx.$$

For a zero boundary layer thickness at $x = 0$ (an assumption which will be discussed later), the latter equation gives

$$\delta^{5/4} = 0.0228 \frac{72}{7} \cdot \frac{5}{4} \left(\frac{v}{U} \right)^{1/4} x,$$

or

$$\delta = 0.375 \left(\frac{v}{U} \right)^{1/5} x^{4/5},$$

(7.68)

$$\frac{\delta}{x} = \frac{0.375}{(N_{RE_x})^{1/5}}.$$

This equation, when compared with Eq. (7.40) for laminar layers, shows that a turbulent boundary layer grows in thickness at a faster rate than a laminar layer under the same conditions. With this relation just established between δ and x, Eq. (7.65) for the local skin friction coefficient may be written in terms of x, the local coordinate. Substitution of Eq. (7.68) into Eq. (7.65) gives

$$C_f = \frac{0.0456}{(0.375)^{1/4}} \left(\frac{v}{Ux} \right)^{1/4} [(N_{RE_x})^{1/5}]^{1/4}$$

$$= 0.0583 \left(\frac{v}{Ux} \right)^{1/5}$$

$$= \frac{0.0583}{(N_{RE_x})^{1/5}}.$$

(7.69)

A comparison between this equation and experimental data for air is shown in Fig. 7.9.

The above analysis assumed that the turbulent layer started, with a zero thickness, at the leading edge of the flat plate. This is not true in practice, although experiment shows that Eq. (7.68) gives reasonable values for δ in spite of the implied contradiction. The actual behavior is illustrated in Fig. 7.10 in which a uniform flow past a flat plate (placed parallel to the approaching flow) first forms a laminar boundary layer which grows according to Eq. (7.40). At some distance along the plate the laminar flow in the boundary layer becomes unstable and a transition to the turbulent state of flow occurs—the resulting turbulent layer then being described by Eq. (7.68). As is the case in several other instances mentioned above, the transition from the laminar

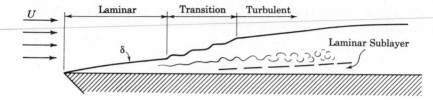

Fig. 7.10. Laminar turbulent transition of a boundary layer on a flat plate.

to the turbulent state of flow does not occur at a definite point, but occurs over a finite length of the plate (Fig. 7.10) called the *transition zone*. As a result of inherent disturbances in the approaching flow and the surface roughness of the plate, oscillations of fairly long wavelengths are observed to appear in the laminar flow. As the flow proceeds along the plate, these oscillations eventually become unstable, depending on the layer thickness, and finally break up into vortices which subsequently degenerate into turbulent fluctuations.

The equations given above for boundary layer thickness [i.e., Eq. (7.40) or (7.68)] apply only to the domains of fully laminar or fully turbulent boundary layers. The transition zone is a region of highly irregular motion, and the knowledge of the flow in this region is quite limited. For most engineering applications it is found that a turbulent layer can be expected to exist for values of the length Reynolds number ($N_{RE_x} = Ux/v$) in excess of 300,000 to 500,000—although under special conditions laminar layers have been maintained up to values of $N_{RE_x} = 3 \times 10^6$, and turbulent layers have been observed for $N_{RE_x} = 80,000$. Picking a typical critical Reynolds number of about 400,000, Eqs. (7.40) and (7.68) (for the laminar and turbulent boundary layer thicknesses, respectively) show that not only does the turbulent layer grow more rapidly in thickness, but there is a considerable thickening in the layer thickness when the laminar turbulent transition takes place.

Prandtl's Mixing Length and the Universal Velocity Distribution

As was pointed out in Sec. 6.13, the principal difficulty associated with the analysis of turbulent boundary layers is the determination of the turbulent shear stress given by either Eq. (6.58) or (6.66):

$$\tau_t = -\rho \overline{v'_x v'_y} = \rho \epsilon \frac{\partial v_x}{\partial y}. \tag{7.70}$$

In other words, one needs expressions relating either the eddy viscosity, ϵ, or the averaged fluctuation product, $\overline{v'_x v'_y}$, to the flow field.

    Prandtl accomplished this by introducing the concept of the *mixing length*. In analogy to the laminar phenomenon in which momentum is transported on the microscopic scale by molecules moving over their mean free paths between collisions, Prandtl assumed that the macroscopic turbulent eddies associated with the velocity fluctuations v_x' and v_y' travel over a finite distance before losing their identity. He called this distance the *mixing length*, l. Figure 7.11 illustrates the concepts involved. A fluid lump, or eddy, coming from $(y + l)$ would have, on the average, a higher v_x than that at y; a lump coming from $(y - l)$ would have a lower v_x. The average of these differences would be, from the definition of the mixing length, the fluctuation v_x':

$$v_x' \propto l \left| \frac{\partial v_x}{\partial y} \right|.$$

If v_y' is of the same order as v_x', the turbulent shear, Eq. (7.70), is

$$\tau_t \propto -\rho l^2 \left(\frac{\partial v_x}{\partial y} \right)^2,$$

$$\tau_t = C_1 \rho l^2 \left(\frac{\partial v_x}{\partial y} \right)^2, \tag{7.71}$$

in which C_1 is a constant.

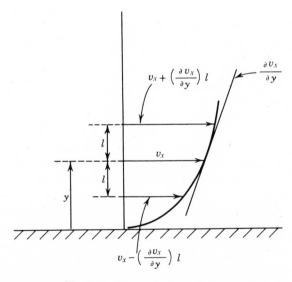

Fig. 7.11. The turbulent mixing length.

From another viewpoint, Prandtl's mixing length hypothesis amounts to the assumption that the eddy viscosity, ϵ, is

$$\epsilon = C_1 l^2 \frac{\partial v_x}{\partial y}.$$

That is, the shear stress is taken to be proportional to the square of the magnitude of the velocity gradient, and the mixing length is introduced to make the resulting expression for the eddy viscosity dimensionally correct.

Prandtl further assumed that the mixing length was directly proportional to the distance from the wall, so

$$l = C_2 y,$$

and Eq. (7.71) becomes

$$\tau_t = C_1 C_2^2 \rho y^2 \left(\frac{\partial v_x}{\partial y}\right)^2$$

$$= C_3^{-2} \rho y^2 \left(\frac{\partial v_x}{\partial y}\right)^2. \tag{7.72}$$

Upon integrating, Eq. (7.72) gives

$$v_x = C_3 \sqrt{\frac{\tau_0}{\rho}} \ln y + C_4, \tag{7.73}$$

in which C_3 and C_4 are constants and τ_0 denotes the wall value of τ_t (the subscript t having been dropped) which has been assumed constant.

As was pointed out earlier, more-sophisticated theories have been developed to eliminate some of the broad assumptions made in Prandtl's development. However, the results of these theories are in close agreement with Eq. (7.73).

Many measurements have validated the form of the velocity distribution given in Eq. (7.73)—except very near the wall, where the laminar sublayer becomes dominant. These results are made "universal" when nondimensionalized in the following fashion. A characteristic velocity, the "shear velocity" v_x^*, is defined:

$$v_x^* = \sqrt{\frac{\tau_0}{\rho}}. \tag{7.74}$$

Then the nondimensional velocity and y coordinate are defined

$$v_x^+ = \frac{v_x}{v_x^*}, \tag{7.75}$$

$$y^+ = y\frac{v_x^*}{v}. \tag{7.76}$$

With these definitions, Eq. (7.73) becomes

$$v_x^+ = C_3 \ln y^+ + C_5. \tag{7.77}$$

The laminar sublayer is sufficiently thin that one may take the velocity distribution there to be linear without much error. If this is done, the laminar viscosity law

$$\frac{\tau_l}{\rho} = \frac{\tau_0}{\rho} = v\frac{\partial v_x}{\partial y}$$

becomes, upon integration and nondimensionalization,

$$v_x^+ = y^+. \tag{7.78}$$

Many experimental results indicate that Eqs. (7.77) and (7.78) will represent the laminar sublayer, the buffer zone, and fully turbulent zone. Von Kármán (Ref. 13) recommends:

For $\ 0 < y^+ < \ \ \ 5: v_x^+ = y^+,$ laminar sublayer.

For $\ 5 < y^+ < \ \ 30: v_x^+ = 5.0 \ln y^+ - 3.05,$ buffer zone.   (7.79)

For $30 < y^+ < 400: v_x^+ = 2.5 \ln y^+ + 5.5,$ fully turbulent.

Figure 7.12 shows the relation between the above *universal velocity distribution* and experimental data. Beyond $y^+ = 400$ the power law of Eq. (7.66) may be used.

The universal distribution expressed in Eq. (7.79) is expressed in terms of the wall shear through the shear velocity $v_x^* = \sqrt{\tau_0/\rho}$. Hence, the wall shear τ_0 must be related to the main stream parameters in order to express the velocity distribution in terms of the main velocity U. This is done by use of the skin friction coefficient given in Eq. (7.69).

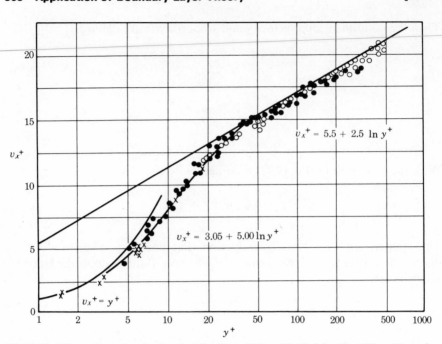

Fig. 7.12. The universal velocity distribution. (From H. Reichardt, "Heat Transfer Through Turbulent Boundary Layers," *N.A.C.A. Tech. Memo. 1047*, Washington, D.C., 1943.)

7.7   Heat Transfer in the Turbulent Boundary Layer on a Flat Plate

The concepts regarding turbulent boundary layers discussed in Secs. 6.13 and 7.6 will now be applied to the analysis of heat transfer on flat surfaces in turbulent flow. A modification of Reynolds' analogy, discussed in Sec. 7.4, will be applied—using both the velocity distributions described in Sec. 7.5. These two examples are perhaps the simplest that may be used to demonstrate the analysis of turbulent heat transfer, but they do illustrate all the essential points.

Prandtl's Modification of Reynolds' Analogy

As noted in Eq. (6.73) of Sec. 6.13 the apparent shear stress and heat flux in a turbulent boundary layer are given by

$$\tau_{\text{app}} = \rho(v + \epsilon)\frac{\partial v_x}{\partial y},$$

$$\left(\frac{q}{A}\right)_{\text{app}} = -\rho c_p(\alpha + \epsilon_H)\frac{\partial t}{\partial y}.$$

$$(6.73)$$

Reynolds' analogy in the laminar form developed in Sec. 7.4 was limited to the laminar regime, in which the eddy coefficients ϵ and ϵ_H vanish. Then if $N_{PR} = 1$, $v = \alpha$ and the analogy results. If now, however, in the *fully turbulent* region of a turbulent layer it is assumed that $\epsilon \gg v$ and $\epsilon_H \gg \alpha$ the apparent stress and heat flux are

$$\tau_{app} = \rho\epsilon\frac{\partial v_x}{\partial y},$$

$$\left(\frac{q}{A}\right)_{app} = -\rho c_p \epsilon_H \frac{\partial t}{\partial y}.$$

Although the eddy coefficients ϵ and ϵ_H are complex functions of the turbulent flow field itself, one may theorize that since both are representations of the turbulent mixing fluctuations, then in the fully turbulent zone $\epsilon \approx \epsilon_H$. If this is so, as many measurements would indicate, the equations above would lead to an identical expression of Reynolds' analogy between momentum and energy transport as obtained for laminar flow in Eq. (7.49):

$$\frac{\tau_r c_p}{U - v_r} = \frac{q_r/A}{t_r - t_f}. \tag{7.49}$$

It is important to remember that this relation is being applied only in the fully turbulent region of the boundary layer and the plane, denoted by the subscript r, located y_r away from the wall must lie within this zone. Also, it is worthy of note that one is *not* assuming necessarily that the Prandtl number $N_{PR} = v/\alpha$, a property of the *fluid*, is 1, only that the ratio ϵ/ϵ_H, a property of the flow field, is 1.

Prandtl's original analysis of turbulent heat transfer on a flat surface consisted of the rather simple representation depicted in Fig. 7.13, in which

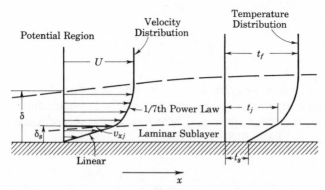

Fig. 7.13. Simplified representation of a turbulent boundary layer on a solid surface.

the boundary layer is presumed to consist of only a laminar sublayer and a fully turbulent zone. In other words, the buffer zone in which ϵ and v are of the same order of magnitude is omitted in this simple representation. The velocity distribution in the fully turbulent region will be represented by the one-seventh power law discussed in Sec. 7.6, and the velocity distribution in the laminar sublayer will be assumed to be linear. This linear variation will be chosen so that its slope at $y = 0$ will yield the wall shear stress given by Eq. (7.65). The velocity distribution in the laminar sublayer will be joined to that in the fully turbulent region at a finite distance from the wall. This distance will be the presumed thickness of the laminar sublayer and will be denoted by δ_s.

The assumption of a linear velocity distribution in the laminar sublayer is actually not as crude an assumption as one might imagine. In reality the laminar sublayer is so thin that even a third-order polynomial will give an almost straight line.

In a manner similar to the above discussion concerning the velocities in the boundary layer, it may be assumed that the temperature distribution through the laminar sublayer is also linear—varying from t_s at $y = 0$ to t_j (the temperature at the "juncture" between the laminar sublayer and the turbulent region) at $y = \delta_s$. This variation is also illustrated in Fig. 7.13. Thus, the heat flow from the surface, per unit area, is

$$\frac{q_0}{A} = \frac{k}{\delta_s}(t_s - t_j). \tag{7.80}$$

If v_{x_j} is used to denote the velocity at the juncture between the laminar sublayer and the fully turbulent zone, and if the reference plane denoted by the subscript r in Eq. (7.49) is taken as this plane of juncture, then Reynolds' analogy for the exchange of heat and momentum between $y = \delta_s$ and the potential free stream is

$$\frac{\tau_j c_p}{U - v_{x_j}} = \frac{q_j/A}{t_j - t_f}.$$

Here τ_j and q_j/A denote the shear stress and heat flux at $y = \delta_s$. Since the velocity and temperature variations through the laminar sublayer are taken to be linear, the shear stress and heat flux at $y = \delta_s$ are equal to their respective values at $y = 0$. So

$$\frac{\tau_0 c_p}{U - v_{x_j}} = \frac{q_0/A}{t_j - t_f}. \tag{7.81}$$

Now, since the heat transfer coefficient is defined as

$$h = \frac{q_0/A}{t_s - t_f},$$

Eqs. (7.80) and (7.81) may each be rewritten as

$$h = \frac{k}{\delta_s} \frac{t_s - t_j}{t_s - t_f}$$

and

$$h = \frac{\tau_0 c_p}{U - v_{x_j}} \frac{t_j - t_f}{t_s - t_f} = \frac{\tau_0 c_p}{U - v_{x_j}} \left(1 - \frac{t_s - t_j}{t_s - t_f} \right).$$

Simultaneous solution of these last two expressions in order to eliminate $(t_s - t_j)/(t_s - t_f)$ gives

$$h = \frac{\tau_0 c_p}{U - v_{x_j}} \left(1 - \frac{h \delta_s}{k} \right).$$

Since the velocity distribution in the laminar sublayer has been assumed linear,

$$\tau_0 = \mu \left(\frac{d v_x}{dy} \right)_{y=0} = \mu \frac{v_{x_j}}{\delta_s}.$$

This equation may be used to eliminate δ_s from the equation for h above, giving

$$h = \frac{\tau_0 c_p}{U + v_{x_j} \left(\dfrac{\mu c_p}{k} - 1 \right)}.$$

Since $N_{PR} = v/\alpha = \mu c_p/k$, this equation becomes

$$h = \frac{\tau_0 c_p/U}{1 + \dfrac{v_{x_j}}{U}(N_{PR} - 1)}. \qquad (7.82)$$

Equation (7.82) is a statement of Prandtl's modification of Reynolds' analogy and may be written in a dimensionless form by multiplying by x/k (x being the local coordinate along the plate surface):

$$N_{NU_x} = \frac{hx}{k} = \frac{\dfrac{\tau_0}{U}\dfrac{c_p x}{k}}{1 + \dfrac{v_{x_j}}{U}(N_{PR} - 1)}$$

$$= \frac{\dfrac{1}{2}\dfrac{\tau_0}{\frac{1}{2}\rho U^2}\dfrac{\mu c_p}{k}\dfrac{Ux\rho}{\mu}}{1 + \dfrac{v_{x_j}}{U}(N_{PR} - 1)}$$

$$N_{NU_x} = \frac{\frac{1}{2}C_f N_{PR} N_{RE_x}}{1 + \dfrac{v_{x_j}}{U}(N_{PR} - 1)}. \tag{7.83}$$

This dimensionless form of Prandtl's analogy relates the local Nusselt number to the skin friction coefficient—in terms of the Prandtl number and the local Reynolds number.

The ratio v_{x_j}/U may be deduced from the assumed velocity distribution in the boundary layer. The velocity distribution of the laminar sublayer gives

$$\tau_0 = \mu \frac{v_{x_j}}{\delta_s},$$

and the empirical law for the skin friction coefficient, Eq. (7.65), gives

$$\tau_0 = \frac{0.0456}{2}\rho U^2 \left(\frac{v}{U\delta}\right)^{1/4}.$$

If these two expressions for τ_0 are equated, the following equation can be obtained

$$\frac{\delta_s}{\delta} = \frac{1}{0.0228}\frac{v_{x_j}}{U}\frac{\mu}{\rho U\delta}\left(\frac{U\delta}{v}\right)^{1/4}.$$

Since the juncture between the sublayer and the turbulent zone must satisfy the one-seventh power law which has been assumed to hold in the turbulent zone,

$$\frac{\delta_s}{\delta} = \left(\frac{v_{x_j}}{U}\right)^7.$$

Thus,

$$\frac{v_{x_j}}{U} = \left(\frac{1}{0.0228}\right)^{1/6} \left(\frac{\nu}{U\delta}\right)^{1/8}. \tag{7.84}$$

Equation (7.84) expresses the ratio v_{x_j}/U in terms of the local boundary layer thickness, but δ is itself related to the local coordinate along the plate by the formula given in Eq. (7.68):

$$\frac{\delta}{x} = 0.375(N_{RE_x})^{-1/5}.$$

Thus,

$$\frac{v_{x_j}}{U} = \left(\frac{1}{0.0228}\right)^{1/6} \left(\frac{\nu}{U}\right)^{1/8} \left(\frac{1}{0.375}\right)^{1/8} \left(\frac{U}{\nu}\right)^{1/40} \left(\frac{1}{x}\right)^{1/10}$$

$$= 2.12 \left(\frac{\nu}{Ux}\right)^{1/10} = 2.12(N_{RE_x})^{-1/10}. \tag{7.85}$$

Equation (7.83) for the local Nusselt number becomes, then,

$$N_{NU_x} = \frac{\frac{1}{2}C_f N_{PR} N_{RE_x}}{1 + 2.12(N_{RE_x})^{-1/10}(N_{PR} - 1)}.$$

Equation (7.69) relates the local skin friction coefficient to N_{RE_x} so that one finally has for the local Nusselt number:

$$N_{NU_x} = \frac{0.0292(N_{RE_x})^{0.8} N_{PR}}{1 + 2.12(N_{RE_x})^{-1/10}(N_{PR} - 1)}. \tag{7.86}$$

This relation is found to give adequate results for turbulent heat transfer coefficients in spite of many simplifications. The degree of agreement between this relation and measured data may be seen in Fig. 7.17, wherein the counterpart of Eq. (7.86) for turbulent pipe flow is compared with experimental data.

Von Kármán's Modification of Reynolds' Analogy

Von Kármán (Ref. 13) improved upon Prandtl's analysis by including the buffer zone through the use of the universal velocity distribution given in Eq. (7.79). The basic assumptions of this analysis may be summarized:
1. In the laminar sublayer $(0 < y^+ < 5)$ the eddy coefficients ϵ and ϵ_H vanish. The velocity and temperature profiles are assumed to be

constant. Thus, the shear stress and heat flux are constant through the sublayer and are equal to their wall values, τ_0 and q_0/A.

2. In the buffer zone ($5 < y^+ < 30$) the eddy coefficients are of the same order as their molecular counterparts: $\epsilon \approx \nu$, $\epsilon_H \approx \alpha$. Further, the shear stress and heat flux are taken to be constant through the zone and, hence, are again equal to their wall values, τ_0 and q_0/A. This assumption determines the variation of ϵ through the zone according to the universal distribution given in Eq. (7.79). Finally, it is assumed that in the buffer zone $\epsilon \approx \epsilon_H$.

3. In the fully turbulent zone ($30 < y$) the molecular coefficients ν and α vanish (or are negligible). Further, it is assumed that $\epsilon \approx \epsilon_H$.

As a result of assumption 3, one may apply Reynolds' analogy between the point denoted by $y^+ = 30$ and the free stream. It is not possible to apply Reynolds' analogy in the buffer zone even with $\epsilon \approx \epsilon_H$ since ν and α are not necessarily equal; hence, Eq. (6.73) does not yield an analogy between momentum and heat transfer. Thus, by the use of the subscript 30 to denote the junction between the buffer and fully turbulent zones, Eq. (7.49) gives

$$\frac{\tau_{30} c_p}{U - v_{x30}} = \frac{q_{30}/A}{t_{30} - t_f},$$

but, by assumption 2 one has

$$\frac{\tau_0 c_p}{U - v_{x30}} = \frac{q_0/A}{t_{30} - t_f}. \tag{7.87}$$

In order to use this equation to evaluate the heat transfer for stated surface and free stream conditions, it is necessary to relate v_{x30} and t_{30} to these conditions.

The temperature at the junction, t_{30}, may be found by integrating the heat flux law in Eq. (6.73):

$$\frac{q_0/A}{\rho c_p} = -(\alpha + \epsilon_H)\frac{\partial t}{\partial y}.$$

First, the dimensionless variable y^+, Eqs. (7.74) and (7.76), is introduced:

$$\frac{q_0/A}{\rho c_p} = -(\alpha + \epsilon_H)\frac{\partial t}{\partial y^+}\frac{dy^+}{dy}$$

$$= -(\alpha + \epsilon_H)\frac{v_x^*}{\nu}\frac{\partial t}{\partial y^+}$$

$$= -\left(\frac{1}{N_{PR}} + \frac{\epsilon_H}{\nu}\right)\sqrt{\tau_0/\rho}\frac{\partial t}{\partial y^+}.$$

Integration of this expression from $y^+ = 0$ (where $t = t_0$) to $y^+ = 30$ gives

$$-(t_{30} - t_0) = \sqrt{\rho/\tau_0}\,\frac{q_0/A}{\rho c_p}\left(\int_0^5 \frac{dy^+}{\frac{1}{N_{PR}} + \frac{\epsilon_H}{v}} + \int_5^{30} \frac{dy^+}{\frac{1}{N_{PR}} + \frac{\epsilon_H}{v}}\right). \quad (7.88)$$

Evaluation of Eq. (7.88) requires finding ϵ_H/v as a function of y^+ in the ranges considered. From assumption 1,

$$\frac{\epsilon_H}{v} = 0 \qquad \text{for } 0 < y^+ < 5. \quad (7.89)$$

The shear stress law of Eq. (7.63) may be written in the following way when the dimensionless variables v_x^+ and y^+ are introduced:

$$\frac{\tau}{\rho} = (v + \epsilon)\frac{\partial v_x}{\partial y}$$

$$= (v + \epsilon)\frac{v_x^{*2}}{v}\frac{\partial v_x^+}{\partial y^+}$$

$$= \frac{\tau_0}{\rho}\left(1 + \frac{\epsilon}{v}\right)\frac{\partial v_x^+}{\partial y^+}.$$

Since assumption 2 takes $\tau/\rho = \tau_0/\rho$ in $5 < y^+ < 30$,

$$\left(1 + \frac{\epsilon}{v}\right)\frac{\partial v_x^+}{\partial y^+} = 1.$$

so Eq. (7.79) gives

$$\frac{\epsilon}{v} = \frac{y^+}{5} - 1,$$

Since $\epsilon = \epsilon_H$ also applies in the buffer zone, one finally has

$$\frac{\epsilon_H}{v} = \frac{y^+}{5} - 1 \qquad \text{for } 5 < y^+ < 30. \quad (7.90)$$

Now when Eqs. (7.89) and (7.90) are introduced into Eq. (7.88),

$$-(t_{30} - t_0) = \sqrt{\rho/\tau_0}\frac{q_0/A}{\rho c_p}\left[\int_0^5 \frac{dy^+}{\frac{1}{N_{PR}}} + \int_5^{30} \frac{dy^+}{\left(\frac{1}{N_{PR}} - 1\right) + \frac{y^+}{5}}\right]$$

$$= \sqrt{\rho/\tau_0}\frac{q_0/A}{\rho c_p}\left[5N_{PR} + 5\ln\frac{\left(\frac{1}{N_{PR}} - 1\right) + 6}{\left(\frac{1}{N_{PR}} - 1\right) + 1}\right]. \qquad (7.91)$$

The velocity $v_{x_{30}}$ required in Eq. (7.87) is readily found from Eq. (7.79).

$$v_{x_{30}} = \sqrt{\tau_0/\rho}(5\ln 30 - 3.05)$$

$$= \sqrt{\tau_0/\rho}(5\ln 6 + 5).$$

If this latter equation and Eq. (7.91) are substituted into Eq. (7.87), considerable algebraic manipulation gives

$$h = \frac{q_0/A}{t_0 - t_f} = \frac{\tau_0 c_p}{U + \sqrt{\tau_0/\rho}\{5(N_{PR} - 1) + 5\ln[1 + \frac{5}{6}(N_{PR} - 1)]\}}.$$

When the Nusselt number and skin friction coefficient are used to non-dimensionalize h and τ_0, there results

$$N_{NU_x} = \frac{\frac{1}{2}C_f N_{PR} N_{RE_x}}{1 + 5\sqrt{\frac{C_f}{2}}\{(N_{PR} - 1) + \ln[1 + \frac{5}{6}(N_{PR} - 1)]\}}. \qquad (7.92)$$

This last expression constitutes von Kármán's analogy for turbulent heat transfer on a flat surface. If the Blasius friction law of Eq. (7.69) is substituted into Eq. (7.92), one has

$$N_{NU_x} = \frac{0.0292(N_{RE_x})^{0.8} N_{PR}}{1 + 0.855(N_{RE_x})^{-1/10}\{(N_{PR} - 1) + \ln[1 + \frac{5}{6}(N_{PR} - 1)]\}}. \qquad (7.93)$$

As in the case of Prandtl's relation, von Kármán's formula has a pipe flow counterpart which is compared with experiment in Fig. 7.17.

As mentioned earlier, there are many more refined theories which more fully explain the physical phenomena taking place in a turbulent layer. Other laws representing the skin friction coefficient and the velocity distribution may be applied to obtain expressions for the local Nusselt number. These other theories are too lengthy and complex to be included here. The main purpose of this chapter is to illustrate the *method* of analysis. The principal results of the other theories will be given in Chapter 8 which will be devoted to a presentation of recommended working formulas for the determination of the heat transfer coefficient for practical engineering problems.

7.8   Viscous Flow in Pipes or Tubes—Fully Developed Flow

The examples given in the foregoing sections illustrated the application of boundary layer theory to the solution of heat convection problems for flow along flat surfaces. Both laminar and turbulent flows were considered. The analyses also implied the absence of any other nearby surfaces.

The next three sections will be devoted to the discussion of flow and heat transfer in pipes or tubes. The engineering significance of such problems should be apparent.

As a viscous fluid, initially of uniform velocity, enters a pipe or tube a boundary layer builds up along the surface of the pipe. This is illustrated in Fig. 7.14. Near the entrance of the pipe this growth of the boundary layer is much like the flow along the flat plate—particularly if the curvature of the pipe is not large. However, the flow cannot be the same as in the flat plate case due to the presence of the opposite wall—where a boundary layer is also developing. As the flow proceeds down the pipe, the boundary layer growing along the pipe wall gets thicker, eventually growing together from opposite sides and filling the pipe with "boundary layer" flow. The state of flow cannot actually be considered as boundary layer flow since the fundamental assumption that the viscous flow region is thin compared to the geometric dimensions of the pipe is no longer satisfied. However, experiment shows that many of the empirical laws quoted in the foregoing sections for flow along a flat surface may also be used in the case of pipe flow when appropriate definitions are made for the undisturbed stream velocity, boundary layer thickness, etc.

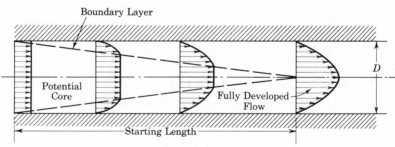

Fig. 7.14. The starting length in pipe flow.

Pipe flow is termed *fully developed flow* once the viscous layers have grown together and filled the pipe cross section. The pipe length required to establish this fully developed flow is called the *starting* or *entrance* length. In the starting length the velocity distribution across a diameter consists of a potential core region near the center of the pipe (the velocity being uniform) which joins the boundary layer region at each surface where the velocity varies from the potential core value to zero at the wall. As one moves along the pipe in the starting region, the viscous or boundary layer portion of the velocity distribution curve increases while the potential core portion decreases. For constancy of mass flow rate the mean velocity across the cross section (mass flow rate divided by density and cross-sectional area) must remain constant—for incompressible flow. Hence, the velocity of the potential core region must increase as the flow proceeds down the starting length. This increase is also shown in Fig. 7.14.

The transition from the laminar to turbulent state of flow may occur either in the boundary layer in the starting length or in the fully developed flow— or not at all—depending on the conditions of the flow (i.e., velocity, viscosity, etc.). Thus, it is possible to have a laminar or turbulent fully developed region. The transition in the starting region can be predicted reasonably well by use of the criterion based on the length Reynolds number, N_{RE_x}, as noted in Sec. 7.6.

For fully developed pipe flow the length Reynolds number loses its significance, and it is customary to employ the *diameter Reynolds number*, R_{RE_D}, defined as

$$N_{RE_D} = \frac{U_m D \rho}{\mu} = \frac{U_m D}{\nu}. \tag{7.94}$$

Here D denotes the pipe diameter and U_m is the mean velocity of the flow in the pipe. Experiments indicate that a critical, or transition, diameter Reynolds number of 2300 may be used to predict reasonably well the existence of fully developed laminar or turbulent pipe flow. For values of $N_{RE_D} < 2300$, laminar flow may be expected, whereas greater values indicate the presence of turbulent flow. As discussed before in connection with transition in boundary layers, this critical value of $N_{RE_D} = 2300$ is not to be treated as a precise value since certain extraneous conditions can cause the transition to occur at other values.

A formula, developed by Langhaar (Ref. 12), which is useful for predicting the length of the starting section of laminar pipe flow is

$$\frac{L_s}{D} = 0.0575 N_{RE_D}. \tag{7.95}$$

This formula must be treated as approximate since the fully developed condition is reached asymptotically, and hence the starting length, L_s, is difficult to define. If the laminar–turbulent transition takes place in the starting region, the starting length decreases from the value indicated by Eq. (7.95). No adequate theory exists which will predict the starting length in turbulent flow since the nature of the pipe entrance, the pipe roughness, etc., will have a serious effect on the growth and transition of the boundary layer in the pipe. Starting lengths of 25 to 40 pipe diameters are typical for turbulent flow.

7.9 Fully Developed Velocity Distributions and Pressure Losses

Laminar Flow

For fully developed laminar flow, the velocity distribution across a pipe cross section is the well-known parabolic form as given by

$$\frac{v}{U_m} = 2\left[1 - \left(\frac{r}{R}\right)^2\right]. \tag{7.96}$$

In Eq. (7.96) v denotes the local axial velocity at radius r, R denotes the pipe radius, and U_m denotes the mean flow velocity of the cross section. Figure 7.15 illustrates these quantities. The viscous shear stress at the pipe surface is given by

$$\tau_R = -\mu\left(\frac{\partial v}{\partial r}\right)_{r=R}.$$

The negative sign appears in this equation since r is measured from the pipe centerline instead of the wall. The symbol τ_R is used to denote the shear at the pipe radius R. This gives

$$\tau_R = 4\mu\frac{U_m}{R}. \tag{7.97}$$

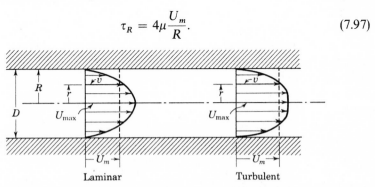

Fig. 7.15. Fully developed profiles for laminar and turbulent pipe flow.

The viscous shear at the pipe wall is balanced by the gradient in pressure along the pipe and, hence (for incompressible flow),

$$-\pi R^2 \frac{dP}{dx} = 2\pi R \tau_R,$$

$$-\frac{dP}{dx} = 8\mu \frac{U_m}{R^2}.$$

It is often customary to define a *friction factor*, f, as

$$f = \frac{-(dP/dx)}{\frac{1}{2}\rho U_m^2/D}$$

$$= \frac{4\tau_R}{\frac{1}{2}\rho U_m^2}. \tag{7.98}$$

The friction factor serves the same purpose for pipe flow that the skin friction coefficient does for flow along a plate. In terms of the wall shear stress, the friction factor differs from the skin friction coefficient by a factor of 4. [See Eqs. (7.15) and (7.98).] In the case of laminar pipe flow being discussed here, f is easily deduced as a simple function of the diameter Reynolds number if Eq. (7.97) is substituted into Eq. (7.98):

$$f = 64\frac{\mu}{U_m D\rho} = \frac{64}{N_{RE_D}}. \tag{7.99}$$

From Eq. (7.98) the following relation for the pressure drop due to laminar flow in a pipe of length L results:

$$-\Delta P = f\frac{L}{D}\frac{1}{2}\rho U_m^2. \tag{7.100}$$

The friction factor just defined here is sometimes called the *Fanning friction factor*. Some workers prefer to use a friction factor which is four times that just defined.

Turbulent Flow

Experiments have shown that many of the relations discussed in Secs. 7.6 and 7.7 with regard to flat plates may be applied also to fully developed flow in smooth tubes. This is done by replacing y with $R - r$, δ with R, and U with U_{max}. The symbol U_{max} is used to denote the maximum tube axial velocity along the centerline.

Power Law Velocity Distribution. The power law velocity distribution given in Eq. (7.66) becomes, for pipe flow,

$$\frac{v}{U_{max}} = \left(\frac{R - r}{R}\right)^{1/7}. \tag{7.101}$$

As in the case of flat plate flow, this relation is only approximate and cannot be expected to describe very accurately the flow situation near the wall. It does serve to give an adequate representation of the gross behavior of turbulent pipe flow. Figure 7.15 illustrates the relative shape of a turbulent and laminar velocity profiles. The turbulent profile is seen to give a more nearly uniform velocity distribution near the center of the pipe than does laminar flow—with a resultant steeper velocity gradient at the wall.

Since it is customary to base pipe flow parameters on the average flow velocity rather than the maximum velocity, Eq. (7.101) may be integrated to obtain a mean

$$U_m = \frac{1}{\pi R^2} \int_0^R v(2\pi r)\, dr$$

$$= \frac{U_{max}}{\pi R^2} \int_0^R \left(\frac{R - r}{R}\right)^{1/7} dr$$

$$= \frac{98}{120} U_{max} = 0.817 U_{max}. \tag{7.102}$$

The symbol U_m is used to denote the mean cross-section velocity.

The Pipe Wall Friction Factor. The Blasius relation for all shear stress may be written for pipe flow by taking Eq. (7.65) and replacing τ_0 by τ_R, δ by $R = D/2$, and U by $U_{max} = U_m/0.817$, giving

$$\frac{\tau_R}{\rho U_m^2} = 0.0396 \left(\frac{v}{U_m D}\right)^{1/4}$$

$$= \frac{0.0396}{(N_{RE_D})^{1/4}}. \tag{7.103}$$

The definition of the friction factor leads to the relation

$$f = \frac{4\tau_R}{\frac{1}{2}\rho U_m^2}$$

$$= \frac{0.316}{(N_{RE_D})^{1/4}}. \tag{7.104}$$

This relation is found to give reasonable results for smooth-walled tubes for the diameter Reynolds numbers from 3000 to 100,000. For the higher Reynolds numbers, better results are obtained using the universal velocity distribution.

Universal Velocity Distribution. The universal velocity distribution given by Eqs. (7.79) may be used as long as the dimensionless variables are defined:

$$v^+ = \frac{v}{v^*},$$

$$y^+ = (R - r)\frac{v^*}{v},$$

$$v^* = \sqrt{\tau_R/\rho}.$$

Using the definition of the friction factor in Eq. (7.98), one finds that

$$f = \frac{4\tau_R}{\frac{1}{2}\rho U_m^2},$$

Prandtl integrated the universal velocity distribution to obtain

$$\frac{1}{\sqrt{f}} = 0.87 \ln (N_{RE_D}\sqrt{f}) - 0.8. \tag{7.105}$$

This relation, derived from the universal distribution, is applicable only to "smooth" tubes—tubes in which the surface irregularities lie well within the laminar sublayer. More detailed analyses due to, mainly, Nikuradse, von Kármán, and Moody (Refs. 3, 4, 5) include the effect of larger surface roughnesses and how they influence the establishment of the turbulent layer. An additional parameter, the *relative roughness*, enters the considera- tion. The relative roughness is defined as ϵ/D, where ϵ is the absolute rough- ness of the pipe surface. These analyses yield the results

$$\frac{1}{\sqrt{f}} = -0.87 \ln \left(\frac{\epsilon/D}{3.7} + \frac{2.51}{N_{RE_D}\sqrt{f}}\right) \tag{7.106}$$

for transitional flow, and

$$\frac{1}{\sqrt{f}} = -0.87 \ln (\epsilon/D) + 1.14 \tag{7.107}$$

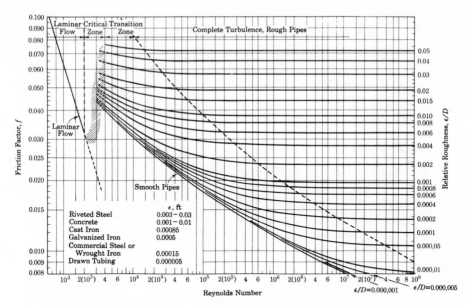

Fig. 7.16. The Moody diagram for pipe friction. (From L. F. Moody, "Friction Factors for Pipe Flow," *Trans. ASME*, Vol. 66, 1944, p. 671.)

for completely turbulent flow. Moody combined the results of Eqs. (7.105), (7.106), and (7.107) into the diagram shown in Fig. 7.16. This figure may be used to avoid the trial and error solution of the above equations.

7.10   Heat Transfer in Fully Developed Pipe Flow

For the calculations of heat transfer coefficients for fully developed pipe flow, it is necessary to define a characteristic temperature of the fluid. In the case of boundary layer flow this was not necessary since the temperature of the free stream flow was the obvious temperature to use. For fully developed pipe flow it is customary to define the "bulk" fluid temperature or the "mixing cup" temperature as the temperature that would result, at a cross section, if the fluid were thoroughly mixed to a uniform temperature. This mixing cup temperature, t_B, may be defined as the flux of enthalpy at a cross section divided by the product of the mass flow rate and the fluid specific heat. Thus, for an incompressible fluid of constant properties,

$$t_B = \frac{\int_0^R c_p t v(2\pi r)\, dr}{c_p \int_0^R v(2\pi r)\, dr} = \frac{\int_0^R t v r\, dr}{\int_0^R v r\, dr}. \tag{7.108}$$

It is apparent that t_B is dependent on the velocity distribution across the pipe cross section. The local heat transfer coefficient is then defined as

$$h = \frac{q/A}{t_B - t_R},$$ (7.109)

where t_R denotes the temperature of the pipe wall (at radius $r = R$).

Laminar Flow

Although the velocity distribution for fully laminar pipe flow is relatively simple—the parabolic expression of Eq. (7.96)—the analysis of the temperature distribution and consequent film coefficient is perhaps one of the most complex problems of heat convection. Nusselt in 1910 (Ref. 6) determined that for very long pipes, the local heat transfer coefficient would tend toward the following limiting value:

$$h = \frac{1.828k}{R}.$$

The constant 1.828 is the result of a numerical integration. Defining a *diameter Nusselt number*, $N_{NU_D} = hD/k$, one obtains

$$N_{NU_D} = \frac{hD}{k} = 3.66.$$ (7.110)

Although it was an achievement in the theoretical analysis of heat convection problems, this results does not represent very well the observed heat transfer coefficients for laminar pipe flow. This is primarily due to two facts. First, the fluid properties were assumed constant (i.e., independent of temperature) in the analysis. Most fluids which will maintain a fully laminar pipe flow are rather viscous oils. These fluids often exhibit a strong temperature dependence in the kinematic viscosities—leading to an actual behavior much different from the assumed constant viscosity flow. Second, in laminar pipe flow the starting length may often be a significant part of the total pipe length, and the assumption of a very long pipe, as used by Nusselt, is not actually realized. Also, there must exist a "thermal starting length" in the same sense that the hydrodynamic starting length exists. This would be the distance from the pipe entrance that is required before the thermal boundary layers at the pipe walls have met at the pipe axis. For fluids with a Prandtl number much different from 1, this thermal starting length may be quite different from the hydrodynamic starting length. Most highly viscous fluids—most likely to be flowing in a laminar fashion—have a Prandtl number much greater than 1 (see Appendix A, Table A.5). Section 7.2 and Fig. 7.4 showed that this leads to a thermal layer that is thinner than the velocity layer. Hence, thermal starting lengths are usually much greater than hydrodynamic starting lengths in laminar flow.

Sellars, Tribus, and Klein (Ref. 14) carried out a very complex analysis of developing laminar flow in circular tubes. The analysis, necessarily two-dimensional, is far too complex to present here, but it involved the simultaneous integration of the two-dimensional energy and momentum equations. The results are interesting inasmuch as they show how the limiting solution of Nusselt is approached as the flow develops. The results of the analysis are given in Table 7.4. In Table 7.4 x denotes the axial distance measured from the pipe entrance, $N_{NU_{D_x}}$ represents the local diameter Nusselt number and $N_{NU_{D_m}}$ represents the *mean* diameter Nusselt number between $x = 0$ and $x = x$.

Table 7.4 Developing Laminar Heat Transfer in a Circular Tube

$\dfrac{x/R}{N_{RE_D}N_{PR}}$	$N_{NU_{D_x}}$	$N_{NU_{D_m}}$
0	∞	∞
0.001	12.86	22.96
0.004	7.91	12.59
0.01	5.99	8.99
0.04	4.18	5.87
0.08	3.79	4.89
0.10	3.71	4.66
0.20	3.66	4.16
∞	3.66	3.66

Because of the complexities of the analytical solutions for developing laminar heat transfer (in particular the influence of temperature dependent viscosity), much use is made of semiempirical formulas for practical applications. Such relations are summarized in Chapter 8.

Prandtl's Analogy for Heat Transfer in Fully Developed Turbulent Pipe Flow

An analogous relation to the Prandtl analogy obtained for flat plate turbulent flow may be derived for heat transfer in the case of fully developed turbulent pipe flow. However, it is possible to deduce Prandtl's analogy for pipe flow directly from that for plate flow by making certain generalizations. Prandtl's analogy for plate flow (Sec. 7.7) is

$$
h = \frac{\tau_0 c_p / U}{1 + \frac{v_{x_j}}{U}(N_{\text{PR}} - 1)}.
\tag{7.82}
$$

This equation was based on an application of Reynolds' analogy to the fully turbulent region of the simplified model assumed to represent the boundary layer. Thus, Eq. (7.82) is based on an assumed similarity in the exchange of momentum and heat between the laminar sublayer and the reservoir of momentum and thermal energy represented by the undisturbed free stream at temperature t_f and velocity U. In the case of fully developed pipe flow this reservoir of the undisturbed free stream is missing, but if one extends Reynolds' analogy to apply to the case of momentum and heat exchange between the laminar sublayer and the main bulk of the fully developed turbulent core at temperature t_B and velocity U_m, the expression obtained for Prandtl's analogy would be the same as Eq. (7.82) with U_m replacing U and τ_R replacing τ_0. Thus, for fully developed pipe flow,

$$
h = \frac{\tau c_p / U_m}{1 + \frac{v_{x_j}}{U_m}(N_{\text{PR}} - 1)}.
$$

Introducing the diameter Nusselt number $N_{\text{NU}_D} = hD/k$, the friction factor $f = 4\tau_R / \frac{1}{2}\rho U_m^2$, and the diameter Reynolds number $N_{\text{RE}_D} = U_m D / v$, one finds that the last equation becomes

$$
N_{\text{NU}_D} = \frac{hD}{k} = \frac{\frac{f}{8} N_{\text{PR}} N_{\text{RE}_D}}{1 + \frac{v_{x_j}}{U_m}(N_{\text{PR}} - 1)}.
\tag{7.111}
$$

Equation (7.84) gave

$$
\frac{v_{x_j}}{U} = \left(\frac{1}{0.0228}\right)^{1/6}\left(\frac{v}{U\delta}\right)^{1/8}.
$$

When one introduces $\delta = D/2$, $U = U_{\text{max}} = U_m/0.817$, this becomes

$$
\frac{v_{x_j}}{U_m} = 2.44\left(\frac{v}{U_m D}\right)^{1/8}.
\tag{7.112}
$$

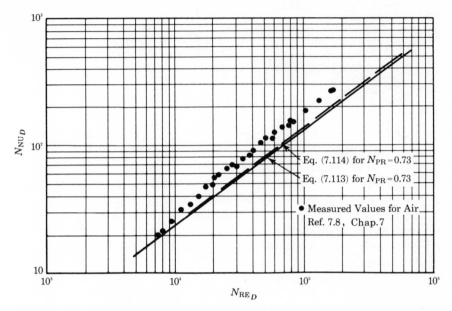

Fig. 7.17. A comparison between measured values for air and the values predicted by von Kármán's and Prandtl's equations for the Nusselt number in the case of turbulent heat transfer in pipes.

When Eqs. (7.104) and (7.112) are introduced into Eq. (7.111), one obtains

$$N_{NU_D} = \frac{0.0396(N_{RE_D})^{3/4} N_{PR}}{1 + 2.44(N_{RE_D})^{-1/8}(N_{PR} - 1)}. \tag{7.113}$$

Experiment shows that this relation works rather well—in spite of the many simplifying assumptions that have been made in its derivation. Figure 7.17 shows a plot of Eq. (7.113) along with some measured values for air. The agreement between theory and experiment is not bad—typical of results obtained in convective heat transfer. Further experiments have shown that the agreement becomes poorer as the Prandtl number becomes very different from 1.

Von Kármán Analogy for Heat Transfer in Fully Developed Turbulent Pipe Flow

Von Kármán's analogy for flat plates given in Eq. (7.92) may be converted to fully developed pipe flow by reasoning and algebraic manipulation identical to that just performed for the Prandtl relation. The result is

$$N_{\mathrm{NU}_D} = \frac{\frac{f}{8} N_{\mathrm{RE}_D} N_{\mathrm{PR}}}{1 + 5\sqrt{f/8}\{(N_{\mathrm{PR}} - 1) + \ln[1 + \frac{5}{6}(N_{\mathrm{PR}} - 1)]\}}$$

$$= \frac{0.0396(N_{\mathrm{RE}_D})^{3/4} N_{\mathrm{PR}}}{1 + 0.995(N_{\mathrm{RE}_D})^{-1/8}\{(N_{\mathrm{PR}} - 1) + \ln[1 + \frac{5}{6}(N_{\mathrm{PR}} - 1)]\}}. \quad (7.114)$$

This relation is also plotted in Fig. 7.17 for $N_{\mathrm{PR}} = 0.73$ for comparison with Prandtl's formula and experimental data.

As mentioned above in the case of laminar flow, the equations are presented mainly to show the method of approach and the results of application of boundary layer theory to the solution of heat convection problems. The complexity of even these simplest situations should illustrate to the reader the necessity of some system of intelligent and logical analysis of empirical information for the purpose of obtaining useful (for engineering design purposes) relations which will predict convective heat transfer coefficients for all manner of complex situations as well as those described above.

References

1. SCHLICHTING, H., *Boundary Layer Theory*, 6th ed., New York, McGraw-Hill, 1968.
2. POHLHAUSEN, E., "Der Wärmeaustausch zwischen festen Körpern und Flüssigkeiten mit kleiner Reibung und kleiner Wärmeleitung," *Zeitschr. für Angew. Math. und Mech.*, Vol. 1, 1921, p. 115.
3. NIKURADSE, J., "Strömungsgesetze in rauhen Rohren," *Forsch. a.d. Geb. d. Ingenieurwes.*, No. 361, 1933.
4. VON KÁRMÁN, T., "On Laminar and Turbulent Friction," *N.A.C.A. Tech. Memo. 1092*, Washington, D.C., 1946.
5. MOODY, L. F., "Friction Factors for Pipe Flow," *Trans. ASME*, Vol. 66, 1940, p. 671.
6. NUSSELT, W., "Die Abhängigkeit der Wärmeübergangszahl von der Rohrlänge," *Zeitschr. V.D.I.*, Vol. 54, 1910, p. 1154.
7. DHAWAN, S., "Direct Measurements of Skin Friction," *N.A.C.A. Tech. Note 2567*, Washington, D.C., 1952.
8. DEISSLER, R. G., and C. S. EIAN, "Analytical and Experimental Investigation of Fully Developed Turbulent Flow of Air in a Smooth Tube with Heat Transfer with Variable Fluid Properties," *N.A.C.A. Tech. Note 2629*, Washington, D.C., 1952.
9. BLASIUS, H., "Boundary Layers in Fluids with Little Friction," *N.A.C.A. Tech. Memo. 1256*, Washington, D.C., 1950.
10. PARMELEE, G. V., and R. G. HUEBSCHER, "Heat Transfer by Forced Convection Along a Smooth Flat Surface," *Heating, Piping and Air Conditioning*, Vol. 19, No. 8, 1947, p. 115.

11. BLASIUS, H., "Das Ähnlichkeitsgesetz bei Reibungsvorgängen in Flüssigkeiten," *Forsch. a.d. Geb. d. Ingenieurwes.*, No. 131, 1913.
12. LANGHAAR, H. L., "Steady Flow in the Transition Length of a Straight Tube," *Trans. ASME*, Vol. 68, 1942, p. A-55.
13. VON KÁRMÁN, T., "The Analogy Between Fluid Friction and Heat Transfer," *Trans. ASME*, Vol. 61, 1939, p. 705.
14. SELLARS, J. R., M. TRIBUS, and J. S. KLEIN, "Heat Transfer to Laminar Flow in a Round Tube or Flat Conduit—The Graetz Problem Extended," *Trans. ASME*, Vol. 78, 1956, p. 441.

Problems

7.1. Find the indicated dimensionless parameters for the conditions indicated. Make sure that the result is truly dimensionless.

(a) Find N_{RE_x} for: $U = 60$ mph
$$x = 2 \text{ ft}$$
air at 80°F

(b) Find N_{RE_D} for: a flow of water at 100°F through a 4-in.-I.D. pipe at the rate of 4000 lb_m/hr.

(c) Find N_{NU_L} for: $h = 2.0$ Btu/hr-ft^2-°F
$$L = 27 \text{ in.}$$
CO_2 at 120°F

(d) Find N_{NU_x} for: $h = 1100$ Btu/hr-ft^2-°F
$$x = 3 \text{ ft}$$
$$k = 0.002 \text{ cal/sec-cm-°C}$$

(e) Find N_{PR} for: $k = 0.0095$ Btu/hr-ft-°F
$$\mu = 0.02 \text{ centipoise}$$
$$c_p = 0.124 \text{ Btu/lb}_m\text{-°R}$$

(f) Find N_{PR} for: $k = 5.02$ Btu/hr-ft-°F
$$\mu = 3.75 \text{ lb}_m/\text{ft-hr}$$
$$c_p = 0.0333 \text{ Btu/lb}_m\text{-°R}$$

7.2. Verify the algebra leading from Eq. (7.2) to Eqs. (7.11).

7.3. Starting with Eq. (7.20), verify the derivation leading to Eq. (7.23). Show that Eq. (7.24) is the solution of Eq. (7.23).

7.4. Air at atmospheric pressure and 40°F flows at 100 mph across the wing of an airplane (to be approximated as a flat plate). Plot a curve showing the thickness of a laminar boundary layer as a function of the distance from the leading edge—up to the distance at about which transition to a turbulent layer is likely to occur. Use either the "exact" solution or the results of the application of the momentum integral method.

7.5. Saturated water at 200°F flows at a velocity of 5 ft/sec past a flat plate maintained at 80°F. Using the results of Sec. 7.2, plot the temperature and velocity profiles at stations 1, 2, and 3 in. from the leading edge of the plate. Also plot the variation in the local film coefficient over the first 3 in. of the plate and determine the average film coefficient for that distance. (Since the fluid properties are tem-perature dependent, evaluate them at the mean of the plate and free stream temperatures.)

7.6. Repeat Prob. 7.5 using results of the integral methods given in Sec. 7.3 and compare the results. Also, using the results of this section, plot the velocity layer thickness and thermal layer thickness as functions of the distance along the plate.

7.7. Using the momentum integral method, find the relation equivalent to Eq. (7.40) when a fourth-degree polynomial is assumed to represent the velocity distribution in a laminar boundary layer.

7.8. Verify the algebra leading from Eq. (7.41) to Eq. (7.45).

7.9. The *drag coefficient*, C_D, of a flat plate in a uniform stream is defined as the ratio of the total drag force, D, to the product of the surface area and the total dynamic pressure. For unit width and plate length L this is

$$C_D = \frac{D}{(\frac{1}{2}\rho U^2)L}.$$

For a velocity distribution in a laminar boundary layer assumed to be represent-able as a third-degree polynomial, find how C_D depends on the length Reynolds number, N_{RE_L}.

7.10. Repeat Prob. 7.9 for a turbulent boundary layer, using the empirical result of Eq. (7.65).

7.11. Water at 100°F flows past a flat plate at 50 ft/sec. Assuming that the boundary layer undergoes the laminar turbulent transition at a length N_{RE} of 400,000, plot the growth of the boundary layer for a plate length of 18 in.

7.12. For the data of Prob. 7.11, plot, for comparison purposes, the velocity distribu-tion in a laminar and turbulent layer at the transition point—assuming that the transition occurred instantaneously.

7.13. Air at 200°F flows over a flat plate at 200 ft/sec. The plate is maintained at 600°F. (a) Assuming that the transition Reynolds number (based on length) is 400,000, find the distance from the leading edge at which the laminar–turbulent transition takes place. (Since the fluid properties are temperature dependent, evaluate them at the mean of the plate and free stream temperatures.) (b) Find the boundary layer thickness, the laminar sublayer thickness, and the temperature gradient at the trailing edge of the plate if the plate is 18 in. long.

7.14. Atmospheric air at 60°F flows past a flat plate at a velocity of 600 ft/sec. (a) Find the recovery temperature of the plate. Find the heat flux (per unit area) at a distance 1 in. from the leading edge if the plate is maintained at (b) 40°F, and (c) 80°F.

7.15. If a first-degree polynomial is used to represent the velocity and temperature distributions in the laminar boundary layer on a flat surface, apply the integral boundary layer equations to obtain expressions for the dependence of the local boundary layer thickness and the local Nusselt number on the local Reynolds number.

7.16. Note that the function

$$g(\eta) = \frac{v_x}{U} = a \sin b\eta$$

automatically satisfies the condition $g(0) = 0$ noted in Eq. (7.38). Using two additional conditions to determine the constants a and b, find expressions relating the boundary layer thickness and skin friction coefficient to local Reynolds number by use of the integral boundary layer equations.

7.17. A flat plate, 6 in. × 9 in., is exposed to an air stream flowing parallel to its surface. The air is at atmospheric pressure and is flowing at 120 ft/sec. A total drag force (for both sides) of 0.3 lb$_f$ is observed to act on the plate. Using Reynolds' analogy, estimate the value of the surface heat transfer coefficient between the plate and the air.

7.18. Verify the derivation leading from Eq. (7.65) to Eq. (7.68).

7.19. Verify the derivation leading from Eq. (7.80) to Eq. (7.82).

7.20. Verify the algebra leading from Eq. (7.87) to Eq. (7.93).

7.21. Water at 150°F flows at the rate of 25 lb$_m$/sec through a pipe with a 3-in. inside diameter. Plot the fully developed velocity distribution for (a) the one-seventh power law, and (b) the universal velocity distribution.

7.22. If the pipe wall in Prob. 7.21 is maintained at 120°F, find
(a) The heat transfer coefficient, using the Prandtl analogy.
(b) The heat transfer coefficient, using the von Kármán analogy.
(c) The eddy viscosity at a point 0.15 in. from the wall. Compare this value with the molecular kinematic viscosity.
Evaluate the properties of the water at the average between the surface and bulk water temperatures.

7.23. Steam, at an average temperature of 600°F and a pressure of 1000 psia, flows in a 5-in., schedule 40 pipe at an average velocity of 20 ft/sec. Find the heat transfer coefficient using both the Prandtl and von Kármán analogies.

7.24. If a dimensionless temperature for turbulent flow past a flat plate is defined as

$$t^+ = \frac{(t_0 - t)\sqrt{\tau_0/\rho}}{\dfrac{q_0/A}{\rho c_p}},$$

deduce a universal temperature distribution based on the universal velocity distribution of Eq. (7.79).

CHAPTER 8

Working Formulas and Dimensionless Correlations for Forced Convection

8.1   Introductory Remarks

The analyses in Chapter 7 and their applications of boundary layer theory to the solution of forced heat convection problems were meant, as stated, to illustrate the methods of approach and to show typical results. A detailed exposition of the derivation and development of all the relationships having engineering importance for the prediction of forced convection heat transfer coefficients is beyond the scope of this book. In fact, some of the most useful engineering formulas for forced convection coefficients have no real theoretical basis. They are the results of extensive experimental analyses.

The purpose of this chapter is to present a collection of the most useful of the existing relations for the more frequently encountered cases of forced convection. Some of these will be relations having theoretical bases, and some will be empirical dimensionless correlations of experimental data. In some instances more than one relation will be given for a particular case. The relations considered to be the preferred—or "recommended"—ones will be marked by a ★. This presentation of working relations for forced convection will, of course, be limited. For sources of relations for cases not covered here, the reader is advised to consult Refs. 1, 2, and 3.

Unless otherwise noted, all the relations to be quoted here will be limited (like those derived in Chapter 7) to the imposed condition of a constant surface temperature. That is, the thermal boundary condition imposed at the solid surface (either analytically or experimentally) is that of a uniform temperature. This is in contrast to other possible conditions—such as uniform heat flux, constant temperature difference, etc. Again, the reader is advised to consult the quoted references for relations based on those other conditions.

The theoretical analyses of Chapter 7 assumed that the physical properties (μ, c_p, k, and ρ) were constants. This is, of course, not true in reality as the discussions of Chapter 2 pointed out. In almost every case there is a significant dependence of these properties on temperature, and in the case of ρ (and v) there is also a pressure effect for gases. The temperature dependence of the properties means that there may be a significant variation in the quantities through the thermal boundary layer. The accuracy of the results of these theoretical relations and of the dimensionless experimental correlations depends on the temperature chosen for the evaluation of the properties. In most cases one of two mean temperatures is used. One is the bulk fluid temperature, or mixing cup temperature, as defined in Eq. (7.108). This is usually applied in the case of forced convection inside a closed duct or pipe and is the mean fluid temperature at a cross section—the temperature that would result if the fluid at a cross section were to be thoroughly mixed. The symbol t_B will be used to denote the *bulk temperature*. The second mean temperature that is often used is the *mean film temperature*. This is the arithmetic mean of the surface temperature, t_s, and (in the case of flow external to a body) the free stream or undisturbed fluid temperature, t_f. Using t_m to denote the mean film temperature, one finds that

$$t_m = \frac{t_s + t_f}{2}. \tag{8.1}$$

Sometimes in the case of internal flow a mean film temperature which is the average of the surface temperature and the bulk temperature will be used in place of the bulk temperature.

In the working formulas to be quoted here, it will be stated in each case which of these mean temperatures should be employed for the evaluation of the fluid properties.

The analysis of the basic governing differential equations of convective heat transfer presented in Sec. 6.8 showed that for forced convection (i.e., buoyant forces neglected) one could expect the sought-for heat transfer coefficient to be represented by a relation of the form [see Eq. (6.33)]

$$N_{NU} = F(N_{RE}, N_{EC}, N_{PR}).$$

The Eckert number, N_{EC}, arises from the dissipation of mechanical energy through viscous friction. When this dissipation is negligible compared with the other energy forms, the form of the heat transfer relation becomes

$$N_{NU} = F(N_{RE}, N_{PR}).$$

The solutions of Chapter 7 were of the above form. In the event analytical solutions are not available, the form of the function F may be deduced experimentally if the data are reduced to the dimensionless groups indicated.

Because of certain effects that were not considered in the development of the above dimensionless representations, additional parameters may enter an experimental correlation of data. Examples of such effects are the temperature dependence of properties and thermal starting lengths. The treatment of these special effects will be illustrated in the following sections.

It should also be pointed out that in the correlation of experimental forced convection data deviations of the order of magnitude of ± 10 per cent are often considered quite satisfactory. Such deviations are due to effects such as those discussed in the above paragraph, difficulty in controlling all variables, experimental error, etc.

8.2   Forced Convection Past Plane Surfaces

Forced convection heat transfer past plane surfaces such as walls, flat air-craft surfaces, etc., may be treated as flow past a flat plate—discussed exten-sively in Chapter 7. The analytical expressions developed in Chapter 7 are found, experimentally, to hold rather well with some modifications in certain instances as noted below.

Laminar Flow. For laminar flow past a flat plate, the analytical expressions of Eqs. (7.30) and (7.32) are found to hold well if the fluid properties are evaluated at the *mean film temperature*, t_m. Thus, for the local film coefficient

$$N_{NU_x} = 0.332(N_{RE_x})^{1/2}(N_{PR})^{1/3}. \qquad (8.2)\bigstar$$

The average heat transfer coefficient for a plate length L may be calculated from

$$N_{NU_L} = 0.664(N_{RE_L})^{1/2}(N_{PR})^{1/3}. \qquad (8.3)\bigstar$$

In the above expressions the local and average Reynolds and Nusselt num-bers, as defined by Eq. (7.29), have been used. The velocity in the undisturbed stream outside the boundary layer should be used in N_{RE}. Since laminar flow is assumed, N_{RE} must be less than 400,000. Equations (8.2) and (8.3) are valid for $N_{PR} \geq 0.6$.

Turbulent Flow. Equations (7.86) and (7.93) for Prandtl's and von Kármán's modifications of Reynolds' analogy for turbulent flow may also be used with confidence for real problems if the fluid properties are evaluated at the mean film temperature and as long as the Prandtl number is not too different from unity. These relations are

$$N_{\text{NU}_x} = \frac{0.0292(N_{\text{RE}_x})^{0.8}N_{\text{PR}}}{1 + 2.12(N_{\text{RE}_x})^{-1/10}(N_{\text{PR}} - 1)},$$

$$N_{\text{NU}_x} = \frac{0.0292(N_{\text{RE}_x})^{0.8}N_{\text{PR}}}{1 + 0.855(N_{\text{RE}_x})^{-1/10}\{(N_{\text{PR}} - 1) + \ln[1 + \frac{5}{6}(N_{\text{PR}} - 1)]\}}.$$

(8.4)

In their present form, however, these equations are difficult to integrate for the more useful average Nusselt number.

Colburn (Ref. 18) finds, on an empirical basis, that for gases the denominator of Eq. (8.4) is nearly constant, since N_{PR} will not be too far removed from 1. For such cases it is recommended that

$$N_{\text{NU}_x} = 0.0292(N_{\text{RE}_x})^{0.8}(N_{\text{PR}})^{1/3}.$$

(8.5)★

Again, the mean film temperature should be used for all properties. For a plate length L, an average Nusselt number may now be evaluated by integration:

$$h_{\text{av}} = \frac{1}{L}\left[0.0292(N_{\text{PR}})^{1/3}\left(\frac{U}{v}\right)^{0.8}k\int_0^L \frac{x^{0.8}}{x}\,dx\right]$$

$$= \frac{1}{L}\left[\frac{5}{4}0.0292(N_{\text{PR}})^{1/3}\left(\frac{U}{v}\right)^{0.8}kL^{4/5}\right]$$

$$N_{\text{NU}_L} = \frac{hL}{k} = 0.036(N_{\text{RE}_L})^{0.8}(N_{\text{PR}})^{1/3}.$$

(8.6)

This latter relation assumes that the boundary layer is fully turbulent from the leading edge of the plate. Actually, as discussed extensively in Sec. 7.6, there must be a certain portion of laminar flow at the front of the plate before the transition length Reynolds number is reached and turbulent flow is established. A better average film coefficient and average Nusselt number would be given by a combination of Eqs. (8.2) and (8.5):

$$h_{\text{av}} = \frac{1}{L}\left[0.332(N_{\text{PR}})^{1/3}k\left(\frac{U}{v}\right)^{1/2}\int_0^{x_{\text{cr}}} \frac{x^{1/2}}{x}\,dx\right.$$

$$\left. + 0.0292(N_{\text{PR}})^{1/3}k\left(\frac{U}{v}\right)^{0.8}\int_{x_{\text{cr}}}^L \frac{x^{0.8}}{x}\,dx\right].$$

The symbol x_{cr} is used to denote the plate length required to reach the transition or "critical" Reynolds number. The assumption that the laminar turbulent transition occurs instantaneously has been made. Integration gives the following expression for the average Nusselt number $N_{RE_{cr}}$, symbolizing the transitional Reynolds number:

$$N_{NU_L} = 0.664(N_{PR})^{1/3}(N_{RE_{cr}})^{1/2} + 0.036(N_{PR})^{1/3}[(N_{RE_L})^{0.8} - (N_{RE_{cr}})^{0.8}],$$

$$N_{NU_L} = 0.036(N_{PR})^{1/3}[(N_{RE_L})^{0.8} - (N_{RE_{cr}})^{0.8} + 18.44(N_{RE_{cr}})^{1/2}]. \qquad (8.7)\bigstar$$

For a transitional Reynolds number of 400,000 this is

$$N_{NU_L} = 0.036(N_{PR})^{1/3}[(N_{RE_L})^{0.8} - 18,700], \qquad (8.8)\bigstar$$

whereas a $N_{RE_{cr}} = 500,000$ gives

$$N_{NU_L} = 0.036(N_{PR})^{1/3}[(N_{RE_L})^{0.8} - 23,100]. \qquad (8.9)$$

EXAMPLE 8.1:   Atmospheric air at 100°F flows at a velocity of 150 ft/sec past a flat plate 2 ft long with its surface maintained at 500°F. At this velocity the air may be treated as incompressible.
(a) Find, for 1-ft width, the heat transferred to the air from the entire plate, from the laminar portion of the boundary layer, and from the turbulent portion of the boundary layer.
(b) What error is involved if the boundary layer is assumed to be entirely turbulent from the leading edge?
(c) Repeat the above if the flow velocity is doubled and all other data remain the same—assuming the flow is still incompressible.

Solution: The mean film temperature is $\frac{1}{2}(100 + 500) = 300°F$, at which value Table A.7 gives

$$k = 0.0203 \text{ Btu/hr-ft-}°F, \quad v = 1.101 \text{ ft}^2/\text{hr}, \quad N_{PR} = 0.687.$$

(a) The N_{RE} for the entire plate is

$$N_{RE_L} = \frac{150 \times 3600 \times 2}{1.101} = 9.80 \times 10^5.$$

Equation (8.8), assuming a transition at $N_{RE_{cr}} = 400,000$, gives

$$N_{NU_L} = 0.036(0.687)^{1/3}[(9.80)^{0.8} \times 10^4 - 18,700] = 1372.$$

So the average film coefficient is

$$h = N_{NU_L}\frac{k}{L} = 1372\frac{0.0203}{2} = 13.9 \text{ Btu/hr-ft}^2\text{-}°F.$$

The total heat transferred is

$$q_{total} = Ah(t_s - t_f) = (1 \times 2)(13.9)(500 - 100) = 11,110 \text{ Btu/hr (3256 W)}.$$

For the laminar portion of the boundary layer, it is necessary to locate the point of transition. For a critical $N_{RE_{cr}} = 400,000$, the critical plate length is

$$x_{cr} = N_{RE_{cr}} \frac{v}{U} = 400,000 \frac{1.101}{150 \times 3600} = 0.817 \text{ ft}.$$

For this length, the average Nusselt number is given by Eq. (8.3):

$$N_{NU} = 0.664(40)^{1/2} \times 10^2 \times (0.687)^{1/3} = 370.$$

Thus,

$$h_{lam} = 370 \frac{0.0203}{0.817} = 9.2 \text{ Btu/hr-ft}^2\text{-°F}.$$

So the heat transferred from the laminar portion is

$$q_{lam} = (1 \times 0.817)(9.2)(500 - 100) = 3010 \text{ Btu/hr (882 W)}, 27.1\% \text{ of the total}.$$

The heat transferred from the turbulent portion is, then,

$$q_{turb} = 11,110 - 3010 = 8100 \text{ Btu/hr (2374 W)}, 72.9\% \text{ of the total}.$$

(b) Upon assuming a turbulent layer for the entire plate, Eq. (8.6) gives, for $N_{RE} = 9.8 \times 10^5$,

$$N_{NU_L} = 0.036(0.687)^{1/3}(9.8)^{0.8} \times 10^4 = 1970.$$

Thus,

$$h = 1970 \times \frac{0.0203}{2} = 20.0 \text{ Btu/hr-ft}^2\text{-°F}.$$

Thus, the total heat transfer is

$$q = (1 \times 2) \times 20 \times (500 - 100) = 16,000 \text{ Btu/hr (4689 W)}.$$

This value is 44% higher than the total found in part (a).

(c) Doubling the flow velocity to 300 ft/sec merely influences the computation of the Reynolds number. Performing the identical calculations will give

$$
\begin{aligned}
q_{lam} &= 3,010 \text{ Btu/hr (882 W), } 13.1\% \text{ of total} \\
q_{turb} &= 19,900 \text{ Btu/hr (5833 W), } 86.9\% \text{ of total} \\
q_{total} &= 23,000 \text{ Btu/hr (6714 W)}.
\end{aligned}
$$

For an assumed turbulent layer for the entire plate, $q = 27{,}800$ Btu/hr, which is 21% high. This shows that the laminar portion has been shortened—the turbulent layer being correspondingly increased.

8.3   Forced Convection Inside Cylindrical Pipes or Tubes

The engineering importance of the forced flow of fluids through pipes and tubes for the purposes of transport, heating, cooling, etc., is apparent. Most heat exchangers involve the heating or cooling of fluids in tubes. Pipes or ducts transporting liquids or gases are invariably at a temperature level different from the surroundings, causing a flow of heat to or from the fluid inside. Both laminar and turbulent conditions are encountered.

Most of the relations to be presented in this section are restricted to fluids with Prandtl numbers in the range, approximately, given by $0.5 < N_{PR} < 100$. If there are any exceptions they will be noted. The case of fluids of very low N_{PR} will be considered later.

Laminar Flow. As discussed in Sec. 7.10, the analytical determination of heat transfer for forced convection by laminar flow in a pipe is one of the most complex problems of heat transfer. As noted by Eq. (7.110), Nusselt showed that for very long tubes, well in excess of the starting length, the Nusselt number based on the pipe diameter will approach

$$N_{NU_D} = 3.66.$$

Hausen (Ref. 5) developed the following empirical relation, which approaches this limiting value as the pipe length becomes very great compared with the diameter:

$$N_{NU_D} = \frac{hD}{k} = 3.66 + \frac{0.0668(D/L)(N_{RE_D})(N_{PR})}{1 + 0.04[(D/L)(N_{RE_D})(N_{PR})]^{2/3}}. \tag{8.10}$$

In this relation the fluid properties are to be evaluated at the *bulk temperature*, t_B. The Reynolds number is based on the pipe diameter and the mean cross-sectional velocity. The pipe diameter and length are D and L, respectively.

Sieder and Tate (Ref. 6) developed the following more convenient empirical formula for short tubes:

$$N_{NU_D} = 1.86(N_{RE_D})^{1/3}(N_{PR})^{1/3}\left(\frac{D}{L}\right)^{1/3}\left(\frac{\mu}{\mu_s}\right)^{0.14} \quad \tag{8.11}$$

The fluid properties are to be evaluated at the *bulk fluid temperature* except for the quantity μ_s. This is the dynamic viscosity evaluated at the temperature of the pipe surface. The last term, $(\mu/\mu_s)^{0.14}$, is included (as discussed in the introductory paragraphs of this chapter) to account for the fact that the boundary layer at the pipe surface (which controls the value of h) is strongly influenced by the temperature dependence of the viscosity. The fluids of engineering importance which are most likely to be in laminar flow are the viscous oils (see Table A.5). These fluids generally exhibit a strong temperature dependence in their viscosities. The term $(\mu/\mu_s)^{0.14}$ includes, then, the influence of whether the fluid is being heated or cooled by the pipe surface. The effect of the starting length is given by the term $(D/L)^{1/3}$.

EXAMPLE 8.2: The engine oil listed in Table A.5 flows at the rate of 3000 lb$_m$/hr through a 3-in.-I.D. pipe. The pipe is maintained at 210°F, and the oil is at 320°F. If the pipe is 50 ft long, compare the film coefficient predicted by Hausen's formula [Eq. (8.10)] and Sieder and Tate's formula [Eq. (8.11)].

Solution: At 320°F, Table A.5 gives

$$\rho = 50.3 \text{ lb}_m/\text{ft}^3,$$
$$\mu = 10.9 \text{ lb}_m/\text{ft-hr},$$
$$k = 0.076 \text{ Btu/hr-ft-°F},$$
$$v = 0.216 \text{ ft}^2/\text{hr},$$
$$N_{PR} = 84.$$

At 210°F, $\mu_s = 41.3 \text{ lb}_m/\text{ft-hr}$.
  The average flow velocity is then

$$U_m = \frac{3000}{50.3\left(\dfrac{9\pi}{4 \times 144}\right)} = 1215 \text{ ft/hr}.$$

So

$$N_{RE_D} = \frac{1215 \times 3/12}{0.216} = 1406.$$

Thus, the flow is laminar, and Eq. (8.10) gives

$$N_{NU_D} = 3.66 + \frac{0.0668\left(\dfrac{3/12}{50}\right)(1406)(84)}{1 + 0.04\left[\left(\dfrac{3/12}{50}\right)(1406)(84)\right]^{2/3}} = 13.96.$$

Thus,

$$h = 13.96\frac{0.076}{3/12} = 4.25 \text{ Btu/hr-ft}^2\text{-°F } (24.1 \text{ W/m}^2\text{-°C}).$$

The formula of Sieder and Tate, Eq. (8.11), gives

$$N_{NU_D} = 1.86\left(\frac{3/12}{50} \times 1406 \times 84\right)^{1/3}\left(\frac{10.9}{41.3}\right)^{0.14}$$

$$= 13.0$$

$$h = 13.0\frac{0.076}{3/12} = 3.95 \text{ Btu/hr-ft}^2\text{-}°\text{F} \ (22.4 \text{ W/m}^2\text{-}°\text{C}).$$

The difference (approximately 10%) between the two methods of calculation is typical of the spread observed in experimentally determined values of the heat transfer coefficient.

Turbulent Flow. The application of boundary layer theory to the analysis of turbulent forced convection in a pipe, as described in Sec. 7.10, involved a great number of simplifying assumptions. Prandtl's analogy led to Eq. (7.113) for the Nusselt number:

$$N_{NU_D} = \frac{0.0396(N_{RE_D})^{3/4}N_{PR}}{1 + 2.44(N_{RE_D})^{-1/8}(N_{PR} - 1)}. \tag{7.113}$$

This equation has been found to be somewhat unreliable for the prediction of turbulent heat transfer coefficients if N_{PR} differs greatly from unity. The main source of the error seems to be due to the constant 2.44, which, as examination of the derivation of Eq. (7.113) will show, is determined by the ratio of the velocity at the edge of the laminar sublayer to the mean pipe velocity. [See Eq. (7.103).] This numerical value resulted from joining the linear sublayer velocity distribution with the one-seventh power law velocity distribution in the turbulent layer. Experiment shows this value to be too large, and Prandtl (Ref. 7) obtained a modification by making the slope of the one-seventh power velocity distribution at its juncture with the laminar sublayer such that it gave a laminar shear stress equal to that required by Eq. (7.65). This also introduces a discontinuity in the velocity profile. Such an analysis modifies the constant by a factor of $(1/7)^{1/6}$, so

$$N_{NU_D} = \frac{0.0396(N_{RE_D})^{3/4}N_{PR}}{1 + 1.74(N_{RE_D})^{-1/8}(N_{PR} - 1)}. \tag{8.12}$$

Hoffman (Ref. 8), on the other hand, suggests that this factor is a function of the Prandtl number:

$$N_{NU_D} = \frac{0.0396(N_{RE_D})^{3/4}N_{PR}}{1 + 1.5(N_{PR})^{-1/6}(N_{RE_D})^{-1/8}(N_{PR} - 1)}. \tag{8.13}$$

Both these relations are based on the determination of the fluid properties at the *bulk fluid temperature*. Von Kármán's analysis, Eq. (7.114), which included the buffer zone, also gives reliable results.

$$N_{NU_D} = \frac{0.0396(N_{RE_D})^{3/4} N_{PR}}{1 + 0.995(N_{RE_D})^{-1/8}\{(N_{PR} - 1) + \ln[1 + \frac{5}{6}(N_{PR} - 1)]\}}. \tag{8.14}$$

The relations expressed in Eqs. (8.12) through (8.14) give results that may be used with confidence for fluids with N_{PR} near 1, but they are awkward to use. As long as the difference between the pipe surface temperature and the bulk fluid temperature is not greater than 10°F for liquids or 100°F for gases, McAdams (Ref. 10) suggests the following empirical relation based on the bulk fluid temperature:

$$N_{NU_D} = 0.023(N_{RE_D})^{0.8}(N_{PR})^n, \tag{8.15}\bigstar$$

where

$$n = 0.4 \text{ for heating, } fluid$$

$$n = 0.3 \text{ for cooling.}$$

Figure 8.1 repeats the experimental data noted in Fig. 7.17, and the predictions of Eqs. (8.12), (8.14), and (8.15) are also plotted for comparison. It is apparent that the simpler expression of Eq. (8.15) gives as good a prediction of the Nusselt number as do the more complicated analytical expressions of Prandtl and von Kármán.

For temperature differences greater than the limits specified for Eq. (8.15) *or* for fluids more viscous than water, the following expression due to Sieder and Tate (Ref. 6) will give better results:

$$N_{NU_D} = 0.027(N_{RE_D})^{0.8}(N_{PR})^{1/3}\left(\frac{\mu}{\mu_s}\right)^{0.14}. \tag{8.16}\bigstar$$

As in the case of Eq. (8.11), all properties are evaluated at the bulk fluid temperature except for μ_s, which is to be evaluated at the pipe surface temperature. Knudsen and Katz (Ref. 25) note that Eq. (8.16) is applicable for $0.7 < N_{PR} < 16,700$. Although Eq. (8.16) is of convenient form, more information is required to evaluate h than in the case of Eq. (8.15), namely the pipe surface temperature. The fact that Eq. (8.15) requires knowledge of only the bulk temperature is the main reason that it is so widely used.

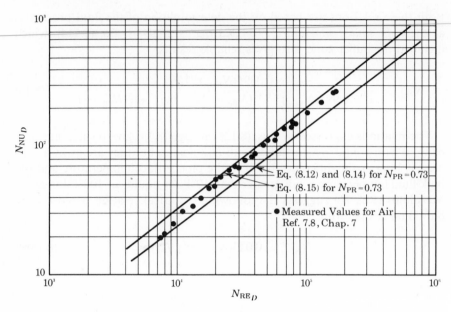

Fig. 8.1. Measured values of the Nusselt number for turbulent air flow in a pipe compared with the equations of Prandtl, von Kármán, and McAdams.

The above equations are based on the assumption that the starting length is not important, and hence the pipe length does not enter as a separate parameter. In a sense, the film coefficient given by these relations is local in that as the fluid proceeds down the pipe the local bulk temperature, upon which the fluid properties are based, will change. Hence, there will be a corresponding change in the resulting film coefficient. This change may, or may not, be small. If the variation is significant, some suitable mean value will have to be used in the application of these relations to design problems. This topic is discussed in Chapter 12.

In reality the starting length will have an effect, since, theoretically, the film coefficient at the entrance to a pipe should be infinite. For short tubes the starting length may significantly affect the value of the local film coefficient but not so much in turbulent flow as in the case of laminar flow. Nusselt (Ref. 11) found the effect to be represented by the following expression:

$$N_{NU_D} = 0.036(N_{RE_D})^{0.8}(N_{PR})^{1/3}\left(\frac{\mu}{\mu_s}\right)^{0.14}\left(\frac{D}{L}\right)^{1/18}. \qquad (8.17)\bigstar$$

EXAMPLE 8.3: Water at 80°F flows at the rate of 1.5 ft/sec through a 1-in., 18-gage, steam condenser tube. Estimate the film coefficient at the inside tube surface.

Solution: Table A.4 gives, at 80°F,

$$k = 0.355 \text{ Btu/hr-ft-°F},$$

$$v = 0.0334 \text{ ft}^2/\text{hr},$$

$$N_{PR} = 5.85.$$

Table B.2 gives the inside diameter of a 1-in., 18-gage tube to be 0.902 in. Thus,

$$N_{RE_D} = \frac{1.5 \times 3600 \times (0.902/12)}{0.0334} = 1.21 \times 10^4.$$

The flow is turbulent, and Eq. (8.15) applies in this case with $n = 0.4$ since the water will be heated in a steam condenser:

$$N_{NU_D} = 0.023(0.121)^{0.8} \times 10^4 \times (5.85)^{0.4}$$

$$= 86.2,$$

$$h = 86.2 \times \frac{0.355}{(0.902/12)} = 407 \text{ Btu/hr-ft}^2\text{-°F} \ (2311 \text{ W/m}^2\text{-°C}).$$

8.4   Forced Convection in Annular Spaces

There are many engineering applications wherein heat is exchanged in a double-tube arrangement—one fluid flowing inside a tube and a second fluid flowing in the annulus formed by another tube concentrically surrounding the first.

The film coefficient for the fluid flowing in the inner pipe is, of course, determined by the relations of Sec. 8.3. The flow in the annulus, however, does not satisfy the conditions applying to the equations and some modification has been sought. Apparently, however, no really reliable results exist for laminar flows in annuli. In the case of turbulent flow in annular spaces, many relations have been proposed (Refs. 1, 2) with varying degrees of success. The most reliable results at the present time seem to be a simple modification of Eq. (8.15) for flow in a cylindrical tube.

For fluid flow in noncircular ducts, it is customary to define an "equivalent diameter" as four times the cross-sectional flow area divided by the wetted perimeter. The equivalent diameter is the diameter of a hypothetical circular pipe which has the same pressure gradient along its axis as the noncircular ducts if the mean flow velocities are the same. If the annulus has an inner diameter (i.e., the O.D. of the inner pipe) of D_1, and an outer diameter (i.e., the I.D. of the outer pipe) of D_2, then the equivalent diameter, D_E, is

$$D_E = 4\frac{\frac{\pi}{4}(D_2^2 - D_1^2)}{\pi(D_2 + D_1)} = D_2 - D_1. \tag{8.18}$$

For forced convection inside an annular space the Nusselt number and Reynolds number are based on this equivalent diameter:

$$N_{NU_{D_E}} = \frac{hD_E}{k},$$

$$N_{RE_{D_E}} = \frac{U_m D_E}{v}.$$

It is recommended that the film coefficient at either surface of the annulus be evaluated by Eq. (8.15) based on the equivalent diameter:

$$N_{NU_{D_E}} = 0.023(N_{RE_{D_E}})^{0.8}(N_{PR})^n,$$

$$n = 0.4, \text{ heating}, \tag{8.19}\bigstar$$

$$n = 0.3, \text{ cooling}.$$

8.5   Forced Convection Inside Cylindrical Tubes for Fluids with Very Low Prandtl Numbers—Liquid Metals

As noted in the introductory remarks of Sec. 8.3, the relations quoted in that section for forced convection inside tubes were to be considered reliable only for fluids with Prandtl numbers greater than about 0.5. This covers the great majority of fluids encountered in engineering practice. The new field of atomic reactor technology, however, often uses liquid metals as heat transfer mediums. Owing to the high thermal conductivity of metals, the Prandtl numbers of liquid metals are in the order of magnitude of 0.01. Table A.11 tabulates the properties of several liquid metals, and the very low values of N_{PR} may be observed.

The analyses of Prandtl and von Kármán relied on neglecting the molecular kinematic viscosity and diffusivity in the fully turbulent core. For fluids of such low conductivity (low Prandtl number) the magnitudes of these molecular coefficients are such that they cannot be neglected.

Martinelli (Ref. 12) derived an analytical expression which more fully accounts for behavior of the fluid in the laminar sublayer and also accounts for the presence of the transition region between the laminar sublayer and the turbulent core. Martinelli's analysis applies to fluids of large Prandtl number as well as low Prandtl number, but its greater application has been made in the cases of liquid metal heat transfer.

Based on an assumed *uniform heat flux* at the wall, Martinelli's analysis may be expressed

$$N_{NU_D} = \frac{N_{RE_D} N_{PR} \sqrt{\dfrac{f}{8} \dfrac{t_s - t_c}{t_s - t_B}}}{5\left[N_{PR} + \ln\left(1 + 5N_{PR}\right) + \dfrac{F}{2} \ln\left(\dfrac{N_{RE_D}}{60} \sqrt{\dfrac{f}{8}}\right) \right]}. \tag{8.20}$$

In this expression f is the friction factor defined in Sec. 7.9. The symbol t_s denotes the pipe surface temperature, t_c the temperature at the centerline, and t_B is the familiar bulk temperature. The factor F is a complex function of the Reynolds number and the Prandtl number, selected values of which are given in Table 8.1 for the case in which the friction factor is assumed to follow the smooth pipe law given in Eq. (7.104):

$$f = \frac{0.316}{(N_{RE_D})^{1/4}}.$$

Substitution of this equation into Eq. (8.20) gives

$$N_{NU_D} = \frac{0.04(N_{RE_D})^{7/8} N_{PR} \dfrac{t_s - t_c}{t_s - t_B}}{N_{PR} + \ln\left(1 + 5N_{PR}\right) + \dfrac{F}{2} \ln\left[\dfrac{(N_{RE_D})^{7/8}}{300}\right]}. \tag{8.21}\bigstar$$

The fluid properties are all evaluated at the bulk temperature t_B, but it should be noted that it is also necessary to have the ratio $(t_s - t_c)/(t_s - t_B)$. Martinelli's analysis relates this ratio to the velocity distribution and temperature

Table 8.1   The Factor F, as a Function of N_{PR} and N_{RE_D}, for Use in Martinelli's Equation*

$\downarrow N_{RE_D} \times N_{PR}$ $\qquad N_{RE_D} \rightarrow$	10^4	10^5	10^6
10^2	0.18	0.098	0.052
10^3	0.55	0.45	0.29
10^4	0.92	0.83	0.65
10^5	0.99	0.985	0.980
10^6	1.00	1.00	1.00

* From W. H. McAdams, *Heat Transmission*, 3rd ed., New York, McGraw-Hill, 1954.

Table 8.2 The Ratio $(t_s - t_B)/(t_s - t_c)$, as Functions of N_{PR} and N_{RE_D}, for Use in Martinelli's Equation*

$N_{PR} \downarrow$ $N_{RE_D} \rightarrow$	10^4	10^5	10^6	10^7
0	0.564	0.558	0.553	0.550
10^{-4}	0.568	0.560	0.565	0.617
10^{-3}	0.570	0.572	0.627	0.728
10^{-2}	0.589	0.639	0.738	0.813
10^{-1}	0.692	0.761	0.823	0.864
1.0	0.865	0.877	0.897	0.912
10	0.958	0.962	0.963	0.966
10^2	0.992	0.993	0.993	0.994
10^3	1.00	1.00	1.00	1.00

* From W. H. MCADAMS, *Heat Transmission*, 3rd ed., New York, McGraw-Hill, 1954.

distribution in the pipe, resulting in another function of the Reynolds and Prandtl numbers. This is given in Table 8.2. These two tables and Eq. (8.21) permit the determination of the heat transfer coefficient when given D, U_m, and t_B.

Although Martinelli's equation is an achievement of analysis, it is awkward to use and may not always be reliable. Seban and Shimazaki (Ref. 13) modified Martinelli's theory for the case in which the wall temperature is constant. Calculation of Nusselt numbers by their analysis involves an iterative procedure. However, for the fluids of very low Prandtl number, their results are well approximated by the formula

$$N_{NU_D} = 5 + 0.025(N_{RE}N_{PR})^{0.8}. \tag{8.22}$$

Lubarsky and Kaufman (Ref. 24), in a comprehensive review of the experimental results of liquid metal heat transfer measurements of several investigators, indicate that the following equation represents very well the expected values of the Nusselt number for most liquid metals—mercury being a notable exception:

$$N_{NU_D} = 0.625(N_{RE_D}N_{PR})^{0.4}. \tag{8.23}★$$

This equation is, as usual, based on the bulk fluid temperature and presumes that the flow is a fully developed turbulent flow.

EXAMPLE 8.4: Liquid sodium at 700°F flows at the rate of 17 ft/sec in a tube of $\frac{1}{2}$-in. inside diameter. Compare the film coefficient predicted by Eqs. (8.21) through (8.23).

Solution: At 700°F, Table A.11 gives the following values for the properties of sodium:

$$\rho = 53.7 \text{ lb}_m/\text{ft}^3,$$
$$k = 41.8 \text{ Btu/hr-ft-°F},$$
$$v = 0.0125 \text{ ft}^2/\text{hr},$$
$$N_{PR} = 0.005.$$

Thus,

$$N_{RE_D} = \frac{17 \times 3600 \times (0.5/12)}{0.0125} = 2.04 \times 10^5.$$

Lubarsky's equation, Eq. (8.23), gives

$$N_{NU_D} = 0.625(2.04 \times 10^5 \times 0.005)^{0.4}$$
$$= 10,$$

$$h = 10 \times \frac{41.8}{0.5/12} = 10,000 \text{ Btu/hr-ft}^2\text{-°F } (56,780 \text{ W/m}^2\text{-°C}).$$

Seban's equation, Eq. (8.22), gives

$$N_{NU_D} = 5 + 0.025(2.04 \times 10^5 \times 0.005)^{0.8}$$
$$= 11.4,$$
$$h = 11,400 \text{ Btu/hr-ft}^2\text{-°F } (64,730 \text{ W/m}^2\text{-°C}).$$

This is in reasonable agreement with the results of Lubarsky's equation, as is Martinelli's equation, Eq. (8.21), although the answer is somewhat greater. From Table 8.1 at $N_{RE_D} = 2 \times 10^5$ and $N_{PR}N_{RE_D} = 10^3$, $F = 0.42$, whereas Table 8.2 gives

$$\frac{t_s - t_c}{t_s - t_B} = \frac{1}{0.65}.$$

Thus,

$$N_{NU_D} = \frac{0.04(2.04 \times 10^5)^{7/8} \times 0.005 \times \dfrac{1}{0.65}}{0.005 + \ln(1.025) + 0.21 \ln\left[\dfrac{(2.04 \times 10^5)^{7/8}}{300}\right]}$$

$$= 12.6,$$
$$h = 12,600 \text{ Btu/hr-ft}^2\text{-°F } (71,550 \text{ W/m}^2\text{-°C}).$$

It might be pointed out here that such agreement as shown by the three equations is considered good for forced convective heat transfer because of the spread observed in actual experimental results.

8.6 Forced Convection in Flow Normal to Single Tubes and Tube Banks

The last four sections have dealt with the flow of fluids past flat surfaces and inside tubes and pipes. Of equal engineering importance is the situation in which a fluid stream flows normal to an obstacle—for instance, a circular cylinder as depicted in Fig. 8.2. Flow around such bluff bodies introduces special complications not encountered in the foregoing situations.

Figure 8.2(a) depicts a circular cylinder placed normal to a fluid stream, the properties of which at points well upstream of the cylinder are denoted by U_∞, t_∞, and ρ_∞. Classical ideal, inviscid, flow theory predicts a pressure distribution around the surface of the cylinder as suggested in Fig. 8.2(b). The pressure at the forward stagnation point equals the sum of the free stream static pressure and the free stream dynamic pressure. Then, the pressure first drops as the fluid speeds up around the forward part of the cylinder. Since the ideal fluid is not retarded by surface friction, it slows around the rear half of the cylinder exactly in reverse of the acceleration around the forward half. Consequently, the pressure rises around the rear half of the cylinder.

The behavior of a boundary layer in a region of increasing pressure, such as on the rear of a cylinder in cross flow, is quite different than those discussed previously for flat surfaces. As a viscous boundary layer moves into a region

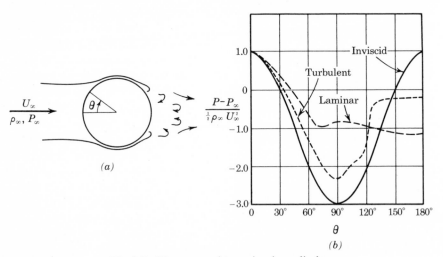

Fig. 8.2. Flow normal to a circular cylinder.

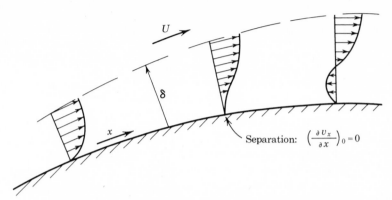

Fig. 8.3. Boundary layer flow in an adverse pressure gradient.

of rising pressure, its velocity profile becomes successively more and more flattened. The retarded layer, deficient in momentum, has increasing difficulty overcoming the rising pressure. Such a process is depicted in Fig. 8.3. Eventually a point is reached at which

$$\left(\frac{\partial v_x}{\partial y}\right)_{y=0} = 0.$$

Downstream of this point reverse flow sets in, and the boundary layer pulls away, or "separates," from the surface. This separation causes a profound change in the pressure distribution along the surface.

Figure 8.2(b) indicates typical pressure distributions in real fluid flows around a circular cylinder compared with the ideal fluid distribution. Significant differences are seen to exist between the laminar and turbulent cases. A laminar layer will separate at a polar coordinate of about 80° from the forward stagnation point; a turbulent layer, having, on the average, more kinetic energy, will remain attached up to about 110°. The separation phenomenon is accompanied by the creation and shedding of eddies in the wake of the cylinder as suggested in Fig. 8.2(a).

The peculiar flow processes just described must have significant influence on the local heat transfer for a cylinder whose surface is maintained at a temperature different from that of the fluid stream. This problem was investigated extensively by Giedt (Ref. 29) for air flowing past heated cylinders. Figure 8.4 shows typical results. At the lower Reynolds numbers the boundary layer is laminar and a minimum value of the local Nusselt number is seen to occur at the separation point. Subsequent to separation, a rise in N_{NU} is seen to occur as transverse mixing is enhanced by the creation of the eddies mentioned above. At higher Reynolds numbers a first minimum in N_{NU}

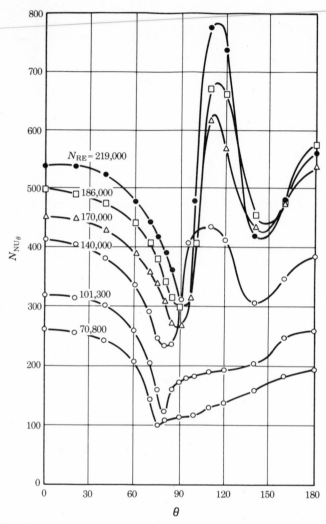

Fig. 8.4. The circumferential variation of a local heat transfer coefficient for the cross flow of air over a cylinder. (From W. H. Giedt, "Variation of Point Unit Heat Transfer Coefficient Around a Cylinder Normal to an Air Stream," *Trans. ASME*, Vol. 71, 1949, p. 375. Used by permission.)

is observed when the laminar turbulent transition takes place and a second when separation occurs.

The above processes are sufficiently complex that an analytical description is virtually impossible. Hilpert (Ref. 14) found that the *average* Nusselt number for gases flowing normal to single cylinders could be represented

Table 8.3   Values of C and m for Hilpert's Equation*

Range of N_{RE_D}	C	m
1–4	0.891	0.330
4–40	0.821	0.385
40–4000	0.615	0.466
4000–40,000	0.174	0.618
40,000–250,000	0.0239	0.805

* From R. HILPERT, "Wärmeabgabe von geheizten Drähten und Rohren im Luftstrom," *Forsch. a.d. Geb. d. Ingenieurwes.*, Vol. 4, 1933, p. 215.

by an empirical relation of the form

$$N_{NU_D} = C(N_{RE_D})^m. \qquad (8.24)\bigstar$$

The diameter used in N_{RE_D} and N_{NU_D} is the external diameter of the cylinder, whereas the velocity used in computing the Reynolds number is that of the gas stream approaching the cylinder. All fluid properties are based on the mean film temperature, t_m. Extensive tests indicate that C and m in Eq. (8.24) are functions of N_{RE_D} and Hilpert recommends the values noted in Table 8.3. Jakob and Hawkins (Ref. 15) recommend that average values of $C = 0.3$, $m = 0.57$ may be used.

Hilpert and also Reiher (Ref. 16) investigated gases flowing normal to shapes other than circular cylinders (elliptical cylinders, rectangular shapes, etc.) such as might be encountered in automotive radiators, cooling coils, etc. Results similar to Eq. (8.24) have been measured for such cases and may be found in Ref. 2.

The case of flow of gases across a bank or bundle of tubes is, naturally, a very complex situation—depending on the arrangement of the tubes in the bank, relative spacing, etc. It is misleading to attempt to quote general expressions for such cases, and one should consult such sources as Refs. 1, 3, and 17 for detailed information on particular arrangements. Such questions will also be considered further in Chapter 12 in connection with the discussion of heat exchangers—an application in which tube banks are frequently used. For a general idea of orders of magnitude, one may use Colburn's (Ref. 18) equation:

$$N_{NU_D} = 0.33(N_{RE_D})^{0.6}(N_{PR})^{1/3}. \qquad (8.25)\bigstar$$

The mean film temperature is advised for the evaluation of the fluid properties, and the Reynolds number is based on the mean velocity in the minimum free cross section for gas flow.

8.7   Forced Convection in High Speed Flow

The correlations for forced convection heat transfer presented thus far in this chapter have been limited to those cases in which the free stream flow velocities were of moderate magnitude. Many applications exist, particularly in connection with modern aerospace problems, in which the flow velocities encountered are sufficiently high that additional effects must be considered. The principal effects that must be included, which have been omitted thus far, are those of viscous dissipation and fluid compressibility.

Viscous Dissipation in Incompressible Flow

The effect of viscous dissipation at high velocities may be significant in certain cases of liquid flow, such as might be encountered in the use of highly viscous oils in lubrication applications, etc. The flow of gases, even though compressible, may usually be treated as incompressible as long as the Mach number of the flow is less than 0.2–0.3.

The effects of viscous dissipation in an incompressible fluid have already been treated for the laminar case in Sec. 7.5. The results of that analysis have been found to agree with experimentally determined heat transfer coefficients rather well. The basic findings of that analysis were that the heat transfer in the presence of dissipation could be found by the use of the recovery temperature, t_r:

$$\frac{q}{A} = h(t_s - t_r), \tag{8.26}$$

using values of the film coefficient found from the relations based on flow without dissipation [such as those given by Eqs. (8.2) and (8.5)]. The recovery temperature is determined from the free stream and surface temperatures according to Eq. (7.64):

$$\frac{t_r - t_f}{t_s - t_f} = r\frac{N_{EC}}{2}, \tag{8.27}$$

where

$$r = \text{recovery factor} = \Psi_0(N_{PR}),$$

$$N_{EC} = \text{Eckert number} = \frac{U^2}{c_p(t_s - t_f)}. \tag{8.28}$$

The analysis of Sec. 7.5 showed that in the laminar case the recovery factor is given, rather accurately, by the approximate relation

$$r = N_{PR}^{1/2}. \tag{8.29}$$

Similar analyses for turbulent flow (Refs. 22, 23, and 26) verify that in this instance

$$r = N_{PR}^{1/3}, \tag{8.30}$$

and that the heat transfer correlations for dissipation free turbulent flow may be used, as in the laminar case, as long as the recovery temperature is used to calculate the heat flux as given in Eqs. (8.26) and (8.27).

In summary, then, the following recommendation is made for heat transfer from flat surfaces, with viscous dissipation:

$$q/A = h(t_s - t_r),$$

$$\frac{t_r - t_f}{t_s - t_f} = r\frac{N_{EC}}{2}. \tag{8.31}$$

The film coefficient is determined by the following relations with the physical properties evaluated at the mean film temperature.

Laminar Flow on Flat Plates

$$r = N_{PR}^{1/2}.$$

Local: $\quad N_{NU_x} = 0.332(N_{RE_x})^{1/2}(N_{PR})^{1/3}. \tag{8.32}\bigstar$

Average: $\quad N_{NU_L} = 0.664(N_{RE_L})^{1/2}(N_{PR})^{1/3}.$

Turbulent Flow on Flat Plates

$$r = N_{PR}^{1/3}.$$

Local: $\quad N_{NU_x} = 0.0292(N_{RE_x})^{0.8}(N_{PR})^{1/3}. \tag{8.33}\bigstar$

Average: $\quad N_{NU_L} = 0.036(N_{RE_L})^{0.8}(N_{PR})^{1/3}.$

In either the laminar or turbulent cases, the general characteristics of the temperature distribution, in particular the gradient at the walls, are as described in the analysis of Sec. 7.5 and depicted in Fig. 7.8. Repetitious discussion here is not justified.

Viscous Dissipation in Compressible Flow

In the consideration of heat transfer on the surfaces of bodies moving at high velocities in gases, consideration must be made of the effects of the fluid compressibility as well as viscous dissipation. Applications in modern aerospace technology for such heat transfer analyses are numerous. Aerodynamic heating effects on high velocity aircraft and missiles, atmospheric entry of spacecraft, etc., are some examples which readily come to mind. At the extreme velocities associated with these applications, recovery (i.e., equilibrium) temperatures in the order of 2000–3000°R and aerodynamic heating rates as high as 10,000–100,000 Btu/hr-ft^2 are encountered.

Compressibility effects in a gas are associated with the Mach number of the free stream flow, M_f, defined as

$$M_f = \frac{U}{c_f},\tag{8.34}$$

where $c_f = \sqrt{\gamma RT}$ is the local *velocity of sound* in the undisturbed stream, γ is the ratio of specific heats of the gas, R is the specific gas constant, and T_f is the absolute temperature of the gas stream. The *stagnation* temperature of the stream is known to be

$$T_0 = T_f + \frac{1}{2}\frac{U^2}{c_p},\tag{8.35}$$

which, for an ideal gas, becomes

$$T_0 = T_f\left(1 + \frac{\gamma - 1}{2}M_f^2\right).\tag{8.36}$$

Since the Eckert number of a flow past a surface at temperature T_s is defined as

$$N_{\text{EC}} = \frac{U^2}{c_p(T_s - T_f)},$$

Eq. (8.35) leads to the alternative formulation:

$$N_{\text{EC}} = 2\frac{T_0 - T_f}{T_s - T_f}.\tag{8.37}$$

Thus, an alternative definition of the *recovery factor*, which is used by aerodynamicists, may be obtained from Eq. (8.27),

$$r = \frac{T_r - T_f}{T_s - T_f} \frac{2}{N_{EC}}$$

$$= \frac{T_r - T_f}{T_0 - T_f}.$$

(8.38)

Experimental evidence (Ref. 27) reveals that the relations $r = N_{PR}^{1/2}$ for laminar flow and $r = N_{PR}^{1/3}$ for turbulent flow are apparently unaffected by compressibility effects and may be used for supersonic flow.

Detailed analyses of the velocity and thermal boundary layers in compressible flow (high subsonic or supersonic speeds), as reported in Refs. 27 and 28, reveal that the main effects of the fluid compressibility are observed in the rather extreme temperature variations which are produced in the boundary layers. Although some analytical studies of high velocity boundary layer flows have been made which include variable fluid properties, a rather simple and reliable technique is recommended by Eckert (Ref. 27). This recommendation is that the constant property heat transfer correlations may be used as long as the properties are evaluated at a reference temperature T^* (rather than the mean film temperature) given by

$$(T^* - T_f) = 0.5(T_s - T_f) + 0.22(T_r - T_f).$$

(8.39)

For low speed flow, $T_r \rightarrow T_f$ and T^* reduces to the mean film temperature. Based on the above, the following correlations are recommended:

Laminar Flow on Flat Plates. The following relations are recommended, where the superscript * denotes the evaluation of properties at T^*:

$$r = N_{PR}^{*1/2}.$$

Local: $N_{NU_x}^* = 0.332(N_{RE_x}^*)^{1/2}(N_{PR}^*)^{1/3}.$

(8.40)★

Average: $N_{NU_L}^* = 0.664(N_{RE_L}^*)^{1/2}(N_{PR}^*)^{1/3}.$

Turbulent Flow on Flat Plates

$$r = N_{PR}^{*1/3}.$$

For $N_{RE}^* \leq 10^7$,

Local: $N_{NU_x}^* = 0.0292(N_{RE_x}^*)^{0.8}(N_{PR}^*)^{1/3}.$

Average: $N_{NU_L}^* = 0.036(N_{RE_L}^*)^{0.8}(N_{PR}^*)^{1/3}.$

(8.41)★

For $10^7 \leq N_{RE}^* \leq 10^9$,

$$\text{Local:} \quad N_{NU_x}^* = \frac{0.185(N_{RE_x}^*)(N_{PR}^*)^{1/3}}{(\log_{10} N_{RE_x}^*)^{2.584}}.$$

$$\text{Average:} \quad N_{NU_L}^* = \frac{0.277(N_{RE_L}^*)(N_{PR}^*)^{1/3}}{(\log_{10} N_{RE_L}^*)^{2.584}}.$$

(8.42)★

Flow Past Cones. The applicability of the cone as a likely geometric shape to be found on aircraft, missiles, and spacecraft, is apparent. For this shape, Johnson and Rubesin (Ref. 4) recommend, for the *local* Nusselt number:

$$\text{Laminar:} \quad N_{NU_x}^* = 0.575(N_{RE_x}^*)^{1/2}(N_{PR}^*)^{1/3}$$

$$\text{Turbulent:} \quad N_{NU_x}^* = 0.0292(N_{RE_x}^*)^{0.8}(N_{PR}^*)^{1/3}.$$

(8.43)★

Stagnation Point Heating

The relations quoted in Eqs. (8.40) through (8.43) are of no use in a very important case—known as *stagnation point flow*. Atmospheric entry of blunted missiles, aircraft, or spacecraft is accompanied by rather high rates of heating at the stagnation point of the blunt surface. Relations for the heat flux at the stagnation point are useful for design purposes. Figure 8.5 illustrates

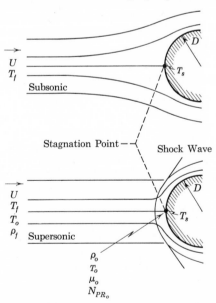

Fig. 8.5. Stagnation point flow.

the basic flow configuration in such an instance. The case of supersonic flow differs from that of a subsonic incident flow in a very significant way—a *bow shock wave* stands in front of the body.

Van Driest (in Ref. 28) makes the following recommendations for calculation of the stagnation point heat flux on blunt bodies:

Incompressible Flow. For fluid properties evaluated at the mean film temperature, and the Reynolds number based on the free stream velocity, U_∞:

$$\frac{q}{A} = h(T_\infty - T_s).$$

Cylinders: $\quad N_{NU_D} = 1.14(N_{RE_D})^{1/2}(N_{PR})^{0.4}.$ $\qquad$ (8.44)★

Spheres: $\quad N_{NU_D} = 1.32(N_{RE_D})^{1/2}(N_{PR})^{0.4}.$

Supersonic Flow. For fluid properties evaluated at the *stagnation state* of the flow *behind* the bow shock wave:

$$\frac{q}{A} = h(T_0 - T_s).$$

Cylinders: $\quad N_{NU_D} = 0.95(N_{RE_D})^{1/2}(N_{PR})^{0.4}(\rho_\infty/\rho_0)^{1/4}.$ $\quad$ (8.45)★

Spheres: $\quad N_{NU_D} = 1.28(N_{RE_D})^{1/2}(N_{PR})^{0.4}(\rho_\infty/\rho_0)^{1/4}.$

In Eqs. (8.45), ρ_∞/ρ_0 is the ratio of the gas density in the free stream to the stagnation density of the stream *behind* the bow shock wave.

EXAMPLE 8.5: A missile travels at $M = 3$ at an altitude of 10,000 ft, where the air has a temperature of 20°F and a density of 0.057 lb_m/ft^3. Find the local heat transfer coefficient and the amount of cooling necessary to maintain a point 1 ft behind the leading edge of a stabilizing fin at a temperature of 300°F.

Solution: At $T_\infty = 20 + 460 = 480°R$, the velocity of sound is (with $\gamma = 1.4$, $R = 53.3$ ft-lb$_f$/lb$_m$-°R for air) $c = \sqrt{\gamma g_c RT} = 1072$ ft/sec. Thus, the flight velocity is $v = Mc = 3 \times 1072 = 3216$ ft/sec. Assume that the flow will be turbulent and that $N_{PR} \approx 0.68$. These assumptions will be checked later. Thus, Eq. (8.36) gives the stagnation temperature to be

$$T_0 = T_\infty\left(1 + \frac{\gamma - 1}{2}M^2\right) = 480(1 + 0.2 \times 9.0)$$

$$= 1345°R = 885°F.$$

The recovery temperature is then, by Eqs. (8.38) and (8.30),

$$(0.68)^{1/3} = \frac{T_r - 480}{1345 - 480}$$

$$T_r = 1240°R = 780°F.$$

Thus, the reference temperature, T^*, is, from Eq. (8.39),

$$T^* = 480 + 0.5(760 - 480) + 0.22(1240 - 480)$$

$$= 788°R = 328°F.$$

At 328°F, Table A.7 gives for air:

$$\mu^* = 0.059 \text{ lb}_m/\text{ft-hr},$$

$$k^* = 0.021 \text{ Btu/hr-ft-}°F,$$

$$N_{PR} = 0.686.$$

The assumed value of N_{PR} is close enough to this value. Now, the Reynolds number, for $x = 1$ ft, is

$$N^*_{RE_x} = \frac{3216 \times 3600 \times 0.057 \times 1}{0.059} = 1.12 \times 10^7.$$

Thus, the flow is turbulent as assumed. Note that v cannot be taken from Table A.7 since that table is based on atmospheric pressure, and v has a strong pressure dependence. Now, since the flow is turbulent, Eq. (8.42) gives the local Nusselt number to be

$$N_{NU} = 0.185(0.686)^{1/3} \frac{1.12 \times 10^7}{[\log_{10}(1.12 \times 10^7)]^{2.584}} = 11,800.$$

The local film coefficient is then

$$h = 11,800 \frac{0.021}{1} = 248 \text{ Btu/hr-ft}^2\text{-}°F \ (1408 \text{ W/m}^2\text{-}°C).$$

The amount of cooling, per square foot, required at this point to maintain the surface at 300°F is then

$$q/A = h(T_r - T_s)$$

$$= 248(780 - 300)$$

$$= 119,000 \text{ Btu/hr-ft}^2 ! \ (3.75 \times 10^5 \text{ W/m}^2).$$

This heat flux is quoted on a per-square-foot basis for the sake of comparison with heat flux values found in the earlier examples for low speed flow. Actually, this magnitude of heat flux does not exist over an entire square foot of the fin surface.

References

1. McAdams, W. H., *Heat Transmission*, 3rd ed., New York, McGraw-Hill, 1954.
2. Jakob, M., *Heat Transfer*, Vol. 1 (1949) and Vol. 2 (1957), Wiley, New York.
3. Knudsen, J. G., and D. L. Katz, *Fluid Dynamics and Heat Transfer*, New York, McGraw-Hill, 1958.
4. Johnson, H. A., and M. W. Rubesin, "Aerodynamic Heating and Convective Heat Transfer," *Trans. ASME*, Vol. 71, 1949, p. 447.
5. Hausen, H., "Darstellung des Wärmeüberganges in Rohren durch verallgemeinerte Potenzbeziehungen," *Zeitschr. V.D.I. Beihefte Verfahrenstechnik*, No. 4, 1943, p. 91.
6. Sieder, E. N., and G. E. Tate, "Heat Transfer and Pressure Drop of Liquids in Tubes," *Ind. Eng. Chem.*, Vol. 28, 1936, p. 1429.
7. Prandtl, L., "Bemerkung über den Wärmeübergang im Rohr," *Phys. Zeitschr.*, Vol. 29, 1928, p. 487.
8. Hoffman, E., "Wärmeübergang bei turbulenter Strömung in Rohren," *Zeitschr. V.D.I.*, Vol. 82, 1938, p. 741.
9. von Kármán, T., "The Analogy Between Fluid Friction and Heat Transfer," *Trans. ASME*, Vol. 61, 1939, p. 705.
10. McAdams, W. H., "Review and Summary of Developments in Heat Transfer by Conduction and Convection," *Trans. A.I.Ch.E.*, Vol. 36, 1940, p. 1.
11. Nusselt, W., "Der Wärmeaustausch zwischen Wand und Wasser im Rhor," *Forsch. a.d. Geb. d. Ingenieurwes.*, Vol. 2, 1931, p. 309.
12. Martinelli, R. C., "Heat Transfer to Molten Metals," *Trans. ASME*, Vol. 69, No. 8, 1947, p. 947.
13. Seban, R. A., and T. T. Shimazaki, "Heat Transfer to a Fluid Flowing Turbulently in a Smooth Pipe with Walls at Constant Temperature," *Trans. ASME*, Vol. 73, 1951, p. 803.
14. Hilpert, R., "Wärmeabgabe von geheizten Drähten und Rohren im Luftstrom," *Forsch. a.d. Geb. d. Ingenieurwes.*, Vol. 4, 1933, p. 215.
15. Jakob, M., and G. A. Hawkins, *Elements of Heat Transfer*, 4th ed., New York, Wiley, 1957.
16. Reiher, H., "Der Wärmeübergang von strömender Luft an Rohr und Rohrbündel im Kreuzstrom," *V.D.I.—Forschungsheft*, No. 269, 1925.
17. Kays, W. M., and A. L. London, *Compact Heat Exchangers*, 2nd ed., New York, McGraw-Hill, 1964.
18. Colburn, A. P., "A Method of Correlating Forced Convection Heat Transfer Data and a Comparison with Fluid Friction," *Trans. A.I.Ch.E.*, Vol. 29, 1933, p. 174.
19. Schlichting, H., *Boundary Layer Theory*, 6th ed., New York, McGraw-Hill, 1968.
20. Shapiro, A. H., *The Dynamics and Thermodynamics of Compressible Fluid Flow*, Vol. 2, New York, Ronald, 1954.

21. POHLHAUSEN, E., "Der Wärmeaustausch zwischen festen Körpern und Flüssig-keiten mit kleiner Reibung und kleiner Wärmeleitung," *Zeitschr. für Angew. Math. und Mech.*, Vol. 1, 1921, p. 115.

22. ECKERT, E. R. G., "Engineering Relations for Friction and Heat Transfer to Surfaces in High Velocity Flow," *J. Aero Sci.*, Vol. 22, 1955, p. 585.

23. WIMBROW, W. R., "Experimental Investigations of Temperature Recovery Factors on Bodies of Revolution at Supersonic Speeds," *N.A.C.A. Tech. Note 1975*, Washington, D.C., 1949.

24. LUBARSKY, B., and S. J. KAUFMAN, "Review of Experimental Investigations of Liquid Metal Heat Transfer," *N.A.C.A. Tech. Note 3336*, Washington, D.C., 1955.

25. KNUDSEN, J. G., and D. L. KATZ, *Fluid Dynamics and Heat Transfer*, New York, McGraw-Hill, 1958.

26. ACKERMANN, G., "Plate Thermometer in High Velocity Flow with a Turbulent Boundary Layer," *Forsch. a.d. Geb. d. Ingenieurwes.*, Vol. 13, 1942, p. 226.

27. ECKERT, E. R. G., "Survey on Heat Transfer at High Speeds," *Aeronautical Res. Lab. Rep. 189*, Office of Aerospace Research, Wright-Patterson Air Force Base, Ohio, 1961.

28. LIN, C. C. (ed.), *Turbulent Flows and Heat Transfer, High Speed Aerodynamics and Jet Propulsion*, Vol. V, Princeton, N.J., Princeton U.P., 1959.

29. GIEDT, W. H., "Investigation of the Variation of Point Unit Heat Transfer Coefficient Around a Cylinder Normal to an Air Stream," *Trans. ASME*, Vol. 71, 1949, p. 375.

Problems

8.1. Atmospheric pressure air at 90°F flows past a plane plate maintained at 600°F. The plate is 3 ft long and the air velocity is 100 ft/sec. Find the heat transferred from the plate, per foot of width.

8.2. Glycerine at 100°F flows at 10 ft/sec past a flat plate 1 ft long which is maintained at 50°F. Find, per foot of width, the heat transferred to the plate. What percentage of this heat is transferred through the laminar portion of the boundary layer?

8.3. A plate has the dimensions of 3 in. × 18 in. It is maintained at 190°F (both sides) and placed in an air stream at 1 atm, 50°F, traveling at 90 ft/sec. Find the total rate of heat transfer from the plate if (a) its 3-in. edge is the leading edge, or (b) its 18-in. edge is its leading edge.

8.4. Air at 70°F, 14.7 psia, flows at 20 ft/sec over a flat plate 10 in. long maintained at 110°F. Find
(a) the local film coefficient at 2.5, 5.0, 7.5, and 10 in. from the leading edge;
(b) the average film coefficient for the entire plate.

8.5. Atmospheric pressure air at 60°F flows in a circular air conditioning duct 8 in. in diameter. The air flows at the rate of 2000 ft³/min. Estimate the film coefficient at the inside duct surface.

8.6. Water at 80°F enters a 2-in., schedule 40 pipe and leaves at 120°F. It flows at the rate of 50,000 lb/hr. Find the heat transfer coefficient at the inner pipe surface for
(a) the entrance conditions;
(b) the exit conditions;
(c) the average of the entrance and exit conditions.

8.7. The engine oil in Table A.5 flows at 1 ft/sec through a 1-in., 18-gage tube. The oil is at 320°F and the tube surface is at 300°F. If the tube is 6 ft long, what is the average film coefficient at the pipe surface? If the tube is 20 ft long, what is the film coefficient?

8.8. Superheated steam at 600 psia, 1000°F flows at the rate of 10,000 lb/hr through an 8-in., schedule 80 pipe. What is the expected heat transfer coefficient at the inner pipe surface?

8.9. Water in a circulating chilled water air-conditioning system enters the chilling unit at 60°F and leaves at 40°F. The tubes of the chilling unit are $1\frac{1}{2}$ in., 10-gage tubes, and a mean water velocity of 2 ft/sec is expected. What is the film coefficient that may be expected at the inner tube surface?

8.10. Air at 500°F flows through the 1-in., 10-gage tubes of an intercooler at the rate of 10 ft/sec. If the intercooler pressure is 40 atm, what is the convective film coefficient at the inner tube surface?

8.11. Natural gas (closely resembling methane) flows at 90°F through a $\frac{3}{4}$-in. pipe at a rate of 10 ft/sec. Calculate the expected heat transfer coefficient at the pipe wall.

8.12. Water at 35°F flows at the velocity of 0.5 ft/sec through a tube $\frac{3}{8}$ in. I.D. The tube surface is at 50°F and is 10 ft long. How much heat is transferred to the water per hour?

8.13. Water flowing at the rate of 350 lb_m/min is heated from 45°F to 70°F as it passes through a 2-in., schedule 40 pipe. The inside pipe wall temperature is 210°F. What is the required length of the pipe?

8.14. Water at the rate of 1.5 lb_m/sec is to be cooled from 120°F to 70°F. Which of the following provides the least pressure loss?
(a) The water flows through a $\frac{1}{2}$-in.-I.D. pipe maintained at 50°F.
(b) The water flows through a 1-in.-I.D. pipe maintained at 60°F.
Base the calculations on an average bulk water temperature of 95°F and presume that the pipes are smooth (i.e., the relative roughness is zero).

8.15. Superheated steam at 1000 psia, 700°F, enters a 1-in.-I.D. pipe at 40 ft/sec. The pipe wall temperature is 950°F. How long must the pipe be to heat the steam to 800°F?

8.16. Superheated steam flows through the annulus formed between a 4-in. and a 6-in. schedule 40 steel pipe. The steam is at 200 psia, 400°F. Its velocity is 5 ft/sec. Water at 120°F is flowing at 2 ft/sec through the inner pipe. How much heat is transferred to the water, per foot of pipe length?

8.17. Liquid sodium at 1000°F flows at the rate of 20 ft/sec through a pipe of 1 in. I.D. Estimate the convective heat transfer coefficient at the pipe surface using all three equations proposed in Sec. 8.5.

8.18. Repeat Prob. 8.17 using mercury at 600°F.

8.19. A 44% potassium, 56% sodium liquid metal mixture is used as the heat transfer medium in an atomic reactor. What film coefficient can be expected at the inner surface of a pipe of 1 in. I.D. if the medium is at 700°F and flows at 10 ft/sec?

8.20. A velocity measuring device consists of a heated wire 0.01 in. in diameter. By measuring the electrical input to the wire, the rate of flow of air across it may be estimated. If the wire is 0.5 in. long and is maintained at 200°F, how much heat does it give up to atmospheric air at 60°F flowing 30 ft/sec normal to it?

8.21. Repeat Prob. 8.20 for the case in which the air velocity is increased to 300 ft/sec.

8.22. A 4-in., schedule 40 pipe is placed in an air stream, the air flowing normal to the pipe. The air is flowing at 15 mph at 80°F, and the pipe surface is at 800°F. What is the estimated film coefficient at the pipe surface?

8.23. Air at 25 psia, 100°F, flows at 50 ft/sec past a 2-in.-O.D. cylinder maintained at 250°F and placed normal to the air flow. Find the surface heat transfer coefficient.

8.24. Compute the local heat transfer coefficient that may be expected to occur at a point 1 ft behind the leading edge of the fin of a supersonic plane traveling at $M = 2$ at
(a) sea level ($P = 14.7$ psia, $t = 59°F$);
(b) 20,000-ft altitude ($P = 6.7$ psia, $t = -12°F$).
The surface is to be maintained at 200°F in both cases. What is the equilibrium surface temperature to be expected in each case if no internal cooling is provided?

8.25. Air at an altitude of 5000 ft ($P = 12.2$ psia, $t = 41°F$) flows past a flat surface, maintained at 50°F, with a velocity of 600 ft/sec. Find the local heat flux at a point 1 in. behind the leading edge if
(a) viscous dissipation is neglected;
(b) viscous dissipation is not neglected.

8.26. A flat plate 2 ft long and 1 ft wide is placed in a wind tunnel where the conditions are $U = 3020$ ft/sec, $p = 0.5$ psia, $t_f = -40°F$. How much cooling must be provided to maintain the plate at 100°F if viscous dissipation is accounted for?

8.27. Air at 5 psia, $-40°F$, flows at Mach 3 past a flat plate 4 in. long and 1 ft wide. The plate is maintained at 400°F. Estimate the total heat transfer to, or from, the surface. Note that the surface is not entirely in either laminar or turbulent flow.

8.28. What is the equilibrium temperature and local heat transfer coefficient that may be expected to occur at a point 1 ft from the tip of the nose cone of a missile traveling at $M = 3$ at an altitude of 60,000 ft ($P = 1$ psia, $t = -67°F$)? For a 20° nose cone one can expect that the resultant shock wave will produce a flow at $M = 2.3$, $P = 3$ psia, $t = 80°F$ past the surface of the cone.

8.29. A body having a spherically blunted nose (diameter, 1 ft) travels at a velocity of 500 ft/sec in air at 40°F, 10 psia. Estimate the heat transfer rate at the stagnation point if the surface is to be maintained at 60°F.

8.30. If the body described in Prob. 8.29 travels at 5000 ft/sec at an altitude of 60,000 ft ($p = 1$ psia, $t = -67°F$), estimate the aerodynamic heating rate at the stagnation point for a surface temperature of 100°F.

8.31. A flat surface travels at 3000 ft/sec in the atmosphere at an altitude of 40,000 ft ($p = 2.7$ psia, $t = -67°F$). (a) What is the recovery temperature for these conditions? (b) The surface described above forms the outer surface of a structure. The over-all conductance of the wall of the structure is 125 Btu/hr-ft²-°F. It is desired to keep the interior surface of the structure at 100°F. What is the cooling rate which must be provided at points 1 in. and 6 in. from the leading edge?

8.32. Water at 200°F flows in a 1-in.-diameter 16-gage copper tube at the rate of 2 ft/sec. The outside surface of the tube is exposed to air at 70°F through a surface film coefficient of 1.5 Btu/hr-ft²-°F. Determine the rate of heat loss from a 1-ft length of the tube.

8.33. Superheated steam at 400 psia, 600°F, flows at the rate of 20 ft/sec through a 6-in., steel, schedule 40 pipe. The outer surface is exposed to a stream of air (14.7 psia, 70°F) flowing at 40 ft/sec normal to the pipe. Estimate the rate of heat loss, per foot of length, from the pipe.

8.34. For fully developed flow in a circular tube with a surface temperature of 300°F, find the local heat transfer coefficient for the circumstances noted below. The mean flow velocity is 20 ft/sec in each case. Discuss the reasons for the difference in the answers.
(a) Air, 200°F, 1-atm pressure, 1-in.-diameter tube.
(b) Air, 200°F, 10-atm pressure, 1-in.-diameter tube.
(c) Air, 200°F, 1-atm pressure, 0.5-in.-diameter tube.
(d) Carbon dioxide, 200°F, 1-atm pressure, 1-in.-diameter tube.
(e) Liquid sodium, 500°F, 1-atm pressure, 1-in.-diameter tube.
(f) Water, 200°F, 10-atm pressure, 1-in.-diameter tube.
(g) Lubricating oil, 200°F, 1-in.-diameter tube.

CHAPTER 9

Heat Transfer by Free Convection

9.1 Introductory Remarks

The examples of convective heat transfer considered up to now have all been examples of *forced* convection between a fluid and a solid body wherein the fluid motion relative to the solid surfaces was caused by an external input of work (i.e., by means of a pump, fan, propulsive motion of an aircraft, etc.). This chapter will be concerned with *free* or *natural* convection in which the fluid velocity at points remote from the body will be essentially zero. Near the body there will be some fluid motion if the body is at a temperature different from that of the free fluid. If such is the case, there will be a density difference between the fluid near the surface and that far removed from the surface. This density difference will produce a positive or negative buoyant force (depending on whether the surface is hotter or colder than the fluid) in the fluid near the surface. The buoyant force results in a fluid motion, substantially in the vertical direction, past the surface—with consequent convective heat transfer taking place. The force of gravity is, then, the driving force which produces the fluid motion and maintains the convective process.

The heating of rooms and buildings by the use of "radiators" is a familiar example of heat transfer by free convective. Heat losses from hot pipes, ovens, etc., surrounded by cooler air are due to free convection, at least in part.

364

9.2 The Governing Equations of Free Convection

For purposes of discussion let the heated surface, at temperature t_s, be a vertical plane. The fluid far removed from the surface is at zero velocity, and its temperature is t_f. If one assumes that $t_s > t_f$, the distribution of the fluid velocity and fluid temperature in a direction normal to the surface will be something like that shown in Fig. 9.1. The temperature decreases (from the value imposed at the surface) asymptotically to the value of the fluid far away. The fluid velocity, on the other hand, must somehow increase from zero at the wall to a maximum value in the immediate neighborhood of the wall and then decrease, asymptotically, to a zero value far away. This distribution of velocity is as shown in Fig. 9.1.

The motion of the fluid is restricted to a region close to the surface. For fluids of low viscosity this region is relatively thin, and it has been found that the simplifications leading to the boundary layer equations of motion and energy [Eqs. (6.46) and (6.47)] are applicable. However, the equation of motion must be modified. Reference to Sec. 6.11 will show that in the development of the equation of motion, Eq. (6.46), the body forces X and Y (defined in Sec. 6.5 in the derivation of the Navier–Stokes equations) were neglected. This was perfectly satisfactory for application to problems of forced convection, wherein the body force due to gravity would be expected to be small in comparison to the inertia forces of the motion. In free convection, however, it is the body force which sustains the fluid motion, and, hence, it certainly must be included. Upon selecting the x-coordinate axis as parallel to the surface, as shown in Fig. 9.1, and the y-coordinate axis as normal to it, the body forces per unit mass will be

$$\text{In the } y \text{ direction, } Y = 0,$$
$$\text{In the } x \text{ direction, } X = \frac{1}{\rho} g(\rho_f - \rho) \tag{9.1}$$

as long as one adopts the convention that the pressure term in the equation of motion represents the difference between the local fluid pressure and the pressure in the fluid, at the same x, far removed from the heated surface. In Eq. (9.1) ρ_f denotes the density of the undisturbed fluid (at temperature t_f) far removed from the surface. The coefficient of volume expansion, β, defined in Sec. 2.6, may be used to replace ρ by temperature:

$$X = \frac{1}{\rho} g(\rho_f - \rho)$$

$$= g\beta(t - t_f). \tag{9.2}$$

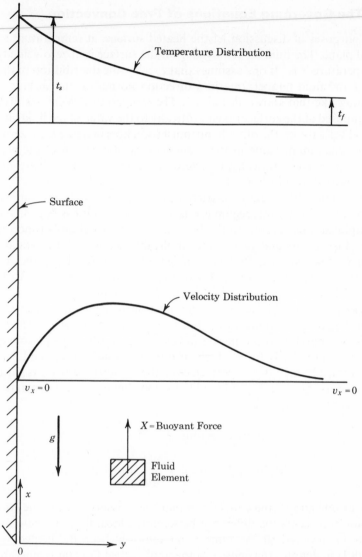

Fig. 9.1. Temperature and velocity distributions near a heated vertical surface in free convection flow.

The boundary layer equation for motion parallel to the surface is then obtained by adding Eq. (9.2) to Eq. (6.46):

$$v_x \frac{\partial v_x}{\partial x} + v_y \frac{\partial v_x}{\partial y} = -\frac{1}{\rho}\frac{dP}{dx} + v\frac{\partial^2 v_x}{\partial y^2} + g\beta(t - t_f). \tag{9.3}$$

The pressure gradient, dP/dx, may, in some cases, be neglected.

The boundary layer equation of energy, Eq. (6.47), needs no alteration other than the readily justified neglection of the viscous dissipation term $(\mu/\rho c_p)(\partial v_x/\partial y)^2$. Thus, one must also satisfy

$$v_x\frac{\partial t}{\partial x} + v_y\frac{\partial t}{\partial y} = \alpha\frac{\partial^2 t}{\partial y^2}. \tag{9.4}$$

For the determination of the velocity and temperature fields near the heated surface, and subsequently the film coefficient h, the last two equations would have to be solved simultaneously to satisfy the conditions

$$\text{At } y = 0: \quad t = t_s, v_x = 0.$$
$$\text{At } y = \infty: \quad t = t_f, v_x = 0. \tag{9.5}$$

The problem as set forth by Eqs. (9.3), (9.4), and (9.5) proves to be quite complex, even for the simple case of a vertical plane surface. The additional complexity involved here over those encountered in the case of forced convection past a plane plate is the additional term just included in the equation of motion. In the case of forced convection this term was missing, and, as was pointed out in Sec. 6.11, this fact made it possible to solve for the velocity distribution in the hydrodynamic boundary layer without consideration of the thermal boundary layer because of the absence of any term in the equation of motion which contained, or was a function of, the fluid temperature. In the problem of free convection this is not the case, and the velocity and temperature distributions must be found, simultaneously.

In like fashion, integral expressions may be written for the free convection boundary layer. The development of the integral momentum equation in Sec. 6.12 neglected the presence of the buoyancy term. Inclusion of this effect in the summation of forces in the development leading to Eq. (6.52) yields the additional term

$$\int_0^\delta g\beta(t - t_f)\, dy$$

on the right side of that equation. If the pressure gradient is neglected and if the fact that $U(x) = 0$ is used, Eq. (6.52) becomes

$$\frac{d}{dx}\int_0^\delta v_x^2\, dy = \int_0^\delta g\beta(t - t_f)\, dy - v\left(\frac{\partial v_x}{\partial y}\right)_0. \tag{9.6}$$

The integral energy equation, Eq. (6.52), remains unchanged:

$$\frac{d}{dx} \int_0^{\delta_t} (t - t_f) v_x \, dy = -\alpha \left(\frac{\partial t}{\partial y} \right)_0.$$

(9.7)

The same boundary conditions, Eq. (9.5), must be satisfied. The solution of the integral equations, Eqs. (9.6) and (9.7), is likewise complicated by the fact that the equations are "coupled." That is, since temperature appears in both the energy and momentum expressions, a simultaneous solution must be accomplished.

The analysis of Sec. 6.8, which determined the various similarity parameters of convection, produced the fact that the significant parameter unique to free convection is the Grashof number, N_{GR},

$$N_{GR} = \frac{l^3 g \beta \, \Delta t}{\nu^2} = \frac{l^3 \rho^2 g \beta \, \Delta t}{\mu^2},$$

(9.8)

in which l represents a characteristic length parameter and Δt represents the surface-fluid temperature difference $(t_s - t_f)$. The Grashof number performs much the same function for free convection flow as the Reynolds number does for forced convection. Under normal conditions one may expect that the laminar turbulent transition will take place at about $N_{GR} \cong 10^9$. In the analysis which follows the Grashof number will be seen to recur frequently.

9.3 Analytical Solutions of Free Convection Past Vertical Plane Surfaces

Extensive work has been carried out in the analytical investigation of free convection problems. Two such solutions for laminar flows will be presented here to illustrate the application of the differential equations [Eqs. (9.3) and (9.4)] and the application of the integral equations [Eqs. (9.6) and (9.7)]. An application of the integral method to the turbulent case will also be shown. Although more complex, the basic ideas in application of these equations are the same as in the forced convection cases. Hence, many details will be omitted in the discussions to follow.

The Solution of the Laminar Case by Differential Equations

Schmidt and Beckmann (Ref. 1) analyzed the problem of laminar free convection past a vertical flat surface, as depicted in Fig. 9.1, by application of Eqs. (9.3), (9.4), and (9.5):

$$v_x \frac{\partial v_x}{\partial x} + v_y \frac{\partial v_x}{\partial y} = v \frac{\partial^2 v_x}{\partial y^2} + g\beta(t - t_f). \tag{9.3}$$

$$v_x \frac{\partial t}{\partial x} + v_y \frac{\partial t}{\partial y} = \alpha \frac{\partial^2 t}{\partial y^2}. \tag{9.4}$$

At $y = 0$: $t = t_s,\ v_x = 0$.

At $y = \infty$: $t = t_f,\ v_x = 0$.
$$\tag{9.5}$$

The x coordinate has been taken as positive measured along the plate in the direction of the buoyant force. The pressure gradient term has been neglected.

The above equations may be made ordinary equations by the introduction of a similarity parameter:

$$\eta = C \frac{y}{x^{1/4}}, \tag{9.9a}$$

$$C = \left[\frac{g\beta(t_s - t_f)}{4v^2} \right]^{1/4}. \tag{9.9b}$$

The reasoning behind the selection of this form for the similarity parameter is rather involved but not unlike that leading to the definition of the Blasius parameter in Sec. 7.2. The parameter η may also be written as

$$\eta = \left(\frac{g\beta\,\Delta t}{4v^2} \right)^{1/4} \frac{y}{x^{1/4}}$$

$$= \left(\frac{1}{4} \right)^{1/4} \left(\frac{g\beta\,\Delta t\,x^3}{v^2} \right)^{1/4} \frac{y}{x}$$

$$= \frac{1}{\sqrt{2}} N_{\mathrm{GR}_x}^{1/4} \frac{y}{x}, \tag{9.10}$$

where N_{GR_x} denotes the *local* Grashof number. The notation

$$\Delta t = t_s - t_f \tag{9.11}$$

will be used throughout the remaining analysis.

The velocity component parallel to the plate surface is next assumed to be of the form

$$v_x = 4vC^2 x^{1/2} F(\eta), \tag{9.12}$$

in which $F(\eta)$ is an as-yet-undetermined function of only η. Continuity considerations then require that v_y satisfy

$$v_y = vCx^{-1/4}[\eta F'(\eta) - 3F(\eta)].$$ (9.13)

Finally, one defines a dimensionless temperature

$$\varphi = \frac{t - t_f}{t_s - t_f} = \frac{t - t_f}{\Delta t}$$ (9.14)

and assumes that it is a function of η *only*. When Eqs. (9.12), (9.13), and (9.14) are substituted into Eqs. (9.3), (9.4), and (9.5), considerable algebra leads finally to

$$F''' + 3FF'' - 2(F')^2 + \varphi = 0,$$

$$\varphi'' + 3N_{PR}F\varphi' = 0,$$

$$F'(0) = 0, \quad \varphi(0) = 1,$$ (9.15)

$$F'(\infty) = 0, \quad \varphi(\infty) = 0.$$

The fact that the system of equations has been converted into a set of ordinary equations involving η only justifies the assumptions made above. However, the equations are still coupled, and they must be solved simultaneously. Presume for the moment that this solution has been accomplished; then one part of the solution would be the dimensionless temperature gradient at the wall, $\varphi'(0)$. This quantity is necessarily a function of only the Prandtl number, since that is the only parameter in Eq. (9.15). Make the definition

$$f(N_{PR}) = -\varphi'(0).$$ (9.16)

Now the local heat transfer coefficient is determined:

$$h_x = -\frac{k\left(\dfrac{\partial t}{\partial y}\right)_0}{t_s - t_f}$$

$$= -k\varphi'(0)\frac{d\eta}{dy}$$

$$= -k\varphi'(0)\frac{1}{\sqrt{2}}N_{GR_x}\frac{1}{x}$$

$$= \frac{1}{\sqrt{2}}\frac{k}{x}N_{GR_x}^{1/4}f(N_{PR}).$$ (9.17)

Integration over a finite plate length L yields an average plate coefficient, h:

$$h = \frac{4}{3} \frac{1}{\sqrt{2}} \frac{k}{L} N_{GR_L}^{1/4} f(N_{PR}), \qquad (9.18)$$

where the plate-length Grashof number is

$$N_{GR_L} = \frac{g\beta \, \Delta t \, L^3}{v^2}. \qquad (9.19)$$

The above expressions may be nondimensionalized:

$$N_{NU_x} = \frac{h_x x}{k} = \frac{1}{\sqrt{2}} N_{GR_x}^{1/4} f(N_{PR}), \qquad (9.20)$$

$$N_{NU_L} = \frac{hL}{k} = \frac{4}{3} \frac{1}{\sqrt{2}} N_{GR_L}^{1/4} f(N_{PR}). \qquad (9.21)$$

The foregoing solutions are still incomplete in that the function $f(N_{PR}) = -\varphi'(0)$ is still not known. Equations (9.15) have been solved by Ostrach (Ref. 6) and the resulting temperature and velocity profiles are given in Figs. 9.2 and 9.3. Values of the function $f(N_{PR})$ may be found in Ref. 7. An empirical fit to the results for $f(N_{PR})$ is

$$f(N_{PR}) = \frac{0.676 N_{PR}^{1/2}}{(0.861 + N_{PR})^{1/4}}. \qquad (9.22)$$

Together with Eqs. (9.20) and (9.21) this enables one to calculate the local and average heat transfer coefficients. For air with $N_{PR} = 0.714$, there results

$$f(N_{PR}) = 0.509,$$

$$N_{NU_x} = 0.360 N_{GR_x}^{1/4}, \qquad (9.23)$$

$$N_{NU_L} = 0.480 N_{GR_L}^{1/4}.$$

These relations yield useful results for practical calculations.

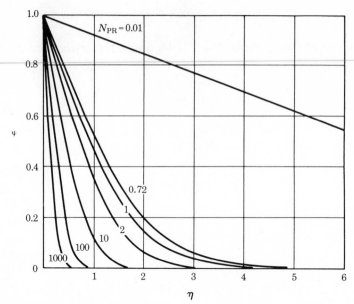

Fig. 9.2. Velocity profile for laminar free convection on a flat plate. (From S. Ostrach, "An Analysis of Laminar Free Convection Flow and Heat Transfer About a Flat Plate Parallel to the Direction of the Generating Body Force," *N.A.C.A. Tech. Note 2635*, Washington, D.C., 1952.)

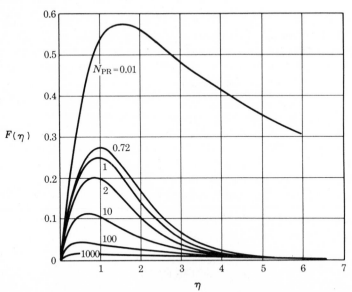

Fig. 9.3. Temperature profile for laminar free convection on a flat plate. (From S. Ostrach, "An Analysis of Laminar Free Convection Flow and Heat Transfer About a Flat Plate Parallel to the Direction of the Generating Body Force," *N.A.C.A. Tech. Note 2635*, Washington, D.C., 1952.)

The Solution of the Laminar Case by Integral Equations

The problem just discussed has been solved also by Eckert (Ref. 8) by use of the integral equations, Eqs. (9.6) and (9.7):

$$\frac{d}{dx} \int_0^\delta v_x^2 \, dy = \int_0^\delta g\beta(t - t_f) \, dy - v\left(\frac{\partial v_x}{\partial y}\right)_0,$$

(9.6)

$$\frac{d}{dx} \int_0^{\delta_t} (t - t_f)v_x \, dy = -\alpha\left(\frac{\partial t}{\partial y}\right)_0.$$

(9.7)

The same basic procedure is used as was used for the integral analysis of the forced convection problem. First, one assumes logical forms of the temperature and velocity profiles. For simplicity in analysis it will be assumed that the thermal and velocity layers are virtually the same (i.e., $\delta \approx \delta_t$). The form of the temperature profile sketched in Fig. 9.1 suggests a dimensionless profile of the form

$$\varphi = \frac{t - t_f}{t_s - t_f} = \left(1 - \frac{y}{\delta}\right)^2.$$

(9.24)

The selection of a form for the velocity profile is more difficult since an obvious characteristic velocity for nondimensionalizing does not exist. The form of the velocity profile of Fig. 9.1 suggests an expression of the form

$$v_x = v_x^* \frac{y}{\delta}\left(1 - \frac{y}{\delta}\right)^2,$$

(9.25)

where v_x^* is an unknown quantity having the dimensions of a velocity. Differentiation of Eq. (9.25) will show that v_x has a maximum value, call it $v_{x_{max}}$, at $y/\delta = \frac{1}{3}$, and this maximum is

$$v_{x_{max}} = \tfrac{4}{27}v_x^*.$$

Thus, the assumed profile in Eq. (9.25) may be written

$$v_x = \tfrac{27}{4}v_{x_{max}} \frac{y}{\delta}\left(1 - \frac{y}{\delta}\right)^2.$$

(9.26)

The quantity $v_{x_{max}}$ is still unknown and would be expected to be a function of the distance from the leading edge, x.

Equations (9.24) and (9.26) may be put in more convenient forms by introducing the parameter

$$\eta = \frac{y}{\delta}, \tag{9.27}$$

so that

$$\varphi = \frac{t - t_f}{t_s - t_f} = (1 - \eta)^2, \tag{9.28}$$

$$\frac{v_x}{v_{x\max}} = \tfrac{27}{4}\eta(1 - \eta)^2. \tag{9.29}$$

In terms of these parameters, Eqs. (9.6) and (9.7) become

$$\frac{d}{dx}\int_0^1 v_{x\max}^2 \left(\frac{v_x}{v_{x\max}}\right)^2 \delta\, d\eta = \int_0^1 g\beta\, \Delta t\, \varphi\delta\, d\eta - v\frac{d}{d\eta}\left(\frac{v_x}{v_{x\max}}\right)_0 \frac{v_{x\max}}{\delta}$$

and

$$\frac{d}{dx}\int_0^1 v_{x\max}\, \Delta t\, \varphi\frac{v_x}{v_{x\max}}\delta\, d\eta = -\alpha\, \Delta t\left(\frac{d\varphi}{d\eta}\right)_0 \frac{1}{\delta}.$$

Substitution of the assumed forms for φ and $v_x/v_{x\max}$, Eqs. (9.28) and (9.29) yields, for the momentum equation,

$$\frac{(27/4)^2}{105}\frac{d}{dx}(v_{x\max}^2\delta) = \frac{g\beta\, \Delta t\, \delta}{3} - \frac{27}{4}\frac{v\, v_{x\max}}{\delta}, \tag{9.30}$$

and for the energy equation,

$$\frac{27/4}{30}\frac{d}{dx}(v_{x\max}\delta) = \frac{2\alpha}{\delta}. \tag{9.31}$$

Equations (9.30) and (9.31) are a coupled set of equations in the unknown functions of x, δ, and $v_{x\max}$. A solution is obtained by assuming a power law of the form

$$v_{x\max} = C_1 x^m,$$

$$\delta = C_2 x^n, \tag{9.32}$$

in which C_1, C_2, m, and n are yet to be determined. When these expressions are introduced into Eqs. (9.30) and (9.31), there results

$$\frac{(27/4)^2}{105} C_1^2 C_2 (2m + n) x^{2m+n-1} = \frac{g\beta \, \Delta t \, C_2}{3} x^n - \frac{27}{4} v \frac{C_1}{C_2} x^{m-n} \qquad (9.33)$$

and

$$\frac{27}{4} \frac{C_1 C_2 (m + n)}{30} x^{m+n-1} = \frac{2\alpha}{C_2} x^{-n}. \qquad (9.34)$$

Since Eqs. (9.33) and (9.34) must be satisfied for all values of x, the exponents of the individual terms in each must be equal:

$$2m + n - 1 = n = m - n,$$

$$m + m - 1 = -n.$$

These are satisfied by

$$m = \tfrac{1}{2},$$

$$n = \tfrac{1}{4},$$

so Eqs. (9.33) and (9.34) may be solved simultaneously for C_1 and C_2:

$$C_1 = \frac{320}{27\sqrt[4]{15}} v \left(\frac{20}{21} + N_{PR}\right)^{-1/2} \left(\frac{g\beta \, \Delta t}{v^2}\right)^{1/2},$$

$$C_2 = (240)^{1/4} \left(\frac{20}{21} + N_{PR}\right)^{1/4} \left(\frac{g\beta \, \Delta t}{v^2}\right)^{-1/4} N_{PR}^{-1/2}.$$

Since $\delta = C_2 x^n$, the following relation for the layer growth is obtained:

$$\frac{\delta}{x} = \frac{C_2}{x^{3/4}} = (240)^{1/4} (\tfrac{20}{21} + N_{PR})^{1/4} N_{PR}^{-1/2} N_{GR_x}^{-1/4}. \qquad (9.35)$$

Now, the local heat transfer coefficient is

$$h_x = -\frac{k\left(\dfrac{\partial t}{\partial y}\right)_0}{\Delta t}$$

$$= -k\varphi'(0)\frac{d\eta}{dy}$$

$$= -k\varphi'(0)\frac{1}{\delta}.$$

From Eq. (9.28), $\varphi'(0) = -2$, so

$$h_x = 2\frac{k}{\delta}, \quad \text{or} \quad N_{NU_x} = \frac{h_x x}{k} = 2\frac{x}{\delta},$$

so Eq. (9.35) gives

$$N_{NU_x} = \frac{2}{(240)^{1/4}}\left(\frac{20}{21} + N_{PR}\right)^{-1/4} N_{PR}^{1/2} N_{GR_x}^{1/4}$$

$$= 0.508(0.952 + N_{PR})^{-1/4} N_{PR}^{1/2} N_{GR_x}^{1/4}. \tag{9.36}$$

The average heat transfer coefficient for a finite plate length L is readily found to be

$$N_{NU_L} = \frac{(4/3)2}{(240)^{1/4}}\left(\frac{20}{21} + N_{PR}\right)^{-1/4} N_{PR}^{1/2} N_{GR_L}^{1/4}$$

$$= 0.678(0.952 + N_{PR})^{-1/4} N_{PR}^{1/2} N_{GR_L}^{1/4}. \tag{9.37}$$

The results indicated by Eqs. (9.36) and (9.37) give reasonably accurate predictions of real free convection processes. For the case of air with $N_{PR} = 0.714$, the above give

$$N_{NU_x} = 0.378 N_{GR_x}^{1/4},$$

$$N_{NU_L} = 0.504 N_{GR_L}^{1/4}.$$

Comparison with the results of the exact solution, Eq. (9.21), shows the close correspondence between the two methods.

Turbulent Free Convection Past Plane Surfaces

For Grashof numbers in excess of about 10^9, free convection boundary layers are observed to be in a turbulent state of flow. Eckert and Jackson (Ref. 9) investigated this case, again using the integral analysis as just described for laminar flow. In fact, the analysis is so similar that only the barest details will be given.

The integral momentum equation must be altered in that the term $-v(\partial v_x/\partial y)_0$ must be replaced by $-\tau_0/\rho$. Also, in the energy equation $-\alpha(\partial t/\partial y)_0$ must be replaced by $(q/A)_0$. Then the following assumptions are made:

1. The thermal and velocity layer thicknesses are the same.
2. The plate is entirely in turbulent flow.
3. The dimensionless temperature and velocity profiles follow laws analogous to those used for laminar flow with the exception that the one-seventh power law is introduced:

$$v_x = v_x^* \left(\frac{y}{\delta}\right)^{1/7} \left(1 - \frac{y}{\delta}\right)^4,$$

$$\varphi = \frac{t - t_f}{t_s - t_f} = 1 - \left(\frac{y}{\delta}\right)^{1/7}.$$

4. The skin friction at the surface follows the Blasius law for forced flow, Eq. (7.65):

$$\frac{\tau_0}{\frac{1}{2}\rho v_x^{*2}} = 0.0456 \left(\frac{v}{v_x^* \delta}\right)^{1/4}.$$

5. Reynolds' analogy may be applied to relate τ_0 to $(q/A)_0$.

With these assumptions, an identical sequence of analytical steps was applied to the integral equations as was just carried out for the laminar case. The details are omitted here, but the final results for the local and average heat transfer were

$$N_{\mathrm{NU}_x} = 0.0295(1 + 0.494N_{\mathrm{PR}}^{2/3})^{-2/5} N_{\mathrm{PR}}^{7/15} N_{\mathrm{GR}_x}^{2/5},$$

$$N_{\mathrm{NU}_L} = 0.0246(1 + 0.494N_{\mathrm{PR}}^{2/3})^{-2/5} N_{\mathrm{PR}}^{7/15} N_{\mathrm{GR}_L}^{2/5}.$$

$$(9.38)$$

These results agree well with experimental data for air and water, but they have not been checked extensively for other fluids.

9.4   Working Formulas for Free Convection

It is the purpose of the next few sections to list some recommended relations for use in the prediction of free convection heat transfer coefficients. Relations will be quoted for the most commonly encountered geometrical shapes—cylinders and plates. Other geometries have been studied and suitable relations may be found in such sources as Refs. 3 and 4.

The results of analyses such as those just discussed may be applied with confidence. However, for convenience of rapid calculation, certain empirical relations will be recommended in their stead. Such empirical relations are developed from experiments based on the similarity analysis of Sec. 6.8, which showed that free convection problems are reducible to the form

$$N_{NU} = F(N_{GR}, N_{PR}).$$

In all free convection correlations it is customary to evaluate the fluid properties at the mean film temperature $t_m = (t_s + t_f)/2$, *except* for the coefficient of volume expansion, β, which is normally evaluated at the temperature of the undisturbed fluid far removed from the surface—namely t_f. Unless otherwise noted, this convention should be used in the application of all relations quoted here.

9.5   Free Convection Around Vertical Plates

The analytical predictions of Sec. 9.3 may be applied for practical calculations of free convection past vertical surfaces. However, for rapid calculation the results quoted below are recommended. On the basis of the correlation of extensive experiments, McAdams (Ref. 3) recommends the use of a relation of the form

$$N_{NU_L} = f(N_{GR_L} N_{PR}).$$

Note that the Grashof and Prandtl numbers appear as a product. This product is sometimes called the *Rayleigh number*,

$$N_{RA} = N_{GR} N_{PR}.$$

Figure 9.4 shows McAdams's recommended relation, along with the supporting experimental data. In certain ranges the curve is well approximated by an equation of the form

$$N_{NU_L} = C(N_{GR_L} N_{PR})^m,$$

where the constants C and m take on different values in the laminar and turbulent ranges.

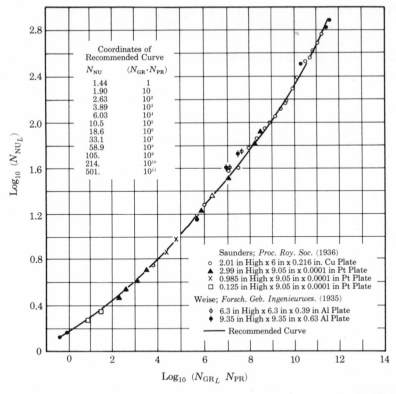

Fig. 9.4. Recommended correlation for free convection around vertical plane surfaces. (From *Heat Transmission*, 3rd ed., by W. H. McAdams. Copyright 1954. McGraw-Hill Book Company, New York. Used by permission.)

Thus, the recommended calculation procedure for vertical surfaces is

$$N_{NU_L} = C(N_{GR_L}N_{PR})^m.$$

$$10^{-1} < (N_{GR_L}N_{PR}) < 10^4: \quad \text{Use Fig. 9.4.}$$

$$10^4 < (N_{GR_L}N_{PR}) < 10^9: \quad C = 0.59, \ m = \tfrac{1}{4}.$$

$$10^9 < (N_{GR_L}N_{PR}) < 10^{12}: \quad C = 0.129, \ m = \tfrac{1}{3}.$$

(9.39)

It is interesting to note that in the turbulent range the exponent $m = \tfrac{1}{3}$. Since the characteristic length appears as L^3 in the Grashof number and as L in the Nusselt number, the significance of this dimension vanishes in the calculation of h.

EXAMPLE 9.1:   A vertical plate 1 ft high maintained at 200°F is immersed in a tank of water at 60°F. Estimate the heat transfer coefficient at its surface.

Solution: For a mean film temperature of $(200 + 60)/2 = 130°F$, Table A.4 gives for water,

$$\rho = 61.54 \, lb_m/ft^3, \qquad\qquad \mu = 1.24 \, lb_m/ft\text{-}hr,$$

$$k = 0.375 \, Btu/hr\text{-}ft\text{-}°F, \qquad N_{PR} = 3.30.$$

At a main fluid temperature of 60°, Table A.4 gives

$$\beta = 0.10 \times 10^{-3} \, 1/°R.$$

Thus,

$$N_{GR_L} = \frac{(1)^3(61.54)^2(32.2)(3600)^2(0.1 \times 10^{-3})(200 - 60)}{(1.24)^2}$$

$$= 1.44 \times 10^{10}.$$

So

$$N_{GR_L} N_{PR} = 4.75 \times 10^{10}.$$

Equation (9.39) suggests that

$$N_{NU_L} = 0.129(4.75 \times 10^{10})^{1/3}$$

$$= 0.129(47.5)^{1/3} \times 10^3$$

$$= 467.$$

Thus,

$$h = 467 \times \frac{0.375}{1}$$

$$= 175 \, Btu/hr\text{-}ft^2\text{-}°F \ (994 \, W/m^2\text{-}°C).$$

9.6   Free Convection Around Horizontal Cylinders

One of the most commonly encountered geometrical shapes as far as free convection is concerned is that of a horizontal cylinder. Heat losses from pipes or wires to an ambient fluid by the process of free convection are often needed in many design calculations. Considerable experimental data have been measured for horizontal cylinders—for cylinders ranging in diameters from 0.002 in. up to as much as 1 ft. A great number of ambient fluids have been investigated.

Figure 9.5 shows some of these data for free convection around horizontal cylinders, along with McAdams's recommended correlation curve. The use

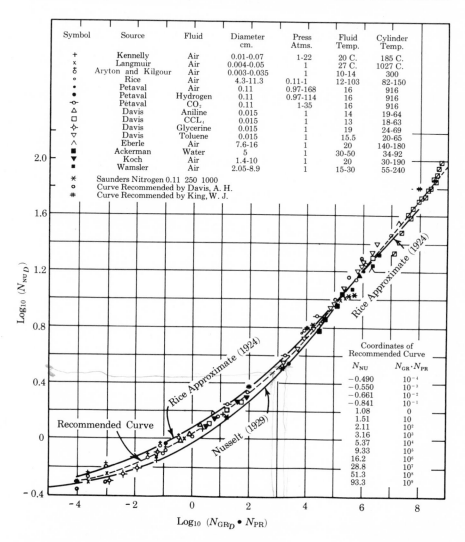

Symbol	Source	Fluid	Diameter cm.	Press Atms.	Fluid Temp.	Cylinder Temp.
+	Kennelly	Air	0.01-0.07	1-22	20 C.	185 C.
x	Langmuir	Air	0.004-0.05	1	27 C.	1027 C.
δ	Aryton and Kilgour	Air	0.003-0.035	1	10-14	300
o	Rice	Air	4.3-11.3	0.11-1	12-103	82-150
•	Petaval	Air	0.11	0.97-168	16	916
•	Petaval	Hydrogen	0.11	0.97-114	16	916
-o-	Petaval	CO_2	0.11	1-35	16	916
△	Davis	Aniline	0.015	1	14	19-64
□	Davis	CCL_4	0.015	1	13	18-63
◇	Davis	Glycerine	0.015	1	19	24-69
▽	Davis	Toluene	0.015	1	15.5	20-65
∧	Eberle	Air	7.6-16	1	20	140-180
■	Ackerman	Water	5	1	30-50	34-92
▼	Koch	Air	1.4-10	1	20	30-190
■	Wamsler	Air	2.05-8.9	1	15-30	55-240

* Saunders Nitrogen 0.11 250 1000
o Curve Recommended by Davis, A. H.
Curve Recommended by King, W. J.

Coordinates of Recommended Curve

N_{NU}	$N_{GR} \cdot N_{PR}$
−0.490	10^{-4}
−0.550	10^{-3}
−0.661	10^{-2}
−0.841	10^{-1}
1.08	0
1.51	10
2.11	10^2
3.16	10^3
5.37	10^4
9.33	10^5
16.2	10^6
28.8	10^7
51.3	10^8
93.3	10^9

Rice Approximate (1924)

Nusselt (1929)

Recommended Curve

$Log_{10} (N_{NU D})$

$Log_{10} (N_{GR_D} \cdot N_{PR})$

Fig. 9.5. Recommended correlation for free convection around horizontal cylinders. (From *Heat Transmission*, 3rd ed., by W. H. McAdams. Copyright 1954. McGraw-Hill Book Company, New York. Used by permission.)

of this correlation is recommended in the following form:

$$N_{NU_D} = C(N_{GR_D}N_{PR})^m.$$

$$0 < (N_{GR_D}N_{PR}) = 10^{-5}: \quad C = 0.4, m = 0.$$

$$10^{-5} < (N_{GR_D}N_{PR}) < 10^4: \quad \text{Use Fig. 9.5.} \qquad (9.40)$$

$$10^4 < (N_{GR_D}N_{PR}) < 10^9: \quad C = 0.525, m = \tfrac{1}{4}.$$

$$10^9 < (N_{GR_D}N_{PR}) < 10^{12}: \quad C = 0.129, m = \tfrac{1}{3}.$$

The subscript D on the Nusselt and Grashof numbers indicates that the characteristic length used in their calculation is the cylinder diameter.

It is interesting to note that for very small N_{GR_D} (i.e., very fine wires or very small Δt), the Nusselt number approaches the limiting constant value of 0.4.

EXAMPLE 9.2: A bare pipe (6-in., schedule 40) has a surface temperature of 950°F. It passes through a room in which the ambient air temperature is 86°F. Find the heat lost, per foot of length, by free convection.

Solution: Table B.3 gives the O.D. of a 6-in. schedule 40 pipe to be 6.625 in. = 0.521 ft. For a mean film temperature of (950 + 86)/2 = 518°F, Table A.7 for air gives

$$\rho = 0.0405 \text{ lb}_m/\text{ft}^3, \qquad \mu = 0.0685 \text{ lb}_m/\text{ft-hr},$$

$$k = 0.0250 \text{ Btu/hr-ft-°F}, \qquad N_{PR} = 0.680.$$

Since air at atmospheric pressure may be treated as an ideal gas, Sec. 2.6 indicates that β is the reciprocal of the absolute temperature. Thus

$$\beta = \frac{1}{86 + 460} = \frac{1}{546}.$$

Thus,

$$N_{GR_D} = \frac{(0.521)^3(0.0405)^2(32.2)(3600)^2(950 - 86)}{(0.0685)^2 \times 546}$$

$$= 3.26 \times 10^7.$$

So

$$N_{GR_D}N_{PR} = 2.22 \times 10^7.$$

Equation (9.40) indicates that the Nusselt number is given by

$$N_{NU_D} = 0.525(2.22 \times 10^7)^{1/4}$$

$$= 0.525(0.222)^{1/4} \times 10^2$$

$$= 36.1.$$

Thus

$$h = 36.1 \times \frac{0.0250}{0.521}$$

$$= 1.73 \text{ Btu/hr-ft}^2\text{-}°F.$$

The heat loss per foot of pipe length is then

$$\frac{q}{L} = (A/L)h(t_s - t_f)$$

$$= \pi \times 0.521 \times 1.73(950 - 86)$$

$$= 2440 \text{ Btu/hr-ft (2346 W/m).}$$

9.7   Free Convection Around Horizontal Flat Surfaces

The data for horizontal plates are varied. After evaluating the properties according to the convention outlined above, the following relations are recommended for *square* plates depending on whether the flow is laminar or turbulent, whether the plate is heated or cooled, and whether the coefficient is sought at the upper surface or at the lower surface.

1. Laminar range $(10^5 < (N_{GR_L}N_{PR}) < 2 \times 10^7)$:
   (a) Upper surface of heated plates or the lower surface of cooled plates (Ref. 4):

   $$N_{NU_L} = 0.54(N_{GR_L}N_{PR})^{1/4}.$$

   (b) Lower surface of heated plates or the upper surface of cooled plates (Ref. 10):

   $$N_{NU_L} = 0.44(N_{GR_L}N_{PR})^{1/5}.$$

2. Turbulent range $[2 \times 10^7 < (N_{GR_L}N_{PR}) < 3 \times 10^{10}]$:
   (a) Upper surface of heated plates or the lower surface of cooled plates (Ref. 4):

   $$N_{NU_L} = 0.14(N_{GR_L}N_{PR})^{1/3}.$$

(b) Lower surface of heated plates or the upper surface of cooled plates:

No data.

9.8   Simplified Free Convection Relations for Air

All the foregoing expressions for cylinders, plates, etc., were of the form

$$N_{NU} = C(N_{GR}N_{PR})^m,$$

at least for values of $N_{GR}N_{PR} > 10^4$. Introduction of the definition of the various dimensionless parameters gives

$$\frac{hl}{k} = C\left(\frac{l^3\rho^2 g\beta \, \Delta t}{\mu^2}\frac{\mu c_p}{k}\right)^m.$$

This equation may be rewritten

$$h = C\left[k\left(\frac{\rho^2 g\beta c_p}{\mu k}\right)^m\right]\frac{(\Delta t)^m}{l^{1-3m}}.$$

For purposes of rapid, rough, estimates of free convection coefficients in air at normal atmospheric conditions, a mean value of the term in the square brackets may be chosen and taken as constant. This then gives an expression for h in the form

$$h = C_1 \frac{(\Delta t)^m}{l^{1-3m}}.$$

In the laminar range where $m = \frac{1}{4}$, this expression for h becomes

$$h = C_1\left(\frac{\Delta t}{l}\right)^{1/4}. \tag{9.41}$$

In the turbulent range, however, $m = \frac{1}{3}$, and the characteristic length disappears:

$$h = C_1(\Delta t)^{1/3}.$$

McAdams (Ref. 3) recommends the relations of Table 9.1 for the cases considered in this chapter.
Obviously the constants in these approximate equations are *not dimensionless*, so care must be taken to use only the dimensions indicated for the quantities involved.

Table 9.1   Approximate Free Convection Equations for Air

Configuration	h in Btu/hr-ft^2-°F, L and D in feet, Δt in °F	
	Laminar $10^9 > N_{GR}N_{PR} > 10^4$	Turbulent $N_{GR}N_{PR} > 10^9$
Horizontal cylinders D = diameter	$h = 0.27\left(\dfrac{\Delta t}{D}\right)^{1/4}$	$h = 0.18(\Delta t)^{1/3}$
Vertical plates and cylinders L = vertical dimension	$h = 0.29\left(\dfrac{\Delta t}{L}\right)^{1/4}$	$h = 0.19(\Delta t)^{1/3}$
Horizontal plates Heated plates, up Cooled plates, down	$h = 0.27\left(\dfrac{\Delta t}{L}\right)^{1/4}$	$h = 0.22(\Delta t)^{1/3}$
Heated plates, down Cooled plates, up L = side dimension	$h = 0.12\left(\dfrac{\Delta t}{L}\right)^{1/4}$	—

It must be kept in mind that these equations are only approximate, and, at times, considerable error may result from their use.

EXAMPLE 9.3:   Repeat Example 9.2 using the approximate formulas for air.

Solution: For $D = 0.521$ ft and $\Delta t = 950 - 86 = 864$°F, and assuming that laminar flow exists, Table 9.1 gives

$$h = 0.27\left(\frac{\Delta t}{D}\right)^{1/4}$$

$$= 0.27\left(\frac{864}{0.521}\right)^{1/4}$$

$$= 1.72 \text{ Btu/hr-ft}^2\text{-°F.}$$

The heat loss, per foot of length, is

$$\frac{q}{L} = \frac{A}{L}h\,\Delta t$$

$$= \pi \times 0.521 \times 1.72 \times 864$$

$$= 2430 \text{ Btu/hr-ft.}$$

This result is 0.5% lower than the value found by the application of Eq. (9.40).

References

1. SCHMIDT, E., and W. BECKMANN, "Das Temperatur und Geschwindigkeitsfeld vor einer Wärmer abgebenden senkrechten Platte bei natürlicher Konvection," *Tech. Mech. und Thermodynamik*, Vol. 1, 1930, p. 341.
2. GOLDSTEIN, S. (ed.), *Modern Developments in Fluid Dynamics*, New York, Oxford U.P., 1938, p. 641.
3. McADAMS, W. H., *Heat Transmission*, 3rd ed., New York, McGraw-Hill, 1954.
4. FISHENDEN, M., and O. A. SAUNDERS, *An Introduction to Heat Transfer*, New York, Oxford U.P., 1950.
5. ECKERT, E. R. G., and R. M. DRAKE, *Heat and Mass Transfer*, New York, McGraw-Hill, 1959.
6. OSTRACH, S., "An Analysis of Laminar Free Convection Flow and Heat Transfer About a Flat Plate Parallel to the Generating Body Force," *N.A.C.A. Rept. 1111*, Washington, D.C., 1953.
7. EDE, A. J., "Advances in Free Convection," *Advances in Heat Transfer*, Vol. 4, New York, Academic, 1967.
8. ECKERT, E. R. G., *Introduction to Heat and Mass Transfer*, New York, McGraw-Hill, 1963.
9. ECKERT, E. R. G., and T. W. JACKSON, "Analysis of Turbulent Free Convection Boundary Layer on a Flat Plate," *N.A.C.A. Tech. Note 2207*, Washington, D.C., 1950.
10. CLIFTON, J. V., and A. J. CHAPMAN, "Natural Convection on a Finite-Size Horizontal Plate," *Intern. J. Heat and Mass Transfer*, Vol. 12, 1969, p. 1573.

Problems

9.1. Verify the algebra leading from Eqs. (9.3) and (9.4) to Eq. (9.15).

9.2. Verify the algebra leading from Eqs. (9.6) and (9.7) to Eqs. (9.30) and (9.31).

9.3. Verify the algebra leading from Eqs. (9.30) and (9.31) to Eqs. (9.36) and (9.37).

9.4. A flat plate 6 in. long and 3 in. wide is placed vertically in air. The plate is maintained at 150°F and the air, far from the plate, is at 90°F, 1 atm pressure. Find the average heat transfer coefficient for the plate by use of
(a) Eqs. (9.21) and (9.22);
(b) Eq. (9.37);
(c) Eq. (9.39).

9.5. A flat plate 4 ft long and 6 in. wide is placed vertically in air. The plate is maintained at 800°F and the air, far from the plate, is at 100°F, 1 atm pressure. Find the average heat transfer coefficient for the plate by use of
(a) Eq. (9.38);
(b) Eq. (9.39).
Explain any differences in results.

9.6. For the plate of Prob. 9.4, find the local heat transfer coefficient, the boundary layer thickness, the maximum velocity, and the location of the maximum velocity at 3 in. from the leading edge and 6 in. from the leading edge.

9.7. A cylinder is 3 in. in diameter and 6 ft long. It is maintained at 200°F and placed in atmospheric air at 100°F. Find the total heat loss by free convection if the cylinder is
(a) horizontal;
(b) vertical.

9.8. A plate 1 ft long is maintained at 300°F in a vertical position. It is placed in air at 100°F. Find the average heat transfer coefficient on the plate surface if the air pressure is
(a) 0.1 atm;
(b) 1.0 atm;
(c) 5.0 atm;
(d) 10.0 atm.

9.9. A horizontal steam pipe passes through a room in which the air is at 150°F. The pipe is a 10-in., schedule 40 pipe with a surface temperature of 800°F. What is the heat loss due to free convection, per foot of pipe length?

9.10. A horizontal 12-in., schedule 40 steam pipe passes through a room in which the air is at 100°F. The pipe surface is 600°F. What is the free convective heat loss per foot of length?

9.11. Find the coefficient of free convective heat transfer between a horizontal 3-in., schedule 40 pipe with a surface temperature of 300°F and carbon dioxide at 125°F and 15 psia.

9.12. What is the coefficient of free convective heat transfer between a horizontal wire 0.1 in. in diameter and air at 75°F? The wire is at 200°F.

9.13. Find the coefficient of free convective heat transfer between a horizontal 5-in. schedule 40 pipe with a surface temperature of 100°F and methane at 0°F and 20 psia.

9.14. If the pipe of Prob. 9.13 is immersed in water at 200°F, what is the convective heat transfer, per foot of length?

9.15. An electrical heating element consists of a wire 0.05 in. in diameter. It is immersed horizontally in a bath of oil (use the oil listed in Table A.5). If its surface temperature is maintained at 300°F and the oil at 100°F, find the coefficient of heat transfer due to free convection.

9.16. An electrical wire is $\frac{1}{8}$ in. in diameter and is covered with insulation $\frac{1}{32}$ in. thick. Four watts of electrical energy are dissipated into heat by the resistance of the wire, per foot of length. If this heat is to be given up to the surrounding air at 80°F, estimate the surface temperature of the wire insulation.

9.17. A large-diameter vertical tank 5 ft high serves as the receiver tank for the output of an air compressor. The air leaving the compressor and entering the tank is at 3000 psia, 600°F. If the inside walls of the tanks are at 200°F, what is the heat transfer by free convection, per square foot, to the vertical side walls?

9.18. A straight cooling fin of uniform thickness of 0.25 in. extends horizontally from a wall maintained at 300°F into still air at 120°F. The fin is 4 in. long. Estimate, per foot of fin width, the heat given up to the surrounding air by free convection if the fin is made of a mild carbon steel. Use the results of Sec. 3.12 to obtain a rational answer.

9.19. Estimate the coefficient of free convection heat transfer between a vertical plate 2 ft high maintained at 250°F and nitrogen at 10°F, 1000 psia.

9.20. A horizontal cylinder with a surface temperature of 400°F is placed in still air at 100°F. Find the free convection film coefficient for diameters of (a) 0.01 in.; (b) 0.1 in.; (c) 1.0 in.; (d) 10 in.

9.21. Repeat Problem 9.20 using the approximate formulas for air and compare the results.

9.22. The heat loss by free convection from a horizontal, electrically heated wire, is known to be 350 Btu/hr-ft^2 when the wire surface is maintained at a temperature of 150°F and the ambient fluid is atmospheric air at 90°F. What is the rate of heat loss from the wire when its surface temperature is raised to 300°F?

9.23. A 12-in.-diameter horizontal cylinder is maintained at a surface temperature of 500°F. It is exposed to atmospheric air at 50°F. Calculate the heat loss, per foot of length, by both the recommended correlations of Eq. (9.40) and the approximate relations of Table 9.1.

9.24. A circular cylinder 6 ft long and 6 in. in diameter is maintained at a surface temperature of 450°F. It is placed in atmospheric air at 75°F. Estimate the total heat loss from the cylinder if it is (a) horizontal or (b) vertical.

9.25. Compare the results of Prob. 9.23 with those obtained for a vertical flat plate (considering heat loss from both sides), the height of which is equal to the half-circumference of the cylinder. Take the surface and air temperatures to be the same as given in Prob. 9.23.

CHAPTER 10

Heat Transfer in Condensing and Boiling

10.1 Introductory Remarks

This chapter will be devoted to processes of convection in which the fluid medium is undergoing a change of phase. A great number of practical engineering processes involving the transfer of heat are concerned with the condensation of a vapor or the boiling of a liquid. When a fluid changes phase from liquid to vapor or from vapor to liquid, there are changes in the fluid properties of density, specific heat, thermal conductivity, viscosity, etc., of such magnitude as to influence greatly the fluid mechanical processes taking place. At the same time there is either a release or consumption of latent heat. This latent heat exchange also has a profound influence on the convection process.

First, the process of condensation of a saturated vapor will be discussed. Subsequent to this discussion the process of heat transfer in boiling will be considered.

10.2 General Remarks Concerning Condensation

The process of condensation of a vapor is usually accomplished by allowing it to come in contact with a surface, the temperature of which is maintained at a value lower than the saturation temperature of the vapor for the pressure at which it exists. The removal of thermal energy from the vapor causes it to release its latent heat of vaporization and, hence, to condense onto the surface.

The appearance of the liquid phase on the cooling surface, either in the form of individual drops or in the form of a continuous film, offers a greater thermal resistance to the further removal of heat from the remaining vapor. In most practical cases of condensation, the condensate is removed by the action of gravity. Naturally, then, the rate of removal of condensate, and with it the rate of heat removal from the vapor, is greater for vertically placed surfaces than for horizontal surfaces. In order to avoid the accumulation of large amounts of liquids at the lower end of a condensing surface it should be short; hence, horizontal cylinders are particularly applicable for such purposes. Thus, most condensing equipment consists of assemblies of horizontal tubes around which the vapor to be condensed is allowed to flow. The cool temperature of the outer tube surface is maintained by circulating a colder medium, usually water, through the inside of the tube.

Two general types of condensation have been observed to occur in practice. The first, called *film condensation*, usually occurs when a vapor, relatively free of impurities, is allowed to condense on a clean surface. Under these conditions, it is found that the condensate will appear as a continuous film all over the surface and will flow off the surface as a film under the action of gravity.

The second type of condensation, less frequently observed, is known as *dropwise condensation*, and has been observed to occur either on highly polished surfaces or on surfaces contaminated with certain fatty acids. In this case the condensate is found to appear in the form of individual drops. These drops increase in size and combine with one another until their size is great enough that their weight causes them to run off the surface, leaving the surface exposed for the formation of a new drop.

Since there will be much less of a liquid barrier on the surface if dropwise condensation is occurring, it should not be surprising to know that heat transfer rates are five to ten times as great for dropwise condensation as for film condensation under similar conditions. For this reason it would be desirable to maintain a condition of dropwise condensation in commercial applications by applying the appropriate contaminants to the condensing surfaces. However, it appears to be rather difficult to maintain dropwise condensation in practice. The presence of noncondensable gases, the nature and composition of the surface, the vapor velocity past the surface, etc., all appear to have some influence on the success of these surface contaminants and on the formation of dropwise condensation. Also, particularly in the case of steam condensers, it is difficult to prevent the removal of the drop-inducing contaminant by the fluid being condensed. Thus, except under carefully controlled conditions film condensation may be expected to occur in most instances. Consequently, for conservative designs in industrial applications, it is usual practice to base the calculation of condensing equipment on the presumption that film condensation will occur. This fact may be unfortunate from the heat transfer point of view, but it so happens that reliable analytical predictions of condensing heat transfer coefficients are much more feasible in the case of film condensation than in the case of dropwise condensation. One such analysis for film condensation is considered in the next section.

10.3   Film Condensation on Vertical Plane Surfaces and Vertical Tubes

Nusselt (Ref. 1) obtained an analytical prediction for the heat transfer coefficient of film condensation on a flat vertical surface by assuming that the film originated at the top of the surface and flowed down it in a laminar fashion. Choosing, as shown in Fig. 10.1, the x coordinate to be parallel to the surface (origin at the top edge) and the y coordinate normal to it, one finds that the film forms a free surface somewhat as shown. The film increases in thickness with x because of flow from above and condensation from the adjacent vapor. A determination of the mass rate of flow past any selected value of x will enable one to determine the rate at which the vapor is being condensed—as a function of the coordinate x. From this it would be easy to evaluate the heat transfer coefficient, h, as a function of x. As long as the radius of curvature is not very small, this analysis may be applied to condensation on the outside of a vertical tube.

The flow in the film is assumed to be laminar, so the laminar law for the viscous shear, $\tau = \mu(\partial v/\partial y)$, may be assumed. Further, it will be assumed

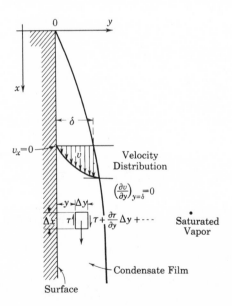

Fig. 10.1. Film condensation on a vertical plane surface.

that the relative velocity between the outer edge of the condensate film and the adjacent vapor is so small that no shear stress is exerted there on the liquid. By using v to denote the velocity in the liquid layer (assumed to be only in the direction parallel to the surface), the assumption of no shear at the liquid–vapor interface means

$$\left(\frac{\partial v}{\partial y}\right)_{y=\delta} = 0, \tag{10.1}$$

where δ is used to denote the local thickness of the liquid layer. Thus, the distribution of the velocity in the liquid layer would be somewhat as sketched in Fig. 10.1.

In order to evaluate the mass rate of flow past any value of x, it is necessary to determine the velocity distribution across the condensate film. This can be obtained from the application of Newton's law to an element of the liquid layer. Choose an element of the liquid film as shown in Fig. 10.1, Δx by Δy in size. A unit depth in the plane of the paper is assumed. For a simple analysis, it is possible, since the velocities are likely to be low, to neglect any normal viscous stress, pressure differences due to elevation, and inertia forces. Thus, only the shear viscous forces and the force of gravity are assumed to be acting on the element. For steady flow these two forces must be equal and opposite. If the shear stress on the element face at y is τ, then that at $y + \Delta y$ is expressible as $\tau + (\partial \tau / \partial y)\,\Delta y + \cdots$. Equating the gravity force to the *net* shearing force, one obtains

$$\frac{g}{g_c}\rho(\Delta x \cdot \Delta y \cdot 1) = (\Delta x \cdot 1)\left[\tau - \left(\tau + \frac{\partial \tau}{\partial y}\Delta y + \cdots\right)\right].$$

The factor g/g_c is included since ρ is normally a *mass* density, whereas the shear is a *force*. Division by the volume of the element gives

$$\frac{g}{g_c}\rho = -\frac{\partial \tau}{\partial y} - \left(\frac{\partial^2 \tau}{\partial y^2}\frac{\Delta y}{2} + \cdots\right).$$

As $\Delta y \to 0$, the following equation, which must be satisfied at each value of x, is obtained:

$$\frac{g}{g_c}\rho = -\frac{\partial \tau}{\partial y}.$$

For laminar flow in the film, $\tau = (\mu/g_c)(\partial v/\partial y)$, g_c being included since μ is usually expressed in mass dimensions. When this fact is substituted into the relation above, one obtains

$$\frac{\partial^2 v}{\partial y^2} = -\frac{\rho g}{\mu}. \tag{10.2}$$

Equation (10.2) can be readily integrated:

$$v = -\frac{\rho g}{\mu}\frac{y^2}{2} + c_1 y + c_2. \tag{10.3}$$

The symbols c_1 and c_2 denote constants of integration. These constants are easily evaluated since, as mentioned above, it is assumed that no shear acts at $y = \delta$ (δ = film thickness) and, also, there is no fluid velocity at the solid surface. In other words,

$$\text{At } y = 0: \quad v = 0.$$

$$\text{At } y = \delta: \quad \frac{\partial v}{\partial y} = 0.$$

These conditions determine that

$$c_1 = \frac{\rho g \delta}{\mu},$$

$$c_2 = 0. \tag{10.4}$$

Thus, finally, the distribution of the velocity through the film is given by the following parabola:

$$v = -\frac{\rho g}{\mu}\left(\frac{y^2}{2} - y\delta\right)$$

$$= -\frac{\rho g \delta^2}{\mu}\left[\frac{1}{2}\left(\frac{y}{\delta}\right)^2 - \frac{y}{\delta}\right]. \tag{10.5}$$

The mass rate of flow, per unit width of plate, $\dot{m}$, is

$$\dot{m} = \int_0^\delta \rho v \, dy.$$

Substitution of Eq. (10.5) for the velocity distribution gives, after evaluation
of the indicated integral,

$$\dot{m} = -\frac{g\rho^2\delta^3}{\mu}\left[\frac{1}{6}\left(\frac{y}{\delta}\right)^3 - \frac{1}{2}\left(\frac{y}{\delta}\right)^2\right]_0^\delta$$

$$= \frac{g\rho^2\delta^3}{3\mu}. \tag{10.6}$$

This mass flow rate is a function of x since the film thickness, δ, is a function
of x.

Since the mass rate of flow down the plate is related to the rate at which
vapor is condensed, $\dot{m}$ may be eliminated from the above expression in terms
of the thermal properties of the vapor, leaving an expression for the film
thickness as a function of x and the fluid properties.

As shown in Fig. 10.2, if one were to select a segment of the plate Δx long,
the amount of vapor condensed between $x = x$ and $x = x + \Delta x$, call this
$\Delta \dot{m}$, would equal the amount of heat removed from the vapor divided by the
latent heat of vaporization of the fluid, λ. Thus,

$$\Delta \dot{m} = \frac{\Delta q}{\lambda}.$$

Since the flow in the film is taken as laminar and parallel to the surface, the
heat removed from the vapor into the surface must pass through the film by
conduction. By assuming the vapor to be saturated, at temperature t_{sat}, and
assuming that the surface is maintained at t_s, the heat conducted through

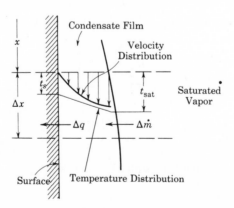

Fig. 10.2

the film is given by the linear law:

$$\Delta q = k(\Delta x \cdot 1)\frac{t_{sat} - t_s}{\delta}. \tag{10.7}$$

Thus, the fluid condensed is

$$\Delta \dot{m} = \frac{k}{\lambda}\frac{\Delta t}{\delta}\Delta x.$$

Here, k denotes the thermal conductivity of the liquid film and Δt is used to signify $(t_{sat} - t_s)$. This analysis will assume that Δt is constant. As the size of the segment becomes small (i.e., $\Delta x \to 0$), the following expression for the rate of change of the mass flow down the plate is obtained:

$$\frac{d\dot{m}}{dx} = \frac{k}{\lambda}\frac{\Delta t}{\delta}.$$

Since δ, the film thickness, is a function of x, so is $d\dot{m}/dx$.

By substituting Eq. (10.6) determined above for the mass flow rate,

$$\frac{g\rho^2}{3\mu}\frac{d(\delta^3)}{dx} = \frac{k}{\lambda}\frac{\Delta t}{\delta}.$$

The following differential expression for δ, the film thickness, as a function of x, is obtained:

$$\delta^3\frac{d\delta}{dx} = \frac{\mu k\,\Delta t}{g\rho^2\lambda}.$$

Upon integrating this equation and noting that, as assumed, $\delta = 0$ at $x = 0$, one finally obtains that δ is related to x as

$$\delta = \left(\frac{4\mu k\,\Delta t\,x}{g\rho^2\lambda}\right)^{1/4}. \tag{10.8}$$

This equation shows that the film grows as the fourth root of the distance down the surface.

Now, Eq. (10.7) shows that for $\Delta x \to 0$ the heat transferred from the vapor to the surface is given by the following function of x:

$$\frac{dq}{dx} = \frac{k\,\Delta t}{\delta}.$$

In combination with Eq. (10.8), this equation gives the following expression for q as a function of x:

$$\frac{dq}{dx} = \left(\frac{g\rho^2\lambda k^3(\Delta t)^3}{4\mu x}\right)^{1/4}.$$

(10.9)

If the *local* film coefficient is defined as

$$h = \frac{dq}{dx}\frac{1}{\Delta t},$$

then

$$h = \left(\frac{g\rho^2\lambda k^3}{4\mu x\,\Delta t}\right)^{1/4}.$$

(10.10)

As has been done before, a *local* Nusselt number, $N_{NU_x} = hx/k$, may be introduced to give

$$N_{NU_x} = \frac{1}{\sqrt{2}}\left(\frac{g\rho^2\lambda x^3}{\mu k\,\Delta t}\right)^{1/4}.$$

(10.11)

Integrating Eq. (10.9) and defining an average film coefficient for a plate of length L, one obtains

$$h = \frac{2}{3}\sqrt{2}\left(\frac{g\rho^2\lambda k^3}{\mu L\,\Delta t}\right)^{1/4}.$$

(10.12)

Thus, the *average* Nusselt number is

$$N_{NU_L} = \frac{hL}{k} = \frac{2}{3}\sqrt{2}\left(\frac{g\rho^2\lambda L^3}{\mu k\,\Delta t}\right)^{1/4}$$

$$= 0.943\left(\frac{g\rho^2\lambda L^3}{\mu k\,\Delta t}\right)^{1/4}.$$

(10.13)

The case of an inclined flat surface may be easily deduced from the results above. The only quantity altered is the magnitude of the acceleration of gravity parallel to the plate. In other words, the above analysis is perfectly applicable to a plate inclined at an angle φ with the horizontal if g is everywhere replaced by $(g \sin \varphi)$. Thus,

$$\text{Local:} \quad N_{\text{NU}_x} = \frac{1}{\sqrt{2}} \left(\frac{g \sin \varphi \rho^2 \lambda x^3}{\mu k\, \Delta t} \right)^{1/4}.$$

$$\text{Average:} \quad N_{\text{NU}_L} = 0.943 \left(\frac{g \sin \varphi \rho^2 \lambda L^3}{\mu k\, \Delta t} \right)^{1/4}. \qquad (10.14)$$

Naturally these relations must be used with caution for small values of φ since some of the assumptions used in their derivation become invalid—an obviously incorrect, absurd, result occurs for $\varphi = 0°$.

When a plate is in a horizontal position, the basic driving force which removes condensate from the surface is that of hydrostatic pressure. Near the center of the plate the condensate film must necessarily be thicker than at the plate edge. This elevation difference causes a flow of the layer toward the edge. This problem was analyzed by Clifton and Chapman (Ref. 8), with the following result:

$$N_{\text{NU}_L} = \frac{(12)^{4/5}}{3} \left(\frac{g\rho^2 \lambda L^3}{\mu k\, \Delta t} \right)^{1/5} \cdot F\left(\frac{k\, \Delta t}{\mu \lambda} \right)$$

$$= 2.43 \left(\frac{g\rho^2 \lambda L^3}{\mu k\, \Delta t} \right)^{1/5} \cdot F\left(\frac{k\, \Delta t}{\mu \lambda} \right). \qquad (10.15)\bigstar$$

The function $F(k\, \Delta t / \mu \lambda)$ is a complex function which is tabulated in Table 10.1. It should be noted that the parameter $(k\, \Delta t / \mu \lambda)$ serves the function of a Prandtl number in this case.

Table 10.1   The Quantity F as a Function of $(k\, \Delta t / \mu \lambda)$ for Use in Eq. (10.15)

$\dfrac{k\, \Delta t}{\mu \lambda}$	F
0.176	0.329
0.381	0.322
0.698	0.320
2.27	0.285
4.08	0.264

Working Formulas and the Effect of Turbulence in the Liquid Film

The equations of Nusselt shown above for condensation on vertical surfaces are found to be valid for short surfaces for which an excessively thick liquid layer does not build up. Most practical applications, however, involve vertical surfaces of such a length that the liquid film becomes thick enough to cause transition to the turbulent state of flow. In order to define the point of such transition, it is necessary to define the Reynolds number of the flow of the condensate film. This definition can be based on the mean velocity in the liquid film, v_m, and the film thickness δ:

$$N_{RE_\delta} = \frac{v_m \rho \delta}{\mu}.$$

It is more convenient, however, to work in terms of the mass flow rate rather than the unknown mean velocity. The numerator of the Reynolds number defined above is recognized as the mass rate of flow per unit of width of the surface. Let this latter quantity be symbolized by $\Gamma = v_m \rho \delta$. (In the case of condensation on a vertical tube Γ is the mass rate of flow of the condensate divided by the cylindrical perimeter, πD). Thus, the Reynolds number based on Γ is

$$N_{RE_\Gamma} = \frac{\Gamma}{\mu}.$$

Experiments have shown that for values of N_{RE_Γ} greater than about 400 to 450, a turbulent liquid film may be expected.

From the examination of many data, McAdams (Ref. 2) recommends that when N_{RE_Γ}, evaluated for Γ at the *lowest point* on the condensing surface, is less than 450, Nusselt's equation, Eq. (10.13), holds well if multiplied by 1.2 —an empirically determined factor. Thus,

$$N_{NU_L} = 1.13 \left(\frac{g\rho^2 \lambda L^3}{\mu k \, \Delta t} \right)^{1/4}. \tag{10.16}\bigstar$$

This correlation is based on the evaluation of the properties at the mean film temperature $(t_s + t_{sat})/2$ except for the latent heat, λ, which is to be taken at the saturation temperature of the condensate. For values of $N_{RE_\Gamma} > 450$ the following empirical relation is recommended:

$$N_{NU_L} = 0.0134 \left(\frac{g\rho^2 L^3}{\mu^2} \right)^{1/3} (N_{RE_\Gamma})^{0.4}. \tag{10.17}\bigstar$$

EXAMPLE 10.1:   Saturated steam at 10 psia condenses on a vertical plate 1 ft high. The plate is maintained at 187°F. Find the average coefficient of heat transfer.

Solution: For a pressure of 10 psia, the steam tables give $t_{sat} = 193°F$, $\lambda = 982$ Btu/lb$_m$. Thus, $\Delta t = 6°F$ and the mean film temperature is 190°F. For 190°F, Table A.4 gives the following data for saturated water:

$$\rho = 60.35 \text{ lb}_m/\text{ft}^3, \qquad \mu = 0.79 \text{ lb}_m/\text{ft-hr}, \qquad k = 0.389 \text{ Btu/hr-ft-°F}.$$

Since the relations which determine h must be selected on the basis of N_{RE_r}, and since the latter quantity depends on the amount of steam condensed, it is necessary first to assume whether film is laminar or not and then to check this assumption. Upon assuming a laminar film, Eq. (10.16) gives

$$N_{NU_L} = 1.13 \left[\frac{32.2 \times (3600)^2 \times (60.35)^2 \times 982 \times 1^3}{0.79 \times 0.389 \times 6} \right]^{1/4} = 6020,$$

$$h = 6020 \times \frac{0.389}{1} = 2340 \text{ Btu/hr-ft}^2\text{-°F } (13{,}290 \text{ W/m}^2\text{-°C}).$$

Now, the amount of steam condensed per foot of width is

$$\Gamma_{max} = 2340 \times \frac{1 \times 6}{982} = 14.3 \text{ lb}_m/\text{ft-hr}.$$

Thus, $N_{RE_r} = 14.3/0.79 = 18.1$. The film is laminar as assumed. If the film had not been laminar, it would have been necessary to use Eq. (10.17). In this case, an iterative solution would be required, since N_{RE_r} is needed to determine h from Eq. (10.17).

The Effect of Noncondensable Gases and the Vapor Velocity

The results quoted above are restricted to cases in which the presence of noncondensable gases has little or no effect on the process of condensation. Also, the analytical approach of Nusselt presumed that the relative velocity at the liquid–vapor interface was so negligibly small that there would be no shear forces acting at the liquid–vapor interface.

Theories exist to explain fully the effect of the presence of noncondensable gases, but are beyond the scope of this book. Experiment shows, as expected, that the presence of such gases tends to decrease materially the heat transfer coefficient. This effect seems to be approximately linear with the weight fraction of the noncondensable gas present.

The effect of having a significant vapor velocity parallel to the direction of motion of the liquid layer is to alter the velocity distribution through the liquid. This effect is caused by the presence of a significant shearing stress at the liquid–vapor interface. Generally speaking, if the vapor flow is upward, the film tends to be thicker than that predicted by the above analyses. For flow downward, in the same direction as the liquid, the reverse is true. The condition of the existence of a significant vapor velocity is most likely to exist for the case of condensation of a vapor under forced flow through a tube, as discussed in the next section.

10.4   Condensation Inside Tubes

The process of the condensation of a vapor flowing inside a cylindrical tube is one of importance in chemical and petrochemical processes. In a comprehensive review of the work in this field, and after additional experimental investigations, Akers, Deans, and Crosser (Ref. 3) report that for condensing vapors in either horizontal or vertical tubes, the following relations correlate the data within about 20 per cent:

$$\text{For } N_{RE_G} < 5 \times 10^4, \; N_{NU_D} = 5.03(N_{RE_G})^{1/3}(N_{PR})^{1/3}.$$

$$\text{For } N_{RE_G} > 5 \times 10^4, \; N_{NU_D} = 0.0265(N_{RE_G})^{0.8}(N_{PR})^{1/3}. \qquad (10.18)\bigstar$$

These expressions appear very similar to those for forced convection without phase change inside of tubes, as discussed in Chapter 7. However, the Reynolds number must be given a new interpretation to account for the existence of the two-phase flow and to express the relative amounts of each phase.

The customary diameter Reynolds number is defined as

$$N_{RE_D} = \frac{U_m \rho D}{\mu}.$$

Instead of the mean velocity, one introduces the *mass velocity*, G, which is the mass rate of flow per unit cross-section area:

$$G = U_m \rho.$$

Then, in terms of the mass velocity, the Reynolds number is

$$N_{RE_G} = \frac{DG}{\mu}. \qquad (10.19)$$

The correlations of Akers et al., given in the equations above, are written in terms of a Reynolds number, N_{RE_G}, based on an *equivalent mass velocity*, G_E, defined as

$$G_E = G_L + G_V \left(\frac{\rho_L}{\rho_V}\right)^{1/2}. \qquad (10.20)$$

In this expression G_L and G_V are the mass velocities, based on the full pipe area, of the liquid and vapor phases, respectively. The symbols ρ_L and ρ_V are the densities of the liquid and vapor phases, respectively. Equation (10.20) results from choosing a model for the flow inside the tube which replaces the vapor core by an analogous liquid which exerts the same shearing stress at the liquid–vapor interface.

10.5   Film Condensation on the Outside of Horizontal Cylinders

The Nusselt analysis shown in Sec. 10.3 for vertical surfaces may be extended to other shapes—such as the very practical case of a condensate film formed on the outside of a horizontal cylinder. Such a case is illustrated in Fig. 10.3(a). For a single tube the film would start with zero thickness at the top of the cylinder. The same method of analysis used for the vertical or inclined plane in Sec. 10.3 would apply in this case, but it would be more convenient to replace the coordinate along the surface, x, by the polar coordinate $\theta = x/(D/2)$ as shown in Fig. 10.3(a). The analysis becomes difficult to carry out in closed form since, in this case, the magnitude of the body force becomes a function of the coordinate θ rather than a constant. That is, the body force is $(g \sin \varphi)$, φ being the angle between the horizontal and the tangent to the tube surface at the point in question—see Fig. 10.3(a). In turn, φ is a function of the polar coordinate θ.

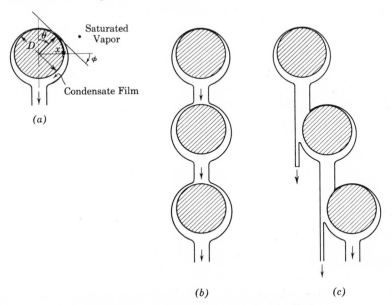

Fig. 10.3. Film condensation on horizontal cylinders.

The carrying out of the analysis to its logical end results in an expression for the film coefficient of exactly the same form as was obtained for the vertical surface, with only the numerical value of the constant being altered. After a numerical integration to obtain this constant, the average diameter Nusselt number is

$$N_{NU_D} = 0.725 \left(\frac{g\rho^2 \lambda D^3}{\mu k \, \Delta t} \right)^{1/4}. \tag{10.21}\bigstar$$

This latter relation is found to give excellent predictions of actual condensing film coefficients on horizontal tubes if, as before, all properties are taken for the condensate at the mean film temperature and λ is evaluated at the saturation temperature. In most practical applications, the tubes used are of such diameter that the film of condensate never becomes thick enough to pass into the turbulent regime.

Equation (10.21) is obtained on the assumption of a zero film thickness at the top of the tube. If, as illustrated in Fig. 10.3(b), the condensate from one tube drips onto the top of a tube just below, the condensate film on the second tube starts with a finite thickness—resulting in a higher film resistance and lower film coefficient. Each successive tube in a vertical array will yield a lower average heat transfer coefficient. Nusselt analyzed such a situation, using the thickness of the film leaving one tube as the initial thickness of the next, and showed the *average* film coefficient for an array of N horizontal tubes placed one above the other is given by

$$N_{NU_D} = 0.725 \left(\frac{g\rho^2 \lambda D^3}{N\mu k \, \Delta t} \right)^{1/4}. \tag{10.22}\bigstar$$

Obviously it is advantageous to stagger the tubes, as shown in Fig. 10.3(c), whenever it is possible.

It is interesting to note the relative effectiveness of horizontal and vertical tubes as condensing surfaces. Taking the ratio of Eq. (10.21) to Eq. (10.16),

$$\frac{N_{NU_D}}{N_{NU_L}} = \frac{h_h D}{h_v L} = \frac{0.725}{1.13} \left(\frac{D^3}{L^3} \right)^{1/4},$$

or

$$\frac{h_h}{h_v} = 0.642 \left(\frac{L}{D} \right)^{1/4}. \tag{10.23}$$

The subscripts h and v denote the horizontal and vertical positions, respectively. In the case of most steam condensers the ratio L/D for the tubes will be in the range 50 to 100, or even more. Thus, Eq. (10.23) indicates that as much as twice the liquid will be condensed in the horizontal position as in the vertical position, other factors being the same.

EXAMPLE 10.2: Saturated steam at 1 psia condenses on a horizontal 1-in. tube, the surface of which is maintained at 95°F. Find the amount of steam condensed per hour, per foot of tube length.

Solution: At 1 psia, the steam tables give the saturation temperature to be 101.7°F and the latent heat, λ, to be 1036 Btu/lb$_m$. For a mean film temperature (95 + 101.7)/2 = 98°F, Table A.4 gives, for saturated water,

$$\rho = 62.1 \text{ lb}_m/\text{ft}^3, \qquad \mu = 1.69 \text{ lb}_m/\text{ft-hr},$$

$$k = 0.363 \text{ Btu/hr-ft-°F}.$$

Thus, Eq. (10.22) yields

$$N_{NU_D} = 0.725 \left[\frac{32.2 \times (3600)^2 \times (62.1)^2 \times (1036) \times (1/12)^3}{1.69 \times 0.363 \times (101.7 - 95)} \right]^{1/4}$$

$$= 0.725(2340)^{1/4}(10^2)$$

$$= 505.$$

So

$$h = 505 \times \frac{0.363}{1/12} = 2200 \text{ Btu/hr-ft}^2\text{-°F}.$$

From this value of h the steam condensed per foot of length is found to be

$$\frac{\dot{m}}{L} = \frac{q}{L}\frac{1}{\lambda}$$

$$= \pi D h \frac{\Delta t}{\lambda}$$

$$= \pi \frac{1}{12}2200\frac{6.7}{1036} = 3.72 \text{ lb}_m/\text{hr-ft } (1.54 \times 10^{-3} \text{ kg/s-m}).$$

It should be noted that condensing film coefficients are much greater, normally, than those found in ordinary forced or free convection.

10.6 Heat Transfer During the Boiling of a Liquid

The process of the purposeful conversion of a liquid into a vapor is one of obvious engineering importance. For many years engineers have been successful in designing and constructing highly efficient devices for, mainly, the purpose of the production of large quantities of steam. Many chemical processes, particularly in the case of organic substances, involve the vaporization of a liquid.

In spite of all this experience and practical success, the process taking place between a heated surface and a boiling liquid remained for many years the least understood of all the convective mechanisms. This subject has been the object of much research during recent years, and a firm understanding of the boiling mechanism appears accomplished.

Experiments have shown that if a liquid, such as water, is distilled and completely degasified under vacuum (i.e., if it contains *no* impurities, not even dissolved gases) it will undergo the liquid–vapor phase change without the appearance of bubbles when it is heated in a clean, smooth vessel. Under normal conditions, however, when the rate of heat input to a saturated liquid is great enough, vapor bubbles are observed to be formed below the free surface of the liquid. These vapor bubbles seem to form at certain favorable nuclei—due either to the presence of solid particles and dissolved gases in the liquid or some irregularities in the surface by which the liquid is being heated. This phenomenon is generally termed *nucleate boiling*.

Other states, or regimes, of boiling, in addition to the nucleate regime, are observed to occur. These regimes are best described in terms of a heating surface submerged in a stagnant (in terms of external agitation) pool of saturated liquid—a set of circumstances described as *pool boiling*. The term "pool boiling" is used in contrast to boiling which occurs simultaneously with forced convection. In pool boiling, the dominant factor determining the heat flux (and, hence, the heat transfer coefficient) is the difference between the heating surface temperature and the saturation temperature of the boiling liquid.

Figure 10.4 illustrates the typical dependence between the heat flux, q/A, and the temperature difference, $\Delta t = t_s - t_{sat}$. The data of Figure 10.4 are

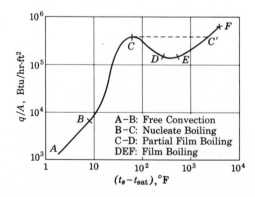

Fig. 10.4. Heat flux versus temperature difference for boiling water heated by a wire.

those obtained for pool boiling of water at atmospheric pressure, using an electrically heated wire which served the dual purpose of a heating surface and a resistance thermometer for measurement of the surface temperature.

This curve is similar in shape to those that would be obtained for other liquids and other pressures. In the low range of values of Δt, the range indicated by A-B in Fig. 10.4, no boiling (i.e., bubble formation) is observed. The liquid is found to be slightly superheated (a phenomenon to be discussed more later) below the free surface. Energy is transported from the heating surface to the liquid–vapor interface by free convection. All vaporization occurs by evaporation at the free surface. The results of Chapter 9 indicate that the free convection coefficient (for laminar flow) is proportional to $(\Delta t)^{1/4}$. Hence, the heat flux in the range indicated by A-B is proportional to $(\Delta t)^{5/4}$.

Continued increase in Δt causes the boiling process to enter in the region B-C in Fig. 10.4, wherein nucleate boiling becomes dominant. Bubbles of vapor form on the heating surface, grow in size, and eventually break off. After breaking away, the bubbles rise to the surface and burst, releasing the vapor. In this regime the fluid around the heating element becomes highly agitated—much more so than in the case of the free convection regime. This increased fluid motion tends, naturally, to intensify the heat transfer process, causing an upturn in the heat flux curve. In the nucleate boiling range the heat flux varies as Δt to the third or fourth power. Since the liquid phase is such a better heat conductor than the vapor phase, most of the heat transferred during the nucleate boiling process is from the surface, into the liquid, and then into the bubble. Only a small fraction of the heat is transferred from the surface directly into the vapor of the bubble.

As Δt is increased further, the number of nucleation centers increases also. Eventually a point is reached at which the increased blanketing of the surface by bubbles offsets the expected increase of heat flux due to the increase Δt. The total heat flux begins to drop off until a peak is reached— point C in Fig. 10.4.

Additional increase in Δt eventually establishes what is called *stable film boiling* (*D-E-F*), wherein the vapor generated has completely covered the heating surface. This covering of the surface with a complete blanket of vapor increases the thermal resistance and, hence, a drop in heat flux is observed. Naturally, the transition from the nucleate boiling regime to the film boiling regime is not a sharp one, and a region of partial film boiling must be passed through, *C-D*, before an established state of stable film boiling can be achieved.

The point of maximum heat flux, point C in Fig. 10.4, at which the transition from nucleate to film boiling occurs is often called the *boiling crisis* or the *Lindenfrost point*, after the investigator who first studied film boiling. The temperature difference associated with this point is of considerable significance, since it would be the desired point of operation of a vapor-producing piece of apparatus, as this would permit the maximum use of surface area. However, certain complications are associated with this idea, as will be seen presently.

In the fully developed region of film boiling (*D-E-F*) the heat flux again rises with increasing Δt, although the rate of increase is not as great as that observed in the nucleate boiling region. As Δt becomes greater and greater in the film boiling region, a considerable portion of the heat exchange may be due to thermal radiation. Eventually a point is reached at which the heating surface melts—point *F* in Fig. 10.4.

The preceding discussion was based on the presumption that the temperature of the heating surface could be maintained at a particular desired value. With Δt thus under control, it is possible to operate a vapor-producing system at the point of maximum heat flux—point *C* in Fig. 10.4. However, in certain applications it is the heat flux and not the surface temperature that is the independent variable. In such cases (e.g., a submerged wire with a fixed voltage applied to it) the heating surface seeks the appropriate temperature required by the relation between the heat flux and Δt symbolized in Fig. 10.4. If such a system is operating at the point of maximum heat flux (i.e., point *C* in Fig. 10.4) and if the heat flux is increased only slightly, the surface temperature will increase until equilibrium between the energy input to the surface and the heat flux away from the surface is established. This equilibrium point is indicated by *C'* in Fig. 10.4. If the material of the heating surface is such that its melting point (point *F*) is less than the temperature corresponding to point *C'*, failure of the heating surface will result.

A significant factor associated with nucleate boiling is the observed liquid superheat referred to above. The amount of superheat depends upon the pressure, surface condition, boiling rate, etc., and is observed to be of the order of about $\frac{1}{2}°$F for water. Physically, the superheat is the result of the action of surface tension on the concave liquid–vapor interface of the bubble. The action of the surface tension requires the pressure of the vapor in the bubble to be greater than that of the bulk liquid. Hence, the temperature of the vapor must be greater than the saturation temperature associated with the pressure of the liquid. Further, if the bubble is to grow, the liquid temperature adjacent to the bubble must be even greater—hence the occurrence of the observed superheat.

The discussion above explains also the observed profound effect of the roughness of the heating surface on the boiling process. If one imagined that a bubble is initiated with zero radius, an infinite pressure to create the bubble would be indicated. Hence, bubbles will tend to form at minute surface imperfections or on foreign particles where the geometry favors the creation of bubbles with a minimum requirement of energy or superheat.

10.7   Working Formulas for Boiling Heat Transfer

Nucleate Pool Boiling

The governing mechanism in any convective heat transfer process is the mixing of hot and cold portions of the fluid which results from the fluid motion. This is certainly true in the case of nucleate pool boiling. As mentioned above, observations prove that very little heat is transferred from the heating surface directly into the vapor of the bubbles forming on it. Most of the heat exchange occurs as a result of transfer directly from the surface to the liquid by the action of convection currents. The convection currents result from the vigorous agitation of the liquid by the motion of the vapor bubbles. Thus, the over-all exchange of heat is related to the rate of growth of the bubbles, their size, the frequency of their release, whether or not the fluid "wets" the heating surface, etc.

The data for nucleate pool boiling are best correlated by the following relation developed by Rohsenow (Ref. 4):

$$\frac{c_{\text{pl}}(t_s - t_{\text{sat}})}{\lambda} = CN_{\text{PR}_l}^{1.7}\left[\frac{q/A}{\mu_l\lambda}\left(\frac{g_c\sigma}{g(\rho_l - \rho_v)}\right)^{0.5}\right]^{0.33}. \qquad (10.24)\bigstar$$

In Eq. (10.24) the subscript l is used to denote the physical properties of the *saturated liquid* and the subscript v denotes those for the *saturated vapor*. The latent heat of vaporization is denoted by λ and the surface tension at the liquid–vapor interface is given by σ. The symbol g represents the acceleration of gravity, while g_c represents the dimensional constant discussed in Sec. 1.7. The constant C is an empirically determined quantity which is dependent upon the way in which the liquid wets the heating surface. A summary of experimentally determined values of C may be found in Ref. 5, but for the purposes of this book the following values for water will suffice:

$$\text{Water with nickel or brass:} \qquad C = 0.006.$$

$$\begin{aligned} &\text{Water with platinum, copper,} \\ &\quad\text{or stainless steel:} \qquad C = 0.013. \end{aligned} \qquad (10.25)$$

Rohsenow (Ref. 6) further recommends that the peak heat flux, point C in Fig. 10.4, may be found from

$$\frac{(q/A)_{\text{max}}}{\rho_v\lambda} = \frac{\pi}{24}\left[\frac{\sigma g g_c(\rho_l - \rho_v)}{\rho_v^2}\right]^{1/4}\left(1 + \frac{\rho_v}{\rho_l}\right)^{1/2}. \qquad (10.26)\bigstar$$

Nucleate Boiling in Forced Convection

The above correlation for nucleate pool boiling will also predict well the heat transfer occurring when nucleate boiling exists simultaneously with forced convection. In such instances, good results are obtained for the total heat transfer by simply superimposing the two effects:

$$\left(\frac{q}{A}\right)_{\text{total}} = \left(\frac{q}{A}\right)_{\text{boiling}} + \left(\frac{q}{A}\right)_{\text{conv}}. \tag{10.27}$$

The boiling heat flux is determined according to Eqs. (10.24) and (10.25), above, while the convective heat flux is determined from the relations of Chapter 8. For instance, for flow in tubes, Eqs. (8.15) and (8.16) may be used for $(q/A)_{\text{conv}}$.

Pool Film Boiling

When the state of full film boiling is established, the heating surface is completely surrounded by a blanket of vapor. The temperature of the surface is quite great in this case since all the heat must be transferred through this film or poorly conducting vapor. The vapor formed in the film rises past the heating surface and forms bubbles which eventually break away, rise to the surface, and burst.

For film boiling on a submerged horizontal cylinder, the convective picture is that of a layer of vapor flowing upward past the cylinder and bounded by a large liquid domain. Bromley (Ref. 7) likened this situation to that of condensation on a horizontal cylinder—a liquid layer flowing downward past the surface and bounded by a large vapor domain. Following the same analysis as Nusselt in the development of Eq. (10.21), the following relation is obtained for the film coefficient:

$$N_{\text{NU}_D} = \frac{hD}{k_v} = 0.725\left[\frac{\rho_v(\rho_l - \rho_v)\lambda D^3 g}{\mu_v k_v(t_s - t_{\text{sat}})}\right]^{1/4}.$$

This relation presumes that no shear exists at the liquid–vapor interface (i.e., that the liquid moves at the same velocity as the vapor). If the liquid is presumed to be stationary, the constant 0.725 becomes 0.512. It is recommended that a mean value be used:

$$N_{\text{NU}_D} = 0.62\left[\frac{\rho_v(\rho_l - \rho_v)\lambda D^3 g}{\mu_v k_v(t_s - t_{\text{sat}})}\right]^{1/4}. \tag{10.28}\bigstar$$

References

1. NUSSELT, W., "Die Oberflächenkondensation des Wasserdampfes," *Zeitschr. V.D.I.*, Vol. 60, 1916, p. 569.
2. MCADAMS, W. H., *Heat Transmission*, 3rd ed., New York, McGraw-Hill, 1954.
3. AKERS, W. W., H. A. DEANS, and O. K. CROSSER, "Condensing Heat Transfer Within Horizontal Tubes," *Chem. Eng. Progr., Symp. Ser.*, Vol. 55, No. 29, p. 171.
4. ROHSENOW, W. M., "A Method of Correlating Heat Transfer Data for Surface Boiling Liquids," *Trans. ASME*, Vol. 74, 1952, p. 969.
5. TONG, L. S., *Boiling Heat Transfer and Two-Phase Flow*, New York, Wiley, 1965.
6. ROHSENOW, W. M., and H. CHOI, *Heat, Mass, and Momentum Transfer*, Englewood Cliffs, N.J., Prentice-Hall, 1961.
7. BROMLEY, L. A., "Heat Transfer in Stable Film Boiling," *Chem. Eng. Progr.*, Vol. 46, 1950, p. 221.
8. CLIFTON, J. V., and A. J. CHAPMAN, "Condensation of a Pure Vapor on a Finite-Size Horizontal Plate," *ASME Paper 67-WA/HT-18*, New York, 1967.
9. ZUBER, N., "On the Stability of Boiling Heat Transfer," *Trans. ASME*, Vol. 80, 1958, p. 711.

Problems

10.1. Ammonia is condensing at 100 psia on the surface of a vertical flat plate 2 ft high and maintained at 50°F. What is the condensing heat transfer coefficient? How long must the plate be in order to just obtain a turbulent film?

10.2. A vertical $\frac{3}{4}$-in. tube is 15 ft long and has its surface maintained at 85°F. Steam at 90°F is condensing on the outer surface of the tube. How much steam is condensed in 1 hr?

10.3. A vertical plate 6 in. wide and 3 ft high is maintained at 150°F while exposed to saturated steam at 1 atm. Find the average heat transfer coefficient and the amount of steam condensed per hour.

10.4. For the data of Prob. 10.3, find the film thickness and maximum film velocity at the bottom edge of the plate and at a point halfway down the plate.

10.5. Saturated steam is exposed to a horizontal 1-in.-diameter cylinder maintained at a surface temperature of 60°F. Find the condensing heat transfer coefficient if the steam pressure is
(a) 14.7 psia;
(b) 50 psia;
(c) 100 psia.

10.6. Freon-12 (dichlorodifluoromethane) at 100°F condenses on a horizontal 1-in. tube, the surface of which is maintained at 94°F. Find the rate at which the Freon is being condensed, per foot of tube length.

10.7. A steam condenser consists of a bundle of $\frac{5}{8}$-in. tubes 6 ft long. Steam at 4 in. Hg abs. is condensing on the outer surface of the tube, maintained at 114°F by cooling water inside. What per cent increase in the amount of steam condensed is obtained by placing the condenser horizontal compared to that obtained in the vertical position?

10.8. Freon-12 at 100 psia is condensing on the inside of a 1-in., 18-gage tube. The tube wall is at 75°F. The total flow rate of Freon is 600 lb/min. Find the rate at which condensate is being produced at the point where there is 50%, by weight, of liquid.

10.9. Steam at 1 psia condenses on a horizontal bank of $\frac{5}{8}$-in. tubes. The bank of tubes is eight tubes high. The tubes are not staggered. The tube surface is maintained at 96°F. Find the average heat transfer coefficient for the bank of tubes. What percentage of the steam condensed is condensed by the top row of tubes?

10.10. Find the heat transfer coefficient for nucleate pool boiling of water on a horizontal stainless-steel surface at atmospheric pressure if the surface is maintained at (a) 230°F; (b) 250°F; and (c) 300°F. The surface tension of water is given by $\sigma = 5.3 \times 10^{-3} (1 - 0.0014T)$, where T is in °F and σ is in lb_f/ft.

10.11. Repeat Prob. 10.10 for a brass surface.

10.12. Find the maximum heat flux and the corresponding surface temperature for the conditions stated in Prob. 10.10.

10.13. What is the rate of heat transfer from a horizontal 3-in. pipe submerged in boiling water at atmospheric pressure if the pipe surface is at 600°F?

10.14. Water at a bulk temperature of 180°F flows through a 2-in., schedule 40 pipe, the inner surface of which is maintained at 300°F. Under these conditions, boiling is likely to occur. If the water pressure is 250 psia, what is the heat flux from the pipe wall to the water?

10.15. Film boiling occurs on a horizontal, 0.01-in. Chromel wire at 1800°F, submerged in water at 212°. What is the rate of heat transfer from the wire, per foot of length?

10.16. A plate of steel ($\frac{1}{4}$ in. thick, 4 in. wide, 12 in. long) is heated to 300°F and then cooled by plunging it vertically into a bath of water at 212°F. What is the expected heat transfer coefficient at the instant the steel is thrust into the water?

CHAPTER 11

Heat Transfer by Radiation

11.1   Introductory Remarks

The fundamental physical phenomenon which forms the basis of all heat transfer studies is the observed fact that the temperature of a body, or a portion of a body, which is hotter than its surroundings tends to decrease with time. This temperature decrease indicates a flow of energy from the body. The entire discussion of the foregoing ten chapters has been limited to cases in which some physical medium was necessary for the transport of the energy from the high temperature source to the low temperature sink, leading to the mechanisms of conduction and convection. Generally speaking, the rate of flow of the thermal energy in these instances was proportional to the difference in temperature between the source and the sink.

If, however, a heated body is physically isolated from its cooler surroundings (i.e., by a vacuum), its temperature is still observed to decrease in time, again showing a loss of energy. In this case an entirely different energy transfer mechanism is taking place, and it is called *thermal radiation*.

A body need not be heated to exhibit the loss of energy by radiation. The "thermal" radiation just referred to is one aspect of a more general phenomenon which might be termed *radiant energy*. The emission of other forms of radiant energy may be caused when a body is excited by such means as an oscillating electrical current, electronic or neutronic bombardment, chemical reaction, etc. Also, when radiant energy strikes a body and is absorbed, it may manifest itself in the form of thermal internal energy, a chemical reaction, an electromotive force, etc., depending on the nature of the incident radiation and the substance of which the body is composed. In either case, emission or absorption, this book will be concerned only with thermal radiation—i.e., radiation produced by or which produces thermal excitation of a body.

Several theories have been proposed to explain the transport of energy by radiation. One theory holds that the body emits discrete *packets*, or *quanta*, of energy and has been successful in explaining the experimental facts observed in such cases as the photoelectric emission of electrons, thermal radiation emission, etc. Another theory asserts that radiation may be represented as an electromagnetic wave motion and has been useful in explaining such phenomena as interference of light, polarization of light, etc. At the present time a dual theory is generally accepted, giving radiant energy the characteristics of a wave motion as well as discontinuous emission.

Whichever theory is used, it is convenient to classify all radiant-energy emissions in terms of the wavelength when considered as electromagnetic wave motions propagating at the velocity of light—186,000 miles/sec. In this way, thermal radiation is usually defined as that portion of the radiant energy spectrum between wavelengths of 1×10^{-1} and 1×10^{2} micron.* Other well-known electromagnetic waves are listed below with their wavelength bands indicated:

Cosmic rays	up to 4×10^{-7} micron
Gamma rays	4×10^{-7} to 1.4×10^{-4} micron
X-rays	1×10^{-5} to 2×10^{-2} micron
Ultraviolet rays	1×10^{-2} to 3.9×10^{-1} micron
Visible light	3.9×10^{-1} to 7.8×10^{-1} micron
Infrared rays	7.8×10^{-1} to 1×10^{3} micron
Heat rays	1×10^{-1} to 1×10^{2} micron
Radio and hertzian waves	1×10^{3} to 2×10^{10} micron

It will be necessary first to define a number of terms and properties which characterize radiant energy emission. The following discussion will apply mainly to thermal radiation, although some of the concepts are valid for other forms of radiant emission.

11.2   Basic Definitions

General definitions will be made first. Further discussion will show a need for the refinement of these definitions in order to account for a multitude of special considerations such as the dependence of physical properties on temperature, directional properties of the incident or emitted radiation, etc.

---

* The unit customarily used to measure wavelength is the micron: 1 micron = 1 micrometer = 1 μm.

As was noted in the introductory discussion, thermal radiation is defined as that electromagnetic radiation between the wavelengths of 1×10^1 and 1×10^2 micron. If the thermal radiation incident upon a surface, or given up by a surface, were to be decomposed into its spectrum over this wavelength band, it would be found that the radiation is not equally distributed over all the wavelengths. The amount of radiant energy is different at each wavelength and the distribution between wavelengths is, generally, different from case to case. The adjective *spectral* is used to denote this dependence upon wavelength. Similarly, the word "spectral" will be used to identify any other phenomenon, surface property, etc., in which a wavelength dependence is observed.

Emissive Power, Radiosity, and Irradiation

Total emissive power, denoted by the symbol W, is the term used to denote the total *emitted* thermal radiation leaving a surface, per unit time, per unit area of emitting surface. Other names often used to denote this quantity are "radiant flux density," "total hemispherical emissive power," or simply "emissive power." The terms *total emissive power* or *emissive power* will be used here, and it should be noted that this quantity is defined to consist only of the *original* emission from a surface. It does not include any energy leaving a surface that may be the result of the reflection of some incident radiation.

The total emissive power is found to be dependent upon the temperature of the emitting surface, the substance of which the surface is composed, and the nature of the surface structure (i.e., roughness, etc.). The emissive power of a given element of a surface may have both a spectral and directional dependence. That is, the emission may be directionally preferential and may also be concentrated in certain wavelength bands, within the thermal radiation spectrum. The *total* emissive power W defined here includes *all* the original radiant emission from a surface element passing into the hemisphere above the element, regardless of its direction or wavelength.

It has been found that the total emissive power of an element of a surface is dependent upon its temperature. For surfaces with temperature-independent physical properties, the emissive power is proportional to the fourth power of the surface temperature. Since W is defined as an energy flux *per unit area*, the total emissive power of a finite-sized surface must be the integrated average of the emissive power of an element taken over the entire surface.

Total Radiosity. *Radiosity* is the term used to indicate the total radiant energy *leaving* a surface, per unit surface area. The symbol J is used to denote the radiosity. This quantity differs from the emissive power in that the radiosity includes reflected energy as well as the original emission. As in the case of the emissive power, the *total* radiosity of a surface element consists of *all* the radiation leaving a surface, regardless of any directional dependence or spectral preference.

Total Irradiation. *Irradiation, G,* is the term used to denote the total radiation *incident* upon a surface per unit time, per unit area of irradiated surface. The irradiation incident upon a surface is the result of emissions and reflections from other surfaces, and, as such, may be directionally and spectrally preferential. The *total* irradiation is, again, the total energy flux incident upon a surface, regardless of these other factors.

Absorptivity, Reflectivity, and Transmissivity

When radiation falls on a surface, part of it may be absorbed by the body, part may be reflected away from the surface, and part may be transmitted through the body. The following symbols are defined:

α = total absorptivity = fraction of radiation absorbed,
ρ = total reflectivity = fraction of radiation reflected,
τ = total transmissivity = fraction of radiation transmitted.

Using these definitions, one has

$$\alpha + \rho + \tau = 1. \tag{11.1}$$

The absorptivity, reflectivity, and transmissivity just defined are *total* values of these quantities. That is, they are defined in terms of the total incident radiation, total reflected radiation, and total transmitted radiation.

In general the absorptivity of a surface is dependent upon the direction of the incident radiation, the spectral distribution of the incident radiation, the composition and structure of the irradiated surface, and the temperature of the surface. In the case of gases, the absorptivity is also dependent upon the geometrical size and shape of the gas bulk through which the radiation passes. Similar dependencies are observed for the reflectivity and transmissivity—plus the additional complications that the reflected or transmitted energy may also have spectral and directional dependency.

Most gases have high values of τ and low values of α and ρ. Air at atmosphere pressure and temperature, for instance, has $\alpha = \rho \approx 0$, and thermal radiation passes through it as though in a vacuum. Other gases, notably water vapor and carbon dioxide, may be highly absorptive to thermal radiation—at least at certain wavelengths. Most solids encountered in engineering practice are opaque to thermal radiation, resulting in $\tau = 0$. The initial parts of this chapter will be devoted to solid surfaces separated by thermally transparent

($\alpha = \rho = 0$, $\tau = 1$) media. The questions which arise when absorbing gases separate the surfaces will be treated in later portions of the chapter.

For thermally opaque solid surfaces one then has

$$\alpha + \rho = 1. \tag{11.2}$$

The definitions of total emissive power W, total radiosity J, and total irradiation G, given earlier lead to the useful relations

$$J = W + \rho G,$$
$$= W + (1 - \alpha)G. \tag{11.3}$$

The Intensity of Radiation

In the consideration of the exchange of radiant energy between surfaces it will be useful to define a quantity known as the *intensity* of radiation. Intensity, to be symbolized by I, is properly defined in terms of a "point source," but it is convenient to discuss the concept in terms of incremental surface elements. Consider, as illustrated in Fig. 11.1, a small emitting surface of area ΔA, the center of which is denoted as the point Q. Let Δq represent the *rate* at which radiant energy leaves ΔA. It will be useful to think in terms of a radiant energy *flux*. The average flux leaving ΔA is defined as

$$f_{av} = \frac{\Delta q}{\Delta A}. \tag{11.4}$$

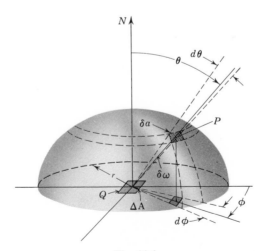

Fig. 11.1

The radiant flux from the *point source* at Q is defined as

$$f_Q = \lim_{\Delta A \to 0} \frac{\Delta q}{\Delta A}. \tag{11.5}$$

It is important to remember that the flux just defined is based on the energy *leaving* a surface (reflected or emitted) and is calculated per unit area of the surface from which it leaves.

Now, the radiant flux from ΔA, or Q, fills the half-space above ΔA. The *intensity* of the radiation at a point in space due to the emission from this point source is defined as the radiant energy passing the spatial point per unit time, per unit solid angle subtended at Q, per unit area of radiating surface projected normal to the direction which the point in space makes with the emitting point. In terms of the drawing in Fig. 11.1, the radiation intensity at some point in space, say P, due to the radiation leaving ΔA may be defined in terms of the radiation falling on an element of the spherical surface (center at Q, radius r) which passes through P. If δa represents an element of the spherical surface surrounding the point P, only a fraction of Δq strikes δa. Let $\delta(\Delta q)$ represent this fraction of the energy leaving ΔA which falls on δa. In reference to Fig. 11.1, N is the normal to the plane containing ΔA, the radiating surface, θ is the angle between N and the line connecting P and Q, while $\delta\omega = \delta a / r^2$ is the solid angle subtended by δa at the point Q.

By definition, the intensity at the element δa due to the radiation from ΔA, call it $I_{\delta a/\Delta A}$, is

$$I_{\delta a/\Delta A} = \frac{\delta(\Delta q)}{\delta\omega(\Delta A \cos \theta)}. \tag{11.6}$$

The intensity at δa due to the radiation coming from the *point source Q* is, by use of Eq. (11.5),

$$I_{\delta a/Q} = \frac{\delta f}{\delta\omega} \frac{1}{\cos \theta},$$
$$= \frac{\delta f}{\delta a} \frac{r^2}{\cos \theta}. \tag{11.7}$$

In Eq. (11.7), the flux f is that flux measured at the *radiating* source, Q, not at the receiving area δa.

The intensity at the *point* P due to the radiation coming from the *point source* Q is found by taking the limit as $\delta a \to 0$:

$$I_{P/Q} = \frac{df}{d\omega} \frac{1}{\cos\theta},$$

$$= \frac{df}{da} \frac{r^2}{\cos\theta}. \tag{11.8}$$

The quantity $df/d\omega$ represents the radiated *flux*, per unit spatial solid angle, while the quantity df/da is the flux at P of the flux leaving Q. That is, df/da is the areal density at P of the flux emanating from Q. Rewriting Eq. (11.8) and dropping the subscripts,

$$\frac{df}{d\omega} = I \cos\theta,$$

$$\frac{df}{da} = I\frac{\cos\theta}{r^2}. \tag{11.9}$$

Equation (11.9) will be useful in finding the radiant energy falling on a *finite* surface A' due to radiation leaving a *finite* surface A, as suggested in Fig. 11.2. Since $f = dq/dA$ (flux from the radiating surface), the fraction of the flux from point Q which actually strikes the finite area A' is

$$\Delta f_{A'/Q} = \int_{A'} I\frac{\cos\theta}{r^2}(dA'\cos\theta'). \tag{11.10}$$

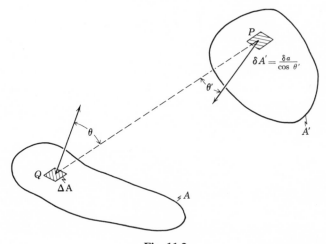

Fig. 11.2

The factor $\cos\theta'$ (where θ' is the angle between the line $\overline{PQ}$ and the normal to dA') appears since δa in Eq. (11.9) was defined as the element of a spherical surface centered at Q and passing through P. If the element of the finite area $\delta A'$ is not an element of such a spherical surface, its projection normal to the connecting line must be used in place of δa.

Similarly, the total radiation falling on a finite surface A' due to emission from the *finite* area A is

$$q_{A'/A} = \int_A \Delta f_{A'/Q} \, dA = \int_A \int_{A'} I \frac{\cos\theta \cos\theta'}{r^2} \, dA' \, dA. \qquad (11.11)$$

In Eqs. (11.10) and (11.11), θ and r are certainly functions of the relative positions of dA' and dA (i.e., the locations of P and Q) on the finite areas A' and A. The intensity, I, may also be a function of position, so the evaluation of the integrals is quite complex.

Return now to the situation depicted in Fig. 11.1. Equation (11.9) gives the energy flux, per unit area of receiving surface, at a point P due to radiation leaving the point Q. If one considers the hemisphere centered on Q, passing through P, and having its equatorial plane coinciding with the plane of the emitting element ΔA, the increment da may be expressed as

$$da = r\, d\theta(r \sin\theta\, d\varphi).$$

Here θ is the colatitude (polar angle) and φ is the longitude (azimuthal angle) of point P. Equation (11.9) or (11.10) gives the total flux passing through the hemisphere to be

$$\Delta f = \frac{dq}{dA} = \int_0^{2\pi} \int_0^{\pi/2} I \sin\theta \cos\theta\, d\theta\, d\varphi.$$

Since *all* the energy leaving Q must be intercepted by the hemisphere, Δf must equal the total flux leaving Q, and this flux is the radiosity of the radiating surface, J. Thus,

$$J = \int_0^{2\pi} \int_0^{\pi/2} I \sin\theta \cos\theta\, d\theta\, d\varphi, \qquad (11.12)$$

or, by use of Eq. (11.3),

$$W + (1-\alpha)G = \int_0^{2\pi} \int_0^{\pi/2} I \sin\theta \cos\theta\, d\theta\, d\varphi. \qquad (11.13)$$

Equations (11.12) and (11.13) relate the radiosity of a point on a surface to the intensity, and its spatial distribution, in the half-space above it. The dependence of I on the angles θ and φ must be known in order to apply these equations.

Diffuse Radiation, Lambert's Law

The radiation emanating from a point on a surface (whether by emission, reflection, or both) is termed *diffuse* if the intensity, I, is constant. It may be seen from Eq. (11.9) that for $I = $ constant,

$$\frac{df}{da} = I\frac{\cos \theta}{r^2},$$

$$\frac{df}{da} \propto \frac{\cos \theta}{r^2}.$$

(11.14)

Thus, the fraction of the flux leaving the surface which is intercepted by an element, df/da, varies as the cosine of the angle θ and inversely as the square of the distance between the radiating point and the receiving element. This is the familiar *Lambert's law* of diffuse radiation.

Another implication of the assumption of diffuse radiation (i.e., $I = $ constant) may be observed from Eqs. (11.12) and (11.13):

$$J = W + (1 - \alpha)G = I\int_0^{2\pi}\int_0^{\pi/2} \sin \theta \cos \theta \, d\theta \, d\varphi,$$

(11.15)

$$= I\pi.$$

Consequently, for a surface of *diffuse* radiosity, the radiosity and the *constant* intensity are related in a very simple way.

The assumption of constant intensity, leading to Lambert's law and Eq. (11.15), implies that the radiation leaving the surface (i.e., the radiosity J) is of a special character. Since the radiosity is composed of an original emission component W, as well as a reflected component, $(1 - \alpha)G$, the diffuse assumption implies that each of these components of the radiosity must leave the surface in a special way. Thus, in order for the radiation leaving a body to be completely diffuse, the surface must be both a *diffuse emitter* and a *diffuse reflector*. Real surfaces, to be discussed in some detail in Sec. 11.4, may deviate from this ideal of diffuse radiosity in either, or both, of these aspects.

All the foregoing discussion concerning radiation intensity has been directed to the characteristics of the radiation *leaving* a surface (by emission or reflection). The flux f was defined as the rate of energy flow per unit *emitting* (or reflecting) surface, while the angles θ and φ were the polar and azimuthal angles of the emitted or reflected beam. In like fashion, the intensity of radiation *incident* upon a surface may be defined. In order to avoid confusion, a subscript i will be used to distinguish quantities associated with incident radiation. Thus, if, as suggested in Fig. 11.3, θ_i and φ_i represent the polar and azimuthal angles of an incident beam, and if f_i represents the energy flux *incident* upon the surface per unit area of *intercepting* surface, the incident beam intensity, I_i, is

$$I_i = \frac{df_i}{d\omega_i} \frac{1}{\cos \theta_i}. \tag{11.16}$$

The irradiation, G, incident on the surface is, then, proceeding as above,

$$dG = I_i \cos \theta_i \, d\omega_i,$$
$$G = \int_0^{2\pi} \int_0^{\pi/2} I_i \sin \theta_i \cos \theta_i \, d\theta_i \, d\varphi_i. \tag{11.17}$$

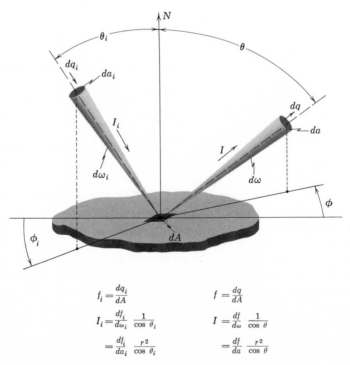

$$f_i = \frac{dq_i}{dA} \qquad\qquad f = \frac{dq}{dA}$$
$$I_i = \frac{df_i}{d\omega_i} \frac{1}{\cos \theta_i} \qquad I = \frac{df}{d\omega} \frac{1}{\cos \theta}$$
$$\quad = \frac{df_i}{da_i} \frac{r^2}{\cos \theta_i} \qquad\quad = \frac{df}{da} \frac{r^2}{\cos \theta}$$

Fig. 11.3. Angular relations for radiation intensity.

Again, if the incident radiation is diffuse (i.e., I_i = constant),

$$G = I_i \pi. \tag{11.18}$$

11.3 Black Body Radiation

In almost every science in which an analytical description of real physical phenomena is attempted, it proves useful to conceive of certain ideal circumstances upon which various definitions and concepts may be based. In thermal radiation this ideal set of circumstances is embodied in the concept known as the *perfect black body*. A perfect black body is one which absorbs *all* incident radiation (i.e., $\alpha = 1$, $\rho = 0$) regardless of the spectral character or directional preference of the incident radiation. The term "black" is used since dark surfaces normally show high values of absorptivity. In practice, a perfect black surface may be closely approximated by the use of a hollow body, say spherical, which is maintained at a uniform inside surface temperature. If a small hole is provided into the interior of the body, radiation entering the hole will be partly absorbed and partly reflected. The reflected portion will strike another portion of the interior, be partly absorbed and partly reflected, and so on. Thus, none, or essentially none, of the incident radiation will ever find its way back out of the hole through which it entered, and the plane of the hole is almost a perfect black surface with respect to the radiation falling on it. The radiation from within the hollow body which escapes from the hole must be of the character of the radiation that would be emitted by a perfect black surface. It is this quantity, the black body emissive power, which is of interest here. Once this concept is established, the emissive powers of nonblack surfaces may be defined in terms of it.

This concept of black body radiation may also be explained in terms of the characteristics of an isothermal enclosure. Figure 11.4 depicts an enclosure, or cavity, the walls of which have been heated to a *uniform* temperature. The surface of the enclosure is not necessarily black, and thus the space is filled

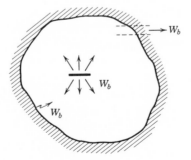

Fig. 11.4

with radiation which consists of reflected energy as well as original emission from the walls. Since the walls are maintained at a uniform temperature, the radiant field inside the cavity is at *thermal equilibrium* with the walls. Now imagine a very small wafer of material to be placed inside the cavity—small enough that it may be assumed that its presence only negligibly affects the radiation within the enclosure. Presume that the wafer has perfect black surfaces, and allow it to come to thermal equilibrium with the walls of the cavity. Thus, the emissive power of the wafer of material must be the emissive power of a black surface, W_b, at the given temperature. However, the wafer is subjected to the irradiation of the radiant field in the cavity, G. Since the wafer is at thermal equilibrium, G is identically equal to W_b. By changing the position and orientation of the perfectly black wafer, it may be concluded that the irradiation at *any* point within the enclosure is equal to the emissive power of a perfect black body maintained at a temperature equal to the temperature of the enclosing walls. This statement is true regardless of the orientation chosen for the evaluation of the irradiation. If a small hole is made in the walls of the enclosure, small enough that the radiation field is negligibly disturbed, the radiation escaping from the hole would be of the character of the radiation which would be emitted from a "black body."

The perfect black body just defined is, then, a surface which absorbs all incident energy—a surface from which emanates only original emission. The emissive power, W_b, of a black body is the same as the irradiation field produced within an isothermal enclosure at the same temperature. It is the amount, the spectral distribution, and the spatial distribution of the emission from a black surface which is of interest here.

The Stefan–Boltzmann Law for Black Body Emissive Power

In the preceding discussion, the radiation escaping from a small hole in an isothermal enclosure was seen to be the same as the emissive power of a black surface at the temperature of the enclosure walls. Since the nature of the interior surface of the cavity did not enter the discussion, the radiant emission from a perfect black surface must be independent of surface properties and be dependent only upon the surface temperature. In 1879 Josef Stefan suggested, on the basis of experimental data, that the total emissive power of a black surface is proportional to the fourth power of the absolute temperature of the surface. Later, Ludwig Boltzmann applied the concepts of the Carnot cycle, using radiant energy as the working medium, and analytically derived the same fact (see Ref. 1). Using the symbol W_b to denote the emissive power of a black body, one may state the *Stefan–Boltzmann law* for black body radiation as

$$W_b = \sigma T^4. \tag{11.19}$$

The Stefan–Boltzmann constant σ has the value

$$\sigma = 0.1714 \times 10^{-8} \text{ Btu/hr-ft}^2\text{-}{}^{\circ}\text{R}^4.$$

The temperature of the black surface, T, is the absolute temperature. It is convenient, for computational purposes, to remember that

$$W_b = \sigma T^4 = 0.1714 \times 10^{-8} T^4$$

$$= 0.1714 \left(\frac{T}{100}\right)^4.$$

The black body emissive power given by the Stefan–Boltzmann law is the *total* emissive power of the black surface, regardless of the spectral or spatial distribution of the emitted energy.

Spatial Characteristics of Black Body Emission

As the earlier discussion explained, black body emission is of the same character as the radiation created in an isothermal enclosure. This radiation was shown to be isotropic. That is, the irradiation incident upon any elemental area placed within the enclosure will be the same, regardless of the location or orientation of the receiving area. From these arguments, the conclusion that the intensity of black body radiation is constant in space follows so that the results of Eqs. (11.14) and (11.15) may be applied. Thus, using the subscript b to denote the black body, one obtains

$$\left(\frac{df}{da}\right)_b \propto \frac{\cos \varphi}{r^2}$$

$$\left(\frac{df}{da}\right)_b = I_b \frac{\cos \varphi}{r^2}, \tag{11.20}$$

$$J_b = W_b = \pi I_b. \tag{11.21}$$

Thus, black body emission obeys Lambert's law of diffuse radiation, and the emissive power and the intensity of the emission are simply related by the factor π. In other words, these facts, together with the definition that $\alpha_b = 1.0$, lead to the observation that black body radiation is an ideal concept so defined as to make it independent of any surface characteristics other than temperature.

The Spectral Distribution of Black Body Emission

The total emissive power of a black body as determined by the Stefan–Boltzmann law represents the *total* thermal radiation emitted regardless of the wavelength of the emission. If the radiant energy emitted by such a black body were to be decomposed into its spectrum over the entire wavelength band, the emission would be found to be unequally distributed over all the wavelengths. A different amount is emitted at each wavelength in the thermal radiation band, and the *total* emissive power, W_b, must be the integrated sum of all the energies at each wavelength. A *monochromatic black body emissive power*, $W_{b\lambda}$, must be defined as the rate of energy emission, per unit area, at a particular wavelength λ. The total emissive power and the monochromatic emissive power are then related by

$$W_b = \int_0^\infty W_{b\lambda} \, d\lambda. \tag{11.22}$$

Since W_b is itself a function of temperature, the monochromatic emissive power, $W_{b\lambda}$, must depend on λ *and* T [i.e., $W_{b\lambda} = fn(\lambda, T)$].

The functional relation among $W_{b\lambda}$, λ, and T was first derived by the thermodynamicist Max Planck and marked the first achievement of his quantum theory of radiation. Planck's law is expressed as

$$W_{b\lambda} = \frac{C_1}{\lambda^5(e^{c_2/\lambda T} - 1)}, \tag{11.23}$$

where

$$C_1 = 1.1870 \times 10^8 \text{ Btu-micron}^4/\text{ft}^2\text{-hr},$$

$$C_2 = 25,896 \text{ micron-}°R,$$

$$\lambda \quad \text{is measured in microns}$$

$$(1 \text{ micron} = 10^{-6} \text{ meter}).$$

Integration of Eq. (11.23) according to Eqs. (11.22) and (11.19) will show that Planck's constants C_1 and C_2 are related to the Stefan–Boltzmann constant according to

$$\sigma = \left(\frac{\pi}{C_2}\right)^4 \frac{C_1}{15}. \tag{11.24}$$

Figure 11.5 shows a plot of Planck's equation, $W_{b\lambda}$ vs. λ, for two different temperatures. The area under these curves represents the total black body

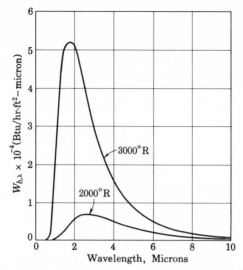

Fig. 11.5. Monochromatic emissive power of a perfect black surface at different temperatures.

emissive power, W_b, and is proportional to T^4 according to the Stefan–Boltzmann law. It may be noted that although the black body emissive power is composed of radiation at *all* wavelengths, the principal contribution is made between the wavelengths of about 1×10^{-7} and 1×10^{-4} meter (0.1 to 100 microns). Also, a peak monochromatic emissive power occurs at a particular wavelength, and this peak shifts to shorter wavelengths as the temperature increases. Wien's law states that this displacement of the maximum monochromatic emissive power follows the equation

$$\lambda_{\text{max}} T = C_3$$

$$= 5215.6 \text{ micron-}°\text{R}.$$

For use in performing calculations, particularly for nonblack surfaces as will be discussed in Sec. 11.4, it is useful to have available numerical values of Planck's equation as well as the emissive power between specific wavelength bands. This latter quantity may be found from tabulated values of the following integral:

$$W_{b(0-\lambda)} = \int_0^\lambda W_{b\lambda} \, d\lambda.$$

The tabulation of numerical values of $W_{b\lambda}$ and $W_{b(0-\lambda)}$ is facilitated by the division of $W_{b\lambda}$ by σT^5 to yield a function of only the product λT:

$$\frac{W_{b\lambda}}{\sigma T^5} = \frac{C_1}{\sigma} \frac{1}{(\lambda T)^5(e^{c_2/\lambda T} - 1)},$$

(11.25)

$$\frac{W_{b(0-\lambda)}}{\sigma T^4} = \frac{W_{b(0-\lambda)}}{W_b} = \int_0^{\lambda T} \frac{C_1 \, d(\lambda T)}{\sigma(\lambda T)^5(e^{c_2/\lambda T} - 1)}.$$

The functions given in Eq. (11.25) are shown graphically in Fig. 11.6 and are tabulated in Table A.14 for use in detailed calculations.

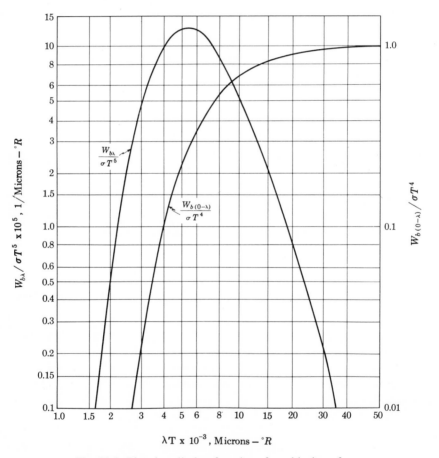

Fig. 11.6. Planck radiation functions for a black surface.

EXAMPLE 11.1:   A black body is maintained at 500°F. What fraction of the thermal radiation given up by the surface is emitted in the wavelength band between 1.0 and 4.0 microns? Repeat the problem for a temperature of 1500°F.

Solution: For $T = 500°F = 960°R$, one has

$$\lambda_1 T = 1.0 \times 960 = 960 \text{ micron-°R},$$

$$\lambda_2 T = 4.0 \times 960 = 3840 \text{ micron-°R}.$$

From Table A.14,

$$\frac{W_{b(0-1.0)}}{W_b} \approx 0,$$

$$\frac{W_{b(0-4.0)}}{W_b} = 0.0890.$$

Thus,

$$\frac{W_{b(1.0-4.0)}}{W_b} = 0.0890.$$

Since $W_b = \sigma T^4 = 0.1714 \times (9.6)^4 = 1456 \text{ Btu/hr-ft}^2$,

$$W_{b(1.0-4.0)} = 129.6 \text{ Btu/hr-ft}^2 \text{ (408.8 W/m}^2).$$

Similarly, for $T = 1500°F = 1960°R$,

$$\frac{W_{b(0-1.0)}}{W_b} = 0.0008,$$

$$\frac{W_{b(0-4.0)}}{W_b} = 0.5416,$$

$$\frac{W_{b(1.0-4.0)}}{W_b} = 0.5408,$$

$$W_{b(1.0-4.0)} = 13,680 \text{ Btu/hr-ft}^2 \text{ (43,150 W/m}^2).$$

11.4   Radiation from Real Surfaces and Ideal Gray Surfaces

The characteristics of the radiation coming from real surfaces, or absorbed by real surfaces, differ in several ways from those just described for an ideal black surface. The emissive power differs in magnitude although the temperature may be the same. The radiosity from real surfaces does not, in general, follow Lambert's law of diffuse radiation. Absorption and reflection may have directional preferences, and almost all real surfaces exhibit spectral preferences with respect to absorption and emission. Although all these effects cannot be discussed here in detail, certain significant characteristics may be pointed out. References 2 and 3 may be consulted for further discussion of radiation from real surfaces.

A proper, rigorous, approach to the description of real surfaces would be that which first defines the *monochromatic directional* properties and then proceeds to the *total* and *hemispherical* properties by appropriate integration. However, for an initial introduction to the subject, a better physical grasp of the concepts involved will be had if the hemispherical quantities are discussed first. Subsequently, some of the concepts introduced will be refined to include directional effects.

The Hemispherical Emissive Power and Absorptivity of Real Surfaces. Kirchhoff's Law of Radiation. The Ideal Gray Surface

If the *total* emissive power of a nonblack surface, W, is measured at a temperature T, then a quantity known as the *total hemispherical emissivity*, ϵ, may be defined as

$$\epsilon = \frac{W}{W_b},$$

$$W = \epsilon \sigma T^4,$$

(11.26)

where the black body emissive power, W_b, is calculated at the same temperature as the real surface. The adjectives *total* and *hemispherical* are applied to this definition of emissivity since W and W_b include all the radiation emitted—regardless of the *spectral* or *spatial* distribution of this energy.

Monochromatic Hemispherical Emissivity. The total hemispherical emissivity just defined described the over-all emission characteristics of a nonblack surface. Since some real surfaces are selective emitters (i.e., emit preferentially at certain wavelengths) a more detailed description of their emission is necessary. Thus, a *monochromatic hemispherical emissivity* is defined as

$$\epsilon_\lambda = \frac{W_\lambda}{W_{b\lambda}},$$

(11.27)

where

ϵ_λ = monochromatic hemispherical emissivity of the nonblack body at wavelength λ,

$W_{b\lambda}$ = monochromatic emissive power of a perfect black body at wavelength λ and at the same temperature as the nonblack body,

W_λ = monochromatic emissive power of a nonblack body at wavelength λ.

Note that no statement has yet been made as to whether ϵ_λ, the emissivity, is greater or less than 1. That is, in this definition of ϵ_λ there is no assumption as to whether a black body is a better or poorer emitter than a nonblack body. The fact that ϵ_λ must be less than 1 will be proved later by use of Kirchhoff's law. It should be emphasized that ϵ_λ is a *hemispherical* emissivity, the emissive powers $W_{b\lambda}$ and W_λ being the *total* energy emitted by the surfaces, in *all* directions, into the hemisphere above the surface.

Most real surfaces exhibit, to some degree, the characteristics of a "selective emitter" in that ϵ_λ, the monochromatic hemispherical emissivity, is different for different wavelengths of emitted energy. In general, ϵ_λ may be a function of the wavelength *and* the surface temperature:

$$\epsilon_\lambda = \epsilon_\lambda(\lambda, T).$$

The temperature dependence of ϵ_λ referred to above enters only to the extent that the physical or chemical nature of the surface are functions of temperature.

A special type of nonblack surface, called a *gray body*, is defined as one for which the monochromatic hemispherical emissivity is independent of the wavelength; i.e., the ratio of W_λ to $W_{b\lambda}$ is the same for all wavelengths of emitted energy at the same temperature. This definition of a gray body does not eliminate the possible dependence of the emissivity on temperature. Thus,

$$\epsilon_{gr} = \epsilon_{gr}(T),$$

where ϵ_{gr} represents the monochromatic, or total, hemispherical emissivity for a gray surface. In some discussions in this chapter it will be convenient to use the concept of a gray surface. This condition is sometimes closely approximated by some materials (e.g., slate, polished aluminum, etc.); however, many materials deviate sufficiently to make it important that the distinction be noted. The assumption of a gray surface, ϵ_λ independent of λ, means that the spectral distribution curves of the monochromatic emissive powers of gray and black surfaces at the same temperature are affine to one another, there being no shift in the peak of the curve as illustrated in Fig. 11.7. Also shown in Fig. 11.7 is the monochromatic emissive power of a typical nonblack, nongray surface.

Monochromatic Absorptivity and Reflectivity. In the same manner in which real surfaces exhibit selective emission, real surfaces also show a selectivity with respect to the absorption of certain wavelengths of incident radiation. If a surface is subjected to a monochromatic irradiation flux of intensity I_{λ_i}, the incident flux is, by Eq. (11.16),

$$df_{\lambda_i} = I_{\lambda_i} \cos \theta_i \, d\omega_i.$$

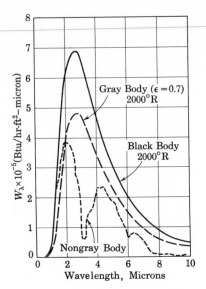

Fig. 11.7. Monochromatic emissive power of a black surface, a gray surface, at 2000°R.

Since, in general, I_{λ_i} may be a function of the polar and azimuthal angles φ_i and θ_i, the total flux incident on the surface at the wavelength λ is found by integrating over the hemisphere above the surface

$$G_\lambda = \int_h I_{\lambda_i} \cos \theta_i \, d\omega_i.$$

In the above, the notation $\int_h$ is used to denote integration over the hemisphere ($0 \leq \theta \leq \pi/2$, $0 \leq \varphi \leq \pi$) with $d\omega_i = \sin \theta_i \, d\theta_i \, d\varphi_i$, as was used in the algebra leading from Eq. (11.14) to (11.15).

A fraction of the incident irradiation is absorbed by the surface, and the *monochromatic hemispherical absorptivity* is defined by

$$\alpha_\lambda = \frac{\int_h (df_{\lambda_i})_{\text{abs}}}{\int_h df_{\lambda_i}}$$

$$= \frac{\int_h (I_{\lambda_i})_{\text{abs}} \cos \theta_i \, d\omega_i}{\int_h I_{\lambda_i} \cos \theta_i \, d\omega_i}. \tag{11.28}$$

This absorptivity, as defined, is seen to be dependent upon the spatial characteristics of the incident radiation as well as the nature of the absorbing surface.

For opaque surfaces the *monochromatic hemispherical reflectivity* may be simply found from

$$\rho_\lambda = 1 - \alpha_\lambda. \tag{11.29}$$

Both α_λ and ρ_λ depend on the wavelength of the incident radiation and the absorbing surface temperature as well as the spatial character of I_{λ_i}.

Kirchhoff's Law. A relation between the monochromatic absorptivity and emissivity may be established by use of Kirchhoff's law. Suppose that a small body is contained within an enclosure—the walls of which are maintained at a fixed temperature T. When equilibrium is established, the small, enclosed, body will be at the same temperature as the walls, and it will give up as much energy as it absorbs. Thus,

$$\alpha G = W, \qquad \text{or} \qquad \frac{W}{\alpha} = G.$$

The irradiation upon the small body, G, is dependent upon the temperature T and the geometrical arrangement of body and the enclosure. If the body is very small compared with the enclosure, it may be assumed that its presence has a negligible effect upon the irradiation field in the enclosure. Thus, G will be the same no matter what the properties of the small body are:

$$\frac{W}{\alpha} = fn(T), \qquad \text{for all bodies.}$$

This statement that the ratio of the total emissive power to the total hemispherical absorptivity is the same for all bodies at thermal equilibrium is known as *Kirchhoff's law*. The absorptivity α must be less than 1, according to Eq. (11.1). Thus, since $W/\alpha = fn(T)$, at a given temperature the emissive power will be the greatest for a body with $\alpha = 1$. Kirchhoff's law, then, states that a perfect black body, defined as a perfect absorber, must also be a perfect emitter. That is, when $\alpha = 1$, $W = W_b$. Since the total emissivity was defined as $\epsilon = W/W_b$, Eq. (11.26), then Kirchhoff's law gives

$$\frac{W}{\alpha} = W_b = \frac{W}{\epsilon},$$

$$\alpha = \epsilon. \tag{11.30}$$

It must be noted that the form of Kirchhoff's law just quoted is subject to the restrictions of *thermal equilibrium* in an *isothermal enclosure*. This also includes the condition of diffuseness, since, as was shown in Eq. (11.21), the radiation in an isothermal enclosure is diffuse. Extensions of Eq. (11.30) to other situations (as attractive as they may be) must be made with caution. Reference 2 presents an excellent and detailed exposition of Kirchhoff's law, including directional and spectral effects. In the case of *monochromatic diffuse* radiation, a generalization of this law may be stated as: If α_λ and ϵ_λ are independent of direction (i.e., φ and θ), they are equal for a given surface temperature:

$$\alpha_\lambda(\lambda_i, T) = \epsilon_\lambda(\lambda_i, T). \tag{11.31}$$

In Eq. (11.31) the symbol λ_i is employed to emphasize the fact that the same wavelength must be used for both α_λ and ϵ_λ. A surface, as above, for which α_λ and ϵ_λ are independent of direction is called a *monochromatically diffuse* surface.

Relations Between the Total and Monochromatic Properties. Still ignoring directional dependencies, one finds it useful to examine relations between the total and monochromatic absorptivity and emissivity (hemispherical in all cases).

In the case of the emissivity, Eqs. (11.22), (11.26), and (11.27) give

$$\epsilon_\lambda = \epsilon_\lambda(\lambda, T_s) = \frac{W_\lambda}{W_{b\lambda}},$$

$$\epsilon = \frac{W}{W_b},$$

$$W_b = \int_0^\infty W_{b\lambda}(\lambda, T_s)\, d\lambda.$$

The symbol T_s is used to denote the dependence of a quantity upon the temperature of the emitting *surface*. From the above, the following relation may be derived for the total emissivity as a function of the monochromatic emissivity and Planck's black body radiation function, $W_{b\lambda}$:

$$\epsilon = \epsilon(T_s) = \frac{\displaystyle\int_0^\infty \epsilon_\lambda(\lambda, T_s) W_{b\lambda}(\lambda, T_s)\, d\lambda}{\displaystyle\int_0^\infty W_{b\lambda}(\lambda, T_s)\, d\lambda}. \tag{11.32}$$

The total hemispherical emissivity is thus seen to be a function only of the emitting surface temperature.

In the case of the absorptivity, the *total* hemispherical absorptivity is, following Eq. (11.28),

$$\alpha = \frac{\displaystyle\int_0^\infty \int_h (df_{\lambda_i})_{abs}\, d\lambda}{\displaystyle\int_0^\infty \int_h df_{\lambda_i}\, d\lambda}.$$

When the definition of the monochromatic hemispherical absorptivity, Eq. (11.28), is introduced:

$$\alpha = \frac{\displaystyle\int_0^\infty \alpha_\lambda [\int_h df_{\lambda_i}]\, d\lambda}{\displaystyle\int_0^\infty \int_h df_{\lambda_i}\, d\lambda}$$

$$= \frac{\displaystyle\int_0^\infty \alpha_\lambda [\int_h I_{\lambda_i} \cos \theta_i\, d\omega_i]\, d\lambda}{\displaystyle\int_0^\infty \int_h I_{\lambda_i} \cos \theta_i\, d\omega_i\, d\lambda}.$$

The above integral expression cannot be evaluated until one knows how α_λ depends on λ and how I_{λ_i} depends on λ, θ_i, and φ_i. Thus, contrary to the case for ϵ, it is seen that the absorptivity α depends on the nature of the incident radiation as well as the absorbing surface.

Now, if the incident radiation is diffuse (i.e., I_{λ_i} = constant), the above expression for the total hemispherical absorptivity becomes

$$\alpha = \alpha(\lambda^*, T^*, T_s) = \frac{\displaystyle\int_0^\infty \alpha_\lambda(\lambda, T_s) I_{\lambda_i}(\lambda^*, T^*)\, d\lambda}{\displaystyle\int_0^\infty I_{\lambda_i}(\lambda^*, T^*)\, d\lambda}. \tag{11.33}$$

The notation $I_{\lambda_i}(\lambda^*, T^*)$ is used to emphasize the fact that the incident radiation has some spectral distribution dependent upon the temperature, T^*, and nature of the source from whence the radiation comes.

Since the total absorptivity depends not only on the temperature of the surface but also on the spectral characteristics of the *incident* radiation, it differs significantly from the total emissivity [see Eq. (11.32)], which is dependent only on the surface temperature. However, for monochromatically diffuse surfaces, for which Kirchhoff's law in Eq. (11.31) holds, one may compare Eqs. (11.32) and (11.33) (for diffuse incidence) to ascertain under what conditions the total emissivity and absorptivity are equal. It is seen that $\epsilon = \alpha$ when either of the following conditions prevail:

(1) If $I_{\lambda_i}(\lambda^*, T^*) = W_b(\lambda, T_s)$. That is, Kirchhoff's law for the total α and ϵ holds if the incident radiation has the spectral distribution of a black body at the temperature of the receiving surface. This is the case in an isothermal enclosure (as used in the above derivations), but this condition is rarely met in most engineering applications.

(2) If $\epsilon_\lambda = \alpha_\lambda$ are *independent of wavelength*, i.e., if the surface is *gray*. Most real materials are not gray, but often a suitable average value of $\epsilon_\lambda = \alpha_\lambda$ may be chosen to enable this assumption to be made. Many materials are nearly gray at moderate temperatures, in which the emission is predominately in the *infrared* portion of the thermal radiation band (see the table on p. 412).

Observed Values of Emissivity and Absorptivity. There is a large amount of experimental information available which gives measured values of the monochromatic emissivity, or absorptivity, of real materials. References 2, 4, 5, and 6 may be consulted in this regard. Figures 11.8 and 11.9 show some typical values. Values of the total emissivity may be measured directly or

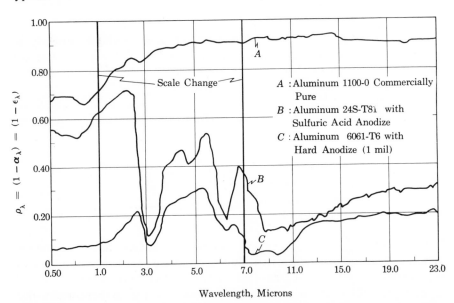

Fig. 11.8. Monochromatic reflectivity of various surfaces. (From D. K. Edwards, K. E. Nelson, R. D. Roddick, and J. T. Gier, "Basic Studies on the Use and Control of Solar Energy," *Univ. Calif. Dept. Eng. Rept. 60–93*, 1960.)

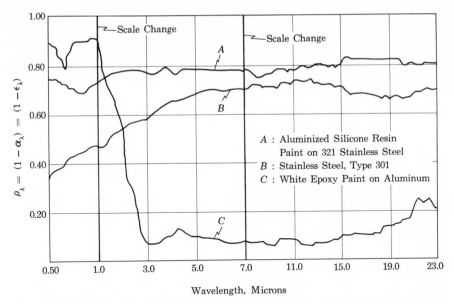

Fig. 11.9. Monochromatic reflectivity of various surfaces. (From D. K. Edwards, K. E. Nelson, R. D. Roddick, and J. T. Gier, "Basic Studies on the Use and Control of Solar Energy," *Univ. Calif. Dept. Eng. Rept. 60–93*, 1960.)

may be calculated from the ϵ_λ vs. λ data by the use of Eqs. (11.32) and (11.25). In terms of the functions given in Eqs. (11.25), Eq. (11.32) may be expressed as

$$\epsilon = \int_0^\infty \epsilon_\lambda \left(\frac{W_{b\lambda}}{\sigma T^5}\right) d(\lambda T)$$

$$\approx \sum \epsilon_\lambda \left(\frac{W_{b\lambda}}{\sigma T^5}\right) \Delta(\lambda T).$$

By selecting increments in λT, the tabulated functions in Table A.1 may be combined with the data supplied in ϵ_λ vs. λ, such as in Figs. 11.8 and 11.9, to evaluate numerically the above integral. Reference 5 may be consulted for detailed examples of such calculations. The procedures just described apply at a single surface temperature. The dependence of the total emissivity on surface temperature must be accounted for whether a surface is gray or not. Table A.12 lists some values of the total emissivity of various surfaces, including the temperature dependency and Fig. 11.10 illustrates, graphically, some typical data.

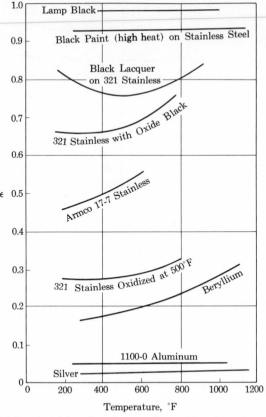

Fig. 11.10. Total emissivities of various surfaces. (Based on data in Table A.12.)

The total absorptivity of a surface may be similarly calculated from Eq. (11.33), presuming $\epsilon_\lambda = \alpha_\lambda$ vs. λ data and the spectral distribution of the *incident* radiation are known. If the incident radiation is known to be of a black body distribution so that Kirchhoff's law holds, the calculations described above may be applied. If the surface is known to be gray, the total emissivities given in Table A.12 may be used for the total absorptivity. In all other cases α must be calculated from Eq. (11.33) for the individual case at hand.

Solar Absorptivity

Since so many applications of current interest exist which are connected with solar phenomena, some specific comments relative to the thermal radiation from the sun seem appropriate. Because the sun's radiation is produced at such a high temperature (about 10,400°R), the spectral distribution of the monochromatic emissive power is significantly different from that of the radiation produced by most surfaces encountered in engineering practice. Even though the spectral distribution of the sun's radiation follows, rather closely, the black body law, the fact that a significant portion of it lies within the visible spectrum means that the difference noted in Eqs. (11.32) and (11.33) between the total emissivity and the total absorptivity becomes quite significant.

In present-day space-technology nomenclature, the total emissivity calculated at moderate surface temperatures is often called the *infrared emissivity* since such surfaces radiate mainly in the infrared band. The absorptivity of a surface irradiated by a source having the intensity and spectral distribution of the sun is called the *solar absorptivity*, α_s. The data given in Table A.12 may be used as "infrared" emissivities. Special values of the "solar absorptivity" are tabulated in Table A.13 for several materials of current importance.

Of particular significance is the ratio α_s/ϵ, and Table A.13 shows that quite a wide range of this ratio may be observed in real materials. A large value of α_s/ϵ indicates that such a surface absorbs a large fraction of the irradiation upon it compared with the amount emitted. This is a desirable property for surfaces to be used in solar energy collectors, solar cells, etc. On the other hand, a surface which is intended for use as a heat rejection surface (such as in a manned spacecraft) should have a low value of α_s/ϵ in order that it maintain a useful heat-rejecting capability even in the solar environment. These facts are sometimes expressed in terms of the equilibrium temperature a surface would achieve if placed, isolated, in a solar-irradiation environment of G_s. In such an instance, the equilibrium temperature is given by

$$T^* = \left(\frac{\alpha_s}{\epsilon} \frac{G_s \cos \beta}{\sigma} \right)^{1/4}$$

in which β represents the angle between the sun's rays and the surface normal. For a fixed orientation and at a fixed distance from the sun, T^* is uniquely determined by the ratio α_s/ϵ. (The variation of G_s with distance from the sun may be deduced from the data given in Prob. 11.1. At the earth's orbit $G_s = 442$ Btu/hr-ft².) The temperature T^* is also referred to as the "equivalent environment temperature" since it is the equilibrium temperature a perfect black surface would achieve if subjected to an external black body radiation equal to σT^{*4}.

EXAMPLE 11.2: For a surface placed normal to the sun's irradiation at the earth's orbit ($G_s = 442$ Btu/hr-ft²), find the equilibrium temperature attained if the surface is composed of (a) 410 stainless steel, a solar collector material, or (b) aluminum coated with white epoxy paint, a heat rejection material.

Solution: Table A.13 gives, for stainless steel, $\alpha_s/\epsilon = 4.24$ and for white-epoxy-painted aluminum, $\alpha_s/\epsilon = 0.28$. Thus, for $G_s = 442$ Btu/hr-ft^2, $\beta = 0$:

$$T^*_{\text{stainless}} = 1015°\text{R} = 555°\text{F} \,(291°\text{C}),$$

$$T^*_{\text{epoxy}} = 517°\text{R} = 57°\text{F} \,(13.9°\text{C}).$$

Directional Emissivity and Reflectivity

The discussion in this section so far has been limited to *hemispherical* properties. That is, the emissive power, absorptivity, emissivity, etc., so far defined have been based on the *total* radiation leaving a surface, or incident upon it, regardless of the directional properties of the emission. The adjective "hemispherical" was used to denote all the radiation in the hemisphere (emitted or incident) above a surface element.

An emissivity may be defined in terms of the *intensity* introduced in Sec. 11.2. The term *directional emissivity* is applied in this case in order to distinguish it from the hemispherical emissivity. The directional emissivity, ϵ_θ, of a surface is defined as the radiant flux emitted at a certain polar angle θ, and azimuthal angle φ, in a solid angle $d\omega$ divided by the same quantity, at the same angle, that would be emitted by a perfect black body at the same temperature. Using the definition of the intensity of *emission* given in Eq. (11.8), one obtains

$$\epsilon_\theta = \frac{I \cos\theta \, d\omega}{I_b \cos\theta \, d\omega} = \frac{I}{I_b}$$

$$= fn(\theta, \varphi, T_s).$$

Since I will, in general, be dependent on both θ and φ, the directional emissivity will depend upon these angles as well as the surface temperature.* For "smooth" surfaces, the dependence on the azimuthal angle φ disappears, and the directional emissivity will depend only upon the polar angle and surface temperature.

Figure 11.11 shows typical distributions of the directional emissivity for several surfaces. Since the black body intensity, I_b, is diffuse and follows Lambert's cosine law (Sec. 11.2), the polar curve of a diffusely emitting surface would be a circle in Fig. 11.11, and any deviation from a circle indicates a departure from diffuse emission. The hemispherical emissivity defined

---

*The question of spectral dependence is being omitted here for simplicity. One could distinguish between a monochromatic and a total directional emissivity, but such questions will not be dealt with in this text. The reader may consult Ref. 2 or 6 in this regard.

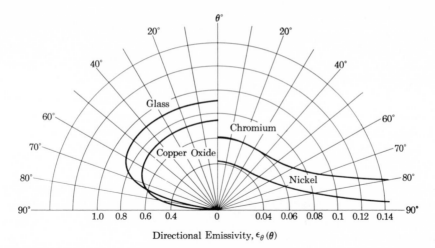

Fig. 11.11. Directional emissivities of various surfaces. (From *Heat and Mass Transfer* by E. R. G. Eckert and R. M. Drake. Copyright 1959. McGraw-Hill Book Company, New York. Used by permission.)

earlier is related to the directional emissivity as

$$\epsilon = \frac{\int_h I \cos \theta \, d\omega}{\int_h I_b \cos \theta \, d\omega}$$

$$= \frac{\int_h \epsilon_\theta \cos \theta \sin \theta \, d\theta \, d\varphi}{\int_h \cos \theta \sin \theta \, d\theta \, d\varphi}$$

$$= \frac{1}{\pi} \int_h \epsilon_\theta \cos \theta \sin \theta \, d\theta \, d\varphi. \tag{11.34}$$

In the above, the notation $\int_h$ is again used to denote integration over the hemisphere. For the "smooth" surface no dependency on φ exists, and

$$\epsilon = 2 \int_0^{\pi/2} \epsilon_\theta \cos \theta \sin \theta \, d\theta.$$

The directional properties of the absorptivity are better explained in terms of the reflectivity. Many different definitions of a reflectivity may be made, depending upon what particular aspect of the reflection phenomenon one is attempting to emphasize. Perhaps the most basic reflectivity that may be defined is the *biangular reflectivity*. Figure 11.12 illustrates the nomenclature

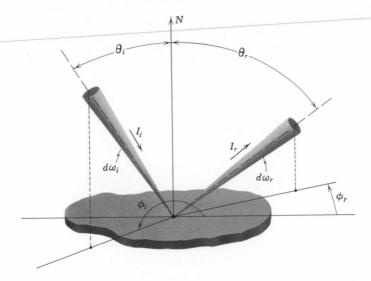

Fig. 11.12. The geometrical relations for biangular reflectivity.

involved in the definition of the biangular reflectivity. An incident beam of intensity I_i, solid angle $d\omega_i$, impinges upon a surface at polar and azimuthal angles of θ_i and φ_i, respectively. This incident beam is reflected in all directions, in general, but the biangular reflectivity gives the fraction of this incident energy which is reflected in a certain direction given by the polar and azimuthal angles θ_r and φ_r. If I_r is the intensity of the reflected beam, the biangular reflectivity, ρ_{ba}, is defined as the intensity of the reflected beam divided by the incident flux of energy. By using the concepts developed in Sec. 11.2, the biangular reflectivity is

$$\rho_{ba} = \frac{I_r}{I_i \cos \theta_i \, d\omega_i} = fn(\theta_i, \varphi_i, \theta_r, \varphi_r, T).$$

(Again, no mention is being made of any spectral dependence, the above definition being applicable to either total or monochromatic reflectivities.) The reflectivity just defined is a function of the direction of the incident beam, the surface temperature, and the particular direction of reflection of interest. *Specular* reflection is the mirrorlike reflection which results in $\rho_{ba} = 0$ for all θ_r and φ_r except when $\theta_r = \theta_i$ and $\varphi_r = \varphi_i + 180°$, in which direction ρ_{ba} is different from 0. *Diffuse* reflection is defined as that reflection for which $\rho_{ba} = $ constant; i.e., the reflectivity is independent of the direction of the incident beam or the reflected azimuthal angle and depends only on the reflected polar angle according to Lambert's cosine law.

Figure 11.13 shows the biangular reflectivity for a ground nickel surface. Plotted in Fig. 11.13 is

$$\frac{\rho_{ba}}{\rho_s} \cos \theta_r \text{ vs. } \theta_r$$

for various values of φ_r with an incident beam at $\theta_i = 10°$, $\varphi_i = 180°$. The symbol ρ_s denotes the reflectivity in the specular direction ($\theta_r = 10°$, $\varphi_r = 0°$), so diffuse reflection would be represented by the cosine curve shown. The deviation of the reflection from the diffuse ideal is apparent in the strong specular preference at certain wavelengths. For analytical purposes, the characteristics of real surfaces are often represented as a combination of a diffuse reflector and a pure specular reflector.

An *angular–hemispherical* reflectivity may also be defined. It expresses the ratio of the total energy reflected in the hemisphere above the surface to the incident energy flux from a given direction. Likewise, there is a *hemispherical–angular* reflectivity which gives the ratio of the intensity of a reflected beam

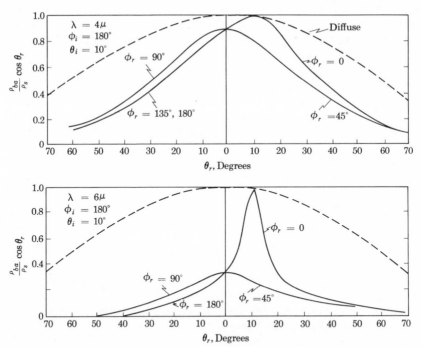

Fig. 11.13. The biangular reflectivity of ground nickel as a function of incident angles, reflected angles, and wavelength. (From R. C. Birkebak and E. R. G. Eckert, "Effects of Roughness of Metal Surfaces on Angular Distribution of Monochromatic Reflected Radiation," *J. Heat Transfer*, Vol. 87C, 1965, p. 85.)

in a particular direction to the total influx of incident energy from the entire hemisphere. These reflectivities may be related to ρ_{ba} by appropriate integrations, but they will not be discussed in detail here. The problems at the end of the chapter may be consulted in this regard.

The *hemispherical* reflectivity introduced at the beginning of this section is defined as the ratio of the total energy reflected into the hemisphere above a surface element to the total energy flux incident from the entire hemisphere. Thus, the relation of the hemispherical reflectivity to the biangular reflectivity is found to be

$$\rho = \frac{\int_h I_r \cos \theta_r \, d\omega_r}{\int_h I_i \cos \theta_i \, d\omega_i}, \tag{11.35}$$

$$\rho = \frac{\int_h \int_h \rho_{ba} I_i \cos \theta_r \, d\omega_r \cos \theta_i \, d\omega_i}{\int_h I_i \cos \theta_i \, d\omega_i}$$

$$= fn(T). \tag{11.36}$$

The directional dependence of the reflection process is eliminated, or obscured, by the introduction of the hemispherical reflectivity. For applications involving only gross energy balances, the hemispherical reflectivity will usually suffice.

Restricting attention to a single wavelength, one finds that the monochromatic, hemispherical reflectivity, ρ_λ, is

$$\rho_\lambda = \frac{\int_h \int_h \rho_{ba_\lambda} I_{i_\lambda} \cos \theta_r \, d\omega_r \cos \theta_i \, d\omega_i}{\int_h I_{i_\lambda} \cos \theta_i \, d\omega_i}.$$

It is this reflectivity, just defined, which, for the opaque surface, is related to the monochromatic emissivity by Kirchhoff's law [Eq. (11.31)] and displayed in Fig. 11.9:

$$\alpha_\lambda = 1 - \rho_\lambda = \epsilon_\lambda.$$

In order to evaluate the hemispherical reflectivity (total or monochromatic) from the above expressions, it is necessary to know the functional dependence of the biangular reflectivity upon θ_i, φ_i, θ_r, and φ_r, in great detail—as well as the directional properties of the incident radiation, I_i. Such detailed information is very rare, and the total, hemispherical, reflectivities that might be inferred from the data of Table A.12, via

$$\rho = 1 - \alpha = 1 - \epsilon,$$

represent only average values that would result when a surface is irradiated uniformly.

Concluding Remarks

The foregoing discussions have attempted to point out the detailed complexity of the radiant behavior of real surfaces. In general, the emissivity of a surface is dependent upon the surface temperature, and the energy emitted may have directional and spectral dependencies. The absorptivity also depends upon the surface temperature as well as the directional and spectral properties of the *incident* radiation. The reflectivity, although related to the absorptivity, may also exhibit directional and spectral preferences as far as the *reflected* energy is concerned.

In the analysis of real, engineering, problems involving radiant heat transfer, there may not be sufficient justification for inclusion of all these detailed effects—or there simply may not be sufficient physical data to account for them correctly. Thus, in the material which follows, only the *hemispherical* values of ϵ, ρ, and α will be used. Also, in most cases, *total hemispherical* values will be used—meaning that detailed directional and specular considerations will be omitted. In addition, all radiation leaving a surface (emitted or reflected) will, generally, be presumed to follow the diffuse law of Lambert, and whenever possible, the simplification offered by Kirchhoff's law ($\alpha = \epsilon$) will be used. This latter case will be true when the idealization of a *gray* surface is used.

11.5   Diffuse Radiant Exchange Between Black or Gray Infinite Planes

The concepts presented thus far provide sufficient background to determine the net radiant exchange between arbitrarily located surfaces of arbitrary shape separated by a nonabsorbing medium. Although this section will consider the exchange between infinite planes, it will be useful to explain first the notation and basic concepts to be used throughout the rest of this chapter. Thus, define

q_i = *net* energy given up by surface i over that which it absorbs,

$q_{i \rightarrow j}$ = energy *leaving* surface i which eventually *strikes* surface j,

q_{i-j} = energy *emitted* by surface i which is *absorbed* at surface j,

q_{ij} = *net* exchange of energy between surface i and surface j.

From the definition of the total hemispherical radiosity and irradiation, a surface which is heated *uniformly* and irradiated *uniformly* gives up the following *net* energy:

$$q_i = A_i(J_i - G_i). \tag{11.37}$$

Applying Eq. (11.3), one may also express the above net heat flow in terms of the total hemispherical emissive power and absorptivity:

$$q_i = A_i(W_i - \alpha_i G_i)$$

$$= A_i\left(\frac{1}{1 - \alpha_i} W_i - \frac{\alpha_i}{1 - \alpha_i} J_i\right). \tag{11.38}$$

For *gray* surfaces, Kirchhoff's law holds for the total emissivity, $\epsilon = \alpha$, so Eq. (11.38) may be written

$$q_i = A_i(W_i - \epsilon_i G_i)$$

$$= A_i\left(\frac{W_i}{1 - \epsilon_i} - \frac{\epsilon_i}{1 - \epsilon_i} J_i\right)$$

$$= A_i\frac{\epsilon_i}{1 - \epsilon_i}(W_{b_i} - J_i). \tag{11.39}$$

In the above equation, W_{b_i} represents the emissive power of a black surface at the same temperature as the surface A_i.

For numerous practical problems, not only is net energy given up by a surface sought (the quantity q_i described above) but also the *net exchange* of energy between two surfaces, q_{ij}, is required. Since q_{i-j} denotes the energy emitted by surface i which is absorbed at surface j,

$$q_{ij} = q_{i-j} - q_{j-i}. \tag{11.40}$$

Thus, evaluation of the net exchange of energy will involve the determination of the amount of energy emitted by a surface which is *eventually* absorbed at the second surface. If other surfaces are present, or if the two surfaces in question are nonblack, or both, the quantity q_{i-j} is composed not only of a *direct* transfer of energy but also of an indirect transfer—the result of multiple reflections, absorption and reradiation, etc. Thus, q_{i-j} will depend upon the geometric shape and orientation of all surfaces present as well as their emissivities and absorptivities. It is the purpose of this and succeeding sections to examine such problems. The simplest geometrical arrangement will be considered first—that of two parallel planes of infinite extent.

Diffuse Radiation Exchange Between Parallel, Infinite Planes

The assumption of an infinite plane implies the neglection of edge effects in the case of finite surfaces. Such a simplification is often justifiable in real problems if the space between the planes is small compared to the area of the plane itself. Of course, it makes sense in such cases to work only in terms of the energy exchange per unit area of surface. Examples of the practical applications of such a simplified configuration are the radiation exchange through the air spaces within large walls and concentric cylindrical shapes of large radii.

Only the case of diffuse radiation will be considered in order that the simplification afforded by Kirchhoff's law can be applied.

If two opposing, parallel, infinite planes (say, 1 and 2) are considered to be mutually irradiating one another (only the exchange between their opposing surfaces being considered), then all the energy leaving one plane must strike the other. Thus,

$$J_1 = G_2,$$

$$J_2 = G_1, \tag{11.41}$$

$$q_1 = q_{12} = -q_{21} = -q_2.$$

The net exchange, per unit area, is, by Eq. (11.37),

$$\frac{q_{12}}{A} = J_1 - G_1.$$

Black Planes. If the two opposing planes are black ($\alpha_1 = \alpha_2 = 1$), Eq. (11.3) gives $J_1 = W_{b_1}$, $J_2 = W_{b_2}$, so that, with Eq. (11.41), the net exchange becomes

$$\frac{q_{12}}{A} = J_1 - G_1$$

$$= W_{b_1} - W_{b_2}$$

$$= \sigma(T_1^4 - T_2^4). \tag{11.42}$$

The fact that the heat exchange is proportional to the difference of the fourth power of the absolute temperature is of considerable importance, as the following example shows.

EXAMPLE 11.3: Two parallel, opposed, infinite, black planes are maintained at 300°F and 400°F, respectively. If the temperature difference is doubled by increasing the 400°F to 500°F, by what factor is the net heat exchange increased?

Solution: For the original case of the 300°F and 400°F temperatures, Eq. (11.42) gives the net heat exchange per unit area to be

$$\frac{q_{12}}{A} = \sigma(T_1^4 - T_2^4)$$

$$= 0.1714\left[\left(\frac{400 + 460}{100}\right)^4 - \left(\frac{300 + 460}{100}\right)^4\right]$$

$$= 366 \text{ Btu/hr-ft}^2 \ (1154 \text{ W/m}^2).$$

When the temperature difference is doubled,

$$\frac{q_{12}}{A} = 0.1714\left[\left(\frac{500 + 460}{100}\right)^4 - \left(\frac{300 + 460}{100}\right)^4\right]$$

$$= 884 \text{ Btu/hr-ft}^2 \ (2788 \text{ W/m}^2).$$

The net heat exchange is increased by a factor of 2.4.

Diffuse Gray Planes. Consider now the same case as above, except let the planes be gray instead of black. The symbols ϵ_1 and α_1 will represent the emissivity and absorptivity of plane 1, and ϵ_2 and α_2 will denote the same for plane 2. Since the planes are gray, the relation among the emittance, radiosity, and irradiation expressed in Eq. (11.3) may be written for each plane:

$$J_1 = W_1 + (1 - \alpha_1)G_1,$$
$$J_2 = W_2 + (1 - \alpha_2)G_2.$$
$$(11.43)$$

As in the case of the black planes, the net radiant heat exchange between the two planes is found by evaluating the net energy given up or absorbed by one of the surfaces. So

$$\frac{q_{12}}{A} = J_1 - G_1. \qquad (11.44)$$

The four expressions given in Eqs. (11.41) and (11.43) may be taken as four equations in the four unknowns—G_1, G_2, J_1, and J_2—since the absorptivities and emissive powers may be treated as known, or given, quantities. Solving for J_1 and G_1, one obtains

$$J_1 = \frac{W_1 + (1 - \alpha_1)W_2}{1 - (1 - \alpha_1)(1 - \alpha_2)},$$

$$G_1 = \frac{W_2 + (1 - \alpha_2)W_1}{1 - (1 - \alpha_1)(1 - \alpha_2)}.$$

Thus, Eq. (11.44) may be expressed as

$$\frac{q_{12}}{A} = \frac{W_1[1 - (1 - \alpha_2)] - W_2[1 - (1 - \alpha_1)]}{\alpha_1 + \alpha_2 - \alpha_2\alpha_1}.$$

However, $W_1 = \epsilon_1\sigma T_1^4$, $W_2 = \epsilon_2\sigma T^4$; and since Kirchhoff's law holds for gray surfaces, $\epsilon_1 = \alpha_1$ and $\epsilon_2 = \alpha_2$. Thus, one finally arrives at the following equation for the net exchange of radiant energy:

$$\frac{q_{12}}{A} = \sigma\frac{\epsilon_1\epsilon_2 T_1^4 - \epsilon_1\epsilon_2 T_2^4}{\epsilon_1 + \epsilon_2 - \epsilon_1\epsilon_2}$$

$$= \sigma\frac{(T_1^4 - T_2^4)}{(1/\epsilon_1) + (1/\epsilon_2) - 1}. \tag{11.45}$$

This expression gives the net radiant exchange between the two planes in terms of their temperatures and emissivities. The denominator of Eq. (11.45) reduces to 1 for the black surface case with $\epsilon_1 = \epsilon_2 = 1$.

EXAMPLE 11.4: Two parallel gray planes have emissivities of 0.8 and 0.7 and are maintained at 500°F and 1000°F. What is the net radiation exchange?

Solution:

$$\frac{q_{12}}{A} = 0.1714\frac{\left(\dfrac{1000 + 460}{100}\right)^4 - \left(\dfrac{500 + 460}{100}\right)^4}{(1/0.8) + (1/0.7) - 1}$$

$$= 3765 \text{ Btu/hr-ft}^2 \text{ (11,876 W/m}^2\text{)}.$$

Had the surfaces been black the net exchange would have been almost double this, 6325 Btu/hr-ft². Note that although the planes are at different temperatures, it does not matter which has $\epsilon = 0.8$ and which has $\epsilon = 0.7$.

Plane Radiation Shields

The above example shows the tremendous influence that a surface with an emissivity different from unity has as far as the net heat exchange is concerned. This fact is often used to reduce the loss or gain or radiant energy from a surface by what is known as a radiation shield.

For simplicity, consider two parallel infinite gray planes of equal emissivity, ϵ. If the two planes are at temperatures T_1 and T_2, the rate of energy exchange per unit area is, by Eq. (11.45),

$$\frac{(q_{12})_0}{A} = \sigma \frac{T_1^4 - T_2^4}{(2/\epsilon) - 1}.$$

The notation $(q_{12})_0$ is used to denote the net radiant exchange between the planes when nothing is placed between them.

Now, what is the reduction in the radiant energy exchange between the two original planes if a third plane is placed between them and allowed to come to thermal equilibrium? If the equilibrium temperature of this third plane is T_3 and if it is assumed that this plane is so thin that it is at the equilibrium temperature on both sides, the following relations give the net exchange of energy between this inserted plane and each of the original two when it is assumed that all the planes have the same emissivity:

$$\frac{q_{13}}{A} = \sigma \frac{T_1^4 - T_3^4}{(2/\epsilon) - 1},$$

$$\frac{q_{32}}{A} = \sigma \frac{T_3^4 - T_2^4}{(2/\epsilon) - 1}.$$

For thermal equilibrium $q_{13}/A = q_{32}/A$, so the equilibrium temperature reached by the inserted plane is

$$T_3^4 = \tfrac{1}{2}(T_1^4 + T_2^4).$$

Thus, one has

$$\frac{q_{13}}{A} = \frac{q_{32}}{A} = \sigma \frac{T_1^4 - \tfrac{1}{2}T_1^4 - \tfrac{1}{2}T_2^4}{(2/\epsilon) - 1}$$

$$= \frac{1}{2}\left[\sigma \frac{T_1^4 - T_2^4}{(2/\epsilon) - 1} \right].$$

The net radiant exchange between the original two planes is also equal to the above expression since the shielding plane placed between them neither increases nor decreases in energy. Hence, the radiant exchange between two planes with one *radiation shield* placed between them is

$$\frac{(q_{12})_1}{A} = \frac{1}{2}\left[\sigma \frac{T_1^4 - T_2^4}{(2/\epsilon) - 1} \right]$$

$$= \frac{1}{2}\frac{(q_{12})_0}{A}.$$

This expression shows that the presence of one radiation shield reduces the net exchange of radiant energy between the source and the sink to 50 per cent of the value obtained when no shield is present. The shield does not give up or absorb energy, and its position between the two planes does not alter its effectiveness as long as it does not touch either plane.

Similarly, for N radiation shields placed between a plane radiation source and sink, the net heat flux is

$$\frac{(q_{12})_N}{A} = \frac{1}{N+1} \frac{(q_{12})_0}{A}. \tag{11.46}$$

Analogous results may be deduced for the more general case in which all the emissivities are not equal; however, the analysis is a bit more involved and the results are less amenable to such a simple interpretation as given in Eq. (11.46).

11.6 Diffuse Radiant Exchange Between Finite Surfaces

The simplicity of the problem discussed above was due primarily to the fact that all the radiation leaving one plane struck the other. Thus, the radiosity of one surface was always equal to the irradiation of the other. In many applications in engineering, such as the radiant exchange between the different walls of a room, various "hot" and "cold" surfaces in spacecraft, etc., this simplification will not hold. It is necessary to consider now the general case of finite surfaces which are not parallel and which are exchanging radiant energy. The surfaces are, generally, not plane, and other surfaces may be present which can interfere with or aid the net exchange of energy.

The Shape Factor

Of basic importance is the *direct* exchange of energy between two surfaces —that fraction of the energy leaving one surface which strikes the second directly, none being considered which is transferred by reflection or reradiation from other surfaces that may be present. This direct flux of energy between one surface and another is expressed in terms of the *shape factor*, which is found by determining the exchange between differential area elements in each surface and then performing simultaneous integrations over both surfaces. When Lambert's law is presumed to hold, the result is dependent solely upon the geometry of the problem.

The geometry of the situation is illustrated in Fig. 11.14. The two surfaces in question are denoted by A_1 and A_2. These surfaces are located arbitrarily in space, are of arbitrary shape, and are not necessarily plane. Elements of each surface are denoted by dA_1 and dA_2 with corresponding normals N_1

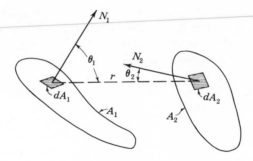

Fig. 11.14

and N_2. The line connecting dA_1 to dA_2 has length r and makes polar angles θ_1 and θ_2 with the two normals N_1 and N_2, respectively. According to Eq. (11.11), which is developed in Sec. 11.2, comparison of the geometry of Fig. 11.2 [upon which Eq. (11.1) is based] with that in Fig. 11.14 reveals that the sought-for radiant energy which leaves A_1 and *strikes* A_2, directly, is

$$q_{1 \to 2} = \int_{A_1} \int_{A_2} I_1 \frac{\cos \theta_1 \cos \theta_2}{r^2} dA_2 \, dA_1.$$

The intensity of the radiation *leaving* the transmitting surface is denoted by I_1.

If the radiosity of the transmitting surface, J_1, is uniform over that surface, the total energy leaving it is $A_1 J_1$, so the *fraction* of the energy leaving A_1 which strikes A_2 is

$$F_{1-2} = \frac{q_{1 \to 2}}{A_1 J_1}. \tag{11.47}$$

If, in addition, the radiant energy leaving A_1 is *diffuse* (both the emissive and reflected portions), then I_1 is constant and Eq. (11.15) states that $J_1 = \pi I_1$. Thus, the two equation above give

$$F_{1-2} = \frac{1}{A_1} \int_{A_1} \int_{A_2} \frac{\cos \theta_1 \cos \theta_2}{\pi r^2} dA_2 \, dA_1. \tag{11.48}$$

The factor F_{1-2} is called the *configuration factor* or *shape factor* of A_1 with respect to A_2—the fraction of the energy leaving A_1 which *directly* strikes A_2. As may be seen from its representation in Eq. (11.48), the shape factor for diffuse radiation is a purely geometric property of the two surfaces involved. The order of the subscripts on F_{1-2} is quite important since F_{2-1} will be used to represent the fraction of energy leaving A_2 which directly strikes A_1. In general, $F_{1-2} \neq F_{2-1}$, since reasoning similar to that given above yields

$$F_{2-1} = \frac{1}{A_2} \int_{A_2} \int_{A_1} \frac{\cos \theta_2 \cos \theta_1}{\pi r^2} dA_1 \, dA_2. \tag{11.49}$$

Some Properties of the Shape Factor

Certain useful properties of the shape factor may be deduced from the definition given by Eqs. (11.47) and (11.48). These properties will be useful for the determination of shape factors for specific geometries and for the analysis of heat exchange between surfaces.

The Reciprocal Property. The most important property of the shape factor is readily deducible from Eq. (11.48) and (11.49). For two surfaces, A_1 and A_2, in general $F_{1-2} \neq F_{2-1}$; however, the following reciprocal relation exists:

$$A_1 F_{1-2} = A_2 F_{2-1}. \tag{11.50}$$

The Additive Property. If one of the two surfaces, say A_i, is divided into subareas $A_{i_1}, A_{i_2}, \ldots, A_{i_n}$ (so that $\Sigma A_{i_n} = A_i$), the application of Eqs. (11.47) and (11.48) will lead to the following relation:

$$A_i F_{i-j} = \sum_n A_{i_n} F_{i_n-j}. \tag{11.51}$$

The receiprocal relation given above will then yield

$$F_{j-i} = \sum_n F_{j-i_n}. \tag{11.52}$$

Thus, if the transmitting surface is subdivided, the shape factor for that surface with respect to a receiving surface is not simply the sum of the individual shape factors, although the AF product is expressed by such a sum. The shape factor from a surface to a subdivided receiving surface *is* simply the sum of the individual shape factors.

If surface A_i is subdivided into n parts ($A_{i_1}, A_{i_2}, \ldots, A_{i_n}$) and surface A_j into m parts ($A_{j_1}, A_{j_2}, \ldots, A_{j_m}$), the above reasoning leads to

$$A_i F_{i-j} = \sum_n \sum_m A_{i_n} F_{i_n-j_m}. \tag{11.53}$$

These additive properties are useful in finding the shape factor for complex shapes by subdividing the surfaces into subsections for which the shape factor is either known or more simply evaluated. This is discussed in detail in Appendix E.

The Enclosure Property. If the interior surface of a completely enclosed space, such as a room, is subdivided into n parts—each part having a finite area $A_1, A_2, \ldots, A_n$, then

$$\sum_{j=1}^{n} F_{i-j} = 1, \qquad i = 1, 2, \ldots, n. \tag{11.54}$$

The above representation admits the shape factors $F_{1-1}, F_{2-2}, \ldots, F_{n-n}$ since some of the surfaces may "see" themselves if they are convex. That is, for example, the definition of the shape factor in Eq. (11.48) does not preclude the possibility that the two incremental areas, dA_1 and dA_2, may be on the same surface.

If a surface A_1, is completely enclosed by a second surface, A_2, and if A_1 does not see itself ($F_{1-1} = 0$), then $F_{1-2} = 1$. Then the reciprocal relation of Eq. (11.50) gives

$$F_{2-1} = \frac{A_1}{A_2}. \tag{11.55}$$

The Calculation of Shape Factors

The definition of the shape factor is given by

$$F_{1-2} = \frac{1}{A_1} \int_{A_1} \int_{A_2} \frac{\cos \theta_1 \cos \theta_2}{\pi r^2} dA_2 \, dA_1. \tag{11.48}$$

The two integrals in Eq. (11.48) are double integrals—each taken over a surface. Calculation of F_{1-2} for a specific geometrical setup will require that a coordinate system be established for each surface. Then the location of dA_1 and dA_2 on A_1 and A_2, respectively, may be expressed in terms of these coordinates. Likewise, the angles θ_1 and θ_2 and the connecting distance r may also be expressed in terms of the coordinate location of dA_1 and dA_2. Finally, the area elements themselves must be expressed in terms of differentials of the coordinates—resulting in a twofold integral for each of the double integrals. Thus, in terms of geometrical coordinates, the expression for F_{1-2} becomes, usually, a fourfold integral.

For even the simplest geometrical shapes the evaluation of the fourfold integral becomes quite involved. Since such integrations are principally mathematical in nature, the details of shape factor evaluation, for a selection of physically significant cases, are given in Appendix E. The resultant expressions for these cases are presented for convenience in Figs. 11.15 through 11.19. Figures 11.15 and 11.16 present the shape factors for directly opposed rectangles and perpendicular rectangles with a common edge; Fig. 11.17 gives F_{1-2} for parallel disks, and Fig. 11.18 presents F_{1-2} for coaxial cylinders of finite length. Two cases of importance in the present-day analysis of radiant exchange between a planetary body and elements of a spacecraft or satellite are given in Fig. 11.19—the shape factors between a large sphere and either a small plane element or a small sphere.

More complex shapes often may be reduced to simpler cases by the application of "shape factor algebra"—i.e., the use of some of the shape factor properties discussed in the preceding section, particularly the additive property. Appendix E illustrates the manner in which this reduction may be carried out. The material which follows in this chapter presumes that the reader is familiar with the methods of Appendix E and has mastered the methods of determining the shape factor for a given geometry.

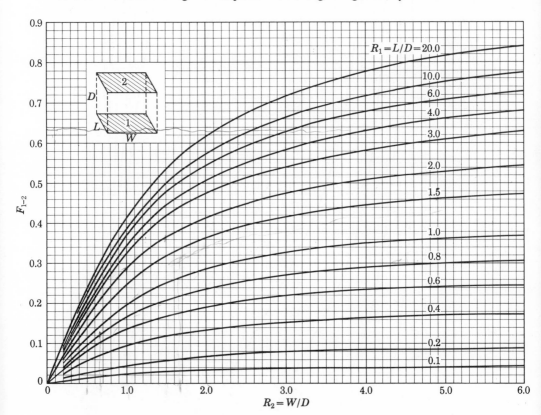

Fig. 11.15. The radiation shape factor for parallel, directly opposed, rectangles.

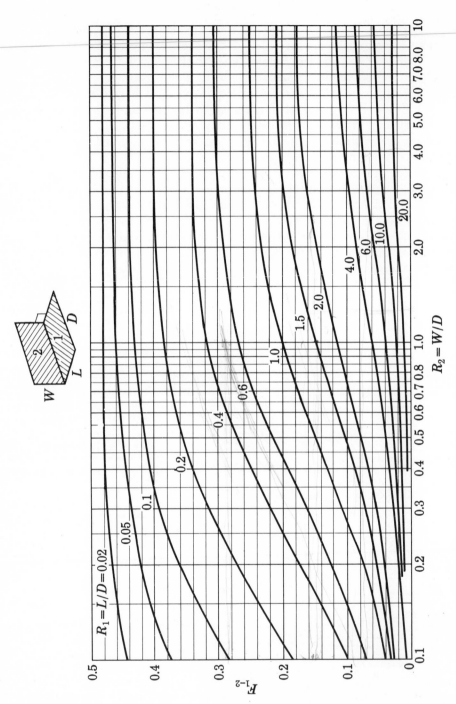

Fig. 11.16. The radiation shape factor for perpendicular rectangles with a common edge.

454

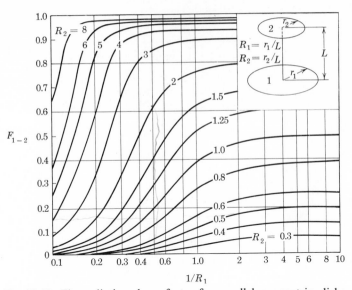

Fig. 11.17. The radiation shape factor for parallel, concentric, disks.

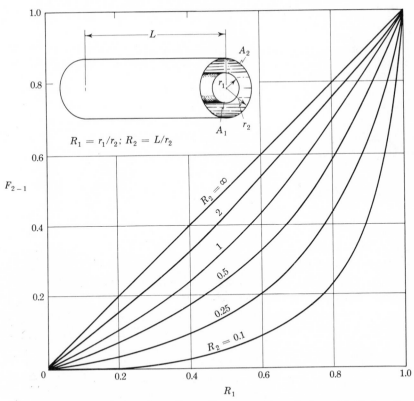

Fig. 11.18. The radiation shape factor for concentric cylinders of finite length.

455

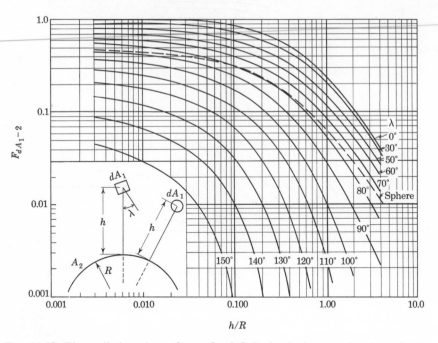

Fig. 11.19. The radiation shape factor for infinitesimal planes and spheres in the presence of a large sphere.

Such mathematical methods as Stokes' theorem may be applied in some cases to reduce the double area integrals to contour integrals (Ref. 7). For many geometries, numerical methods utilizing digital computers may be advantageous (Ref. 8).

Radiant Exchange Between Black Surfaces

Consider, now, a system of two or more black surfaces, $A_1, A_2, \ldots, A_n$, each maintained at a fixed temperature, $T_1, T_2, \ldots, T_n$. The surfaces may form a complete enclosure, or they may not. If they do not, imagine them to be situated in an irradiation free space. That is, any radiation leaving one surface which does not strike another may be presumed lost and no irradiation is incident upon any surface from sources which are not one of the given black surfaces. In general, two quantities are sought for such an instance: the net exchange between any two surfaces and the net energy given up by any one of the surfaces.

Since the surfaces are black, the energy striking a surface is all absorbed; and the radiosity of a surface is solely its emissive power. Thus,

$$q_{i-j} = q_{i \to j},$$

where the meanings of the symbols q_{i-j} and $q_{i \to j}$ are as given at the beginning of Sec. 11.5. By virtue of Eq. (11.47) and the fact that all the surfaces are black so that the radiosity is solely the black body emissive power, W_b,

$$q_{i-j} = A_i F_{i-j} J_i$$

$$= A_i F_{i-j} W_{b_i}.$$

Similarly, the amount of radiation leaving surface A_j which is *absorbed* at surface A_i is

$$q_{j-i} = A_j F_{j-i} W_{b_j}.$$

The net *exchange* of energy between the two surfaces is

$$q_{ij} = q_{i-j} - q_{j-i}$$

$$= A_i F_{i-j} W_{b_i} - A_j F_{j-i} W_{b_j}.$$

Upon employing the reciprocal relation of Eq. (11.50),

$$q_{ij} = A_i F_{i-j}(W_{b_i} - W_{b_j}), \tag{11.56}$$

or

$$q_{ij} = A_i F_{i-j} \sigma(T_i^4 - T_j^4). \tag{11.57}$$

Thus, the shape factor provides a very convenient way of expressing the exchange of radiant energy between two black surfaces. Quite apparently the exchange based on the surface A_j is

$$q_{ji} = -q_{ij} = A_j F_{j-i} \sigma(T_j^4 - T_i^4).$$

The net energy given up by one of the n surfaces, say A_i, is

$$q_i = A_i W_i - \sum_{p=1}^{n} q_{p-i}$$

$$= A_i W_i - \sum_{p=1}^{n} q_{p \to i}$$

$$= A_i W_i - \sum_{p} A_p F_{p-i} W_{b_p}$$

$$= A_i \left(W_i - \sum_{p} F_{i-p} W_{b_p} \right). \tag{11.58}$$

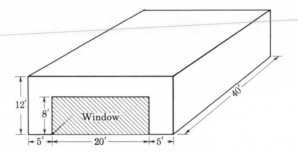

Fig. 11.20. Geometry for Example 11.5.

EXAMPLE 11.5: An outlet shoe store with a display window in the front is shown, with dimensions, in Fig. 11.20. The store is to be heated by making the floor a radiant heating panel at 100°F. If the glass window acts as a black plane at 40°F and all the other walls and the ceiling act as black planes at 80°F, find the net exchange between the floor and the window and the net heat given up by the floor. The representation of the window as a black surface is reasonable since practically none of the radiation passing through it will ever return to the room.

Solution: The floor and window form a configuration as noted in Fig. 11.21. The floor is denoted by 1 and the window by 2, and 3 through 6 are subdivisions to find F_{1-2}. Application of the principles of shape factor algebra illustrated in Appendix E gives

$$A_1 F_{1-2} = A_5 F_{5-2} + 2A_6 F_{6-2}$$

$$= A_5 F_{5-2} + [A_{(5,6)} F_{(5,6)-(2,3)} - A_5 F_{5-2} - A_6 F_{6-3}]$$

$$= A_{(5,6)} F_{(5,6)-(2,3)} - A_6 F_{6-3}.$$

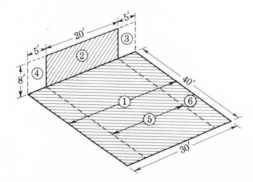

Fig. 11.21

Use of Fig. 11.16 gives

$$F_{(5,6)-(2,3)} = 0.075,$$

$$F_{6-3} = 0.04,$$

so

$$F_{1-2} = \frac{1}{30 \times 40}(25 \times 40 \times 0.075 - 5 \times 40 \times 0.04)$$

$$= 0.056.$$

Since the room forms an enclosed space, it is not necessary to find, individually, the shape factor of the floor with respect to all the other walls since they are at the same temperature. Letting F_{1-w} denote the shape factor between the floor and all the walls (and ceiling) except the window,

$$F_{1-w} = 1 - 0.056$$

$$= 0.944.$$

By Eq. (11.57) the exchange between the floor and the window is

$$q_{12} = A_1 F_{1-2} \sigma(T_1^4 - T_2^4)$$

$$= (30 \times 40)(0.056)(0.1714)\left[\left(\frac{560}{100}\right)^4 - \left(\frac{500}{100}\right)^4\right]$$

$$= 4130 \text{ Btu/hr } (1210 \text{ W}).$$

By Eq. (11.58) the net heat given up by the floor is

$$q_1 = A_1[W_{b_1} - (F_{1-2}W_{b_2} + F_{1-w}W_{b_w})]$$

$$= (30 \times 40)(0.1714)\left[\left(\frac{560}{100}\right)^4 - 0.056\left(\frac{500}{100}\right)^4 - 0.944\left(\frac{540}{100}\right)^4\right]$$

$$= 29,980 \text{ Btu/hr } (8787 \text{ W}).$$

11.7 Diffuse Radiant Exchange in Enclosures

This section and the following section will be devoted to the analysis of the radiant exchange between an assembly of isothermal surfaces which form an enclosure. Applications in engineering practice for such analyses are numerous. The radiant exchange between the walls of a room or furnace, the exchange between a body and a surrounding room, etc., are examples of such applications. In addition, the radiant exchange between a surface, or a collection of surfaces, and a "space environment" may also be treated by the enclosure theory to be presented here. Numerous applications exist in modern space technology wherein a surface, or surfaces, exchange radiant energy with an irradiation from space (a special case of which may be that of zero irradiation—i.e., "deep space"). In such instances it will be only necessary to represent the environs as an imaginary surface of large extent—with a specified, or known, radiosity (perhaps zero) and a unit absorptivity. All other surfaces will be presumed to be isothermal, of uniform radiosity, and uniform irradiation. These assumptions of uniformity may be approached in practice by sufficient subdivision of the enclosure (with an associated increase in computational labor). Integral theories of enclosures exist (Refs. 2 and 9) which treat the cases of nonuniformly irradiated surfaces. In addition, the analysis of this and the subsequent section will presume that the incident and emitted radiation is diffuse in every instance.

Generally two different quantities may be sought in the analysis of an enclosure. It is often necessary to find, for specified surface temperatures (i.e., emissive power), the *net* heat given up or absorbed by any one of the surfaces, q_i. In addition, the net exchange of energy (based on emissive power) between any two of the surfaces $q_{ij} = q_{i-j} - q_{j-i}$ may be desired. It should be expected that the net exchange, q_{ij}, will be proportional to the difference of the emissive powers,

$$q_{ij} = A_i \varphi_{i-j} (W_{b_i} - W_{b_j}),$$

where φ_{i-j} is an exchange coefficient dependent upon the geometry of the enclosure. The treatment which follows investigates the form of φ_{i-j} for various cases of practical significance.

In the analyses which follow, various kinds of imposed conditions will be considered—i.e., surfaces of specified temperature, adiabatic surfaces, etc. One fundamental relation may be developed for the energy given up by any surface in terms of the radiosities of the other surfaces in the enclosure. Consider an enclosure of n surfaces. Let $\beta = 1, 2, \ldots, \alpha, \ldots, n$ represent an index indicating each of the surfaces, and let their uniform, diffuse, radiosities be denoted by $J_1, J_2, \cdots, J_n$. For one of the surfaces in question, call it α, the *net* energy given up by the surface is, by Eq. (11.37),

$$q_\alpha = A_\alpha J_\alpha - A_\alpha G_\alpha.$$

The irradiation on surface A_α, G_α, is the sum of the fraction of the radiosities of each of the other surfaces which reaches A_α:

$$A_\alpha G_\alpha = \sum_{\beta=1}^{n} A_\beta F_{\beta-\alpha} J_\beta,$$

so

$$q_\alpha = A_\alpha J_\alpha - \sum_{\beta=1}^{n} A_\beta F_{\beta-\alpha} J_\beta$$

$$= A_\alpha J_\alpha - \sum_{\beta=1}^{n} A_\alpha F_{\alpha-\beta} J_\beta.$$

The reciprocity property of the shape factor, Eq. (11.50), has been used. The enclosure property, Eq. (11.54), may be stated

$$\sum_{\beta=1}^{n} F_{\alpha-\beta} = 1,$$

so

$$q_\alpha = A_\alpha J_\alpha \sum_{\beta=1}^{n} F_{\alpha-\beta} - \sum_{\beta=1}^{n} A_\alpha J_\beta F_{a-\beta}$$

$$= \sum_{\beta=1}^{n} A_\alpha F_{\alpha-\beta}(J_\alpha - J_\beta). \qquad (11.59)$$

This equation is the relation being sought. It will be applied in several applications in what follows. It should be emphasized that all the radiosities J_β are not necessarily known—in fact, finding them is the fundamental question of enclosure theory.

Enclosures Consisting of Black Surfaces Only

Consider first an enclosure consisting solely of black surfaces. The analysis of the preceding section has already yielded enough information to consider the enclosure problem for black surfaces solved. The sought-for exchange coefficient φ_{i-j} is already known to be F_{i-j}. However, it will be useful for further developments to consider the black enclosure from a different point of view.

In a system of n surfaces of specified emissive power, the exchange of energy between any two of the surfaces has been shown to be

$$q_{ij} = A_i F_{i-j}(W_{b_i} - W_{b_j}).$$

The net heat flow from any one of the surfaces is readily found from Eq. (11.59) since the surface radiosities are known to be the specified emissive powers inasmuch as the surfaces are black and no reflection occurs. Thus,

$$q_i = \sum_{p=1}^{n} A_i F_{i-p}(W_{b_i} - W_{b_p}), \qquad i = 1, 2, 3, \ldots, n. \qquad (11.60)$$

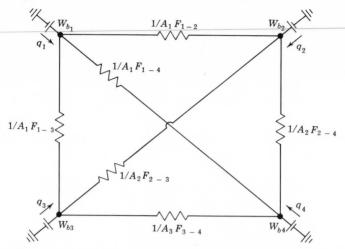

Fig. 11.22. The network representation of an enclosure of four black surfaces.

If one of the n surfaces represents an irradiation environment from space, then W_{b_p} takes on the value of the space irradiation, J_s (perhaps zero).

The form of Eq. (11.60) suggests the network analogy (Ref. 10) illustrated in Fig. 11.22 for an enclosure of, say, four surfaces. An electrical network analogy may be established in which the surfaces of the enclosure are represented by nodal points maintained at potentials equal to the known black body emissive powers. The nodes are connected by resistances which are $\mathcal{R}_{ij} = 1/A_i F_{i-j} = 1/A_j F_{j-1}$. The summation of Eq. (11.60) symbolizes the net current flow at each node.

The Effect of the Presence of Reradiating Surfaces

Frequently the situation occurs in which one, or more, of the surfaces of an enclosure acts as an *adiabatic reradiating surface.* Such a surface is one for which there is no net influx or outflux of radiant energy. In other words, a reradiating surface either reflects or absorbs and reemits all energy falling on it. That is, $J = G$ for a reradiating surface. Thus, a reradiating surface reaches an equilibrium temperature as determined by the other surfaces with which it communicates. As a result of its definition, a reradiating surface is insensitive to its value of α or ϵ. The presence of such reradiating surfaces may affect profoundly the exchange of energy between the "active" surfaces.

Imagine an enclosure consisting of n active block surfaces ($i = 1, 2, \ldots, n$) and m reradiating surfaces denoted by $r1, r2, \ldots, rm$. Then for each active surface the radiosities are known to be equal to the specified emissive power, $J_i = W_{b_i}$, while the reradiating surface radiosities, J_{rk}, are not known. Equation (11.59) may be written for each surface in the enclosure, and it is desirable to split each such expression into one term involving the known radiosities and another involving the unknown. For each of the active surfaces,

$$q_i = \sum_{p=1}^{n} A_i F_{i-p}(W_{b_i} - W_{b_p}) + \sum_{k=1}^{m} A_i F_{i-rk}(W_{b_i} - J_{rk}),$$

$$i = 1, 2, \ldots, n. \quad (11.61)$$

For each of the reradiating surfaces

$$\sum_{p=1}^{n} A_{rk} F_{rk-p}(J_{rk} - W_{b_p}) + \sum_{l=1}^{m} A_{rk} F_{rk-rl}(J_{rk} - J_{rl}) = 0,$$

$$k = 1, 2, \ldots, m. \quad (11.62)$$

The equivalent electrical network for an enclosure of three active and two reradiating surfaces is shown in Fig. 11.23(a). The reradiating surfaces are

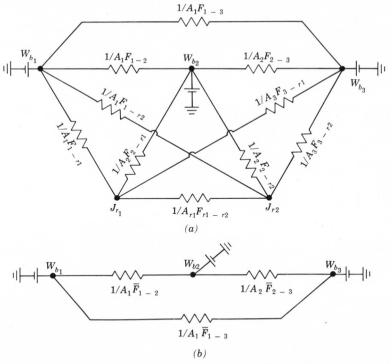

Fig. 11.23. The network representation of an enclosure of three black surfaces and two reradiating surfaces.

represented as floating, ungrounded, nodes of unknown potential. As before, the above summations represent the total current flow at each node.

For given geometry and specified values of the W_{b_i}'s of the active surfaces, there are n unknown q_i's and m unknown radiosities, J_{rk}, of the reradiating surfaces. Equations (11.61) and (11.62) give $m + n$ equations in these unknowns. General methods, based on the solution of these equations by determinates (Refs. 9, 10, 11), have been developed. It is sufficient here to simply state that an answer is obtained by simultaneous solution of the equations.

EXAMPLE 11.6:   Figure 11.24 depicts an enclosure consisting of a room. The ceiling, A_1, is a black surface maintained at 200°F; the floor, A_2, is a black surface at 300°F; and the right wall, A_3, is a black surface at 400°F. The left wall, A_{r2}, is one reradiating zone, and the two ends are to be treated as a single second reradiating zone, A_{r1}. Find the net heat given up, or absorbed, by each of the active surfaces.

Solution: By the methods of Appendix E, the necessary shape factors are found to be $F_{1-2} = 0.360$, $F_{1-3} = F_{2-3} = F_{1-r2} = F_{2-r2} = 0.185$, $F_{1-r1} = F_{2-r1} = 0.270$, $F_{3-r1} = 0.270$, $F_{3-r2} = 0.170$, and $F_{r1-r2} = 0.180$. Since $W_b = \sigma T^4$, $W_{b_1} = 325$, $W_{b_2} = 572$, and $W_{b_3} = 938$ Btu/hr-ft^2. The various surface areas may be deduced easily from Fig. 11.24.

Equations (11.61) and (11.62) give

$$q_1 = A_1 F_{1-2}(W_{b_1} - W_{b_2}) + A_1 F_{1-3}(W_{b_1} - W_{b_3})$$
$$+ A_1 F_{1-r1}(W_{b_1} - J_{r1}) + A_1 F_{1-r2}(W_{b_1} - J_{r2}),$$

$$q_2 = A_1 F_{1-2}(W_{b_2} - W_{b_1}) + A_2 F_{2-3}(W_{b_2} - W_{b_3})$$
$$+ A_2 F_{2-r1}(W_{b_2} - J_{r1}) + A_2 F_{2-r2}(W_{b_2} - J_{r2}),$$

$$q_3 = A_1 F_{1-3}(W_{b_3} - W_{b_1}) + A_2 F_{2-3}(W_{b_3} - W_{b_2})$$
$$+ A_3 F_{3-r1}(W_{b_3} - J_{r1}) + A_3 F_{3-r2}(W_{b_3} - J_{r2}),$$

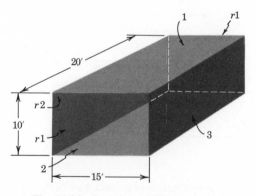

Fig. 11.24. Geometry of Example 11.6.

$$0 = A_1 F_{1-r1}(J_{r1} - W_{b_1}) + A_2 F_{2-r1}(J_{r1} - W_{b_2})$$
$$+ A_3 F_{3-r1}(J_{r1} - W_{b_3}) + A_{r1} F_{r1-r2}(J_{r1} - J_{r2}),$$
$$0 = A_1 F_{1-r2}(J_{r2} - W_{b_1}) + A_2 F_{2-r2}(J_{r2} - W_{b_2})$$
$$+ A_3 F_{3-r2}(J_{r2} - W_{b_3}) + A_{r1} F_{r1-r2}(J_{r2} - J_{r1}).$$

The reciprocal relation $A_i F_{i-j} = A_j F_{j-i}$ has been used. When the known values of the F's and W_b's are introduced:

$$q_1 = -16,335 - 81.0 J_{r1} - 55.5 J_{r2},$$

$$q_2 = 84,441 - 81.0 J_{r1} - 55.5 J_{r2},$$

$$q_3 = 136,879 - 54.0 J_{r1} - 55.5 J_{r2},$$

$$0 = 123,309 - 270 J_{r1} + 54.0 J_{r2}$$

$$0 = 81,676 - 199 J_{r2} + 54.0 J_{r1}.$$

Solution of these equations yields

$$q_1 = -93,840 \text{ Btu/hr} (-27,500 \text{ W}),$$

$$q_2 = 6940 \text{ Btu/hr} (2034 \text{W}),$$

$$q_3 = 86,900 \text{ Btu/hr} (25,470 \text{ W}),$$

$$J_{r1} = 570 \text{ Btu/hr-ft}^2 (167 \text{ W}),$$

$$J_{r2} = 565 \text{ Btu/hr-ft}^2 (166 \text{ W}),$$

The closeness of the J_{r1} and J_{r2} suggests that the two reradiating zones could be considered as one without incurring much error.

The above analysis determines the net energy given up by each active surface, but it does not yield the *net exchange* between any two of them. This latter quantity may be found from $q_{ij} = q_{i-j} - q_{j-i}$, in which q_{i-j}, the amount of the emission from A_i which is absorbed at A_j, is calculated by noting that the linearity of Eq. (11.61) or (11.62) in the W_b's leads to

$$q_{i-j} = W_{b_i}(-q_j^{(i)}).$$

The symbol $q_j^{(i)}$ denotes the net flow of heat from surface A_j when W_{b_i} is maintained at unity and all other active surfaces are maintained at $W_b = 0$. In this instance surface A_j receives heat only from A_i (directly or via the reradiating surfaces).

It is customary to express the heat flow q_{i-j} as a fraction of the original emission from A_i, so a modified shape factor, $\bar{F}_{i-j}$, may be found in this case to account for the presence of the reradiating surfaces:

$$q_{i-j} = A_i W_{b_i} \bar{F}_{i-j},$$

$$\bar{F}_{i-j} = \frac{1}{A_i}(-q_j^{(i)}).$$

$$(11.63)$$

Equation (11.61) for $W_{b_i} = 1$ (all other emissive powers being zero) gives

$$-q_j^{(i)} = A_j F_{j-i} + A_j \sum_k F_{j-rk} J_{rk}^{(i)} - \delta_{ij} A_i,$$

or

$$\bar{F}_{i-j} = F_{i-j} + \frac{A_j}{A_i} \sum_k F_{j-rk} J_{rk}^{(i)} - \delta_{ij}. \qquad (11.64)$$

The operator δ_{ij} ($\delta_{ij} = 0$ for $i \neq j$, $\delta_{ij} = 1$ for $i = j$) has been introduced, so Eq. (11.64) will apply when $j = i$. The radiosities, J_{rk}, of the reradiating surfaces when $W_{b_i} = 1$ and all other active surfaces are zero may be found from Eq. (11.62):

$$\sum_p A_{rk} F_{rk-p} J_{rk}^{(i)} - A_{rk} F_{rk-i} + \sum_l A_{rk} F_{rk-l} (J_{rk}^{(i)} - J_{rl}^{(i)}) = 0. \qquad (11.65)$$

Equation (11.65) represents a system of m equations ($k = 1, 2, \ldots, m$), one for each reradiating surface. This system may be solved for the unknown $J_{rk}^{(i)}$'s and the results substituted into Eq. (11.64) to find the factor $\bar{F}_{i-j}$.

This direct, or *brute-force*, approach is recommended for the majority of engineering applications, and this is illustrated in Example 11.7. However, one may carry out the indicated algebra, using the methods of determinants, to develop a generalized expression for $\bar{F}_{i-j}$ in terms of the conventional shape factors of the system. Such an analysis is too complex for this text, and Ref. 11 should be consulted for further details. One result of such an analysis is the fact that the modified shape factor $\bar{F}$ satisfies the reciprocal relation

$$A_i \bar{F}_{i-j} = A_j \bar{F}_{j-i}. \qquad (11.66)$$

Since $q_{i-j} = A_i \bar{F}_{i-j} W_{b_i}$, $A_i \bar{F}_{i-j} = A_j \bar{F}_{j-i}$, and $q_{ij} = q_{i-j} - q_{j-i}$, the *net heat exchange* between any two active surfaces is

$$q_{ij} = A_i \bar{F}_{i-j} (W_{b_i} - W_{b_j}). \qquad (11.67)$$

This is the sought-after form in which the net exchange is expressed in terms of the emissive powers of the surfaces and an exchange coefficient. In other words, the electrical network of Fig. 11.23(a) has been reduced to the equivalent network shown in Fig. 11.23(b) based on resistors involving the $\bar{F}$'s.

It is worth noting from Eq. (11.64) that since the J's are all positive, $\bar{F}_{i-j}$ is always greater than the direct exchange factor F_{i-j}. Thus, the effect of presence of the reradiating surfaces is the enhancement of the exchange between the active surfaces. Also, when reradiating surfaces are present, two active surfaces will exchange energy even though they do not "see" each other, since $\bar{F}_{i-j} \neq 0$ when $F_{i-j} = 0$.

In summary, then, the exchange factor $\bar{F}_{i-j}$ is found from Eq. (11.64) after finding the reradiator radiosities from the system of m equations given in Eq. (11.65). The net exchange is then found from Eq. (11.67). Normally this direct approach is adequate for most applications.

The case in which there is only *one* reradiating zone, call it simply A_r, is of sufficient interest to warrant special note. In this case an algebraic expression for $\bar{F}$ may be found. Equation (11.65) gives in this case

$$J_r^{(i)} = \frac{F_{r-i}}{\sum\limits_{p} F_{r-p}}.$$

Since there is only one reradiator, the enclosure rule gives

$$J_r^{(i)} = \frac{F_{r-i}}{1 - F_{r-r}},$$

F_{r-r} being the self-shape factor of the enclosing reradiating surface. Equation (11.64) yields

$$\bar{F}_{i-j} = F_{i-j} + \frac{A_j}{A_i} F_{j-r} J_r^{(i)}$$

$$= F_{i-j} + \frac{A_j}{A_i} \frac{F_{j-r} F_{r-i}}{1 - F_{r-r}}.$$

Using $A_j F_{j-r} = A_r F_{r-j}$ and $A_r F_{r-i} = A_i F_{i-r}$, one finally has

$$\bar{F}_{i-j} = F_{i-j} + \frac{F_{i-r} F_{r-j}}{1 - F_{r-r}}. \tag{11.68}$$

It will be left as an exercise to show that if, in addition, only *two* active surfaces are present, then

$$\bar{F}_{1-2} = F_{1-2} + \cfrac{1}{\cfrac{A_1}{A_2}\cfrac{1}{F_{2-r}} + \cfrac{1}{F_{1-r}}}, \tag{11.69}$$

and if, further, $F_{1-1} = F_{2-2} = 0$ (i.e., the active surfaces do not see themselves),

$$\bar{F}_{1-2} = \frac{(A_2/A_1) - F_{1-2}^2}{(A_2/A_1) + (1 - 2F_{1-2})}. \tag{11.70}$$

In this instance only the direct exchange factor for the two active surfaces is needed to evaluate $\bar{F}_{1-2}$, and the specific geometry of the reradiating surface does not enter.

EXAMPLE 11.7:   For the data given in Example 11.6, find the net heat exchanged by surfaces A_1 and A_2, and A_1 and A_3.

Solution: For $W_{b_1} = 1$, $W_{b_2} = W_{b_3} = 0$, the equations for q_2 and q_3 in Example 11.6 give

$$q_2^{(1)} = 180(-1) + 0 + 81(-J_{r1}^{(1)}) + 55.5(-J_{r2}^{(1)}),$$
$$q_3^{(1)} = 55.5(-1) + 0 + 54(-J_{r1}^{(1)}) + 34(-J_{r2}^{(1)}).$$

The heat balances of the two reradiating surfaces give

$$270J_{r1}^{(1)} = 81 + 54J_{r2}^{(1)},$$
$$199J_{r2}^{(1)} = 55.5 + 54J_{r1}^{(1)}.$$

Thus, $J_{r1}^{(1)} = 0.3762$, $J_{r2}^{(1)} = 0.3810$, so

$$-q_2^{(1)} = 159.6,$$
$$-q_3^{(1)} = 88.8.$$

Consequently, Eq. (11.63) gives

$$\bar{F}_{1-2} = \frac{159.6}{300} = 0.532,$$

$$\bar{F}_{1-3} = \frac{88.8}{300} = 0.296.$$

When compared with $F_{1-2} = 0.360$ and $F_{1-3} = 0.186$, the effect of the reradiating walls is apparent. The net heat exchanges are

$$q_{12} = A_1 \bar{F}_{1-2}(W_{b_1} - W_{b_2})$$

$$= 159.6(325 - 572) = -39{,}430 \text{ Btu/hr} \ (-11{,}560 \text{ W}),$$

$$q_{13} = A_1 \bar{F}_{1-3}(W_{b_1} - W_{b_3})$$

$$= 88.8(325 - 938) = -54{,}410 \text{ Btu/hr} \ (-15{,}950 \text{ W}).$$

The net heat given up by surface A_1 may be determined by summing q_{12} and q_{13}, as shown, yielding the same result as Example 11.6. In the example considered here, this method may prove simpler for the determination of q_1; however, for a large number of surfaces, the approach of Example 11.6 may prove better, particularly if the net exchanges, q_{12}, q_{13}, etc., are not desired.

EXAMPLE 11.8: If all the walls other than the window and floor of the shoe store described in Example 11.5 are considered as a single reradiating wall, and if the floor and windows are at the same temperatures given in that example, find the net heat exchange between the floor and the window. Also, what will be the equilibrium temperature of the reradiating walls?

Solution: If the floor is denoted as A_1 and the windows as A_2, Example 11.5 gives $F_{1-2} = 0.056$. Since there is only one reradiating surface and only two plane radiant sources, Eq. (11.70) holds for $\bar{F}_{1-2}$:

$$\bar{F}_{1-2} = \frac{A_2 - A_1 F_{1-2}^2}{A_1 + A_2 - 2A_1 F_{1-2}}$$

$$= \frac{20 \times 8 - 30 \times 40 \times (0.056)^2}{30 \times 40 - 20 \times 8 - 2 \times 30 \times 40 \times 0.056} = 0.127.$$

The net heat exchange in this case is

$$q_{12} = A_1 \bar{F}_{1-2}\sigma(T_1^4 - T_2^4)$$

$$= 0.1714 \times 30 \times 40 \times 0.127 \left[\left(\frac{560}{100}\right)^4 - \left(\frac{500}{100}\right)^4 \right]$$

$$= 9366 \text{ Btu/hr} \ (2745 \text{ W}).$$

The equilibrium temperature of the walls is found from the fact that the walls exchange the same amount of energy with A_1 as with A_2. Thus,

$$\sigma A_1 F_{1-r}(T_1^4 - T_r^4) = \sigma A_r F_{r-2}(T_r^4 - T_2^4)$$

$$= \sigma A_2 F_{2-r}(T_r^4 - T_2^4).$$

The equilibrium temperature may be found by solving the equation above:

$$A_1(1 - F_{1-2})(T_1^4 - T_r^4) = A_2(1 - F_{2-1})(T_r^4 - T_2^4)$$

$$= (A_2 - A_1 F_{1-2})(T_r^4 - T_2^4).$$

$$\frac{T_1^4 - T_r^4}{T_r^4 - T_2^4} = \frac{(A_2/A_1) - F_{1-2}}{1 - F_{1-2}},$$

$$\frac{(560)^4 - T_r^4}{T_r^4 - (500)^4} = \frac{(2/15) - 0.056}{0.944},$$

$$T_r = 555°\text{R} = 95°\text{F}\,(35°\text{C}).$$

11.8   Diffuse Radiant Exchange in Gray Enclosures

The foregoing discussions related to enclosures involving black surfaces will now be extended to such enclosures in which the active surfaces are gray. Reradiating surfaces may, or may not, be present. Consider, then, a system of n gray surfaces (areas $A_1, A_2, \ldots, A_n$; emissivities $\epsilon_1, \epsilon_2, \ldots, \epsilon_n$) maintained at fixed, known, temperatures so that the *equivalent black body* emissive powers are known (i.e., $W_{b_1}, W_{b_2}, \ldots, W_{b_n}$). Presume that these active surfaces form an enclosure together with m reradiating surfaces. These reradiating surfaces attain equilibrium values of their radiosities ($J_{r_1}, J_{r_2}, \ldots, J_{rm}$), which must be determined. The number of reradiating surfaces may actually be zero, the enclosure consisting solely of gray, active, surfaces.

The net heat flow from any of the surfaces is given by Eq. (11.37) or (11.39):

$$q_s = A_s(J_s - G_s), \tag{11.37}$$

$$q_s = A_s \frac{\epsilon_s}{1 - \epsilon_s}(W_{b_s} - J_s)$$

$$= A_s E_s(W_{b_s} - J_s). \tag{11.39}$$

The notation

$$E_s = \frac{\epsilon_s}{1 - \epsilon_s} \tag{11.71}$$

has been introduced for simplicity of expression. The index s has been used to denote *all* the surfaces of the enclosure, active or reradiating. The indices i and p are reserved for the n active surfaces and the indices k and l for the reradiating surfaces, as before.

The general relation of Eq. (11.59) must be satisfied by all the surfaces, active or reradiating. Thus,

$$q_s = \sum_t A_s F_{s-t}(J_s - J_t), \qquad s, t = 1, 2, \ldots, n, r1, r2, \ldots, rm. \quad (11.72)$$

For the n active surfaces the unknown q_s may be replaced with Eq. (11.39), so

$$\sum_t A_i F_{i-t}(J_i - J_t) + A_i E_i(J_i - W_{b_i}) = 0,$$

$$t = 1, 2, \ldots, n, r1, r2, \ldots, rm; \qquad i = 1, 2, \ldots, n, \quad (11.73)$$

while the remaining reradiating surfaces are adiabatic:

$$\sum_t A_{rk} F_{rk-t}(J_{rk} - J_t) = 0,$$

$$t = 1, 2, \ldots, n, r1, r2, \ldots, rm, \qquad k = 1, 2, \ldots, m. \quad (11.74)$$

Equations (11.73) and (11.74) are sufficient to solve for the $m + n$ unknowns in a given enclosure problem. These equations are expressed graphically in terms of the electrical network illustrated in Fig. 11.25(a) for an enclosure

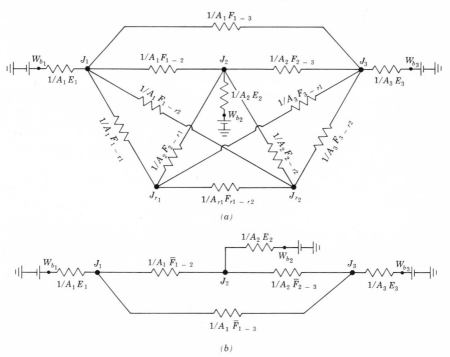

(a)

(b)

Fig. 11.25. The network representation of an enclosure of three gray surfaces and two reradiating surfaces.

of three active surfaces and two reradiating surfaces. The radiosities are represented by floating nodes connected with resistors as in the black enclosure case. Nodes representing the active surfaces' emissive powers are connected to the corresponding radiosity nodes with resistors of value $\mathcal{R} = 1/A_iE_i$. Equations (11.73) and (11.74) express the summation of the current flows at each radiosity node. Once the radiosities are found from these equations, the desired net heat flows from each surface may be found from Eq. (11.37).

Particular matrix inversion or determinate techniques have been developed for solution of this particular set of equations (Refs. 10, 11).

EXAMPLE 11.9: Repeat Example 11.6 except for the modification that the active surfaces are to be considered gray with $\epsilon_1 = 0.8$, $\epsilon_2 = 0.7$, $\epsilon_3 = 0.6$.

Solution: For the given emissivities, $E_1 = 4.0$, $E_2 = 2.33$, $E_3 = 1.50$. For the data given, Eqs. (11.73) and (11.74) yield

$$108(J_1 - J_2) + 55.5(J_1 - J_3) + 81(J_1 - J_{r1})$$
$$+ 55.5(J_1 - J_{r2}) + 4.0 \times 300 \times (J_1 - 325) = 0,$$
$$108(J_2 - J_1) + 55.5(J_2 - J_3) + 81(J_2 - J_{r1})$$
$$+ 55.5(J_2 - J_{r2}) + 2.33 \times 300(J_2 - 572) = 0,$$
$$55.5(J_3 - J_1) + 55.5(J_3 - J_2) + 54(J_3 - J_{r1})$$
$$+ 34(J_3 - J_{r2}) + 1.50 \times 200(J_3 - 938) = 0,$$
$$81(J_{r1} - J_1) + 81(J_{r1} - J_2) + 54(J_{r1} - J_3) + 54(J_{r1} - J_{r2}) = 0,$$
$$55.5(J_{r2} - J_1) + 55.5(J_{r2} - J_2) + 34(J_{r2} - J_3) + 54(J_{r2} - J_{r1}) = 0.$$

Simultaneous solution (either a matrix inversion or solution by determinants is recommended) yields

$$J_1 = 378 \text{ Btu/hr-ft}^2, \qquad J_2 = 557,$$
$$J_3 = 761, \qquad\qquad J_{r1} = 540,$$
$$J_{r2} = 537,$$

from which the net heat flows are found:

$$q_1 = A_1E_1(W_{b_1} - J_1)$$
$$= 300 \times 4.0(325 - 378)$$
$$= -63,600 \text{ Btu/hr} (-18,640 \text{ W}),$$
$$q_2 = 10,500 \text{ Btu/hr} (3077 \text{ W}),$$
$$q_3 = 53,100 \text{ Btu/hr} (15,560 \text{ W}).$$

The net heat exchange between two active surfaces, say A_i and A_j, may be found by the same scheme as that described for enclosures involving black surfaces—namely ascertaining the quantity $-q_j^{(i)}$ which represents the heat exchange between surfaces A_i and A_j when $W_{b_i} = 1$ and all other active-surface emissive powers are set equal to zero. This quantity led to the definition of an exchange coefficient, $\bar{F}_{i-j} = -q_j^{(i)}/A_i$. This procedure will be followed again. In this instance the exchange coefficient will be given a different symbol in order to emphasize the fact that gray surfaces are involved. Thus, define

$$\mathscr{F}_{i-j} = -q_j^{(i)}/A_i$$

whenever gray active surfaces are involved—whether reradiating surfaces are present or not. The procedure to be followed in determining $\mathscr{F}_{i-j}$ is simply that of solving Eqs. (11.73) and (11.74) for the radiosity $J_j^{(i)}$ when $W_{b_i} = 1$ and all other W_b's are zero. Then

$$\mathscr{F}_{i-j} = -q_j^{(i)}/A_i = \frac{A_j}{A_i}E_j J_j^{(i)}, \qquad (11.75)$$

according to Eq. (11.39).

With $\mathscr{F}_{i-j}$ thus defined, the energy flow from A_i to A_j is

$$q_{i-j} = A_i \mathscr{F}_{i-j} W_{b_i},$$

and the *net exchange* is

$$\begin{aligned} q_{ij} &= q_{i-j} - q_{j-i} \\ &= A_i \mathscr{F}_{i-j} W_{b_i} - A_j \mathscr{F}_{j-i} W_{b_j}. \end{aligned}$$

The two previously defined exchange coefficients satisfy the reciprocal relation: $A_i F_{i-j} = A_j F_{j-i}$; $A_i \bar{F}_{i-j} = A_j \bar{F}_{j-i}$. Similarly, it may be demonstrated that the $\mathscr{F}$ factor (see Ref. 11) satisfies the same relation:

$$A_i \mathscr{F}_{i-j} = A_j \mathscr{F}_{j-i}, \qquad (11.76)$$

so that, for *gray enclosures*,

$$q_{ij} = A_i \mathscr{F}_{i-j}(W_{b_i} - W_{b_j}). \qquad (11.77)$$

Proof of the reciprocal relation for $\mathscr{F}$ will not be given here, although it is easily demonstrated for some of the special cases to be quoted later.

An alternative formulation to that just presented in Eqs. (11.73) and (11.74) and illustrated in Fig. 11.25(a) may be made. If, when reradiating surfaces are present, the coefficients $\bar{F}$ defined earlier have been already evaluated, the equivalent electrical circuit may be reduced to that shown in Fig. 11.25(b), and the energy balances given in Eqs. (11.73) and (11.74) are replaced by the single expression

$$\sum_p A_i \bar{F}_{i-s}(J_i - J_p) + A_i E_i(J_i - W_{b_i}) = 0, \qquad p, i = 1, 2, \ldots, n. \quad (11.78)$$

The summation is to be made over *only* the active surfaces. Also, if the $\bar{F}$'s are known, Eq. (11.78) may be used in conjunction with Eq. (11.75) to find the exchange coefficient $\mathscr{F}_{i-j}$.

In general, $\mathscr{F}_{i-j}$ must be calculated for each case at hand. However, certain special cases of physical significance lend themselves to algebraic solutions. Without detailed algebra being given, these cases are summarized below.

Two Gray Surfaces

If a complete enclosure consists of only two gray surfaces, A_1 and A_2, then Eq. (11.73) yields only two expressions and none results from Eq. (11.74). Manipulation of these expressions along with the definition of $\mathscr{F}$ given in Eq. (11.75) gives

$$\frac{1}{\mathscr{F}_{1-2}} = \frac{1}{F_{1-2}} + \frac{A_1}{A_2}\left(\frac{1}{\epsilon_2} - 1\right) + \left(\frac{1}{\epsilon_1} - 1\right). \quad (11.79)$$

Two Convex Gray Surfaces in a Radiation-Free Space

If two gray surfaces do not form a complete enclosure but are located in space wherein no external irradiation is present, an enclosure may be envisioned by defining a third enclosing surface for which $J = W = 0$. If $F_{1-1} = F_{2-2} = 0$, then Eqs. (11.73), (11.74), and (11.75) will yield

$$\mathscr{F}_{1-2} = \frac{F_{1-2}}{\dfrac{1}{\epsilon_1 \epsilon_2} - \left(\dfrac{1}{\epsilon_1} - 1\right)\left(\dfrac{1}{\epsilon_2} - 1\right)\dfrac{A_1}{A_2}F_{1-2}^2}. \quad (11.80)$$

Two Gray Surfaces Enclosed by a Single Reradiating Surface

In this instance the formulation based on $\bar{F}$ given in Eq. (11.78) provides the easiest solution. That equation, written only for the two active surfaces, gives two expressions which are formally the same as those mentioned above which led to Eq. (11.79). The solution in this instance becomes

$$\frac{1}{\mathscr{F}_{1-2}} = \frac{1}{F_{1-2}} + \frac{A_1}{A_2}\left(\frac{1}{\epsilon_2} - 1\right) + \left(\frac{1}{\epsilon_1} - 1\right). \tag{11.81}$$

The expressions for $\bar{F}$ in terms of F given in Eqs. (11.68), (11.69), and (11.70) are applicable here.

One Gray Surface Completely Enclosing a Second

If a gray surface, A_2, completely encloses a second gray surface, A_1, which does not see itself (i.e., $F_{1-1} = 0$, $F_{1-2} = 1$, $F_{2-1} = A_1/A_2$), then Eq. (11.79) reduces to

$$\frac{1}{\mathscr{F}_{1-2}} = \frac{1}{\epsilon_1} + \frac{A_1}{A_2}\left(\frac{1}{\epsilon_2} - 1\right). \tag{11.82}$$

If, as may often be the case, a convex body, A_1, is enclosed in a very *large* room (or space), then $A_1/A_2 \to 0$ and

$$\mathscr{F}_{1-2} \to \epsilon_1. \tag{11.83}$$

EXAMPLE 11.10: Find the net heat exchange between A_1 and A_2 in Example 11.9.

Solution: If, in the heat balance equations given in Example 11.9, the emissive power $W_{b_1} = 325$ is replaced by $W_{b_1} = 1$ and the other emissive powers (namely 572 and 938) are replaced by 0, the resulting set of equations may be solved to find

$$J_2^{(1)} = 0.162.$$

Then Eq. (11.75) gives

$$\mathscr{F}_{1-2} = \frac{A_2}{A_1}\frac{\epsilon_2}{1 - \epsilon_2}J_2^{(1)}$$

$$= \frac{300}{200} \times \frac{0.7}{1 - 0.7} \times 0.162$$

$$= 0.378.$$

Thus,

$$q_{12} = A_1\mathscr{F}_{1-2}(W_{b_1} - W_{b_2})$$

$$= 300 \times 0.378 \times (325 - 572)$$

$$= -28,010 \text{ Btu/hr} (-8209 \text{ W}).$$

EXAMPLE 11.11: A rectangle, 5 ft × 10 ft, is maintained at $T_1 = 500°F$ and has an emissivity of $\epsilon_1 = 0.7$. It is directly opposed, 5 ft away, by a second rectangle of identical size with $T_2 = 900°F$, $\epsilon_2 = 0.9$. Find the net heat exchange between the two surfaces if (a) they are placed in a radiation free space; and (b) they are connected by a single reradiating surface.

Solution: For $T_1 = 500°F = 960°R$, $W_{b_1} = 1456$ Btu/hr-ft²; $T_2 = 900°F = 1360°R$, $W_{b_2} = 5864$ Btu/hr-ft². The equivalent electrical networks for the two cases are shown in Fig. 11.26. Based on $F_{1-2} = 0.285$, as determined from Fig. 11.15, and $F_{1-r1} = 1 - F_{1-2}$, the appropriate resistances are shown. Although the required information may be found by use of the network and Eqs. (11.73) and (11.74), the special relations just given will be used here.

(a) In the case that the surfaces are exposed to a radiation free environment, Eq. (11.80) gives

$$\mathscr{F}_{1-2} = \frac{0.285}{\dfrac{1}{0.7 \times 0.9} - \left(\dfrac{1}{0.7} - 1\right)\left(\dfrac{1}{0.9} - 1\right)(0.285)^2} = 0.180.$$

Thus,

$$q_{12} = A_1 \mathscr{F}_{1-2}(W_{b_1} - W_{b_2}) = -39{,}700 \text{ Btu/hr} (-11{,}640 \text{ W}).$$

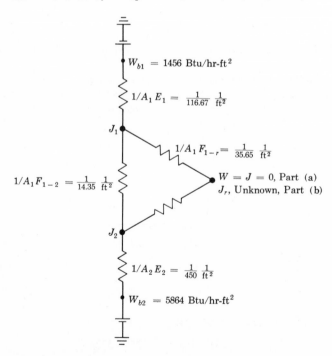

Fig. 11.26. The network representation for Example 11.11.

(b) If the surfaces are connected by a reradiating surface, Eq. (11.81) applies. First $\bar{F}_{1-2}$ must be found, and since just two, convex, active surfaces are involved, Eq. (11.70) applies for $\bar{F}_{1-2}$:

$$\bar{F}_{1-2} = \frac{1 - (0.285)^2}{1 + (1 - 2 \times 0.285)} = 0.643,$$

so

$$\mathscr{F}_{1-2} = \frac{1}{\dfrac{1}{0.643} + \left(\dfrac{1}{0.9} - 1\right) + \left(\dfrac{1}{0.7} - 1\right)} = 0.477.$$

Thus,

$$q_{12} = -105{,}100 \text{ Btu/hr} \, (-30{,}800 \text{ W}).$$

11.9   Radiation in the Presence of Absorbing and Emitting Gases

For other than the very broad definitions given in Sec. 11.2, all discussions in this chapter have been devoted to the characteristics of solid surfaces (emitting, absorbing and reflecting characteristics) and the exchange of radiation between solid surfaces separated by nonabsorbing media. The solid surfaces considered were taken as opaque to thermal radiation ($\tau = 0$) while the media between surfaces were taken as transparent and nonemitting ($\tau = 1, \epsilon = \alpha = 0$).

Elementary gases with symmetrical molecules are, indeed, transparent to thermal radiation. Many gases with more complex molecules, however, do emit and absorb thermal radiation—at least in certain wavelength bands. Water vapor, carbon dioxide, ammonia, and most hydrocarbons are examples of the latter.

The Absorptivity and Emissivity of Gases

Figure 11.27 shows, as an example, the absorption characteristics of carbon dioxide. Similar data are available for other gases of engineering significance (Ref. 12). Examination of Fig. 11.27 reveals that the absorptivity depends upon the thickness of the gas layer as well as upon the wavelength of the incident radiation. The thermodynamic state of the gas is also a determining factor for the absorptivity. The dependence of the gas absorption upon wavelength and thickness is described by the equation

$$dI_\lambda = -I_\lambda a_\lambda \, ds,$$

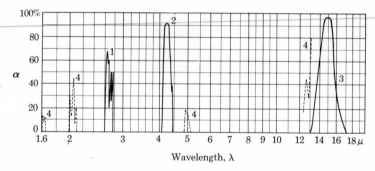

Fig. 11.27. Absorptivity of carbon dioxide: (1) 5 cm thick, (2) 3 cm, (3) 6.3 cm, and (4) 100 cm. (From *Heat and Mass Transfer* by E. R. G. Eckert and R. M. Drake. Copyright 1959. McGraw-Hill Book Company, New York. Used by permission.)

where I_λ represents the monochromatic intensity of the incident beam, s the path length of the beam, and a_λ the *absorption coefficient*. The absorption coefficient is dependent upon wavelength and the thermodynamic state of the gas. To a first approximation, a_λ varies linearly with pressure, at constant temperature. Thus, in a mixture of an absorbing and nonabsorbing gas the absorption coefficient should be proportional to the partial pressure of the absorbing gas.

An integration of the above equation yields

$$I_\lambda = I_{\lambda 0}\, e^{-a_\lambda s},$$

where $I_{\lambda 0}$ represents the intensity of the radiant beam as it enters the gas at $s = 0$. Since the reflection of thermal radiation at a gas-to-gas interface is generally negligible, the above result yields the following for the monochromatic transmissivity, emissivity, and absorptivity of a gas:

$$\tau_\lambda = e^{-a_\lambda s},$$
$$\alpha_\lambda = 1 - e^{-a_\lambda s}. \tag{11.84}$$

If Kirchhoff's law holds, then a monochromatic emissivity of a gas is

$$\epsilon_\lambda = 1 - e^{-a_\lambda s} = \frac{I_\lambda}{I_{b\lambda}}. \tag{11.85}$$

As indicated in Eq. (11.85) the emissivity of a gas has to be interpreted as the intensity of radiation arriving at a point, divided by the equivalent black body intensity. As such, the emissivity given above is associated with the radiation arriving at a point from a given direction and through a given thickness of gas. In order to account for radiant exchange between a gas mass and an element of its boundary surface, consideration must be made for *all* the radiation arriving at the element from *all* directions. The geometry of such a situation is depicted in Fig. 11.28.

According to Eq. (11.34), the hemispherical emissivity of the entire gas mass with respect to dA is then

$$\bar{\epsilon}_\lambda = \frac{1}{\pi} \int_0^{2\pi} \int_0^{\pi/2} (1 - e^{-a_\lambda s}) \sin\theta \cos\theta \, d\theta \, d\varphi.$$

The integration in the above equation can be performed only when s is related to θ and φ—meaning that the geometry of the gas mass must be known. Then a second integration must be performed, over the finite boundary surface.

In general, calculations like those just described are quite complex. However, in the case of a hemispherical mass of gas, the emissivity for radiation to the center of the base may be found easily. Hottel and Sarofim (Ref. 11) have shown that other shapes of practical interest may be reduced to equivalent hemispherical masses. The radii of these equivalent hemispheres are referred to as the *mean beam length*, customarily denoted by L_e. Table 11.1 gives recommended mean beam lengths for some shapes of interest. Since for the hemispherical shape, the mean beam length is constant, the associated value of $\bar{\epsilon}_\lambda$ may be integrated over all wavelengths, to obtain finally a total gas emissivity. Thus, the mean beam length becomes a useful engineering concept, giving the following simpler relation for emissive power of a given gas shape:

$$W = \bar{\epsilon}(L_e)\sigma T^4. \tag{11.86}$$

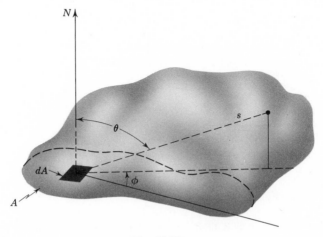

Fig. 11.28

Table 11.1   Mean Beam Lengths for Various Gas Shapes

Shape	L_e
Sphere (radiation to surface)	0.65 × diameter
Circular cylinder, infinite length (radiation to curved surface)	0.95 × diameter
Circular cylinder, length = diameter (radiation to base center)	0.77 × diameter
Infinite planes (radiation to surfaces)	1.80 × distance between planes
Cube (radiation to surfaces)	0.66 × edge length
Arbitrary shape (radiation to surface)	≈ 3.6 × (volume/surface)

In Eq. (11.86), $\bar{\epsilon}$ represents the total gas emissivity described above. Quite apparently, $\bar{\epsilon}$ is a function of the gas composition, its thermodynamic state, *and* the mean beam length, L_e.

Values of $\bar{\epsilon}(L_e)$ may be found from experimental data such as those shown in Figs. 11.29 and 11.30 for carbon dioxide and water vapor mixed with air. As noted in the earlier discussion, the gas emissivity is dependent upon density, and in mixtures of emitting and nonemitting gases, $\bar{\epsilon}$ should depend upon the partial pressure of the absorbing component—as Figs. 11.29 and 11.30 show.

EXAMPLE 11.12:   Find the emissivity to be used for calculating the radiant exchange between a spherical mass of water vapor–air mixture and its surface if the mass has a diameter of 10 ft, a temperature of 1500°F, and a total pressure of 1 atm. The partial pressure of the water vapor is 0.2 atm.

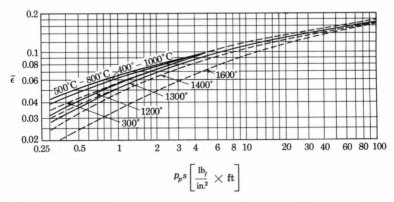

$$p_p s \left[\frac{lb_f}{in.^2} \times ft \right]$$

Fig. 11.29. Emissivity of carbon dioxide in mixture with air or nitrogen. The total pressure is 1 atm, and p_p represents the partial pressure of the carbon dioxide. (From *Heat and Mass Transfer* by E. R. G. Eckert and R. M. Drake. Copyright 1959. McGraw-Hill Book Company, New York. Used by permission.)

$$p_p s \left[\frac{\text{lb}_f}{\text{in.}^2} \right] \times \text{ft}$$

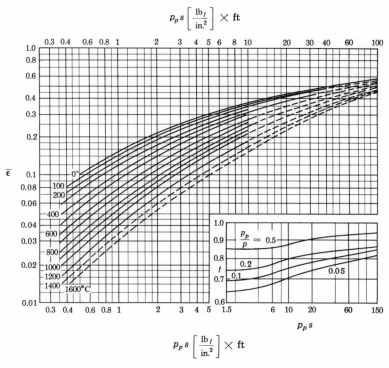

$$p_p s \left[\frac{\text{lb}_f}{\text{in.}^2} \right] \times \text{ft}$$

Fig. 11.30. Emissivity of mixtures of water vapor and air or nitrogen for a total pressure of 1 atm, p_p representing the partial pressure of the water vapor. When p_p differs from 1, the values from the large chart must be multiplied by f from the small chart. (From *Heat and Mass Transfer* by E. R. G. Eckert and R. M. Drake. Copyright 1959. McGraw-Hill Book Company, New York. Used by permission.)

Solution: For a diameter of 10 ft, Table 11.1 gives the mean beam length to be $L_e = 0.65 \times 10 = 6.5$ ft. Thus, $p_p L_e = 0.2 \times 6.5 = 1.30$. At $1800°F = 815°C$, for $p_p L_e = 1.30$, $p_p/p = 0.2$, Fig. 11.30 gives

$$\bar{\epsilon} = 0.74 \times 0.06 = 0.044.$$

Radiation Exchange in Gas-Filled Enclosures

In the event that the enclosures in Secs. 11.7 and 11.8 are filled with an emitting and absorbing gas, the procedures given in those sections must be modified. The radiating gas must be treated as an individual "surface" itself. Since Eq. (11.86) gives the emissive power of the gas relative to surface i as $\bar{\epsilon}_{g-i} W_{b_g}$, the gas may be represented by a node maintained at a potential W_{b_g} connected to each active radiating surface through a resistor of $1/A_i \bar{\epsilon}_{g-i}$.

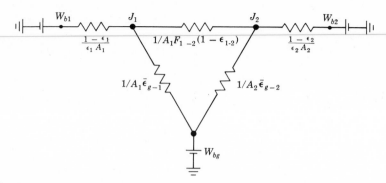

Fig. 11.31. The network representation of two gray surfaces separated by an absorbing, emitting gas.

The direct radiant exchange between active surfaces, $A_i F_{i-j}(J_i - J_j)$, when no gas is present, must be decreased by the amount absorbed by the intervening gas. Thus, the resistors connecting active nodes have to be of the form $1/A_i F_{i-j}(1 - \bar{\epsilon}_{i-j})$, in which $\bar{\epsilon}_{i-j}$ is the gas absorptivity evaluated for all rays traveling between A_i and A_j. This amounts to the evaluation of the equation for $\bar{\epsilon}$ over the geometrical angles admitted by F_{i-j}. This calculation can become quite complex. References 10 and 12 may be consulted in this respect. Figure 11.31 illustrates the equivalent network for a two-surface enclosure filled with an emitting and absorbing gas.

References

1. ZEMANSKY, M. W., *Heat and Thermodynamics*, 4th ed., New York, McGraw-Hill, 1957.
2. SIEGEL, R., and J. R. HOWELL, *Thermal Radiation Heat Transfer*, New York, McGraw-Hill, 1972.
3. JAKOB, M., *Heat Transfer*, Vol. 1 (1949) and Vol. 2 (1957), New York, Wiley.
4. DUNKLE, R. V., "Thermal Radiation Tables and Applications," *Trans. ASME*, Vol. 65, 1954, p. 549.
5. KRIETH, FRANK, *Radiation Heat Transfer*, Scranton, Pa., International Textbook, 1962.
6. EDWARDS, D. K., D. E. NELSON, R. D. RODDICK, and J. T. GIER, "Basic Studies on the Use and Control of Solar Energy," *Univ. Calif. Dept. Eng. Rept. 60–93*, Los Angeles, 1960.
7. SPARROW, E. M., "A New and Simpler Formulation for Radiative Angle Factors," *J. Heat Transfer*, Vol. C85, 1963, p. 81.
8. TOUPS, K. A., "Confac II, A General Computer Program for the Determination of Radiant Interchange Configuration and Form Factors," *Tech. Documentary Rept. FDL-TDR-64-43*, Air Force Flight Dynamics Laboratory, Wright-Patterson Air Force Base, Ohio, 1964.

9. SPARROW, E. M., "Radiation Heat Transfer Between Surfaces," *Advances in Heat Transfer*, Vol. 2, New York, Academic, 1965.

10. OPPENHEIM, A. K., "Radiation Analysis by the Network Method," *Trans. ASME*, Vol. 78, 1956, p. 725.

11. HOTTEL, H. C., and A. F. SAROFIM, *Radiative Heat Transfer*, New York, McGraw-Hill, 1967.

12. MCADAMS, W. H., *Heat Transmission*, 3rd ed., New York, McGraw-Hill, 1954.

Problems

11.1. The average rate at which radiant energy is incident on the earth's surface has been measured to be about 350 Btu/hr-ft^2 at noon on a clear day. Approximately 90 Btu/hr-ft^2 is absorbed in the earth's atmosphere. Estimate the temperature of the surface of the sun if its radius is known to be 430,000 miles. Take the radius of the earth's orbit to be 93×10^6 miles.

11.2. What is the intensity of radiation, I, of the sun? Use the data of Prob. 11.1 and assume that Lambert's law holds.

11.3. Starting with Eqs. (11.19), (11.22), and (11.23), verify Eq. (11.24).

11.4. A black body is maintained at 400°F. Find (a) the wavelength at which the maximum monochromatic emissive power occurs; (b) the emissive power at that wavelength; (c) the total emissive power; and (d) the fraction of the total emission which occurs between the wavelengths of 1.0 and 4.0 micron.

11.5. Repeat Prob. 11.4 for a black body temperature of 2000°F.

11.6. According to the tables of Appendix A, an aluminum-alloy plate coated with silicon could be expected to have a total emissivity of $\epsilon = 0.12$ and a "solar absorptivity" of $\alpha_s = 0.52$. Find the equilibrium temperature which the plate would achieve and the rate of heat absorption (or emission) if it is isolated and subjected to an external normal irradiation of 400 Btu/hr-ft^2 that has a spectral distribution which is (a) that for black body radiation at the same temperature as the surface or (b) that of the sun's emission.

11.7. Repeat Prob. 11.6 for the case in which the aluminum is coated with white epoxy paint.

11.8. Using the data of Fig. 11.9, find the *total hemispherical emissivity* at 450°R and the solar absorptivity (sun at 10,400°R) of white epoxy paint on aluminum and compare the results with the data of Table A.13.

11.9. Compute the radiant exchange between two parallel, infinite black planes at 100°F and 800°F, respectively.

11.10. The air space within the wall of a house is only 3.75 in. thick, so the boundary surfaces may be treated as infinite planes. If the inner surface of the outside brick ($\epsilon = 0.93$) is at 100°F and the opposite surface of the air space is building paper ($\epsilon = 0.91$) at 80°F, what is the radiant heat flux through the wall?

11.11. If, in Prob. 11.10, a layer of aluminum foil ($\epsilon = 0.09$) is placed on top of the building paper, to what value is the radiant flux reduced if the aluminum assumes the temperatures of the paper?

11.12. If two aluminum sheets (taken as infinite planes with $\epsilon = 0.09$) face each other at 500°F and 200°F, respectively, what is the radiant flux between the planes? If a third sheet is placed between the first, what will be its equilibrium temperature and what will be the radiant flux between the planes?

11.13. Two infinite parallel planes, one at 500°F with $\epsilon = 0.5$ and the other at 100°F with $\epsilon = 0.8$, are to be shielded by placing a third plane with $\epsilon = 0.9$ in between. What is the radiant heat flux before and after the insertion of the shield? What is the equilibrium temperature of the shield?

11.14 through 11.21. Find the shape factors, F_{1-2}, for the configurations shown in the accompanying figures.

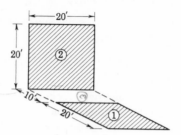

Problem 11.14

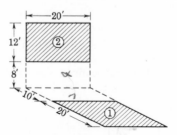

Problem 11.15

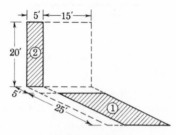

Problem 11.16

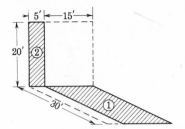

Problem 11.17

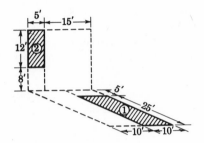

Problem 11.18

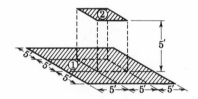

Problem 11.19

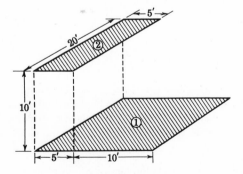

Problem 11.20

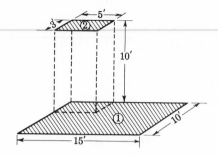

Problem 11.21

11.22. Write integral expressions which relate the *angular–hemispherical* reflectivity ρ_{ah}, to the biangular reflectivity, ρ_{ba}, and the *hemispherical* reflectivity, ρ_h, to the *hemispherical–angular* reflectivity, ρ_{ha}.

11.23. A circular disk of radius r_1 is placed parallel and directly opposite another disk of radius r_2. The distance between the disks is L. Show that the shape factor, F_{1-2}, is

$$F_{1-2} = \frac{1}{2}\left[x - \sqrt{x^2 - 4\left(\frac{R_2}{R_1}\right)^2}\right],$$

where

$$x = 1 + (1 + R_2^2)\frac{1}{R_1^2},$$

$$R_1 = \frac{r_1}{L}, \qquad R_2 = \frac{r_2}{L}.$$

11.24. Two perfect black rectangles, 8 ft × 10 ft, are parallel and directly opposed—spaced 6 ft apart. If their temperatures are 500°F and 300°F, respectively, what is the rate of radiant heat exchange between them? What is the rate at which the 500°F rectangle is losing energy?

11.25. Two perfect black squares, 5 ft × 5 ft, are spaced 10 ft apart, parallel and directly opposed. What is the rate of heat exchange between the two if their temperatures are 100°F and 300°F, respectively?

11.26. A perfect black rectangle, 8 ft × 10 ft, is located normal to a second black rectangle, 6 ft × 10 ft. The two rectangles have their 10 ft sides in common. Find the rate of heat exchange between the two if the 8 ft × 10 ft rectangle is at 500°F and the other at 1000°F. What is the rate at which the 1000°F rectangle is losing energy?

11.27. What is the rate at which the rectangles shown in Prob. 11.15 exchange radiant energy if A_1 is held at 500°F and A_2 is held at 100°F?

11.28. A cubical room, 10 ft on a side, is composed of black surfaces. One wall is maintained at 70°F and the opposite wall at 90°F. All the other walls are at 80°F. Find the heat lost by the 90°F wall.

11.29. The accompanying figure depicts an artist's studio with a skylight and door in one wall. The floor is to be used as a radiant heating panel and acts as a black plane at 150°F. The skylight and door act as black planes at 40°F, whereas the rest of the walls and the ceiling act as black planes at 90°F. Find the net radiant energy given up by the floor.

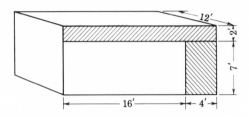

Problem 11.29

11.30. If, for the geometry given in Prob. 11.29, the floor acts as a black surface at 200°F, the door as a black surface at 100°F, and the skylight as a black plane at 40°F, find the net energy given up by the floor if all the other walls act as a single reradiating surface.

11.31. Assume that an enclosure consists of a parallelepiped 5 ft × 10 ft × 20 ft. One of the 5 ft × 10 ft surfaces acts as a black plane at 500°F while the other acts as a black plane at 1000°F. The other surfaces act as *individual* reradiating surfaces; however, symmetry considerations permit the treatment of the two 5 ft × 20 ft surfaces as a single reradiating surface and the two 10 ft × 20 ft surfaces as another. Find the net *exchange* of energy between the two active surfaces.

11.32. Repeat Prob. 11.31 if all the nonactive surfaces act as a single reradiating surface.

11.33. If two opposite surfaces of a rectangular parallelepiped are *black* active surfaces maintained at different temperatures, what error is incurred in calculating the *net* heat exchange between them if the remaining four walls are taken to be a single reradiating zone as compared to the result that would be obtained if the reradiators were taken to be four individual zones?

11.34. If the planes given in Prob. 11.24 are enclosed by a single reradiating surface, what is the net heat exchanged between them?

11.35. If the planes in Prob. 11.27 are enclosed by a single reradiating surface, what is the net heat given up by the plane at 500°F?

11.36. An enclosure consists of a circular cylinder (1 ft in diameter, 1 ft long) closed at the ends by circular disks. The circular ends act as reradiating surfaces and the cylindrical surface as a black surface at 200°F. What is the equilibrium temperature of the ends? Can an intuitive answer be given? Verify this intuitive answer by examination of the appropriate heat balance equations.

11.37. Using the results of Prob. 11.23 or Fig. 11.17, find the net heat exchange between two disks, both 3 ft in diameter and spaced 6 ft apart, if they are maintained at 300°F and 600°F, respectively, and if
(a) they are black, unenclosed, surfaces placed in radiation-free surroundings at 0°R;
(b) they are black surfaces enclosed by a single reradiating surface;
(c) they are gray surfaces (the 300°F surface with $\epsilon = 0.6$, the other with $\epsilon = 0.8$) placed in surroundings as in (a);
(d) they are gray, as in (c), enclosed by a single reradiating surface.

11.38. Verify Eqs. (11.69) and (11.70).

11.39. Repeat Prob. 11.31 for the case in which the 500°F plane is gray with $\epsilon = 0.7$ and the 1000°F plane is also gray with $\epsilon = 0.5$. All other data remain the same.

11.40. The floor of a furnace acts as a plane at 1500°F, $\epsilon = 0.6$, and the ceiling acts as a gray plane at 800°F, $\epsilon = 0.8$. The furnace is 10 ft wide, 12 ft long, and 15 ft high. If all other surfaces behave as a single reradiating wall, what is the rate of radiant heat exchange between the floor and ceiling, and what is the equilibrium temperature of the walls?

11.41. Find the *net* exchange between two 3-ft × 3-ft rectangles, directly opposed, spaced 6 ft apart if they are maintained at 400°F and 700°F and if they are
(a) black, unenclosed, in a radiation-free environment;
(b) black, enclosed by a single reradiating surface;
(c) gray (400°F surface with $\epsilon = 0.6$; 700°F surface with $\epsilon = 0.8$) and placed in a radiation-free environment;
(d) gray, as in (c), and enclosed by a single reradiating surface.

11.42. Two parallel, directly opposed, square planes (4 ft × 4 ft) are spaced 4 ft apart. One plane, at 500°F, has an emissivity of 0.5 while the other is at 1000°F with $\epsilon = 0.8$. What is the net exchange of energy between the planes if
(a) they are in a radiation-free environment?
(b) they are enclosed by a single reradiating wall?

11.43. Two 4-ft × 4-ft directly opposed squares are spaced 4 ft apart. One, with $\epsilon = 0.5$, is maintained at 500°F, while the other, with $\epsilon = 0.8$, is maintained at 100°F. A third, thermally isolated, 4-ft-square plate with $\epsilon = 0.9$ on both sides is placed equidistant between the two planes. It is allowed to come to equilibrium with the two original plates.
(a) What is the net heat exchange between the plates before the third is inserted?
(b) What is the net heat exchange after the third plane is inserted?
(c) What is the equilibrium temperature of the third plate?
Compare these results with the answer of Prob. 11.13.

11.44. If in Prob. 11.26 the 500°F plane is gray with $\epsilon = 0.8$ and the other has $\epsilon = 0.6$, find the net radiant exchange between them if (a) they are placed in a radiation-free environment or (b) they are enclosed by a single reradiating surface.

11.45. A cylindrical cavity is 1 ft in diameter and 1 ft long. The ends are closed by circular disks. One of the ends acts as a gray plane ($\epsilon = 0.7$, 200°F), the other end acts as a gray surface ($\epsilon = 0.5$, 400°F), and the cylindrical wall acts as a third gray surface ($\epsilon = 0.8$, 1000°F). Find the energy given up by each surface.

11.46. Two infinitely long cylinders are placed coaxially. The outer cylinder has an inside diameter of 6 in., an emissivity of 0.4, and is maintained at 400°F. The inner cylinder has an outside diameter of 2 in., an emissivity of 0.2, and is maintained at 800°F. A third cylindrical shield (very thin so that its temperature is approximately the same on both sides) is placed concentrically midway between the first two and allowed to come to thermal equilibrium. It has an emissivity of 0.6. Find
  (a) net heat exchange, per foot of length, between the two cylinders before the shield is put in place;
  (b) net heat exchange, per foot of length, between the two cylinders after the shield is put in place;
  (c) the equilibrium temperature of the shield.
  (d) What is the optimum position of the shield (i.e., diameter) to give the maximum shielding effect?

11.47. Starting with Eq. (11.73), use Eq. (11.75) to obtain Eq. (11.79).

11.48. Starting with Eqs. (11.73) and (11.74), use Eq. (11.75) to obtain Eq. (11.80).

11.49. Starting with Eq. (11.78), use Eq. (11.75) to obtain Eq. (11.81).

11.50. A large piece of steel (1 ft × 3 ft × 6 ft) is heated to 1800°F and then placed in a large room at 100°F. If the emissivity of the steel is 0.5, what is the initial rate at which radiant energy is lost?

11.51. A peephole in a furnace wall is 2 in. in diameter. If the interior temperature of the furnace is 200°F and the furnace is quite large, estimate the radiation that is lost through the peephole if it acts as a gray plane, $\epsilon = 0.5$, at 80°F.

11.52. A 3-in. nominal, bare, wrought iron pipe has a surface temperature of 300°F. It passes through a furnace at 800°F. What is the radiant heat loss per foot of length?

11.53. A cylindrical cavity has a diameter of 6 in. and a length of 8 in. It is capped on one end by a circular disk and on the other end by a circular disk provided with a 3-in.-diameter, concentric, hole. The surroundings with which the 3-in.-diameter hole communicates are radiation free and at 0°R. Find the rate of radiant heat loss out the hole if
  (a) all interior surfaces are black at 1000°R;
  (b) the base of the cavity (the 6-in.-diameter end) is gray at 1000°R, $\epsilon = 0.5$, and other surfaces are reradiating surfaces;
  (c) all interior surfaces are gray at 1000°R, $\epsilon = 0.5$.

11.54. An annealing furnace is so arranged that the steel (representable as a gray plane, $\epsilon = 0.8$, 1500°F) does not "see" the fire (another gray plane, $\epsilon = 0.7$, 2500°F). If all the other enclosing walls act as a single reradiating surface, find the rate of heat exchange between the fire and the steel.

11.55. At the distance of the moon from the sun, the irradiation of the sun on a surface placed normal to the rays is known to be 442 Btu/hr-ft². Of the sun's energy striking the moon, 93 % is absorbed and reradiated as by a black body at 700°F, the remainder is reflected diffusely with the same spectral distribution as the sun. (This fraction of the sun's irradiation that is reflected by a planetary body is usually referred to as the *albedo* of that body.) For a spacecraft surface with $\epsilon = 0.88$, $\alpha_{solar} = 0.248$, find the adiabatic equilibrium temperature it would achieve if (a) the surface directly faces the sun or if (b) the surface faces directly the fully lighted side of the moon, at an altitude of 50 miles above the moon's surface.

11.56. Calculate $\bar{\epsilon}_\lambda$ for a plane layer of an absorbing and emitting gas between two plates and verify that $L_e = 1.8 \times d$, as given in Table 11.1.

11.57. An infinite gray plane ($\epsilon = 0.8$) is maintained at 2000°F. It is placed parallel to a second gray plane ($\epsilon = 0.6$) at 1500°F, 5 ft away. The space between the planes is filled with an air–water vapor mixture at 1 atm (0.5 atm partial pressure of water vapor). Find the net rate of heat exchange between the plates.

CHAPTER 12

Heat Transfer by Combined Conduction and Convection

12.1 Introductory Remarks

The title of this chapter may be misleading in that it seems to imply that all the foregoing chapters dealt with situation in which only one mode of heat transfer was taking place. Strictly speaking, this is not true. The previous chapters on the various aspects of conduction and convection certainly emphasized one or the other of these modes of heat exchange.

However, in much of the work presented on heat conduction, particularly Chapters 3 and 4, heat convection was included as a boundary condition at the surface of a conducting solid. In these cases it was presumed that the convection film coefficient, h, was known ; and from this, and other given facts, much could be deduced about the conduction process within the body— including the determination of the temperature of the solid surface exposed to the convecting fluid. On the other hand, all of Chapters 6 through 10, which considered the hydrodynamical and thermodynamical aspects of convective heat transfer, presumed knowledge of the surface temperature for the determination of the film coefficient, h.

In real examples encountered in practice, however, conduction may be taking place in the interior of a solid body while convection is taking place from its surface. In this case, neither the surface temperature nor the film coefficient is known. For instance, imagine a hot fluid that is separated from a cold fluid by a solid wall. If in this case the temperatures of both fluids are known, the rate of heat exchange could be found if the film coefficients of convection were known at each boundary of the separating wall—along with the thermal properties of the wall. In order to find these film coefficients, it is usually necessary to know the temperature of the solid boundaries. This requirement is certainly true for free convection and laminar forced convection, although in certain cases of turbulent forced convection only the bulk fluid temperature is needed—e.g., see Eq. (8.15). However, if the surface temperatures of the solid are known, it is not necessary to find the film coefficients to obtain the heat flux through the wall since the boundary temperatures alone provide sufficient information, theoretically at least, to determine the heat flow rate. Hence, in such cases where *neither* the surface temperatures *nor* the film coefficients are known, some iterative procedure must be employed to find these quantities and the associated rate of heat flow.

The situation described above, that of a solid wall separating convecting fluids of different temperatures, is encountered frequently in practical applications. In particular, an analysis of the performance of a heat exchanger (a device for transferring heat from one fluid to another without mixing) will involve such a problem and will be treated in some detail in this chapter.

The case of extended surfaces is even more complex since the analyses of Chapter 3 showed that the surface temperature of a fin is not constant all over its exposed boundary. At the same time, the analyses of Chapter 3 presumed a constant surface film coefficient. These two facts are not compatible with the facts learned in the chapters on convection, where it is shown that a variable local film coefficient is likely to occur under such conditions. In some cases it may be possible to attack the problem of an extended surface wherein the film coefficient is not specified by using an iterative procedure based on an average constant film coefficient computed from an average fin surface temperature.

12.2   The Over-all Heat Transfer Coefficient

Problems of the type discussed above which are most often encountered in practical situations are ones in which a solid wall (usually plane or cylindrical) separates two convecting fluids. The wall may be single- or multi-layered, and extended surface fins may be present on either one or both of its exposed sides. The over-all heat transfer coefficient, U, defined in Sec. 3.10 is applicable in such cases.

The over-all coefficient is defined as that factor, for given geometric and hydrodynamic conditions, which, when multiplied by the difference between the two fluid temperatures, yields the heat flux through the wall:

$$\frac{q}{A} = U \, \Delta t.$$

The over-all coefficient depends on the geometric shape of the separating wall, its thermal properties, and the convective coefficients at the two surfaces. For other than plane walls, the above definition of U shows that it is also dependent upon the particular surface area A, at which the heat flux is calculated. Thus, in many instances it is more convenient to work with the UA product, which is the same at any surface:

$$UA = \frac{q}{\Delta t}.$$

The UA product depends, then, on the selection of the locations at which the temperature difference Δt is selected.

In many applications the separating surface is equipped with fins, or extended surfaces. Such a situation is depicted schematically in Fig. 12.1. The subscripts c and h are used to denote the "cold" and "hot" (in a relative sense only) fluids on either side of the separating wall. If $\mathcal{R}_w$ denotes the total thermal resistance of the base wall (i.e., without extended surfaces) and $\mathcal{R}_c$ and $\mathcal{R}_h$ denote the cold- and hot-side surface resistances, respectively, the over-all resistance between the two fluid streams is

$$\mathcal{R} = \mathcal{R}_c + \mathcal{R}_w + \mathcal{R}_h.$$

The sought-for UA product is simply the reciprocal of the over-all resistance,

$$\frac{1}{UA} = \mathcal{R}_c + \mathcal{R}_w + \mathcal{R}_h.$$

By definition of thermal resistance,

$$\frac{1}{UA} = \frac{t_{cw} - t_c}{q} + \mathcal{R}_w + \frac{t_h - t_{hw}}{q},$$

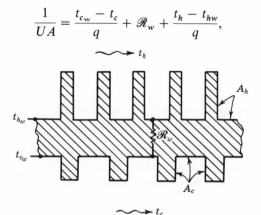

Fig. 12.1

where t_{c_w} and t_{h_w} denote the temperatures of the base wall (i.e., primary surface at the foot of the fins) on the cold and hot sides, respectively.

Since it is customary to "credit" an entire finned surface with the temperature of the fin base by introducing the surface effectiveness of Sec. 3.18, the equation above becomes

$$\frac{1}{UA} = \frac{t_{c_w} - t_c}{A_c \eta_c h_c (t_{c_w} - t_c)} + \mathscr{R}_w + \frac{t_h - t_{h_w}}{A_h \eta_h h_h (t_h - t_{h_w})}.$$

In the above η_c and η_h denote the cold- and hot-surface effectiveness, h_c and h_h denote the cold- and hot-side heat transfer coefficients, and A_c and A_h denote the *total* exposed surface areas, including the fins. Thus,

$$\frac{1}{UA} = \frac{1}{h_c \eta_c A_c} + \mathscr{R}_w + \frac{1}{h_h \eta_h A_h}. \tag{12.1}$$

Frequently, it is desirable to express an over-all conductance based on the hot side,

$$\frac{1}{U_h} = \frac{1}{h_h \eta_h} + A_h \mathscr{R}_w + \frac{A_h / A_c}{h_c \eta_c}, \tag{12.2}$$

or, for the cold side,

$$\frac{1}{U_c} = \frac{1}{h_c \eta_c} + A_c \mathscr{R}_w + \frac{A_c / A_h}{h_h \eta_h}. \tag{12.3}$$

In the event that either one of the exposed sides of the wall has no extended surface, the corresponding surface effectiveness, η, becomes unity. In such instances Eqs. (12.2) and (12.3) reduce to Eq. (3.44) or (3.46) for the special cases of the plane and cylindrical walls when the correct relations for the wall resistance are used, as developed in Chapter 3.

12.3 Examples of Trial and Error Solutions of Combined Conduction and Convection Cases

Some examples of the problems discussed above will now be presented. The simple geometrical cases of plane walls and hollow cylinders without fins will be used, but the methods may be applied to other shapes as long as the solutions of the individual conduction and convection parts of the problem are known. Likewise, not all the possible conditions of the convecting fluid (i.e., forced laminar convection, forced turbulent convection, free convection, condensation, etc.) will be illustrated—only typical cases will be considered, the other possibilities being left as problems at the end of the chapter.

Since the solution of such problems is mainly a matter of procedure, much of the illustration of this solution will be by numerical example. In order to illustrate the procedure in the simplest instance, the first case to be considered will be that in which the conditions are such that one film coefficient will be virtually independent of the surface temperature.

One Coefficient Independent of the Surface Temperature. As a first example, consider the case of a pipe (perhaps insulated) which carries a hot fluid flowing fast enough for turbulent flow to exist. If this pipe is located in, say, air of a known temperature, there will be a loss of heat from the pipe into the air by free convection. If the fluid inside the pipe is of a known temperature and is flowing turbulently, the film coefficient at the inner pipe wall is readily found since the dimensionless correlation, Eq. (8.15), for such a case is based on the bulk fluid temperature—not the inside surface temperature. To find the convective heat loss from the pipe with the outside air temperature and the inside fluid bulk temperatures given, it is necessary to obtain the over-all coefficient, U, and to apply Eq. (12.3). This means that the free convection coefficient at the outside surface is needed, but this coefficient depends on the outside surface temperature—a quantity that is dependent on the rate of heat flow through the pipe!

Thus, it becomes necessary to resort to a trial and error procedure. Assuming that the film coefficient at the inner pipe wall is known, or easily found, the assumption of a value of the outer surface temperature, t_s, enables one to calculate the outside surface film coefficient and, hence, the over-all coefficient U. Now, since the inside fluid temperature, t_i, and outside air temperature, t_o, are known, the heat flux is

$$\frac{q}{A_o} = U_o(t_i - t_o),$$

where the subscript o on A and U indicates that the over-all coefficient is based on the outside surface area. However, the heat flux through the outside area may be found a second way. The definition of the film coefficient at the outside surface, h_o, gives

$$\frac{q}{A_o} = h_o(t_s - t_o).$$

Upon equating these two relations,

$$h_o(t_s - t_o) = U_o(t_i - t_o). \tag{12.4}$$

This expression may be used to compute t_s. Then the result may be compared with the assumed value. When the computed result does not check with the assumed value, a revised assumption is made and the calculations are repeated. This process is continued until a satisfactory agreement is obtained between the assumed and derived values. The extent to which this iterative calculation is carried depends on the accuracy desired.

Obviously, this iterative process is shortened if an assumption of the surface temperature close to the correct answer is made initially. Previous experience helps in making a good initial assumption. In cases in which the surface condition is that of free convection in air at normal atmospheric conditions, a reasonable starting assumption may be obtained from the fact that free convection film coefficients in atmospheric air are of the order of magnitude of 1 Btu/hr-ft^2-°F. These points are best illustrated by a numerical example.

EXAMPLE 12.1:   Steam (400 psia, 600°F) flows at the rate of 1000 ft/min through a 6-in., schedule 40 steam pipe ($k = 27$ Btu/hr-ft-°F) which is covered with 1.5 in. of 85% magnesia insulation ($k = 0.04$ Btu/hr-ft-°F). If the pipe is located in a horizontal position in a room in which the ambient air temperature is 70°F, find the temperature of the outer surface of the insulation, the over-all heat transfer coefficient, and the rate of heat loss per foot of pipe length.

Solution: For convenience in notation, Fig. 12.2 shows the configuration under consideration and the numbers to be used to designate important locations. Table B.3 gives the following diameters:

$$D_2 = 6.07 \text{ in.} = 0.505 \text{ ft,}$$

$$D_3 = 6.63 \text{ in.} = 0.552 \text{ ft,}$$

$$D_4 = 9.63 \text{ in.} = 0.802 \text{ ft.}$$

The given data show that $t_1 = 600°F$ and $t_5 = 70°F$. First, the film coefficient at the inside surface (i.e., h_{12}) is determined. For steam at 400 psia, 600°F, Table A.6 gives

$$\rho = 0.677 \text{ lb}_m/\text{ft}^3, \qquad k = 0.0255 \text{ Btu/hr-ft-°F,}$$

$$\mu = 0.0494 \text{ lb}_m/\text{ft-hr,} \qquad N_{PR} = 1.12.$$

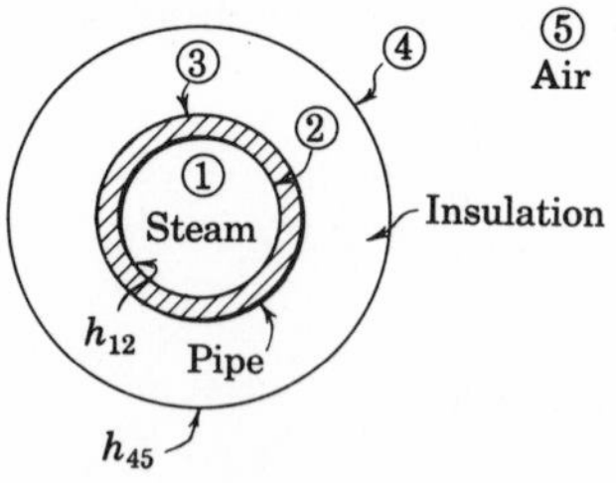

Fig. 12.2

Thus, the Reynolds number of the flow is found to be

$$N_{RE_D} = \frac{vD\rho}{\mu} = \frac{1000 \times 60 \times 0.505 \times 0.677}{0.0494} = 4.15 \times 10^5.$$

The flow is turbulent and Eq. (8.15) applies with $n = 0.3$ since the steam is being cooled:

$$N_{NU} = 0.023(N_{RE_D})^{0.8}(N_{PR})^{0.3}$$
$$= 0.023 \times (4.15)^{0.8} \times 10^4 \times (1.12)^{0.3} = 742.5.$$

This value of the Nusselt number gives the inside film coefficient to be

$$h_{12} = 742.5 \times \frac{0.0255}{0.505} = 37.5 \text{ Btu/hr-ft}^2\text{-°F.}$$

This coefficient is not dependent on the surface temperature t_2 and may thus be treated as constant in the calculations that follow.

To obtain a starting guess for t_4, the outside surface temperature, let h_{45} (the outside film coefficient) be 1 Btu/hr-ft^2-°F. Thus, the over-all heat transfer coefficient, U_4, based on the outside surface area, is by Eq. (12.3),

$$U_4 = \cfrac{1}{\cfrac{r_4}{r_2 h_{12}} + \cfrac{r_4 \ln (r_3/r_2)}{k_{23}} + \cfrac{r_4 \ln (r_4/r_3)}{k_{34}} + \cfrac{1}{h_{45}}}$$

$$= \cfrac{1}{\cfrac{0.802}{0.505 \times 37.5} + \cfrac{0.802}{2 \times 27} \ln \left(\cfrac{0.552}{0.505}\right) + \cfrac{0.802}{2 \times 0.04} \ln \left(\cfrac{0.802}{0.552}\right) + \cfrac{1}{1}}$$

$$= \frac{1}{0.0423 + 0.00132 + 3.75 + 1}$$

$$= 0.209 \text{ Btu/hr-ft}^2\text{-°F.}$$

It is immediately noticed that the controlling factors in this case are the outside film and the insulation—the thermal resistance of the pipe wall and inside film being negligible compared to the others. In many cases it is perfectly permissible to omit these last two items entirely—simplifying the computations somewhat. For the sake of completeness these items will be retained in this example.

Now, Eq. (12.4) permits the initial estimate of the surface temperature to be found:

$$h_{45}(t_4 - t_5) = U_4(t_1 - t_5),$$
$$1(t_4 - 70) = 0.209(600 - 70),$$
$$t_4 = 181\text{°F.}$$

Using this value of t_4—or more conveniently, $t_4 = 180°F$—one may compute the film coefficient, h_{45}, due to free convection. For $t_4 = 180°F$, one has

$$\Delta t = 110°F, \qquad t_m = 125°F, \qquad \beta = \frac{1}{530°R}.$$

Table A.7 gives, for air,

$$\rho = 0.0678 \text{ lb}_m/\text{ft}^3, \qquad\qquad \mu = 0.0476 \text{ lb}_m/\text{ft-hr},$$

$$k = 0.0163 \text{ Btu/hr-ft-}°F, \qquad N_{PR} = 0.702.$$

Thus, the Grashof number is

$$N_{GR_D} = \frac{D^3 \rho^2 g \beta \, \Delta t}{\mu^2}$$

$$= \frac{(0.802)^3 \times (0.0678)^2 \times 32.2 \times (3600)^2 \times 110}{(0.0476)^2 \times 530}$$

$$= 0.905 \times 10^8.$$

This means that $N_{GR_D} \times N_{PR} = 0.635 \times 10^8$. Thus Eq. (9.40) indicates that under these conditions for a horizontal cylinder:

$$N_{NU_D} = 0.525(N_{GR_D}N_{PR})^{1/4}$$

$$= 46.9.$$

So

$$h_{45} = 46.9 \times \frac{0.0163}{0.802} = 0.953 \text{ Btu/hr-ft}^2\text{-}°F.$$

Now U_4 may be computed as above to give a new value: $U_4 = 0.206$ Btu/hr-ft^2-°F. This then, as before, yields a new value of the outside surface temperature:

$$h_{45}(t_4 - t_5) = U_4(t_1 - t_5),$$

$$0.953(t_4 - 70) = 0.206(600 - 70),$$

$$t_4 = 185°F.$$

Compared to the starting value of $t_4 = 180°F$, this result may be close enough for some applications—particularly in view of the fact that U_4 did not change much because of the dominance of the insulation resistance. However, carrying the process to its logical end, another iteration should be attempted with $t_4 = 185°F$ as the initial assumption. When this is done, the following results are obtained for $t_4 = 185°F$ and $t_5 = 70°F$:

$$t_m = 127.5°F, \qquad \Delta t = 115,$$

$$N_{GR_D} = 0.937 \times 10^8,$$

$$N_{GR_D} N_{PR} = 0.658 \times 10^8,$$

$$N_{NU_D} = 47.2,$$

$$h_{45} = 0.960 \text{ Btu/hr-ft}^2\text{-}°F,$$

$$U_4 = 0.207 \text{ Btu/hr-ft}^2\text{-}°F \ (1.18 \text{ W/m}^2\text{-}°C).$$

One final check with $U_4 = 0.207$ Btu/hr-ft^2-°F and $h_{45} = 0.960$ Btu/hr-ft^2-°F gives $t_4 = 184°F$—close enough to the initial $t_4 = 185°F$. Thus, the heat loss is

$$\frac{q}{L} = \frac{A_4}{L} U_4(t_1 - t_5)$$

$$= \pi \times 0.802 \times 0.207(600 - 70)$$

$$= 276 \text{ Btu/hr-ft } (265 \text{ W/m}).$$

As was pointed out earlier, this example is just one of many instances in which such an iterative type of solution must be employed. The condition at the outer surface could be that of a condensing vapor, for instance. This situation will be the case to be encountered later in the analysis of certain types of heat exchangers. In such cases the iterative procedure noted above will constitute a single step in a much longer trial and error solution, wherein allowance is made for the fact that the bulk fluid temperatures (t_1 and t_5 in the above example—taken as constant) will change as the fluid moves through the pipe or tube.

In any of these iterative—or trial and error—solutions, the agreement between the assumed and derived surface temperatures that must be obtained before stopping the iteration depends on the circumstances of the problem at hand. Usually the given data are not precise enough to justify the carrying of the solution as far as was done in the example above. This is particularly true if one is seeking the over-all coefficient, U, which the example above showed to be relatively insensitive to small changes in the surface temperature—at least in the case of free convection.

Two Film Coefficients Dependent on the Surface Temperature. If the flow in the pipe of the above example had been found to be laminar, an examination of the relations for laminar pipe flow film coefficients [Eq. (8.11)] will show that it is also necessary to find the inside pipe temperature by the same kind of iterative procedure because the dimensionless correlations for laminar flow are based in part on this surface temperature in addition to the bulk fluid temperature. Thus, such a problem—or any other situation for which two unknown surface temperatures are needed—must be solved by a double iterative scheme. This is illustrated in the next example.

EXAMPLE 12.2:   The masonry wall of a building consists of a layer of facing brick ($k = 0.76$ Btu/hr-ft-°F) 4 in. thick followed by a 6-in.-thick layer of common brick ($k = 0.40$ Btu/hr-ft-°F). Free convection takes place at the outer surface into still air at 100°F, while free convection takes place at the inner surface to still air at 70°F. Find the over-all heat transfer coefficient of this arrangement and the temperatures of the two exposed surfaces. Assume that the portion of the wall under consideration is high enough so that the free convection boundary layer is turbulent—thus eliminating the need of a specific characteristic length.

Solution: The configuration and the associated notation are shown in Fig. 12.3. The thermal conductance of the solid wall is

$$\left(\frac{\Delta x_{23}}{k_{23}} + \frac{\Delta x_{34}}{k_{34}}\right)^{-1} = \left(\frac{4}{12 \times 0.76} + \frac{6}{12 \times 0.4}\right)^{-1} = 0.59 \text{ Btu/hr-ft}^2\text{-°F.}$$

If this value is compared to likely film conductances at the surfaces in the order of 1 Btu/hr-ft^2-°F, it seems that a reasonable initial assumption would be that the temperature drop through the solid wall would be half that through either of the convective films. That is, a starting assumption would logically be that $t_1 - t_2 = 12$°F, $t_2 - t_4 = 6$°F, $t_4 - t_5 = 12$°F.
   The procedure to follow is: Select a value for, say, t_2 and treat it as known and constant. Then under this assumption assume successive values of t_4 and follow the same iterative procedure of the proceeding example until a satisfactory check is obtained. Then check the initially assumed value of t_2 and adjust it accordingly. With this adjusted value of t_2, the iterative process is applied to finding t_4 once again. This double iterative procedure is continued until a satisfactory check of both t_2 and t_4 is obtained.

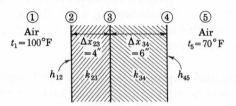

Fig. 12.3

(a) Take $t_2 = 88°F$. For this value of t_2 and the given value of $t_1 = 100°$, the outside film coefficient is found as follows:

$$\Delta t = 12°F, \qquad\qquad\qquad t_m = 94°F,$$

$$\rho = 0.0716 \; lb_m/ft^3, \qquad\qquad \mu = 0.0456 \; lb_m/ft\text{-}hr,$$

$$k = 0.0156 \; Btu/hr\text{-}ft\text{-}°F, \qquad N_{PR} = 0.707.$$

The Grashof number, in terms of an unspecified characteristic length, l, is then

$$N_{GR} = 2.21 \times 10^7 \times l^3$$

and

$$N_{GR}N_{PR} = 1.56 \times 10^7 \times l^3.$$

In the turbulent free convection range, Eq. (9.39) gives

$$N_{NU} = 0.129(N_{GR}N_{PR})^{1/3}$$

$$= 32.2 \times l.$$

Thus,

$$h_{12} = N_{NU}\frac{k}{l} = 32.2 \times 0.0156$$

$$= 0.502 \; Btu/hr\text{-}ft^2\text{-}°F.$$

Treating this value as constant, one finds that two successive iterations determine t_4 after starting with $t_4 = 82°F$.

(1)

$$t_4 = 82°F, t_5 = 70°F, \Delta t = 12°F, t_m = 76°F,$$

$$N_{GR}N_{PR} = 1.89 \times 10^7 l^3,$$

$$N_{NU} = 34.3l,$$

$$h_{45} = 0.518 \; Btu/hr\text{-}ft^2\text{-}°F,$$

$$U = \cfrac{1}{\cfrac{1}{0.502} + \cfrac{4}{12 \times 0.76} + \cfrac{6}{12 \times 0.4} + \cfrac{1}{0.518}} = 0.178 \; Btu/hr\text{-}ft^2\text{-}°F.$$

To check t_4:

$$U(t_1 - t_5) = h_{45}(t_4 - t_5),$$

$$0.178(30) = 0.518(t_4 - 70),$$

$$t_4 = 80.4°F.$$

(2) Repeating, with $t_4 = 80°F$, one obtains:

$$N_{GR}N_{PR} = 1.58 \times 10^7 l^3,$$

$$h_{45} = 0.488 \text{ Btu/hr-ft}^2\text{-°F},$$

$$U = 0.174 \text{ Btu/hr-ft}^2\text{-°F},$$

$$t_4 = 70 + \frac{0.174 \times 30}{0.488}$$

$$= 80.7°F.$$

Compared with the assumed value of $t_4 = 80°$, this last computation is sufficiently accurate. The assumed value of $t_2 = 88°F$ may now be checked:

$$U(t_1 - t_5) = h_{12}(t_1 - t_2),$$

$$0.174 \times 30 = 0.502(100 - t_2),$$

$$t_2 = 89.6°F.$$

This value of t_2 is probably close enough to the assumed value of 88°F, but the complete procedure would be to take a new t_2.

(b) Take $t_2 = 90°F$. For this value of t_2 the steps of part (a) must be repeated. Summarized, these calculations give

$$h_{12} = 0.473 \text{ Btu/hr-ft}^2\text{-°F},$$

$$t_4 = 80°F \text{ (assumed)},$$

$$h_{45} = 0.488 \text{ Btu/hr-ft}^2\text{-°F},$$

$$U = 0.171,$$

$$t_4 = 70 + \frac{0.171 \times 30}{0.488} = 80.4°F \text{ (computed)}.$$

This value is sufficiently close to the original value—only one iteration being required because of the knowledge gained in step (a).

Now upon checking t_2:

$$t_2 = 100 - \frac{0.171 \times 30}{0.473}$$

$$= 89.2°F,$$

sufficiently close to the assumed 90°F. Thus, the final results are

$$t_1 = 100°F\ (37.8°C),$$

$$t_2 = 89°F\ (31.7°C),$$

$$t_4 = 80°F\ (26.7°C),$$

$$t_5 = 70°F\ (21.1°C),$$

$$U = 0.171\ Btu/hr\text{-}ft^2\text{-}°F\ (0.971\ W/m^2\text{-}°C).$$

This example was probably carried to too great a degree of refinement—the given data usually not justifying such accuracy. However, the details are given to illustrate the method of attack.

Many other situations may give rise to a computational approach similar to that given above. For example, laminar flow in a pipe or tube with condensing or free convection occurring at the outer surface yields such a case—with two unknown, but necessary, surface temperatures.

The examples of the preceding sections were presented to establish a procedure that will be used later in this chapter. As discussed earlier, the case of a solid wall separating two convecting fluids of different temperatures, as analyzed above, forms the basic thermal circuit of typical heat exchangers. These devices are discussed in the next section.

12.4   Heat Exchangers—The Various Types and Some of Their General Characteristics

The process of exchanging heat between two different fluids is one of the most important and frequently encountered processes found in engineering practice. Boilers, condensers, water heaters, automobile radiators, air heaters or coolers, etc., are examples of processes in which heat is exchanged between a "cold" and a "hot" fluid. (The terms "cold" and "hot" are used in a relative sense.) The modern petrochemical industry is based on innumerable processes involving the use of devices to exchange heat between two fluids. Such devices are generally termed *heat exchangers*. This section will be devoted to descriptions of the more commonly encountered forms of heat exchangers.

Ordinary heat exchangers may be divided into two general classes, depending on the relative orientation of the direction of flow of the two fluid streams. If the two fluid streams cross one another in space—usually, but not necessarily, at right angles—the exchanger is termed a *cross-flow* heat exchanger. An automobile radiator or the cooling unit in an air-conditioning duct are simple examples of cross-flow heat exchangers.

Heat exchangers in which the two fluid streams move in parallel directions in space may be termed *unidirectional* heat exchangers. The usual *shell and tube* heat exchanger is the most frequently encountered form of a unidirectional heat exchanger, although the concentric pipe (or *double-tube*) exchanger is also of this type. The shell and tube, unidirectional, heat exchanger may be further subclassified and will be extensively treated in this chapter because of its engineering importance.

Unidirectional Heat Exchangers

The "classical" form of a unidirectional shell and tube heat exchanger is indicated by the simplified drawing in Fig. 12.4(a). The exchanger consists of a bundle of tubes (usually many more than shown in Fig. 12.4) secured at either end in tube sheets which are large drilled plates into which the tubes are either rolled or soldered. The entire tube bundle is placed inside a closed shell, which seals around the tube sheets to form the two domains for the hot and cold fluids. One fluid circulates through the tubes—entering and leaving via the water boxes formed between the tube sheets and shell heads. The other fluid flows around the outside of the tube—in the space between the tube sheets and enclosed by the outer shell. Baffle plates are located normal to the tube bundle to ensure thorough distribution of the shell-side fluid.

Figure 12.4(a) is to be interpreted only as a simplified illustration of the principal components of a typical shell and tube exchanger. There are many refinements in the design and construction of a heat exchanger which are not

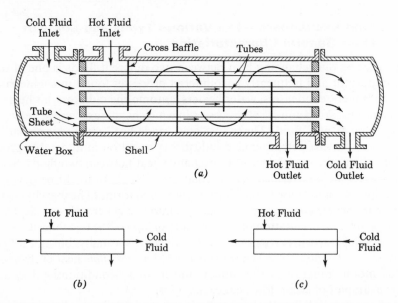

Fig. 12.4. Schematic drawing of (a) one-shell-pass, one-tube-pass heat exchanger; (b) parallel flow; and (c) counterflow.

shown. Such factors as thermal expansion stresses, tube fouling due to con-taminated fluids, ease of assembly and disassembly, etc., must be provided for in a detailed design of a heat exchanger.

The unidirectional heat exchanger shown in Fig. 12.4(a) is designated as having one shell pass and one tube pass since both the shell-side fluid and the tube-side fluid make a single traverse through the heat exchanger. Schemati-cally, a one-shell-pass, one-tube-pass heat exchanger will be indicated as shown in Figs. 12.4(b) and 12.4(c). Figure 12.4(b) indicates the flow arrange-ment for what is termed a *parallel-flow*, one-shell-pass, one-tube-pass, uni-directional heat exchanger. This name is derived from the fact that the two fluid streams travel in the same direction as they pass through the heat ex-changer.

Figure 12.4(c) indicates, schematically, a *counterflow*, one-shell-pass, one-tube-pass, unidirectional heat exchanger in which the two fluid streams flow in opposite directions through the exchanger.

Although they are formally quite similar, the parallel-flow and counter-flow heat exchangers differ greatly in the manner in which the fluid tempera-tures vary as the fluids pass through. This difference is illustrated in the diagrams in Fig. 12.5, which plot the temperature of the fluid stream against the distance along the heat exchanger.

As indicated in Fig. 12.5, the parallel-flow exchanger brings the hottest portion of the hot fluid into communication with the coldest portion of the cold fluid. This provides a large *potential difference* initially, but as the fluids move through the exchanger and transfer heat, one to the other, the temperature difference drops rapidly. The temperatures of the two streams approach one another asymptotically, and it would theoretically require an infinitely long heat exchanger to raise the cold fluid to the temperature of the hot fluid.

On the other hand, the counterflow exchanger places the hottest portion of the hot fluid in communication with the hottest portion of the cold fluid—and the coldest portion of the hot fluid in communication with the coldest portion of the cold fluid. This provides a more nearly constant temperature difference between the two fluids all along the heat exchanger and allows the possibility of heating the cold fluid to a temperature greater than the outlet temperature of the hot fluid. In general the counterflow arrangement results in a greater rate of heat exchange than in the corresponding parallel-flow case.

Figure 12.5 also illustrates the importance of another parameter—the product of the mass flow rate and the specific heat of each of the fluids. If $\dot{m}$ is the mass flow rate and c_p is the specific heat of the fluids, the product, $\dot{m}c_p$, is called the *heat capacity rate*.

The subscript c will be used to denote the cold fluid, and the subscript h will be used to denote the hot fluid. The subscript i will be used to indicate the inlet fluid conditions and o will be used to indicate the outlet fluid conditions. If one uses these subscripts and the definition of the heat capacity rate, an over-all heat balance of the heat exchanger gives

$$\dot{m}_c c_{p_c}(t_{c_o} - t_{c_i}) = \dot{m}_h c_{p_h}(t_{h_i} - t_{h_o}),$$

$$\frac{\dot{m}_c c_{p_c}}{\dot{m}_h c_{p_h}} = \frac{t_{h_i} - t_{h_o}}{t_{c_o} - t_{c_i}}.$$

The sketches of Fig. 12.5 show how the relative variation of the two fluid temperatures through the heat exchanger is influenced by whether $\dot{m}_c c_{p_c}$ is greater or less than $\dot{m}_h c_{p_h}$.

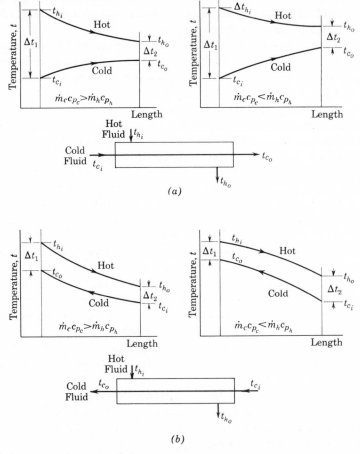

Fig. 12.5. Temperature variations in (a) parallel-flow and (b) counterflow heat exchangers.

The relative magnitudes of the heat capacity rates of the hot and cold fluids determine the maximum amount of heat that may be transferred by a given heat exchanger. In particular, in counterflow the maximum amount of heat that could be transferred by an exchanger of infinite length is limited to that value which would result when the cold fluid is heated to the inlet temperature of the hot fluid—provided that $\dot{m}_c c_{p_c} < \dot{m}_h c_{p_h}$. When $\dot{m}_c c_{p_c} > \dot{m}_h c_{p_h}$, the maximum possible heat transfer is attained when the hot fluid is cooled to the inlet temperature of the cold fluid. These last two statements may be easily verified by examining the sketches of Fig. 12.5.

In other words, the maximum possible heat that may be exchanged is $\dot{m}_c c_{p_c}(t_{h_i} - t_{c_i})$ when $\dot{m}_c c_{p_c} < \dot{m}_h c_{p_h}$, or it is $\dot{m}_h c_{p_h}(t_{h_i} - t_{c_i})$ when $\dot{m}_c c_{p_c} > \dot{m}_h c_{p_h}$. In future sections of this chapter it will be desirable to express the heat transferred by an exchanger in terms of the maximum possible heat transfer. In such cases the above distinction based on the relative magnitudes of the heat capacity rate will be important.

It is not necessary for all unidirectional heat exchangers to be of the one-shell-pass, one-tube-pass form just discussed. In fact, there are many other arrangements which are used and which are just as frequently encountered in practice. In Fig. 12.6(a) a simplified drawing of another heat exchanger is shown. This is the same in all respects as the exchanger depicted in Fig. 12.4(a)

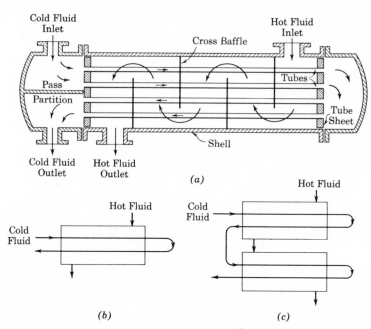

Fig. 12.6. Multiple pass heat exchangers: (a), (b) one-shell pass and two-tube passes; (c) two-shell passes and four-tube passes.

except that a pass partition has been added in the water box at one end and the tube fluid outlet has been relocated. This arrangement causes the tube fluid to make one pass through half of the tubes, reverse its direction of flow, and make a second pass through the remaining half of the tubes. This type of exchanger is referred to as a *one-shell-pass, two-tube-pass* heat exchanger and is typical of many steam condensers. This arrangement is shown schematically in Fig. 12.6(b) and it is apparent that the exchanger is neither a parallel-flow nor counterflow heat exchanger—it is often called a parallel-counterflow heat exchanger.

Many other possible flow arrangements exist and are used. By including longitudinal baffles within the heat exchanger shell, it is possible to cause the shell-side fluid to make two or more passes through the shell. Proper positions of pass partitions in the tube fluid circuit will provide as many tube passes per shell pass as may be desired. A two-shell-pass, four-tube-pass heat exchanger is depicted schematically in Fig. 12.6(c).

Cross-Flow Heat Exchangers

As defined earlier, a cross-flow heat exchanger is one in which the two fluid streams are oriented at an angle to one another. For purposes of discussion, Fig. 12.7 depicts a simplified plate type of cross-flow heat exchanger. As is often the case in the cross-flow heat exchangers used in aircraft or spacecraft applications, Fig. 12.7 shows the separating plate, or wall, to be provided

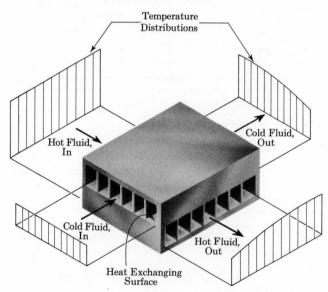

Fig. 12.7. Temperature variation of unmixed fluids in a cross-flow heat exchanger.

with fins. Although simple in principle, the analysis of this type of heat exchanger is extremely complex, since the fluid temperatures may vary both in the direction of flow and normal to the direction of flow if no fluid mixing is allowed in the normal direction. This produces inlet and outlet fluid temperature distributions as indicated in Fig. 12.7—resulting in the fact that the fluid temperature difference varies with the two coordinate fluid stream directions. The designations "mixed" and "unmixed" are used to denote whether or not mixing of either fluid is allowed in the direction transverse to the direction of the flow stream. The exchanger illustrated in Fig. 12.7 is, as noted, one in which there is no mixing in either fluid stream. Other cross-flow heat exchangers may allow for mixing in one or both of the fluid streams.

Over-all Heat Transfer Coefficients for Heat Exchangers

One of the basic parameters to be used in the design of, or the analysis of the performance of, heat exchangers is the over-all heat transfer coefficient of a given exchanger.

In the usual shell and tube heat exchanger, it is customary to base the definition of the over-all coefficient on the total exposed tube surface area —including any exterior fins. In most instances in which a shell and tube exchanger is used, the cold fluid flows inside the tubes and no extended surfaces are provided internally. Under these circumstances, the over-all coefficient for a shell and tube heat exchanger is, from Eq. (12.2),

$$U = \frac{1}{\dfrac{A/2\pi r_i L}{h_c} + \dfrac{A \ln (r_o/r_i)}{2\pi L k} + \dfrac{1}{h_h \eta_h}},$$

in which r_o and r_i are the outside and inside tube radii; k is the tube wall conductivity; h_c and h_h are the inside and outside film coefficients, respectively; and A denotes the total outside exposed surface area. The total temperature effectiveness of the outside surface, η_h, is found according to Eq. (3.86). If there are no fins on the outside tube surface,

$$U = \frac{1}{\dfrac{r_o/r_i}{h_c} + \dfrac{r_o \ln (r_o/r_i)}{k} + \dfrac{1}{h_h}}. \tag{12.5}$$

In the case of cross-flow heat exchangers, such as depicted in Fig. 12.7, the separating wall is usually a plate. Most applications of such exchangers are found in aircraft or spacecraft systems involving the exchange of heat between two gas streams. As a result, both surfaces of the plate are often provided with extended surface fins. There being no preference of one exposed surface over the other for the bases of the over-all coefficient, one may best express the proper relation to use as

$$(UA)_{h,c} = \frac{1}{\dfrac{1}{h_h \eta_h A_h} + \dfrac{\Delta x}{k A_w} + \dfrac{1}{h_c \eta_c A_c}}. \tag{12.6}$$

The notation $(UA)_{h,c}$ is used to denote the UA product of either the hot or cold side of the exchanger. Again, η_h and η_c are the surface temperature effectiveness and are found as described in Sec. 3.18. The symbol A_w is used to denote the wall area available for heat flow—i.e., the plate area with the fins, if any, removed; and Δx represents the plate thickness.

12.5 Heat Exchanger Mean Temperature Differences

As the foregoing discussions have indicated, the difference between the temperatures of the two fluids of a heat exchanger vary from point to point throughout the exchanger. Considering an incremental portion of the length of the heat exchanger, one finds that the heat transferred from the hot to the cold fluid is

$$dq = U \, dA \, \Delta t. \tag{12.7}$$

Here U denotes the over-all heat transfer coefficient (based on the total outside tube surface) of the tubes within the increment chosen. The exposed tube surface area (including fins) for heat transfer in this increment is denoted by dA, and Δt is the temperature difference between the hot and cold fluids. The temperature difference Δt will be considered constant over the increment dA.

As the above descriptions of the various heat exchangers have shown, the temperature difference between the fluids, Δt, varies throughout the exchanger. This variation in Δt is dependent upon the arrangement of the shell and tube passes of the heat exchanger. For convenience in the analysis of the performance of a given heat exchanger, it is desirable to ascertain a mean temperature difference, Δt_m, so that the total heat exchanged throughout the heat exchanger may be expressed by a relation of the form

$$q_{\text{total}} = U_m A \, \Delta t_m. \tag{12.8}$$

The symbol U_m is used to symbolize a suitable mean over-all heat transfer coefficient for the heat exchanger as a whole, whereas A denotes the total surface area. As the examples worked out in Sec. 12.3 show, the over-all coefficient depends, in part, on the fluid temperatures. Since the fluid temperatures vary as the fluids pass through the heat exchanger, U would be expected to vary along the length of the tubes. Often this variation is slight enough for U to be treated as constant when evaluated for suitable mean values of fluid temperatures.

The next few sections will illustrate methods of finding the mean temperature difference, Δt_m, for certain important heat exchanger arrangements.

Parallel-Flow and Counterflow Heat Exchangers. The Log-Mean Temperature Difference

First consider the one-tube-pass, one-shell-pass heat exchanger described in Sec. 12.4. For purposes of discussion, a parallel-flow heat exchanger will be analyzed here, but, as will be readily apparent, the analysis will apply equally well to a counterflow exchanger. The following assumptions are made:

(1) The over-all U is constant throughout the heat exchanger.
(2) If either fluid undergoes a change of phase (e.g., as in a steam condenser), then this phase change occurs throughout the heat exchanger —not in just a portion of the exchanger.
(3) The specific heats and mass flow rates of both fluids are constant.
(4) No heat is lost to the surroundings.
(5) There is no conduction of heat *along* the tubes of the heat exchanger.
(6) At any cross section in the heat exchanger, each of the fluids may be characterized by a single temperature.

The notation used in Sec. 12.4 will be used here. That is, the subscripts h and c will be used to denote the hot and cold fluids, respectively, whereas 1 and 2 will denote the two ends of the heat exchanger. The subscripts i and o will denote the inlet and outlet conditions. Thus, t_{h_i}, t_{h_o}, t_{c_i}, and t_{c_o} denote the inlet and outlet temperatures of the heat-exchanging fluids, as shown in Fig. 12.8. For the case of parallel flow, the temperature differences at the two ends of the heat exchanger are

$$\Delta t_1 = t_{h_i} - t_{c_i},$$

$$\Delta t_2 = t_{h_o} - t_{c_o}.$$

These fluid temperatures are presumed to be known. Now, if one selects an incremental length of the heat exchanger (surface area dA), the heat exchanged between the two fluids may be expressed in the following three ways:

$$dq = U\, dA\, \Delta t, \tag{12.7}$$

$$dq = \dot{m}_h c_{p_h} dt_h, \tag{12.9}$$

$$dq = \dot{m}_c c_{p_c} dt_c. \tag{12.10}$$

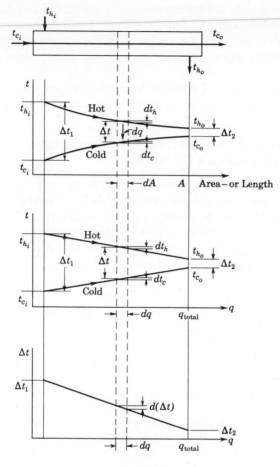

Fig. 12.8

The difference in the temperatures of the two fluids over the increment dA is denoted by Δt, and dt_h and dt_c represent the changes in the temperatures of the fluids as they pass through the increment under consideration. See Fig. 12.8.

Now, Eqs. (12.9) and (12.10) show that if one were to plot the fluid temperatures against the amount of heat transferred up to the point in question, the resulting curve would be a straight line, as illustrated in Fig. 12.8. This means, also as shown in Fig. 12.8, that the fluid temperature difference, Δt, is linear with q—varying from Δt_1 at one end of the exchanger to Δt_2 at the other end. Thus,

$$\frac{d(\Delta t)}{dq} = \frac{\Delta t_2 - \Delta t_1}{q_{\text{total}}}.$$

Equation (12.7) gives $dq = U \, dA \, \Delta t$, so

$$\frac{d(\Delta t)}{\Delta t} = \frac{U(\Delta t_2 - \Delta t_1)}{q_{total}} dA. \tag{12.11}$$

Integrating from one end (i.e., $A = 0$, $\Delta t = \Delta t_1$) of the heat exchanger to the other end (i.e., $A = A$, $\Delta t = \Delta t_2$), one obtains

$$\ln \Delta t \Big]_{\Delta t_1}^{\Delta t_2} = U \frac{\Delta t_2 - \Delta t_1}{q_{total}} A \Big]_o^A,$$

or

$$q_{total} = U A \frac{\Delta t_1 - \Delta t_2}{\ln (\Delta t_1 / \Delta t_2)}.$$

This latter expression is of the form sought—see Eq. (12.8). That is, for a parallel-flow or counterflow, heat exchanger the total heat transferred is expressed in terms of a constant over-all coefficient U, the total tube surface area A, and a mean temperature difference Δt_{lm}:

$$q = U A \, \Delta t_{lm}, \tag{12.12}$$

$$\Delta t_{lm} = \frac{\Delta t_1 - \Delta t_2}{\ln (\Delta t_1 / \Delta t_2)}. \tag{12.13}$$

This mean temperature difference is called the *log-mean temperature difference.*

EXAMPLE 12.3:   In a one-shell-pass, one-tube-pass exchanger, the hot fluid enters at 800°F and leaves at 500°F, while the cold fluid enters at 100°F and leaves at 300°F. Find the log-mean temperature difference if the heat exchanger is arranged for (a) parallel flow, and (b) counterflow.

Solution: (a) For parallel flow the temperature differences are

$$\Delta t_1 = 800 - 100 = 700°F,$$

$$\Delta t_2 = 500 - 300 = 200°F,$$

$$\Delta t_{lm} = \frac{700 - 200}{\ln (7/2)} = 399°F \, (222°C).$$

(b) For counterflow,

$$\Delta t_1 = 800 - 300 = 500°\text{F},$$

$$\Delta t_2 = 500 - 100 = 400°\text{F},$$

$$\Delta t_{lm} = \frac{500 - 400}{\ln(5/4)} = 448°\text{F} \,(249°\text{C}).$$

Note that for equal values of U and A, the heat exchanger operating on the counterflow principle will transfer 12.3 % more heat than the exchanger operating on the parallel-flow principle.

In the event that the over-all heat transfer coefficient, U, varies greatly from one end of a heat exchanger to the other, it may not be possible to represent U by an average, constant value. If one presumes that the variation of U may be represented as a linear function of the temperature difference, then

$$U = a + b\,\Delta t.$$

The symbols a and b denote constants. Equation (12.11) then gives

$$\frac{d(\Delta t)}{\Delta t\,(a + b\,\Delta t)} = \frac{dA\,(\Delta t_2 - \Delta t_1)}{q_{\text{total}}}.$$

Upon integration and evaluation at the limits, this equation gives

$$(\Delta t_2 - \Delta t_1)\frac{A}{q_{\text{total}}} = \frac{1}{a}\ln\left(\frac{\Delta t}{a + b\,\Delta t}\right)_{\Delta t_1}^{\Delta t_2}$$

$$= \frac{1}{a}\ln\left[\frac{\Delta t_2(a + b\,\Delta t_1)}{\Delta t_1(a + b\,\Delta t_2)}\right].$$

Since $U_1 = a + b\,\Delta t_1$ and $U_2 = a + b\,\Delta t_2$, the constant a may be found to be

$$a = \frac{U_1\,\Delta t_2 - U_2\,\Delta t_1}{\Delta t_2 - \Delta t_1},$$

$$q = A\left[\frac{U_2\,\Delta t_1 - U_1\,\Delta t_2}{\ln\left(\dfrac{U_2\,\Delta t_1}{U_1\,\Delta t_2}\right)}\right]. \tag{12.14}$$

This equation would then replace Eqs. (12.12) and (12.13).

Multiple Pass Shell and Tube Heat Exchangers. Cross-Flow Heat Exchangers

The above analysis and the derived log-mean temperature difference applies only to heat exchangers having just one shell pass and one tube pass. In the event that a multiple pass heat exchanger is under consideration, it is necessary to obtain a different expression for the mean temperature difference, dependent on the arrangement of the shell and tube passes.

Consider, for instance, the one-shell-pass, two-tube-pass heat exchanger discussed earlier and shown again in Fig. 12.9. The subscripts h and c are used as in the above example, and again the subscripts i and o will denote inlet and outlet fluid conditions. The subscripts a and b will be used to denote the first and second tube passes, respectively.

For an incremental length of the heat exchanger—involving a total tube surface area of dA—the following expression may be written for the heat transferred:

$$dq = \dot{m}_h c_{p_h} \, dt_h,$$

$$dq = \dot{m}_c c_{p_c}(dt_{c_a} - dt_{c_b}),$$

$$dq = U \, dA[(t_h - t_{c_a}) + (t_h - t_{c_b})].$$

These three equations may be combined to eliminate any two of the three unknown temperatures—say t_{c_a} and t_{c_b}—resulting in a differential equation in

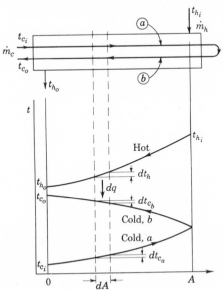

Fig. 12.9. Temperature variation in a one-shell-pass, two-tube-pass heat exchanger.

t_h. This differential equation may be solved along with the following fact, derived from an over-all heat balance of the exchanger:

$$\dot{m}_h c_{p_h}(t_{h_i} - t_{h_o}) = \dot{m}_c c_{p_c}(t_{c_o} - t_{c_i}).$$

(12.15)

The following expression for the total heat transfer then results:

$$q = U A \, \Delta t_m,$$

in which the mean temperature difference, Δt_m, is

$$\Delta t_m = \frac{\sqrt{(t_{h_i} - t_{h_o})^2 + (t_{c_o} - t_{c_i})^2}}{\ln\left[\dfrac{(t_{h_i} + t_{h_o}) - (t_{c_i} + t_{c_o}) + \sqrt{(t_{h_i} - t_{h_o})^2 + (t_{c_o} - t_{c_i})^2}}{(t_{h_i} + t_{h_o}) - (t_{c_i} + t_{c_o}) - \sqrt{(t_{h_i} - t_{h_o})^2 + (t_{c_o} - t_{c_i})^2}}\right]}.$$

(12.16)

The actual algebra leading to this expression has been omitted due to its length and complexity.

Similar expressions may be developed for each individual shell and tube pass arrangement—or, in a much more complex analysis, for cross-flow heat exchangers. This is done in some detail in Ref. 1. In each case the mean temperature difference is expressed in terms of the four terminal fluid temperatures: t_{h_i}, t_{h_o}, t_{c_i}, and t_{c_o}. The expressions for the mean temperature difference may be simplified (see Ref. 2) by introducing the following dimensionless ratios:

$$\text{Capacity ratio:} \quad R = \frac{t_{h_i} - t_{h_o}}{t_{c_o} - t_{c_i}} = \frac{\dot{m}_c c_{p_c}}{\dot{m}_h c_{p_h}}.$$

(12.17)

$$\text{Effectiveness:} \quad P = \frac{\dot{m}_c c_{p_c}(t_{c_o} - t_{c_i})}{\dot{m}_c c_{p_c}(t_{h_i} - t_{c_i})} = \frac{t_{c_o} - t_{c_i}}{t_{h_i} - t_{c_i}}.$$

(12.18)

The capacity ratio, R, is so named since it is the ratio of $\dot{m}_c c_{p_c}$ to $\dot{m}_h c_{p_h}$, above. The effectiveness is the ratio of the total heat transferred to the cold fluid to the heat that would be transferred to the cold fluid if it were raised to the inlet temperature of the hot fluid. With these definitions it is possible to relate the mean temperature difference to P, R, and the log-mean temperature difference calculated for *counterflow*. That is, one may define a function of P and R, call it $F(P, R)$, so that

$$\Delta t_m = F(P, R) \cdot \Delta t_{lm}.$$

(12.19)

Here

$$\Delta t_{lm} = \frac{(t_{h_i} - t_{c_o}) - (t_{h_o} - t_{c_i})}{\ln\left(\dfrac{t_{h_i} - t_{c_o}}{t_{h_o} - t_{c_i}}\right)} \qquad (12.20)$$

and $F(P, R)$ is determined by the shell and tube arrangement of the heat exchanger.

For instance, in the case just discussed, it is not hard to combine Eqs. (12.16) through (12.20) to obtain

$$F(P, R) = \frac{\sqrt{R^2 + 1}}{R - 1} \cdot \frac{\ln\left(\dfrac{1 - P}{1 - PR}\right)}{\ln\left[\dfrac{2 - P(R + 1 - \sqrt{R^2 + 1})}{2 - P(R + 1 + \sqrt{R^2 + 1})}\right]}. \qquad (12.21)$$

With F thus defined, it may be considered to be a *correction factor* to be applied to the usual, and readily calculated, log-mean temperature difference for counterflow according to Eq. (12.20). It can be noted in Eq. (12.21) that a solution for F does not exist for all possible combinations of R and P. The reason that there are certain areas of impossible solutions is due to the definition of the effectiveness P. The effectiveness is defined as the ratio of $\dot{m}_c c_{p_c}(t_{c_o} - t_{c_i})$ to $\dot{m}_c c_{p_c}(t_{h_i} - t_{c_i})$. In Sec. 12.4 it was noted that the *maximum* possible heat that could be transferred in a *counterflow* heat exchange with *infinite* surface area was equal to $\dot{m}_c c_{p_c}(t_{h_i} - t_{c_i})$ for $R < 1$. This quantity is the denominator in the definition of the effectiveness [see Eq. (12.18)]. The two-tube-pass heat exchanger described by Eq. (12.21) exchanges some heat in parallel flow as well as counterflow. For this reason there are certain values of the effectiveness which cannot be achieved—even for an exchanger of infinite area. The maximum effectiveness would, of course, be dependent on the capacity ratio R.

In addition to the limitations just described, it is economically poor practice to employ a heat exchanger for conditions under which $F < 0.75$. This criterion further restricts the maximum effectiveness of a heat exchanger. For applications in which the specified values of the terminal temperatures are such that the associated values of P and R give no solution for F, or if F is less than 0.75, some other shell and tube arrangement must be employed. Some other arrangements will be discussed below.

Values of F for a one-shell-pass, two-tube-pass heat exchanger have been computed from Eq. (12.21) by Bowman, Mueller, and Nagle (Ref. 3). The results are shown in Fig. 12.10. Similar results have been obtained by similar

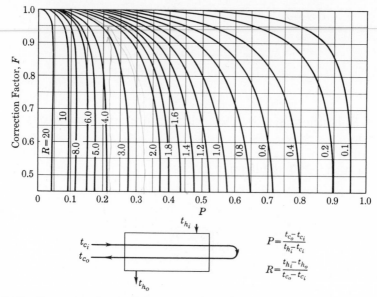

Fig. 12.10. Correction factor F for a one-shell-pass, two-tube-pass heat exchanger. (From *Standards of Tubular Exchanger Manufacturers Association*, 3rd ed., Tubular Exchanger Manufacturers Association, Inc., New York, 1952.)

analyses for a number of shell and tube arrangements and for various cross-flow heat exchangers. These results may be found in Ref. 2, and some of them are presented here. Figure 12.11 gives F for a heat exchanger with two shell passes and two tube passes per shell pass. Figure 12.12 present the results of Bowman, Mueller, and Nagle (Ref. 3) for a cross-flow heat exchanger in which there is no mixing of either fluid stream in a direction transverse to their direction of flow. The outlet fluid temperature in this case is interpreted as the "mixed mean" outlet temperature of the fluids—i.e., transverse mixing being allowed after exit from the heat exchanger.

One important item to be noted about this manner of presentation of the mean temperature of a multiple pass heat exchanger is the unfortunate shape of the curves shown in Figs. 12.10, 12.11, and 12.12. As is readily apparent, considerable error may result in the value of F read from these charts if only a small error is made in the value of the effectiveness P. Another approach to this problem will be considered in Sec. 12.7, in which this trouble is circumvented—at least partially.

EXAMPLE 12.4: A one-shell-pass, two-tube-pass exchanger is used to heat 20,000 lb_m/hr of water from $100°F$ to $250°F$ by the use of 10,000 lb_m/hr of water entering the exchanger at $600°F$. The total heat exchange surface area is $50 ft^2$. Find the average over-all heat transfer coefficient of the heat exchanger.

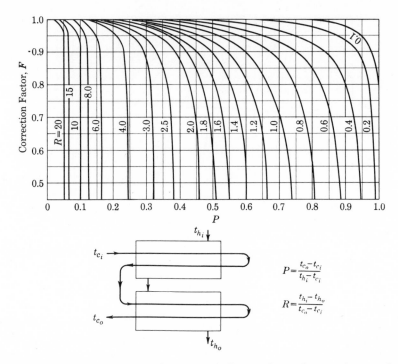

Fig. 12.11. Correction factor F for a two-shell-pass, four-tube pass heat exchanger. (From *Standards of Tubular Exchanger Manufacturers Association*, 3rd ed., Tubular Exchanger Manufacturers Association, Inc., New York, 1952.)

Solution: The given data include $t_{c_i} = 100°F$, $t_{c_o} = 250°F$, $t_{h_i} = 600°F$. Thus, an over-all heat balance gives $(600 - t_{h_o}) = (250 - 100)(20{,}000/10{,}000)$, so the hot water leaves the exchanger at $t_{h_o} = 300°F$. The log-mean temperature difference for counterflow is, by Eq. (12.20),

$$\Delta t_{\text{lm}} = \frac{(600 - 250) - (300 - 100)}{\ln\left(\dfrac{600 - 250}{300 - 100}\right)} = 268°F.$$

The capacity ratio and effectiveness defined in Eqs. (12.17) and (12.18) are

$$R = \frac{600 - 300}{250 - 100} = 2,$$

$$P = \frac{250 - 100}{600 - 100} = 0.30.$$

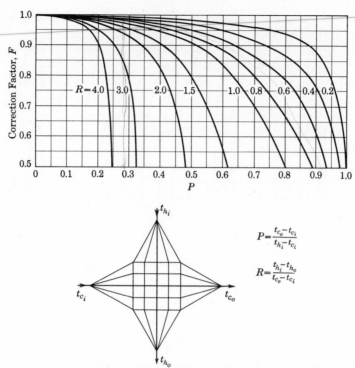

Fig. 12.12. Correction factor F for a cross-flow heat exchanger with both fluids un-mixed. (From R. A. Bowman, A. C. Mueller, and W. M. Nagle, "Mean Temperature Difference in Design," *Trans. ASME*, Vol. 62, 1940, p. 288.)

Figure 12.10 gives $F = 0.88$, so the mean temperature difference is given by

$$\Delta t_m = 268 \times 0.88 = 236°F.$$

Since

$$q = UA\,\Delta t_m = \dot{m}_h c_{p_h}(t_{h_i} - t_{h_o}),$$

$$U = \frac{10,000 \times 1 \times (600 - 300)}{236 \times 50} = 254 \text{ Btu/hr-ft}^2\text{-°F} \ (1442 \text{ W/m}^2\text{-°C}).$$

12.6 Calculation of the Performance of a Given Heat Exchanger

A problem which is frequently encountered in engineering practice is the estimation of the *performance* of a given heat exchanger. That is, given the complete physical description of a heat exchanger and the inlet fluid conditions, what will be the outlet fluid temperatures and the rate of heat exchange?

The complete physical description of the exchanger involves knowing the type of exchanger (counterflow, cross-flow, etc.), the number and size of the tubes (if a shell and tube exchanger), or the size of the separating wall and any associated fins (if a cross-flow exchanger).

In order to find these quantities, it is necessary to determine the over-all transfer coefficient for the heat exchanger. This will involve, in part, trial and error solutions of the same nature as those discussed in Sec. 12.3 since the tube surface temperatures upon which the local film coefficient depend are not known.

Most of the examples of the remainder of this chapter will consider the shell and tube type exchanger since the correlations for the film coefficients in such instances have been discussed in Chapters 7, 8, 9, and 10. The more complex case of finned cross-flow exchangers will not receive as much attention here since the determination of the surface coefficients has not been treated in detail. The basic methods of approach are identical in either case, however, and the examples will suffice to show how calculations are carried out. Reference 4 considers the cross-flow exchanger in some detail and contains numerous correlations for determination of the surface film coefficient for the most-often-encountered practical cases.

Before investigating this general problem, it will be instructive to consider first a simplified example in which it is assumed that the over-all heat transfer coefficient, U, is known. This simplification will permit the illustration of the method of determining the outlet fluid temperatures without the additional complication of the simultaneous determination of U.

If $t_{c_i}, t_{h_i}, \dot{m}_c, \dot{m}_h, c_{p_c}, c_{p_h}, U$, and A are all specified, what are the values of the outlet fluid temperatures (t_{c_o} and t_{h_o}) and what is the total amount of heat transferred? The heat transferred may be expressed in three ways:

$$q = UA\,\Delta t_m,$$

$$q = \dot{m}_c c_{p_c}(t_{c_o} - t_{c_i}),$$

$$q = \dot{m}_h c_{p_h}(t_{h_i} - t_{h_o}).$$

Formally, these three equations can be solved simultaneously for the three unknowns: q, t_{h_o}, t_{c_o}. However, the mean temperature difference—found according to the methods of Sec. 12.5—cannot be solved explicitly for the terminal temperatures. Thus, an iterative solution must be carried out to find values of q, t_{h_o}, and t_{c_o} which will simultaneously satisfy the above three equations.

Usually, the iterative procedure starts with the assumption of either t_{c_o} or t_{h_o}. The other of these two temperatures can then be found from $\dot{m}_c c_{p_c}(t_{c_o} - t_{c_i}) = \dot{m}_h c_{p_h}(t_{h_i} - t_{h_o})$. With the four terminal temperatures now known, the mean temperature difference, Δt_m, may be found for the particular heat exchanger by the methods of Sec. 12.5. Then the assumed value of t_{c_o} (or t_{h_o}) may be checked against values computed from $U A \Delta t_m = \dot{m}_c c_{p_c}(t_{c_o} - t_{c_i}) = \dot{m}_h c_{p_h}(t_{h_i} - t_{h_o})$. The computations are then repeated with revised values of t_{h_o} and t_{c_o} until a satisfactory agreement is obtained.

EXAMPLE 12.5: A one-shell-pass, two-tube-pass heat exchanger has 50 ft^2 of surface area, and its over-all heat transfer coefficient is known to be 250/Btu hr-ft^2-°F. If 10,000 lb$_m$/hr of water enters the shell at 600°F while 20,000 lb$_m$/hr of water enters the tubes at 100°F, find the outlet water temperatures. Take the specific heat of water to be 1 Btu/lb$_m$-°F.

Solution: The capacity ratio is

$$R = \frac{\dot{m}_c c_{p_c}}{\dot{m}_h c_{p_h}} = 2.0.$$

As a first assumption, take $t_{h_o} = 350°F$. Thus, t_{c_o} may be found to be

$$20,000(t_{c_o} - 100) = 10,000(600 - 350),$$

$$t_{c_o} = 225°F.$$

Thus, the effectiveness is

$$P = \frac{t_{c_o} - t_{c_i}}{t_{h_i} - t_{c_i}} = \frac{225 - 100}{600 - 100} = 0.25,$$

and the log-mean Δt for counterflow is $\Delta t_{lm} = 308$. Figure 12.10 gives $F = 0.93$, so $\Delta t_m = 308 \times 0.93 = 286$. Since $U A \Delta t_m = \dot{m}_h c_{p_h}(t_{h_i} - t_{h_o})$, the hot fluid outlet temperature may be calculated:

$$250 \times 50 \times 286 = 10,000(600 - t_{h_o})$$

$$t_{h_o} = 240°F.$$

The last value is considerably lower than the assumed $t_{h_o} = 350°F$. Thus, the calculation must be repeated for a new assumption of t_{h_o}. A final check is obtained for this example. Assume

$$t_{h_o} = 302°F.$$

Thus,

$$t_{c_o} = 249°F,$$

$$R = 2.0, P = 0.3, F = 0.88,$$

$$\Delta t_{lm} = 270°F, \Delta t_m = 224°F,$$

$$t_{h_o} = 303°F.$$

Example 12.5 illustrates the iterative solution necessary to determine the outlet fluid temperatures when it is assumed that the over-all heat transfer coefficient for the exchanger is known. Generally U is not known—it depends on the unknown outlet fluid temperatures as well as the inlet fluid temperatures. Thus, it is often necessary to determine U simultaneously while performing the iterative calculation described above. The determination of U is itself a trial and error calculation—as illustrated in Sec. 12.3.

The simultaneous determination of the outlet fluid temperatures and the over-all heat transfer coefficient will now be illustrated. Using the same notation as Sec. 12.3, one may best define the important parameters of a typical heat exchanger tube by use of Fig. 12.13. The symbol t_1 denotes the inside bulk temperature, t_4 the surrounding fluid temperature, and h_{12} the inside film coefficient, whereas h_{34} denotes the outside film coefficient. With this notation the over-all U is given by Eq. (12.5).

In the example of Sec. 12.3 in which similar situations were encountered, the unknown surface temperature t_3 was assumed. Then h_{34} and the over-all U were computed from this, and the assumed t_3 was checked by making a heat balance of the type

$$h_{34}(t_4 - t_3) = U(t_4 - t_1).$$

However, in the case of a heat-exchanger tube the fluid temperatures t_1 and t_4 will be variable along the length of the tube. (If the exchanger is a condenser, t_4 may be constant.) Thus, h_{34} and U will vary along the tubes. Usually it is adequate to treat the over-all coefficient U as constant at a value computed for the average fluid temperatures. That is, t_1, above, is taken as the average between the inlet and outlet tube fluid temperatures and t_4 is taken as the average of the inlet and outlet shell fluid temperatures. (In some instances, if the variation in one of the fluid temperatures is large, better results may be obtained by choosing the mean values t_1 and t_4, so they differ by the Δt_m for the heat exchanger under consideration.) In either case, the outlet fluid temperatures are not known—if they were, the over-all U could easily be found from the mean temperature difference as was done in Example 12.4.

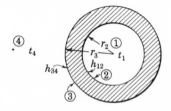

Fig. 12.13

Thus, as in the example given above, an outlet temperature has to be assumed for one of the fluids, an over-all U computed on the basis of this (by the iteration procedure of Sec. 12.3), and then an outlet fluid temperature can be calculated from this result and checked with the original assumption. This may be summarized as below. For purposes of discussion, the typical case in which the cold fluid flows inside the tubes is assumed. It is assumed that all the geometric dimensions of the exchanger are known and that $t_{c_i}, t_{h_i}, \dot{m}_c, \dot{m}_h, c_{p_h}$, and c_{p_c} are specified. The procedure is:

A. Assume either t_{c_o} or t_{h_o} and compute the other from the over-all heat balance:

$$\dot{m}_h c_{p_h}(t_{h_i} - t_{h_o}) = \dot{m}_c c_{p_c}(t_{c_o} - t_{c_i}).$$

1. Calculate the mean temperature difference, Δt_m, from the four terminal temperatures according to the methods of Sec. 12.5.

2. For $t_1 = (t_{c_o} + t_{c_i})/2$, calculate h_{12}—this assumes, as is usually true, that the inside film coefficient is independent of the inside surface temperature.

3. Assume an average outside tube surface temperature, t_3.
    (a) Find h_{34} based on t_3.
    (b) Find U based on h_{12} and h_{34}.
    (c) Check the assumed t_3 by

    $$h_{34}(t_4 - t_3) = U(t_4 - t_1),$$

    using

    $$t_4 = \frac{t_{h_o} + t_{h_i}}{2}.$$

4. Repeat step 3 until a satisfactory check of t_3 is obtained.

5. Compute, say, t_{c_o} from the over-all heat balance

    $$\dot{m}_c c_{p_c}(t_{c_o} - t_{c_i}) = U A \Delta t_m,$$

    in which A is the total heat transfer surface area.

B. Compare this computed value of t_{c_o} (or t_{h_o}) to the assumed value and repeat computations, starting with step A.

The actual carrying out of such a computational procedure is best shown by the use of a numerical example.

EXAMPLE 12.6: A steam condenser is made of $\frac{5}{8}$-in. brass ($k = 64$ Btu/hr-ft-°F) tubes, 18 gage. The exchanger has one shell pass and two tube passes with 129 tubes per pass. The length of the tubes is 6 ft. Estimate the over-all heat transfer coefficient and the pounds per hour of saturated steam at 4 in. Hg. abs. that will be condensed when the cooling water enters the tubes at 70°F and has an average velocity within the tubes of 4 ft/sec.

Solution: A $\frac{5}{8}$-in., 18-gage tube has the following inside and outside diameters:

$$D_2 = 0.527 \text{ in.} = 0.0439 \text{ ft,}$$

$$D_3 = \tfrac{5}{8} \text{ in.} = 0.052 \text{ ft.}$$

The total tube surface area for heat exchange is

$$A = (\pi \times 0.052) \times 6 \times 129 \times 2 = 254 \text{ ft}^2.$$

The total cross-sectional area for water flow, per tube pass, is

$$A_x = \frac{\pi(0.0439)^2}{4} \times 129 = 0.195 \text{ ft}^2.$$

The saturation temperature of steam at 4 in. Hg. is 125°F. Thus, $t_{h_i} = 125$°F. Cooling water temperature rises of about 25–30°F are typical for steam condensers. Thus, following the procedure outlined above with $t_{c_i} = 70$°F:

A. Assume that $t_{c_o} = 95$°F. Since the hot fluid is condensing steam, its outlet temperature equals its inlet temperature—$t_{h_o} = t_{h_i} = 125$°F.

1. Although this is a two-tube-pass exchanger, the mean temperature difference is equal to the log mean since the hot fluid is always at the same temperature. That is, since the hot fluid is a condensing vapor, $\dot{m}_h c_{p_h} \to \infty$. Thus, $R \to 0$, and $F = 1$ regardless of the value of the effectiveness. Consequently,

$$\Delta t_m = \Delta t_{lm} = \frac{(125 - 70) - (125 - 95)}{\ln\left(\dfrac{125 - 70}{125 - 95}\right)} = 41°\text{F}.$$

2. For $t_1 = (95 + 70)/2 = 82.5$°F, Table A.4 of the Appendix gives, for water, $\rho = 62.1$ lb$_m$/ft^3, $\mu = 2.02$ lb$_m$/ft-hr, $k = 0.356$ Btu/hr-ft-°F, and $N_{PR} = 5.67$. (One might also use as a mean value of the tube water temperature the result given by $t_1 = t_4 - \Delta t_m$. In this case, the resulting values of the fluid properties are almost the same as those found here by use of the arithmetic mean.) Thus, with a velocity of 4 ft/sec, $N_{RE_{D2}} = 0.194 \times 10^5$. The flow is turbulent, and Eq. (8.15) gives $N_{NU_{D2}} = 0.023(0.194)^{0.8} \times 10^4 (5.67)^{0.4} = 125$, for which $h_{12} = 1018$ Btu/hr-ft^2-°F.

3. Assume an average tube surface temperature $t_3 = 113°F$. Thus, the average film temperature is $(113 + 125)/2 = 119°F$ and $\Delta t = 12°$. Thus, the property values from Table A.4 give

$$\frac{g\rho^2\lambda D_3^3}{\mu k\,\Delta t} = 3.78 \times 10^{10}.$$

Hence, Eq. (10.21) gives

$$N_{NU_D} = 0.725(378)^{1/4} \times 10^2 = 320.$$

Thus, $h_{34} = 2290$ Btu/hr-ft²-°F. With $h_{12} = 1018$ Btu/hr-ft²-°F and $h_{34} = 2290$ Btu/hr-ft²-°F, Eq. (12.5) gives

$$U = \cfrac{1}{\cfrac{0.0520}{0.0439 \times 1018} + \cfrac{0.0520}{2 \times 64}\ln\left(\cfrac{0.0520}{0.0439}\right) + \cfrac{1}{2290}}$$

$$= 598 \text{ Btu/hr-ft}^2\text{-°F}.$$

To check the assumed t_3:

$$U(t_4 - t_1) = h_{34}(t_4 - t_3),$$
$$598(125 - 82.5) = 2290(125 - t_3),$$
$$t_3 = 114°F.$$

4. This computed value of t_3 is sufficiently close to the assumed value of 113°F.

5. Now, to check the assume outlet water temperature of 95°F, the mass flow rate of the cooling water must be found. For an average water density of 62.1 lb$_m$/ft³, the mass flow rate is

$$\dot{m}_c = 0.195 \times 4 \times 62.1 \times 3600$$
$$= 175,000 \text{ lb}_m/\text{hr}.$$

Thus,

$$q = U A\,\Delta t_m = \dot{m}_c c_{p_c}(t_{c_o} - t_{c_i}),$$
$$598 \times 254 \times 41 = 175,000 \times 1 \times (t_{c_o} - 70).$$

This equation gives $t_{c_o} = 105°F$.

B. This last computed value of t_{c_o} is 10°F greater than the assumed value of 95°F. The difference is great enough to require a second calculation at a new value of

t_{c_o}. A final checked value of $t_{c_o} = 103°F$ eventually results, although it is probably not justifiable to carry the computations too far since U is not very sensitive to changes in t_{c_o}. A summary of the final, checked, results is

$$t_{c_o} = 103°F \ (39.4°C),$$

$$\Delta t_m = 36°F \ (20°C),$$

$$h_{12} = 1020 \ \text{Btu/hr-ft}^2\text{-}°F \ (5792 \ \text{W/m}^2\text{-}°C),$$

$$h_{34} = 2390 \ \text{Btu/hr-ft}^2\text{-}°F \ (13,570 \ \text{W/m}^2\text{-}°C),$$

$$U = 603 \ \text{Btu/hr-ft}^2\text{-}°F \ (3420 \ \text{W/m}^2\text{-}°C).$$

Since the heat of vaporization of steam at 4 in. Hg. abs. is 1023 Btu/lb_m (steam tables), the steam condensed per hour is

$$\text{Condensate rate} = \frac{q}{\lambda} = \frac{U A \Delta t_m}{\lambda} = \frac{603 \times 254 \times 36}{1023}$$

$$= 5380 \ \text{lb}_m/\text{hr} \ (0.678 \ \text{kg/s}).$$

This example illustrates a typical performance calculation for a typical heat exchanger. It should not be taken as a precise prediction of how the actual heat exchanger will perform under actual operating conditions because of the simplifying assumptions that have been made. No allowance was made for probable fouling of the tube surfaces due to contaminants in the water nor for the effects of other factors, such as the cross baffles in the shell, noncondensable gases in the steam, subcooling of the condensate, etc. Such effects may be accounted for, and the reader is referred to Refs. 1, 2, and 5 for the necessary information.

In the event that the assumption of a constant over-all U cannot be made, or if the use of average hot-fluid and cold-fluid temperatures cannot be tolerated for the computation of the inside and outside tube surface film coefficients, the computation procedure becomes more complex. In such an instance, one might have to compute the film coefficients and U at each end of the heat exchanger after assuming an outlet water temperature. Then Eq. (12.14) could be used in place of Eq. (12.13) when checking the assumed temperature and for the final computation of the total heat transferred.

12.7 Heat Exchanger Effectiveness and the Number of Transfer Units

A relatively simple case, a steam condenser, was chosen in the above example, making the computation of the mean temperature difference quite simple and rather accurate. If a shell and tube arrangement had been chosen that would have required the use of the log-mean correction factors discussed in Sec. 12.5, considerable error in the predicted performance might have resulted because of the difficulty of reading values of F from the charts presented in Figs. 12.10 through 12.12. In addition, it should be noted that the above example involved two trial and error solutions—one involving the average tube surface temperature needed to find U and one involving the outlet fluid temperatures needed to find the mean temperature difference. As the first example of this section showed, even if a value for the over-all coefficient U had been assumed known and taken as constant, it still would have been necessary to perform the trial and error determination of the outlet fluid temperatures. This complication is caused by the fact that the total rate of heat exchanged is expressible in three ways:

$$q = UA\,\Delta t_m,$$

$$q = \dot{m}_c c_{p_c}(t_{c_o} - t_{c_i}),$$

$$q = \dot{m}_h c_{p_h}(t_{h_i} - t_{h_o}).$$

These expressions all involve the exit fluid temperatures, so even if U, A, t_{c_i}, and t_{h_i} are known, a trial and error solution must be obtained for t_{c_o} and t_{h_o} which will satisfy all three simultaneously. The use of a concept known as the *number of transfer units* (NTU) eliminates this difficulty.

This concept of the *number of transfer units* was first suggested by Nusselt and has been developed extensively by Kays and London (Ref. 4). The heat-exchanging capacity, per degree of mean temperature difference, of a heat exchanger is given by the product UA. The heat transferred, per degree of temperature rise, to or from either fluid, is given by the products $\dot{m}_h c_{p_h}$ or $\dot{m}_c c_{p_c}$. The product UA may be nondimensionalized by division by $\dot{m}c_p$—giving a number which denotes the heat transfer capacity of the exchanger. The number of transfer units, NTU, is defined as the ratio of UA to the smaller of $\dot{m}_h c_{p_h}$ and $\dot{m}_c c_{p_c}$:

$$\text{NTU} = \frac{UA}{(\dot{m}c_p)_{\min}}. \tag{12.22}$$

The definitions of the capacity ratio, R, and the heat exchanger effectiveness, P, introduced in Sec. 12.6, must be generalized in the following way:

$$\text{Capacity ratio:}\quad C_R = \frac{(\dot{m}c_p)_{\min}}{(\dot{m}c_p)_{\max}}. \tag{12.23}$$

$$\text{Effectiveness:}\quad \epsilon = \begin{cases} \dfrac{t_{c_o} - t_{c_i}}{t_{h_i} - t_{c_i}}, & \dot{m}_c c_{p_c} < \dot{m}_h c_{p_h}, \\[2ex] \dfrac{t_{h_i} - t_{h_o}}{t_{h_i} - t_{c_i}}, & \dot{m}_c c_{p_c} > \dot{m}_h c_{p_h}. \end{cases} \tag{12.24}$$

This latter quantity, the effectiveness, is the ratio of the actual heat transferred to the maximum heat that could be transferred in a counterflow heat exchanger of infinite area—i.e., ϵ is $\dot{m}_h c_{p_h}(t_{h_i} - t_{h_o}) = \dot{m}_c c_{p_c}(t_{c_o} - t_{c_i})$ divided by $\dot{m}_c c_{p_c}(t_{h_i} - t_{c_i})$ for $\dot{m}_c c_{p_c} < \dot{m}_h c_{p_h}$ or by $\dot{m}_h c_{p_h}(t_{h_i} - t_{c_i})$ for $\dot{m}_c c_{p_c} > \dot{m}_h c_{p_h}$. (See the discussion of Sec. 12.4.)

These three parameters (NTU, C_R, ϵ) do not involve any new quantities but simply replace the three parameters (Δt_m, R, P) used in the mean temperature method of Sec. 12.6. The utility of the NTU-ϵ-C_R approach lies in the fact that ϵ may be determined as an explicit function of C_R and NTU for a given heat exchanger. Thus, when the exchanger geometry, its over-all coefficient, and the fluid flow rates and heat capacities are given, the outlet fluid temperature may be directly obtained from ϵ for any specified set of inlet temperatures without a trial and error solution.

The actual determination of the ϵ-NTU-C_R relations will be illustrated by considering a parallel-flow heat exchanger.

The Parallel-Flow Exchanger

For convenience in discussion, let it be assumed that $\dot{m}_h c_{p_h} < \dot{m}_c c_{p_c}$, so that ϵ, C_R, and NTU become [see Eqs. (12.22), (12.23), and (12.24)]

$$\text{NTU} = \frac{UA}{\dot{m}_h c_{p_h}}, \tag{12.25}$$

$$C_R = \frac{\dot{m}_h c_{p_h}}{\dot{m}_c c_{p_c}}, \tag{12.26}$$

$$\epsilon = \frac{t_{h_i} - t_{h_o}}{t_{h_i} - t_{c_i}}. \tag{12.27}$$

Now, in an incremental length of the heat exchanger, the heat transferred, dq, may be expressed in the following ways:

$$dq = U \, dA(t_h - t_c), \tag{12.28}$$

$$dq = \dot{m}_h c_{p_h}(-dt_h) = \dot{m}_c c_{p_c} \, dt_c. \tag{12.29}$$

Equation (12.29) may be rewritten as

$$\dot{m}_h c_{p_h} \, dt_h = \dot{m}_c c_{p_c} \, dt_c,$$

$$\dot{m}_h c_{p_h}(dt_h - dt_c) = -dt_c(\dot{m}_h c_{p_h} + \dot{m}_c c_{p_c}),$$

$$= \left(\frac{-dq}{\dot{m}_c c_{p_c}}\right)(\dot{m}_h c_{p_h} + \dot{m}_c c_{p_c}).$$

Thus,

$$d(t_h - t_c) = \frac{-dq}{\dot{m}_h c_{p_h}}(C_R + 1).$$

Substitution of Eq. (12.28) into the last expression and subsequent integration gives

$$\ln\left(\frac{t_{h_o} - t_{c_o}}{t_{h_i} - t_{c_i}}\right) = -\text{NTU}(1 + C_R),$$

$$\frac{t_{h_o} - t_{c_o}}{t_{h_i} - t_{c_i}} = e^{-\text{NTU}(1 + C_R)}.$$

(12.30)

However, the effectiveness ϵ and the capacity ratio C_R also relate the terminal temperatures:

$$\epsilon = \frac{t_{h_i} - t_{h_o}}{t_{h_i} - t_{c_i}},$$

$$C_R = \frac{\dot{m}_h c_{p_h}}{\dot{m}_c c_{p_c}} = \frac{t_{c_o} - t_{c_i}}{t_{h_i} - t_{h_o}}.$$

These equations combine to give

$$\epsilon = \frac{1 - (t_{h_o} - t_{c_o})/(t_{h_i} - t_{c_i})}{1 + C_R}.$$

So, Eq. (12.30) gives

$$\epsilon = \frac{1 - e^{-\text{NTU}(1 + C_R)}}{1 + C_R}.$$

(12.31)

Exactly the same relation would have been obtained if one had assumed $\dot{m}_h c_{p_h} > \dot{m}_c c_{p_c}$ and if the corresponding changes had been made in ϵ, NTU, and C_R as required by their definitions in Eqs. (12.22), (12.23), and (12.24). The results of Eq. (12.31) for the effectiveness as a function of the number of transfer units and the capacity ratio are shown in Fig. 12.14. It must be noted that since $C_R = (\dot{m}c_p)_{min}/(\dot{m}c_p)_{max}$, only values of $0 < C_R < 1$ are necessary. The limiting case of $C_R = 0$ applies to condensers since the specific heat of a condensing vapor is infinite.

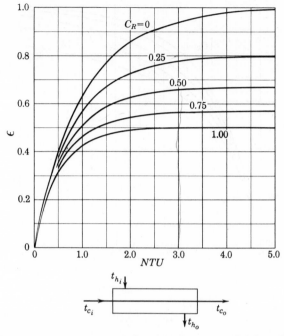

Fig. 12.14. The effectiveness–NTU relationship for a parallel-flow heat exchanger. (Adapted by permission from *Compact Heat Exchangers* by W. M. Kayes and A. L. London. Copyright 1958. McGraw-Hill Book Company, New York.)

The Counterflow Exchanger

An analogous derivation for a counterflow heat exchanger gives the following expression for the exchanger effectiveness:

$$\epsilon = \frac{1 - e^{-\text{NTU}(1 - C_R)}}{1 - C_R e^{-\text{NTU}(1 - C_R)}}. \tag{12.32}$$

This result is presented graphically in Fig. 12.15.

The Multiple Pass and Cross-Flow Heat Exchangers

Expressions similar to those given above in Eqs. (12.31) and (12.32) for parallel-flow and counterflow heat exchangers have been derived and presented graphically for a great number of other shell and tube exchangers and for cross-flow heat exchangers. This information is available in Ref. 4. The effectiveness for a one-shell-pass, two-tube-pass exchanger is shown in Fig. 12.16, and that for a cross-flow exchanger is given in Fig. 12.17.

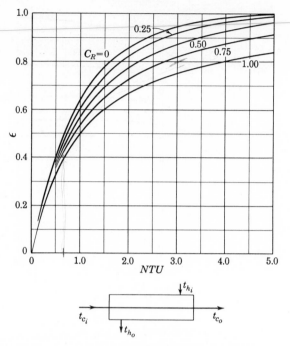

Fig. 12.15. The effectiveness–NTU relationship for a counterflow heat exchanger. (Adapted by permission from *Compact Heat Exchangers* by W. M. Kayes and A. L. London. Copyright 1958. McGraw-Hill Book Company, New York.)

EXAMPLE 12.7:   Repeat Example 12.5 by use of the ϵ-NTU relationships.

Solution: No trial and error solution is required since the NTU and the capacity ratio may be determined from the given data. Taking the c_p of water to be 1 Btu/lb$_m$-°F:

$$(\dot{m}c_p)_{\min} = \dot{m}_h c_{p_h} = 10,000 \text{ Btu/hr-°F},$$

$$(\dot{m}c_p)_{\max} = \dot{m}_c c_{p_c} = 20,000 \text{ Btu/hr-°F},$$

$$C_R = 0.5,$$

$$\text{NTU} = \frac{UA}{(\dot{m}c_p)_{\min}} = \frac{250 \times 50}{10,000} = 1.25.$$

Figure 12.16 gives the effectiveness to be $\epsilon = 0.60$. Since $(\dot{m}_h c_{p_h}) < (\dot{m}_c c_{p_c})$, Eq. (12.24) gives the effectiveness to be

$$\epsilon = \frac{t_{h_i} - t_{h_o}}{t_{h_i} - t_{c_i}}.$$

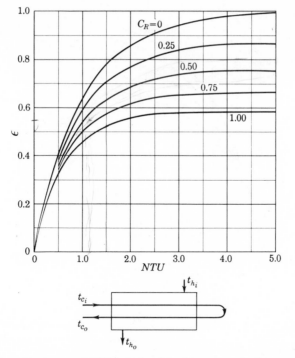

Fig. 12.16. The effectiveness–NTU relationship for a one-shell-pass, two-tube-pass heat exchanger. (Adapted by permission from *Compact Heat Exchangers* by W. M. Kayes and A. L. London. Copyright 1958. McGraw-Hill Book Company, New York.)

Thus,

$$0.60 = \frac{600 - t_{h_o}}{600 - 100}.$$

The exit temperature of the hot fluid is then

$$t_{h_o} = 300°F.$$

The outlet temperature of the cold fluid is found now from the capacity ratio

$$C_R = \frac{(\dot{m}c_p)_{min}}{(\dot{m}c_p)_{max}} = \frac{\dot{m}_h c_{p_h}}{\dot{m}_c c_{p_c}} = \frac{t_{c_o} - t_{c_i}}{t_{h_i} - t_{h_o}},$$

$$0.5 = \frac{t_{c_o} - 100}{600 - 300},$$

$$t_{c_o} = 250°F.$$

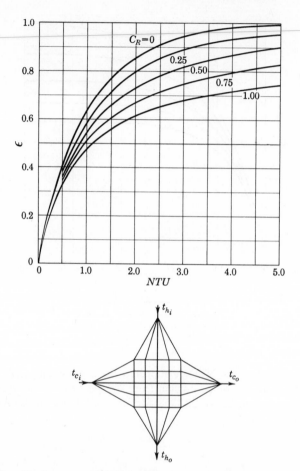

Fig. 12.17. The effectiveness–NTU relationship for a cross-flow heat exchanger with fluids unmixed. (Adapted by permission from *Compact Heat Exchangers* by W. M. Kayes and A. L. London. Copyright 1958. McGraw-Hill Book Company, New York.)

EXAMPLE 12.8: A compact cross-flow, aircraft heat exchanger is depicted in Fig. 12.18. The exterior dimensions are shown. The basic construction of the exchanger is of the "plate-fin" type. Plates, 0.012 in. thick, spaced 0.25 in. apart, from the separating walls between the two fluids. The surface of each plate is covered with longitudinal fins 0.006 in. thick. Both the plate and fins have a $k = 15$ Btu/hr-ft-°F. By alternating the direction of the fins on successive plates, a compact cross-flow pattern is obtained as shown.

The cold fluid is air ($c_p = 0.25$ Btu/lb$_m$-°F) entering at 0°F and flowing at the rate of 800 lb$_m$/hr. The hot fluid consists of combustion gases ($c_p = 0.27$ Btu/lb$_m$-°F) entering at 800°F and flowing at the rate of 200 lb$_m$/hr. Manufacturer's data yield the geometric facts that the total, unfinned, wall area for heat exchange is $A_w = 4.0$ ft^2,

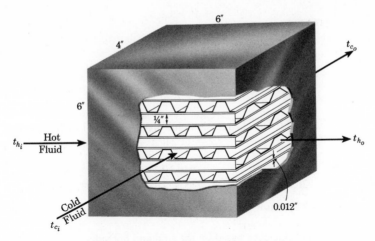

Fig. 12.18. A compact cross-flow heat exchanger.

while the total fin area exposed to the cold fluid is $A_{f_c} = 6.5\,\text{ft}^2$ and that exposed to the hot fluid is 7.2 ft^2 (there being a different spacing of the fins in the two streams). Based on either the manufacturer's predictions or on methods such as outlined in Ref. 4, the surface film coefficients on the cold and hot surfaces may be expected to be $h_c = 13.0$ and $h_h = 20.0\ \text{Btu/hr-ft}^2\text{-}°\text{F}$, respectively.

Estimate the mixed outlet temperatures of the two fluids.

Solution: In order to find the outlet temperatures according to the method of the above example, the over-all coefficient of the exchanger must be found. For the application of Eq. (12.6), the total temperature effectiveness of the hot and cold surfaces must be found first. The fins may be taken to be straight fins of uniform thickness with lengths, by symmetry, equal to one-half of the plate spacing. Thus, for the cold side, Eqs. (3.82) and (3.86), or Eq. (3.87), give

$$(mL)_c = \sqrt{\frac{2h_c}{kw}}\,L$$

$$= \sqrt{\frac{2 \times 13.0}{15.0 \times 0.006/12}} \times \frac{0.25/2}{12} = 0.615,$$

$$\kappa_c = \frac{1}{(mL)_c}\tanh{(mL)_c}$$

$$= \frac{1}{0.615}\tanh 6.15 = 0.892,$$

$$\eta_c = 1 - \frac{A_{f_c}}{A}(1 - \kappa_c)$$

$$= 1 - \frac{6.5}{4.0 + 6.5}(1 - 0.892) = 0.93.$$

Similarly, for the hot side, one finds

$$(mL)_h = 0.765,$$

$$\kappa_h = 0.842,$$

$$\eta_h = 0.90.$$

The UA product for the exchanger is found from Eq. (12.6) with $A_w = 4.0$, $A_c = 4.0 + 6.5 = 10.5$, and $A_h = 4.0 + 7.2 = 11.2 \text{ ft}^2$:

$$\frac{1}{UA} = \frac{1}{20.0 \times 0.90 \times 11.2} + \frac{0.012/12}{15 \times 4.0} + \frac{1}{13.0 \times 0.93 \times 10.5},$$

$$UA = 77.8 \text{ Btu/hr-}°\text{F}.$$

Thus, the over-all coefficients, based on the cold and hot sides, are

$$U_c = 7.41 \text{ Btu/hr-ft}^2\text{-}°\text{F},$$

$$U_h = 6.95 \text{ Btu/hr-ft}^2\text{-}°\text{F}.$$

The heat capacity rates are

$$(\dot{m}c_p)_{\min} = (\dot{m}c_p)_h = 200 \times 0.27 = 54.0 \text{ Btu/hr-}°\text{F},$$

$$(\dot{m}c_p)_{\max} = (\dot{m}c_p)_c = 800 \times 0.25 = 200.0 \text{ Btu/hr-}°\text{F},$$

so

$$C_R = 0.27,$$

$$\text{NTU} = \frac{UA}{(\dot{m}c_p)_{\min}} = \frac{77.8}{54.0} = 1.44.$$

Finally, then, Fig. 12.17 yields $\epsilon = 0.70$, and Eq. (12.24) gives

$$\epsilon = \frac{t_{h_i} - t_{h_o}}{t_{h_i} - t_{c_i}},$$

$$0.70 = \frac{800 - t_{h_o}}{800 - 0},$$

$$t_{h_o} = 240°\text{F},$$

$$t_{c_o} = 151°\text{F}.$$

The advantage of this approach to the problem given above is apparent. Generally speaking, however, the ϵ-NTU approach has a decided advantage over the F-log-mean approach only when a value of the over-all U is known—

or can be approximated fairly well. In other words, if t_{c_i}, t_{h_i}, $\dot{m}_h$, $\dot{m}_c$, c_{p_c}, c_{p_h}, and U are given, specification of the total transfer area A enables one to find quickly the outlet temperatures t_{c_o} and t_{h_o} without trial and error—or, conversely, one can find the necessary transfer area to give desired terminal temperatures.

When the over-all U is not known—and this is often the case since U is dependent on the terminal temperatures or the average fluid temperatures—then a trial and error solution is still required. Values of the terminal temperatures must be assumed to calculate U, as in the second example of Sec. 12.5, from which the terminal temperatures can be recomputed, using the ϵ-NTU approach. In a great many applications, however, the over-all U is not too sensitive to changes in the terminal temperatures so that a reasonable first assumption will give a value of U which is adequate for most engineering purposes.

Another advantage of the ϵ-NTU method is that the curves plotted for ϵ in Figs. 12.14 through 12.17 are not as sensitive to errors as are the F-factor curves of Figs. 12.10, 12.11, and 12.12.

12.8 The Design or Selection of Heat Exchangers for Specific Performance Characteristics

The discussions of the two preceding sections were limited to cases in which it was desired to ascertain the performance characteristics of a given heat exchanger of fixed geometry. That is, the total heat transfer area and the shell and tube arrangement were known, and it was desired to determine the exit temperatures of the fluids for various entrance conditions.

The problem of the selection or design of a heat exchanger of unspecified dimensions to accomplish a certain desired transfer of heat between fluids of specified terminal temperatures is more complex. The problem of design is particularly complex since there is no single answer to the problem. Several different heat exchangers may be able to accomplish the same exchange of heat equally well. The decision of which exchanger is to be used for a particular set of conditions depends on many factors other than the heat transfer aspects discussed above—such as cost, space requirements, personal opinions of the designer, etc. Also, it is usually desirable to adhere to certain standard practices—such as the use of tubes of standard diameter, lengths, etc.

If the flow rates, specific heats, and terminal temperatures of the two heat-exchanging fluids are specified, one can easily calculate the total heat to be transferred and the mean temperature difference—or the NTU. In either way, the required UA product of the heat exchanger is easily determined. The question then becomes, what heat exchanger will have the proper area A, and over-all coefficient U, to satisfy this? The value of the over-all heat transfer coefficient cannot be found until the velocity of the fluid in the tubes is known.

The selection, based on experience, of a tube size and the average tube fluid velocity will determine the number of tubes and their length (once U is calculated) since the flow rate of the tube fluid is known, or a tube length and size may be selected from which the number of tubes may be determined. In actual practice either approach may be used. Usually several possible designs are computed, and the decision of which is to be used is based on the economic factors, such as operating costs as determined by the pumping requirements due to pressure losses, initial investment, etc.

A very elementary example will be illustrated below to show the various factors to be included. This example is to be taken only as an illustration. Other, perhaps more practical, designs might be obtained for the conditions stated. It must be remembered that there is never a unique answer for any engineering design problem.

EXAMPLE 12.9: It is desired to design a feedwater heater to supply a boiler producing 20,000 lb_m of steam per hour. The raw feedwater temperature is 70°F, and the desired temperature to which the water is to be heated is 150°F. The feedwater is to be heated in a horizontal shell and tube heat exchanger by condensing steam at 20 psia. Because of space limitations the tubes should not be longer than 8 ft.

Solution: As a starting assumption, pick 1-in., 16-gage, brass tubes. In order to reduce pressure losses due to tube friction, a low water velocity of, say, 1 ft/sec might be used. A 1-in., 16-gage tube has the following geometric properties:

$$\text{O.D.} = 1 \text{ in.,}$$

$$\text{I.D.} = 0.870 \text{ in.,}$$

$$\text{Surface area/ft} = 0.262 \text{ ft}^2/\text{ft,}$$

$$\text{Cross-sectional area} = 0.00413 \text{ ft}^2.$$

At an average water temperature of $(70 + 150)/2 = 110°F$,

$$\rho = 61.84 \text{ lb}_m/\text{ft}^3, \qquad \mu = 1.49 \text{ lb}_m/\text{ft-hr,}$$

$$k = 0.368 \text{ Btu/hr-ft-°F}, \qquad N_{PR} = 4.04.$$

Since the total flow rate is 20,000 lb/hr and the water velocity is 1 ft/sec, the number of tubes required, per tube pass, is

$$\text{No. tubes} = \frac{20,000}{61.8 \times 1 \times 3600 \times 0.004130}$$

$$= 21.8.$$

Thus, 22 tubes must be used, per tube pass.

The Reynolds number of the tube side water is

$$N_{RE_D} = \frac{3600 \times 1 \times 0.870 \times 61.8}{1.49 \times 12}$$

$$= 0.108 \times 10^5.$$

The flow is turbulent, so the film coefficient at the inner surface is given by

$$N_{NU_D} = 0.023(N_{RE_D})^{0.8}(N_{PR})^{0.4}$$

$$= 0.023(0.108)^{0.8} \times 10^4(4.04)^{0.4}$$

$$= 67.7.$$

Thus,

$$h_{12} = 67.7 \times \frac{0.368}{0.870} \times 12 = 343 \text{ Btu/hr-ft}^2\text{-}°\text{F.}$$

The outside surface condensing film coefficient must be found by assuming values of the outside tube surface temperature in the manner discussed in Sec. 12.3. A satisfactory check is obtained when the tube surface temperature, t_3, is taken to be 215°F. That is, let $t_3 = 215°$F. Steam temperature $= 228°$F. Thus,

$$t_m = \frac{215 + 228}{2} = 222°\text{F.}$$

$$\Delta t = 13°\text{F}, \qquad\qquad \lambda = 960 \text{ Btu/lb}_m,$$

$$\rho = 59.6 \text{ lb}_m/\text{ft}^3, \qquad\qquad \mu = 0.65 \text{ lb}_m/\text{ft-hr},$$

$$k = 0.395 \text{ Btu/hr-ft-}°\text{F}, \qquad N_{PR} = 1.66.$$

So

$$\frac{g\rho^2 \lambda D^3}{uk\,\Delta t} = 2.48 \times 10^{11},$$

$$N_{NU_D} = 0.725(2480)^{1/4}10^2 = 512,$$

$$h_{34} = 512 \times \frac{0.395}{1/12} = 2420 \text{ Btu/hr-ft}^2\text{-}°\text{F.}$$

The over-all heat transfer coefficient may now be calculated:

$$U = \frac{1}{\dfrac{1.0}{0.870 \times 343} \quad \dfrac{1.0}{2 \times 12 \times 64}\ln\left(\dfrac{1.0}{0.870}\right) + \dfrac{1}{2420}}$$

$$= 261 \text{ Btu/hr-ft}^2\text{-}°\text{F.}$$

To check the assumed surface temperature:

$$2420(228 - t_3) = 261(228 - 110)$$

$$t_3 = 215.3°F.$$

The correspondence between the assumed and calculated values of t_3 is satisfactory. Thus, a value of $U = 261$ Btu/hr-ft²-°F may be used. The total heat required to be transferred is $20,000(150 - 70) = 1,600,000$ Btu/hr. The mean temperature difference is the log mean regardless of the number of tube passes eventually used since the hot fluid remains at a constant temperature. Thus,

$$\Delta t_m = \frac{(228 - 70) - (228 - 150)}{\ln\left(\dfrac{228 - 70}{228 - 150}\right)} = 113°F.$$

Since $q = U A \Delta t_m$, the required total surface area is

$$A = \frac{1,600,000}{261 \times 113} = 54.2 \text{ ft}^2.$$

Since 22 tubes are required per pass to accommodate the water flow, the length of tubes required in a single pass exchanger will be

$$L = \frac{54.2}{22 \times 0.262} = 9.42 \text{ ft.}$$

This is longer than the specified 8-ft maximum, so two tube passes must be used. Thus, the final design would be

> 1 shell pass
> 2 tube passes   } Design 1
> 22 tubes per pass, 1-in. O.D., 16 gage
> 5 ft long

As in the previous example, this design does not allow for other factors, such as tube fouling, shell-side fluid distribution, etc. Nonetheless, these calculations serve to illustrate method and the effect of certain parameters.

The effect of changing the water velocity can be seen if all quantities are retained the same except the velocity—which will be raised to 1.5 ft/sec. Under these conditions, the following computations result:

> No. tubes = 15 tubes/pass,
> Inside film: $h_{12} = 475$ Btu/hr-ft²-°F,
> Outside film: $h_{34} = 2160$ Btu/hr-ft²-°F,
> Over-all coefficient: $U = 339$ Btu/hr-ft²-°F,
> Temperature difference: $\Delta t_m = 113°F$,
> Surface area: $A = 41.8$ ft²,
> Length of a single pass = 10.6 ft.

This design—that of fewer tubes with a higher velocity—did give a greater over-all U, but longer tubes are required since there are fewer tubes in a pass. The final design would be

$$
\left.\begin{array}{l}
\text{1 shell pass} \\
\text{2 tube passes} \\
\text{15 tubes per pass, 1-in. O.D., 16 gage} \\
5\tfrac{1}{2}\text{ ft long}
\end{array}\right\} \text{Design 2}
$$

This design would be less expensive to build than the first since there are fewer tubes. However, the operating cost of the second design would probably be greater than the first design since the work required to pump the tube side fluid would be greater because the frictional pressure loss of the flow through the tubes will be greater for the higher tube velocity. Application of the pressure-loss formulas of Eq. (7.100) and Fig. 7.16 gives an approximate figure for comparison purposes. Application of these formulas shows that a pressure loss of 0.017 psia occurs through the tubes in the first case, whereas in the second case it is 0.036 psia. Thus, the operating cost in the second case, because of the increased pumping load, will be about twice that in the first case

The effect of changing the tube size is shown by returning to a water velocity of 1 ft/sec and by decreasing the tube size to, say, a $\frac{3}{4}$-in., 14-gage tube. In this case the following quantities are obtained:

> No. tubes = 43 tubes/pass,
> Inside film: $h_{12} = 338$ Btu/hr-ft^2-°F,
> Outside film: $h_{34} = 2650$ Btu/hr-ft^2-°F,
> Over-all coefficient: $U = 246$ Btu/hr-ft^2-°F,
> Temperature difference: $\Delta t_m = 113$°F,
> Surface area: $A = 57.6$ ft^2,
> Length of a single pass = 6.8 ft.

These data lead to the final specifications:

$$
\left.\begin{array}{l}
\text{1 shell pass} \\
\text{1 tube pass} \\
\text{43 tubes per pass, } \tfrac{3}{4}\text{-in. O.D., 14 gage} \\
\text{7 ft long}
\end{array}\right\} \text{Design 3}
$$

This design requires only 1 tube pass, but the total number of tubes and their length make this design comparable in initial cost to the first design, which required 22 tubes per pass, 2 passes 5 ft long. The pressure loss in the tube fluid is comparable to that of the first design—0.018 psia.

The three heat exchangers in the designs illustrated above are certainly not the only acceptable heat exchangers for the specified requirements. They are presented to illustrate the effect of varying certain of the parameters which are under the control of the designer. The two comparable designs, the first and third, would have to be examined further, on the basis of a more detailed economic analysis, to enable one to decide which should be built.

References

1. JAKOB, M., *Heat Transfer*, Vol. 2, New York, Wiley, 1957.
2. *Standards of Tubular Exchanger Manufacturers Association*, 3rd ed., New York, Tubular Exchanger Manufacturers Association, Inc., 1952.
3. BOWMAN, R. A., A. C. MUELLER, and W. M. NAGLE, "Mean Temperature Difference in Design," *Trans. ASME*, Vol. 63, 1940, p. 283.
4. KAYS, W. M., and A. L. LONDON, *Compact Heat Exchangers*, 2nd ed., New York, McGraw-Hill, 1964.
5. KERN, D. Q., *Process Heat Transfer*, New York, McGraw-Hill, 1950.

Problems

12.1. Steam at 500 psia, 800°F, flows at the rate of 1500 ft/min through a 6-in., schedule 40 pipe which is covered with 1.5 in. of 85% magnesia insulation. If the pipe is located horizontally in a room in which the ambient air temperature is 80°F, find the temperature of the outer surface of the insulation, the over-all heat transfer coefficient, and the rate of heat loss from a 100-ft length of the pipe.

12.2. Steam at 600°F, 800 psia, flows at 20 ft/sec through a horizontal 5-in., schedule 40 pipe covered with 1 in. of insulation ($k = 0.5$ Btu/hr-ft-°F). The pipe is exposed to stagnant atmospheric air at 100°F. Find (a) the temperature of the outer surface of the insulation, and (b) the over-all heat transfer coefficient. Use the approximate relations of Table 9.1 to obtain the free convection coefficient.

12.3. A condenser tube (1-in. diameter, 16 gage) is made of brass. It condenses steam at 3 psia, in a horizontal position, by the use of cooling water flowing at 70°F, 4 ft/sec inside. Estimate the over-all heat transfer coefficient.

12.4. A condenser tube made of brass is a $\frac{3}{4}$-in., 18-gage tube. Saturated steam at 5 psia is condensing on the outside of the tube, in a horizontal position, and cooling water at 80°F flows through the tube with an average velocity of 2 ft/sec. Find the over-all heat transfer coefficient of the tube and the outside tube surface temperature.

12.5. A 10-in., nominal, schedule 40 steel pipe is bare and carries superheated steam at 500 psia, 1000°F. It flows at the rate of 20,000 ft/min. The outer surface of the pipe is exposed to air at 180°F. Find the over-all heat transfer coefficient and the temperature of the pipe surface.

12.6. Steam at 100 psia, 800°F, flows at the rate of 1000 ft/min through a horizontal 5-in., schedule 80 steel pipe covered with 3 in. of 85% magnesia insulation. If the pipe is surrounded by air at 95°F, find the temperature of the outer surface of the insulation and the over-all heat transfer coefficient.

12.7. A brass condenser tube (1 in., 16 gage) has steam at 10 in. Hg abs. condensing on its outer surface and water at 85°F flowing through it at 5 ft/sec. Find the temperature of the tube surface and the over-all heat transfer coefficient.

12.8. The engine oil listed in Table A.5 flows at 1 ft/sec through a 1-in., 16-gage brass tube 6 ft long. The oil is at 120°F. The outer surface of the tube is surrounded by air at 70°F. If the tube is in a horizontal position, find the over-all heat transfer coefficient and the inner and outer tube surface temperatures.

12.9. Superheated steam flows at a velocity of 20 ft/sec through the annulus formed between a 4-in. and a 2-in. pipe. The steam is at 150 psia, 500°F. The inner pipe carries oil (Table A.5) at 320°F flowing at 0.5 ft/sec. The outer pipe is covered with 1 in. of 85% magnesia insulation and is exposed to still air at 70°F. If the pipes are 12 ft long, find the rate of heat loss to the air. How much heat is transferred to the oil?

12.10. A cast iron rod, $\frac{3}{4}$ in. in diameter and 10 in. long, protrudes horizontally from a heat source at 300°F into still air at 90°F. Estimate the rate of heat loss from the rod.

12.11. In a one-shell-pass, one-tube-pass heat exchanger, one fluid enters at 100°F and leaves at 500°F, whereas the other fluid enters at 800°F and leaves at 600°F. Find the log-mean temperature difference for (a) parallel flow and (b) counterflow.

12.12. The cold fluid entering a one-shell-pass, one-tube-pass heat exchanger at 100°F leaves at 400°F. The hot fluid enters at 700°F and leaves at 300°F. What is the log-mean temperature difference?

12.13. If the exchanger in Prob. 12.11 had been a one-shell-pass, two-tube-pass exchanger, what would be the mean temperature difference?

12.14. In a cross-flow exchanger one fluid enters at 100°F and leaves at 300°F, whereas the other fluid enters at 800°F and leaves at 500°F. Find the mean temperature difference.

12.15. A 1-in., 14-gage heat exchanger tube is equipped with 20 equally spaced straight fins placed longitudinally along the tube. The fins extend 1 in. in the radial direction. The tube and fins are made of steel, the fins $\frac{1}{16}$ in. thick. For inside and outside film coefficients of 200 and 45 Btu/hr-ft²-°F, respectively, what is the over-all heat transfer coefficient, U?

12.16. In a parallel-flow heat exchanger 400,000 lb/hr of water is heated from 250°F to 310°F by the use of hot gases entering at 750°F and leaving at 500°F. If the exchanger has a total heat transfer surface area of 22,000 ft², find the over-all heat transfer coefficient for the exchanger.

12.17. A steam condenser is made of $\frac{5}{8}$-in., 14-gage, brass tubes. The cooling water enters at 75°F and has an average velocity of 4.5 ft/sec. If there are two tube passes of 125 tubes/pass—6 ft long—find the number of pounds of saturated steam at 2 psia that will be condensed if the outside tube surface film coefficient is known to be 2000 Btu/hr-ft²-°F. Use the mean temperature difference method and then repeat using the NTU method.

12.18. The over-all heat transfer coefficient for a one-shell-pass, two-tube-pass heat exchanger is known to be 300 Btu/hr-ft^2-°F. The shell side fluid ($c_p = 1.0$ Btu/lb$_m$-°R) flows at the rate of 20,000 lb/hr, entering at 500°F. The tube side fluid ($c_p = 0.85$ Btu/lb$_m$-°R) flows at the rate of 60,000 lb/hr, entering at 90°F. For a total surface area of 100 ft^2, find the outlet fluid temperature for (a) the mean temperature difference approach, and (b) the NTU approach.

12.19. A heat exchanger is to be used to heat 10,000 lb/hr of water from 100°F to 300°F by the use of 5000 lb/hr of water being cooled from 750°F to 350°F. If the over-all coefficient is to be 320 Btu/hr-ft^2-°F, find the surface area required for (a) parallel-flow; (b) counterflow; and (c) a one-shell-pass, two-tube-pass exchanger.

12.20. A water-to-water heat exchanger is to be made of $\frac{3}{4}$-in., 16-gage, brass tubes. The tube water enters at 90°F, leaves at 120°F, and flows at the rate of 30,000 lb$_m$/hr. The shell water flows at the rate of 20,000 lb$_m$/hr, entering at 200°F. The over-all heat transfer coefficient is 250 Btu/hr-ft^2-°F. The exchanger has one shell pass, and the average tube water velocity is 1 ft/sec. If the length of the tubes is not to exceed 8 ft, find
(a) number of tubes per pass;
(b) number of tube passes;
(c) length of the tubes.

12.21. A water-to-water heat exchanger is to be made of 1-in., 12-gage tubes. The tube water enters at 90°F, leaves at 130°F, and flows at the rate of 40,000 lb/hr. The shell water flows at the rate of 20,000 lb/hr, entering at 200°F. The over-all heat transfer coefficient is 300 Btu/hr-ft^2-°F. The exchanger has one shell pass, and the average tube water velocity is 1 ft/sec. If the length of the tubes is not to exceed 10 ft, find
(a) number of tubes per pass;
(b) number of tube passes;
(c) length of the tubes.

12.22. A steam condenser has two tube passes, 180 tubes per pass, 10 ft long. The tubes are $\frac{5}{8}$-in., 17-gage brass tubes. There is one shell pass. The average tube-water velocity is 4 ft/sec and the cooling water enters at 75°F. How many Btu/hr will be transferred when the condenser is condensing saturated steam at 4 in. Hg abs.? What is the outlet temperature of the cooling water?

12.23. A 3-in., schedule 40, steel pipe carries hot water at the rate of 25 gal/min. The water enters the pipes at 180°F. The pipe is 50 ft long and it passes through a large room where it loses heat to the ambient air at 95°F by free convection. Estimate the outlet temperature of the water.

12.24. A steam condenser is built to the following specifications:

> One shell pass
> Two tube passes
> Tubes: $\frac{3}{4}$ in., 17 gage
> > Brass
> > Length = 8 ft
> > 220 tubes/pass
> > Average water velocity = 5 ft/sec
> > Inlet water temperature = 70°F

How many pounds per hour of saturated steam at 3 in. Hg abs. will be condensed?

12.25. It is desired to design a steam condenser of a 25,000 lb/hr turbine. The turbine exhaust pressure is to be 5 psia, and the quality of the exhaust steam is 90%. The condenser is to be supplied with cooling water at 80°F, and the maximum water temperature rise is to be 25°F. The water velocity is to be 7 ft/sec. The tubes are not to be longer than 12 ft and are to be $\frac{7}{8}$-in., 16-gage brass tubes. Find the number of tubes per pass, the number of passes, and the length of the tubes.

12.26. A heat exchanger is to be built of 1-in., 12-gage, brass tubes not longer than 8 ft. It is to heat 15,000 lb/hr of water, entering at 70°F and flowing at 1.4 ft/sec, to a temperature of 100°F. The water is heated by saturated steam condensing at 35 psia. Calculate the number of tubes per pass, the number of passes, and the length of the tubes.

12.27. Design a water heater that will heat 12,000 lb/hr of water from 70°F to 130°F by the use of condensing steam at 30 psia. Set the maximum tube length at 8 ft.

12.28. Derive Eq. (12.32).

12.29. Show that the effectiveness of a one-shell-pass, two-tube-pass heat exchanger is given by

$$\epsilon = \frac{2}{(1 + C_R) + \sqrt{1 + C_R^2}\dfrac{1 + e^{-x}}{1 - e^{-x}}},$$

$$x = \text{NTU}\sqrt{1 + c_R^2}.$$

12.30. Find the limiting form of Eq. (12.32) when $C_R \to 1$.

12.31. A one-shell-pass, one-tube-pass, heat exchanger is built with 60 tubes ($\frac{5}{8}$ in., 0.1-in. wall thickness). Water enters the shell side at 300°F and leaves at 100°F, flowing at the rate of 20,000 lb$_m$/hr. The tube water enters at 80°F and leaves at 150°F. The inside and outside film coefficients of the tubes are 300 and 1500 Btu/hr-ft^2-°F, respectively. Find the required length of the tubes.

12.32. A one-shell-pass, one-tube-pass heat exchanger is to be used to heat 60,000 lb$_m$/hr of water from 185°F to 210°F by using steam condensing at 50 psia. There are 30 1-in., 16-gage, brass tubes, and the over-all heat transfer coefficient is known to be 500 Btu/hr-ft^2-°F. Find the length of the tubes.

12.33. A steam condenser is made of brass ($\frac{3}{4}$-in., 14-gage) tubes. There are two tube passes, 150 tubes per pass. The cooling water flows at 5 ft/sec, enters at 80°F, and leaves at 120°F. The condensing film coefficient is known to be 1800 Btu/hr-ft²-°F, and the coefficient inside the tube is 1200 Btu/hr-ft²-°F. Steam is being condensed at 6 psia. Find the length of the tubes.

12.34. A one-shell-pass, one-tube-pass heat exchanger operates in counterflow. The exchanger uses 15,000 lb$_m$/hr of water entering at 75°F, to cool 30,000 lb$_m$/hr of oil ($c_p = 0.5$ Btu/lb$_m$-°F) entering at 210°F. The total exchanger surface area is 150 ft², and the over-all heat transfer coefficient is 65 Btu/hr-ft²-°F. What are the outlet temperatures of the two fluids?

12.35. A steam condenser is made of brass tubes ($\frac{7}{8}$-in., 14 gage) 5 ft long. There are two tube passes, 115 tubes per pass, and the cooling water enters at 70°F, 5 ft/sec. The condensing fllm coefficient is known to be 1250 Btu/hr-ft²-°F. The thermal resistance of the tube wall is negligible. The steam is being condensed at 5 psia. Find (a) the outlet water temperature, and (b) the rate at which steam is condensed.

12.36. A one-tube-pass, one-shell-pass heat exchanger operates in counterflow. The exchanger uses 10,000 lb$_m$/hr of water, entering at 60°F, to cool 20,000 lb$_m$/hr of oil ($c_p = 0.5$ Btu/lb$_m$-°F) entering at 200°F. The total tube surface area is 115 ft², and the over-all heat transfer coefficient is 50 Btu/hr-ft²-°F. What is the outlet temperature of the two fluids?

12.37. Repeat the problem of Example 12.8, assuming that the plates and fins are made of aluminum with $k = 73$ Btu/hr-ft-°F. Compare the results with those of Example 12.8.

12.38. A cross-flow, plate-fin heat exchanger, similar to that described in Example 12.8, is used to heat 300 lb$_m$/hr of air, entering at 70°F, by the use of 100 lb$_m$/hr of hot gases entering at 500°F. The plates of the exchanger are 5 in. × 5 in., spaced 0.375 in. apart, and are stacked 6 in. high. The plates are 0.0012 in. thick and the fins are 0.006 in. thick. Both the plates and the fin matrix are made of stainless steel with $k = 12$ Btu/hr-ft-°F. The following data are known for the two fluid streams. Hot side: $h_h = 37$ Btu/hr-ft²-°F, $A_f/A = 0.84$; cold side: $h_c = 47$ Btu/hr-ft²-°F, $A_f/A = 0.76$. Assuming that the heat gases have the same heat capacity as air, estimate the outlet temperature of the two fluids.

12.39. A cross-flow heat exchanger consists of a bundle of 30 copper tubes ($\frac{1}{2}$ in., 18 gage) placed normal to an air stream in a square duct. The tubes are 2.0 ft long, extending across the duct which is 2 ft × 2 ft. Atmospheric air, entering at 50°F, flows at the rate of 2500 ft³/min in the duct. Hot water enters the tubes at 300°F, flowing with a velocity of 1.5 ft/sec. A film coefficient of 40 Btu/hr-ft²-°F exists at the outer surface of the tubes. Find the fluid outlet temperatures and the rate of heat transfer.

12.40. Repeat Prob. 12.39 if the tubes are covered with annular fins, made of copper, 0.50 in. long, 0.01 in. thick, spaced 0.125 in. apart on centers. Assume that the same surface film coefficient applies for all exposed surfaces.

CHAPTER 13

Additional Cases of Combined Heat Transfer

13.1 Introductory Remarks

The examples of combined heat transfer considered in Chapter 12 were restricted to cases in which only the conduction and convection modes were present, and primary emphasis was placed on the application of these cases to the analysis and design of heat exchangers. This chapter will be devoted to other examples of combined heat transfer, some of which will include thermal radiation as one of the simultaneous modes taking place. Some applications of these principles for the estimation of thermometric errors will be illustrated.

One of the newest developments in heat transfer, the *heat pipe*, is an example of combined modes—conduction, convection, vaporization, condensation. This development is discussed in this chapter.

13.2 Simultaneous Convective and Radiant Heat Losses from Completely Enclosed Bodies

The instances of heat transfer to or from the surface of a solid body treated thus far in this book have been restricted either to cases in which convection was considered to be the only heat transfer mechanism taking place (Chapters 7 through 10) or to cases in which it was assumed that only thermal radiation was occurring at the surface (Chapter 11). In reality, the surface of any heated body that is not located in a vacuum must lose heat to the surroundings by the

547

simultaneous action of both the convection and radiation mechanisms. In many instances the temperature level and emissivity of a surface are such that the radiant emission from the surface is negligible, whereas in other cases the radiant exchange may be of the same order of magnitude as the convective exchange—or even greater.

When the fluid surrounding the solid surface is a nonabsorbing medium as far as radiation is concerned, one may treat the convection heat transfer and radiant heat transfer as independent, simultaneous mechanisms. This enables one to define a radiation heat transfer coefficient independent of the familiar convective coefficient.

In Eq. (11.83) it was shown that the heat loss by radiation from a body of area A, emissivity ϵ, and surface temperature T_s to ambient surroundings at temperature T_a is given by the following formula as long as the area of the surface of the surroundings is very large compared with A:

$$q_r = \sigma A \epsilon (T_s^4 - T_a^4). \tag{13.1}$$

In Eq. (13.1) q_r is used to denote the heat transfer rate due to radiation. Even though the loss of radiant energy is proportional to the difference of the fourth powers of the temperatures, it has been found expedient to define a *radiation coefficient*, h_r, on the basis of the linear temperature difference. That is, h_r is defined so that

$$q_r = A h_r (T_s - T_a). \tag{13.2}$$

This definition is made in analogy to the convective film coefficient. The definition of h_r in terms of the linear temperature difference is convenient for many computational and design purposes, but it obscures the real nature of the radiant exchange mechanism, since h_r must now be some function of σ, ϵ, *and* the temperatures T_s and T_a. A combination of Eqs. (13.1) and (13.2) shows that h_r is given by

$$h_r = \frac{T_s^4 - T_a^4}{T_s - T_a} \sigma \epsilon.$$

Factoring the numerator leads to the following expression for h_r:

$$h_r = \sigma \epsilon \frac{(T_s^2 - T_a^2)(T_s^2 + T_a^2)}{(T_s - T_a)}$$

$$= \sigma \epsilon (T_s + T_a)(T_s^2 + T_a^2). \tag{13.3}$$

Now, the *convective* heat transfer from the surface to the ambient fluid is given by Newton's law of cooling:

$$q_c = Ah_c(T_s - T_a). \tag{13.4}$$

Equation (13.4) is based on the assumption that the ambient fluid is at the same temperature, T_a, as the walls of the enclosure. Since the two mechanisms of convection and radiation take place simultaneously and independently, the total rate of heat loss from the surface is obtained by the addition of Eqs. (13.2) and (13.4):

$$q = A(h_c + h_r)(T_s - T_a). \tag{13.5}$$

Sometimes a combined coefficient, h_{cr}, is defined so that

$$q = Ah_{cr}(T_s - T_a),$$
$$h_{cr} = h_c + h_r. \tag{13.6}$$

The coefficient h_r is given by Eq. (13.3) and depends on the surface emissivity as well as the temperatures of the surface and the surroundings. The coefficient h_c is found by the methods of Chapters 6 through 10 and is dependent on the fluid properties, the surface geometry, and the temperatures.

The above representation of the combined radiant and convective heat transfer from a body in an enclosure is used to estimate heat losses from pipes, as illustrated in Example 13.1.

EXAMPLE 13.1: A horizontal, oxidized, iron pipe (6-in. schedule 40) has a surface temperature of 210°F and is exposed to air at 70°F. Find the convective and radiant heat transfer coefficients and the combined coefficient, h_{cr}.

Solution: The outside pipe diameter is 6.625 in. (Table B.3), and Table A.12 gives the emissivity of oxidized wrought iron to be $\epsilon = 0.94$.

The convective coefficient is found by the methods of Chapter 9 for free convection. At an average film temperature of $(210 + 70)/2 = 140°F$, Table A.7 gives

$$\rho = 0.0661 \text{ lb}_m/\text{ft}^3, \qquad \mu = 0.0484 \text{ lb}_m/\text{ft-hr},$$
$$k = 0.0166 \text{ Btu/hr-ft-}°F, \qquad N_{PR} = 0.700.$$

Thus,

$$N_{GR} = 3.46 \times 10^7,$$
$$N_{NU_D} = 36.8,$$
$$h_c = 1.11 \text{ Btu/hr-ft}^2\text{-}°F \ (6.30 \text{ W/m}^2\text{-}°C).$$

Equation (13.3) gives h_r, the radiation coefficient, to be

$$h_r = 0.1714 \times 10^{-2} \times 0.94 \left(\frac{530}{100} + \frac{670}{100} \right) \left[\left(\frac{530}{100} \right)^2 + \left(\frac{670}{100} \right)^2 \right]$$

$$= 1.42 \text{ Btu/hr-ft}^2\text{-}°F \text{ (8.06 W/m}^2\text{-}°C).$$

Thus, the combined coefficient is

$$h_{cr} = 1.11 + 1.42$$

$$= 2.53 \text{ Btu/hr-ft}^2\text{-}°F \text{ (14.4 W/m}^2\text{-}°C).$$

Note that for even the modest surface temperature used, the heat loss by radiation is of the same order of magnitude as that by free convection. If this example is repeated for a surface temperature of, say, 900°F, the convective coefficient remains practically unchanged at $h_c = 1.88$ Btu/hr-ft^2-°F whereas the radiation coefficient increases to $h_r = 6.50$—about three times the convective coefficient.

The method of calculation shown in Example 13.1 forms the basis for the various tables and charts available (see, for example, Ref. 1) in the literature which give values of the combined coefficient for pipes as a function of the diameter and temperature difference. The results for insulated pipes are substantially the same as those illustrated above for a bare pipe (except that lower surface temperatures are usually involved) because of the fact that the emissivity of the canvas insulation wrapping is about the same as that of oxidized steel—about 0.8.

An instance of a transfer of heat by a combination of all three modes of heat transfer is encountered, for example, in the case of heat loss from a pipe (or wall) when the surface temperature is not specified as it was in the above example. If, instead, the bulk temperature of the fluid flowing inside the pipe is specified, one must account for the inside convective film coefficient and the thermal resistance of the pipe and insulation (if any) as well as the combined coefficient. The determination of the rate of heat loss from the pipe then necessitates a trial and error solution just like those presented in Sec. 12.3, for cases where radiation was neglected, except that the computation of h_{cr} illustrated above replaces the step of the calculations wherein the outside convective coefficient was computed. The iterative solution of such problems converges less rapidly when the thermal radiation is included than when it is neglected, because the radiation coefficient is much more sensitive to changes in the surface temperature than is the convective coefficient.

13.3   Combined Convection and Radiation in Air Spaces

Another example of combined heat transfer by conduction, convection, and radiation is found in the case of air spaces which are included in the walls, ceilings, etc., of structures such as buildings, furnaces, houses, etc. Ideally, an air space is provided in such structures to increase the thermal resistance by enclosing a stagnant layer of air with a very low thermal conductivity. In reality, the air spaces of many house walls are wide enough so that a certain amount of air movement will occur—increasing the ratio of heat exchange over that which would be obtained by pure conduction alone.

Experiments reported in Ref. 2 indicate that if

$$d^3(\Delta t) < 3 \text{ in.}^3\text{-}^\circ\text{F},$$

one may assume that no convection is taking place through the air. In the above expression d is used to denote the air space thickness, and Δt is the difference between the temperatures of the two surfaces. If it is true that no convection is taking place, the heat that is transferred is exchanged by radiation and conduction through the air. The combination of the conduction and the radiant exchange between the two walls gives the following expression for the heat transfer through the air space:

$$\frac{q}{A} = \frac{k}{d}(T_1 - T_2) + \sigma\frac{T_1^4 - T_2^4}{(1/\epsilon_1) + (1/\epsilon_2) - 1}.$$

The radiation term in this expression comes from Sec. 11.5, wherein the radiant exchange between large, parallel gray planes was analyzed. The subscripts 1 and 2 denote the two bounding surfaces of the air space, k is the thermal conductivity of the air, and d is the air space width. If a combined conductance, C, is defined so that

$$\frac{q}{A} = C(T_1 - T_2), \tag{13.7}$$

then C is given by

$$C = \frac{k}{d} + \frac{\sigma(T_1 + T_2)(T_1^2 + T_2^2)}{(1/\epsilon_1) + (1/\epsilon_2) - 1}. \tag{13.8}$$

When the air space width is such that pure conduction through the air cannot be assumed, some allowance must be made for additional heat transfer by convection. Reference 2 recommends the values for C given in Table 13.1 as suitable averages for air spaces between $\frac{3}{4}$ in. and 4 in. thick

Table 13.1 Air Space Conductances

Mean Temperature of Air Space, °F	Δt, °F	C, Btu/hr-ft^2-°F		
		$\epsilon = 0.167$	$\epsilon = 0.67$	$\epsilon = 0.9$
50	10	0.42	0.69	0.99
50	30	0.52	0.79	1.08
90	10	0.46	0.80	1.16

when it is assumed that both the surface emissivities have the value indicated. The use of the mean values given in this table is usually adequate for most purposes—less than 5 per cent error resulting in most cases.

13.4 The Thermocouple Lead Error in Surface Temperature Measurements

Some of the principles of the various modes of heat transfer may be combined to analyze, approximately, the error that may be expected when the temperature of a body or fluid is measured with one of the usual thermometric devices—a thermometer or a thermocouple. Since the introduction of a temperature-measuring device into a system in which heat transfer is taking place will alter the thermal conditions existing there, one must expect that the device will indicate a value of the sought-after temperature that differs from the temperature that would actually exist had no measurement been made. This error is due to the thermal effects of the measuring device itself on the system and is not a calibration error of the thermometer. Some typical cases will be discussed in the following sections.

If a thermocouple is attached to the surface of a solid body in the manner illustrated in Fig. 13.1(a), heat will be conducted along the thermocouple leads and dissipated into the surroundings by convection from the wires. This

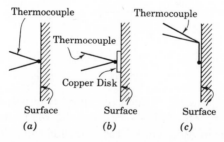

Fig. 13.1. The measurement of surface temperatures by the use of thermocouples.

conduction of heat sets up temperature gradients within the body in the vicinity of the point of attachment—a condition that would not exist had the thermocouple not been attached. The thermocouple will read a value corresponding to the local depression in the surface temperature at the point of attachment.

Since the thermocouple wires are usually long and thin, the conduction of the heat along them—and its eventual dissipation into the surroundings— may be represented as heat flow along a spine of uniform cross section and infinite length. This is treated in Prob. 3.44. To simplify the system under consideration, let the thermocouple be represented as a single wire of radius R and thermal conductivity k_t. The surface film coefficient at the surface of the wire will be denoted by h. Let t_a denote the temperature of the ambient air and let t_t denote the depressed temperature of the surface under the point of attachment of the thermocouple. Then Prob. 3.44 shows that the rate of heat flow from the surface into the wire is

$$q = k_t \sqrt{\frac{hC}{k_t A}} \times A(t_t - t_a).$$

Here C and A are the perimeter and cross-sectional areas of the wire, respectively. Thus,

$$q = \pi \sqrt{2k_t R^3 h} \times (t_t - t_a). \tag{13.9}$$

The heat noted in Eq. (13.9) must be supplied from within the body. The body may be represented as a semi-infinite solid maintained at a temperature t_s at points far removed from the point of the thermocouple attachment. The point of attachment may be approximated as an area of radius R maintained at temperature t_t. Thus, t_s is the true surface temperature, t_t is the temperature indicated by the thermocouple, and $(t_s - t_t)$ is the error introduced by the attachment of the thermocouple.

One of the classical solutions of the mathematical theory of heat conduction is that for a semi-infinite solid of a uniform temperature, say t_s, at points far removed from a heat sink on the surface which has a finite radius, say R, and a temperature t_t. The solution is not given here but may be found in Ref. 3. The rate of heat flow into the sink is given by this solution to be

$$q = 4Rk_s(t_s - t_t). \tag{13.10}$$

In Eq. (13.10) k_s is used to denote the thermal conductivity of the surface material. Combination of Eqs. (13.9) and (13.10) gives the measured error to be

$$t_s - t_t = \frac{\frac{\pi}{k_s}\sqrt{\frac{k_t R h}{8}}(t_s - t_a)}{1 + \frac{\pi}{k_s}\sqrt{\frac{k_t R h}{8}}}.$$

Now, h is determined by the free convection relations of Chapter 9. Generally, the diameter of a thermocouple wire is quite small, and the average temperature difference between the wire and the ambient air would not be expected to be very great. Hence, as an approximation, one may presume that the Grashof number for the free convection around the wire will be small. According to the experimental correlations noted in Eq. (9.40), one may take the Nusselt number for the wire equal to $N_{NU_D} = 0.4 = h(2R)/k_a$ if $N_{GR}N_{PR} < 10^{-5}$. The symbol k_a is used to denote the thermal conductivity of the ambient fluid. Thus,

$$t_s - t_t = \frac{\frac{\pi}{\sqrt{40}}\sqrt{\frac{k_a k_t}{k_s^2}}(t_s - t_a)}{1 + \frac{\pi}{\sqrt{40}}\sqrt{\frac{k_a k_t}{k_s^2}}}. \tag{13.11}$$

The relation gives an approximation to the expected error in the surface temperature indicated by a thermocouple when it is attached in the manner shown in Fig. 13.1(a). Equation (13.11) is only an approximation in view of the many simplifying assumptions that were made in its derivation, but it is valuable for the purpose of estimating the order of magnitude of the error likely to be encountered. It is interesting to note that the error is independent of the diameter of the wire. Also, the greater the thermal conductivity of the wire, k_t, the greater the error is, since heat will be more readily conducted away. Similarly, the smaller the value of the thermal conductivity of the surface material, k_s, the greater is the error, since the heat conducted away by the wire must be supplied from within the body. In some instances, the error involved may be quite significant, as the following example shows.

EXAMPLE 13.2:   A furnace wall of magnesite has a surface temperature of 400°F. The ambient air is at 100°F. What error may be expected to be indicated by a thermocouple of copper wire attached to the surface?

Solution: From the tables of Appendix A,

$$k_a = 0.0157 \text{ Btu/hr-ft-°F},$$

$$k_t = 216 \text{ Btu/hr-ft-°F},$$

$$k_s = 2.2 \text{ Btu/hr-ft-°F}.$$

The error is then

$$t_s - t_t = \frac{\dfrac{\pi}{\sqrt{40}}\sqrt{\dfrac{0.0157 \times 216}{(2.2)^2}}(400 - 100)}{1 + \dfrac{\pi}{\sqrt{40}}\sqrt{\dfrac{0.0157 \times 216}{(2.2)^2}}}$$

$$= 88°F \ (49°C)!$$

Had the surface been made of copper the error would be much less,

$$t_s - t_t = 1.3°F \ (0.7°C).$$

The error in the reading of a thermocouple attached to a surface may be reduced by one of the schemes indicated in Fig. 13.1(b) or 13.1(c). Figure 13.1(b) shows the thermocouple attached to a thin disk of material of high thermal conductivity (say copper). The disk is then placed in contact with the surface. The arrangement is advantageous since it provides a large area at the surface from which the heat conducted away by the thermocouple is withdrawn, producing a less severe local depression in the surface temperature.

If the leads of the thermocouple are placed along the surface for a few inches before being led away, as shown in Fig. 13.1(c), the junction of the thermocouple does not lie at the point of departure of the leads from the surface—the point at which the greatest local depression in temperature is likely to occur.

13.5   Thermometer Well Errors Due to Conduction

The temperature of a fluid flowing in a closed conduit is often measured by the use of a thermometer or thermocouple placed in a thermometer well inserted into the fluid stream. Such an arrangement is indicated schematically in Fig. 13.2.

If the thermometer well is assumed to have good thermal contact with the pipe wall, at temperature t_s, then it may be treated as a spine of uniform cross

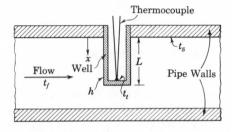
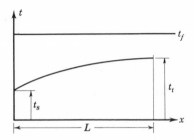

Fig. 13.2

section and finite length L. The thermometer or thermocouple placed in the well will be assumed to have perfect contact with the bottom of the well, so the indicated temperature, t_t, may be assumed to be that of the end of the spine.

The finite length spine was treated in Sec. 3.12, and when losses out of the end of the spine are neglected, Eq. (3.54) shows that the temperature of the end is given by

$$t_t - t_f = \frac{t_s - t_f}{\cosh ml},$$

$$(13.12)$$

$$m = \sqrt{\frac{hC}{kA}}.$$

In Eq. (13.12) t_f is the fluid temperature—the desired quantity—so that $(t_t - t_f)$ is the error in the reading given by the thermocouple. Also, in Eq. (13.12) A is the cross-sectional area for heat flow in the thermometer well wall and k is the thermal conductivity of the well material. If one assumes that little convection takes place at the inner well surface (the well is usually filled with a stagnant liquid), C denotes the perimeter of the outer well surface, and h is the surface film coefficient there. Usually a thermometer well is of circular shape, so the outside film coefficient h may be determined by the relation given in Eq. (8.24):

$$N_{NU_D} = C'(N_{RE_D})^{m'}.$$

$$(13.13)$$

Values of C' and m' are tabulated in Table 8.3 as functions of N_{RE_D}.

In order to use Eq. (13.12) to estimate the error of the temperature reading in a thermometer well, one must determine the pipe surface temperature t_s. The error is directly proportional to the difference between the pipe surface temperature and the fluid temperature $(t_s - t_f)$. Thus, the error may be reduced by insulating the outside pipe surface so that t_s will be nearer to t_f. Also, any means of making the product mL as great as possible will reduce the error.

13.6 Radiation Effects in the Measurement of Gas Temperatures

If the temperature of a high-temperature gas stream is to be measured by the insertion of a thermometer or thermocouple, the effects of the radiant exchange between the pipe walls and the temperature-sensing element may be of such an order of magnitude as to influence markedly the value indicated.

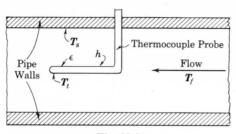

Fig. 13.3

Conduction Effects Negligible. If the thermocouple (perhaps contained in a housing) is inserted into the gas stream in the manner illustrated in Fig. 13.3—with a significant length of the thermocouple parallel to the flow—then one might be able to neglect the effects of the heat which is conducted along the wire to the pipe wall. If this is so, any loss in heat by radiation from the thermocouple-sensing element is balanced by the gain in heat due to convective heat transfer between the fluid and the element. On a unit area basis, this balance of heat gives

$$h_c(T_f - T_t) = \sigma\epsilon(T_t^4 - T_s^4). \tag{13.14}$$

where T_f = fluid temperature, T_t = thermocouple temperature, T_s = pipe surface temperature, h_c = convective film coefficient, and ϵ = emissivity of thermocouple housing. Absolute temperatures are used since radiant heat transfer is involved. The expression on the right side of Eq. (13.14) is based on Eq. (11.83) for the radiant exchange between a surface (the thermocouple) completely enclosed by a second surface (the pipe walls).

The coefficient of convective heat transfer, h_c, is to be found by the methods of Chapters 7 through 9, depending on the geometry of the thermocouple housing. Since the correlations for convection are generally based on an average film temperature $[(T_t + T_f)/2$ in this case], it is necessary to perform a trial and error solution of Eq. (13.14) in order to obtain the error $(T_f - T_s)$.

EXAMPLE 13.3: A large duct conveys a high temperature stream of air at a velocity of 20 ft/sec. The duct wall temperature is measured to be 300°F and a thermocouple ($\epsilon = 0.5$) is placed at right angles to the flow. If conduction effects are negligible, what is the temperature of the gas stream when the thermocouple reads 1000°F? The diameter of the thermocouple housing is $\frac{1}{2}$ in.

Solution: A trial and error solution is necessary, but for the sake of space just the final iteration will be given. Assume that $t_f = 1300°F$, $T_f = 1760°R$. Then since $t_t = 1000°F$ ($T_t = 1460°R$), the mean film temperature is 1150°F, and Table A.7 gives

$$\rho = 0.0246 \text{ lb}_m/\text{ft}^3, \qquad \mu = 0.094 \text{ lb}_m/\text{ft-hr},$$
$$k = 0.036 \text{ Btu/hr-ft-°F}, \qquad N_{PR} = 0.697.$$

Thus, based on the diameter of the thermocouple housing,

$$N_{RE_D} = 785.$$

Using Eq. (8.24) and Table 8.3, one obtains

$$N_{NU_D} = 0.615(785)^{0.466}$$

$$= 13.7.$$

So

$$h = 13.7 \frac{0.036}{0.5} \times 12 = 11.85 \, \text{Btu/hr-ft}^2\text{-}°\text{F}.$$

Equation (13.14) may be used to verify the assumed value of t_f:

$$11.85(t_f - 1000) = 0.1714 \times 0.5 \left[\left(\frac{1460}{100} \right)^4 - \left(\frac{760}{100} \right)^4 \right],$$

$$t_f = 1305°\text{F} \ (747°\text{C}).$$

Thus, an error of about 300°F is observed in the gas temperature!

The radiation error involved in the measurement of the temperature of a very hot gas may be reduced considerably by the use of radiation shields. The use of plane radiation shields was discussed in Sec. 11.5, and it was noted in that case that the use of one shield reduced the radiant energy exchange to one-half of its original, two shields reduced it to one-third, etc. The same sort of advantage can be deduced for a radiation shield around a thermocouple housing. If T_{sh} denotes the equilibrium temperature of a radiation shield placed around a thermocouple, and if one assumes that the shield is large compared to the thermocouple but small compared to the duct walls, the equilibrium temperature is given by

$$\sigma \epsilon_t (T_t^4 - T_{sh}^4) = \sigma \epsilon_{sh} (T_{sh}^4 - T_s^4).$$

If one can say that the emissivities of the thermocouple and pipe walls are practically the same,

$$T_{sh}^4 = \frac{T_s^4 + T_t^4}{2}.$$

Thus, Eq. (13.14) may be replaced by

$$h_c(T_f - T_t) = \sigma\epsilon(T_{sh}^4 - T_s^4),$$

$$= \frac{\sigma\epsilon}{2}(T_t^4 - T_s^4). \tag{13.15}$$

The error is reduced to one-half of its former value. It is not hard to show, as in the case of plane radiation shields, that if N shields are placed around the thermocouple probe, the error will be reduced to $1/(N + 1)$ of its value when no shield is used.

Conduction Effects Not Negligible. Section 13.5 considered the error involved in the use of a thermometer well in a gas stream when the effects of radiation were neglected. The discussion above considered the case in which conduction errors were negligible and radiant effects were dominant. This section will consider the case of the error likely to be involved in the measurement of the temperature of a fluid stream by means of a thermometer well when all three modes of heat transfer must be taken into account.

Figure 13.4 illustrates the simplified system assumed to represent the thermometer well. The temperatures of the pipe surface, flowing gas, and thermometer well end are denoted by T_s, T_f, and T_t, respectively. By assuming no losses in the interior of the well, T_t will be the temperature indicated by the thermometer or thermocouple. The symbol h_c denotes the convective film coefficient on the surface of the well, C denotes its perimeter, and A is the cross-sectional area for heat flow. Following the concept introduced in Sec. 13.2, one uses h_r to denote the radiant coefficient defined in Eq. (13.3):

$$h_r = \sigma\epsilon(T_s + T_f)(T_s^2 + T_f^2). \tag{13.16}$$

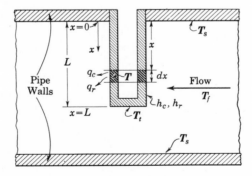

Fig. 13.4

If the thermometer well is treated as a straight spine maintained at T_s on one end, the methods of Sec. 3.12 may be applied with the inclusion of the radiation loss at the surface. If x denotes the coordinate measured along the length of the thermometer well, a heat balance on an element dx thick gives

$$kA\frac{d^2T}{dx^2}\,dx = h_cC\,dx(T - T_f) + h_rC\,dx(T - T_s).$$

This expression may be written as

$$\frac{d^2T}{dx^2} - m_1^2T = -m_2^2\left[T_f + \left(\frac{m_1^2}{m_2^2} - 1\right)T_s\right], \tag{13.17}$$

where

$$m_1 = \sqrt{\frac{(h_c + h_r)C}{kA}}, \tag{13.18}$$

$$m_2 = \sqrt{\frac{h_cC}{kA}}. \tag{13.19}$$

Equation (13.17) is an ordinary, linear, nonhomogeneous differential equation, the solution to which is

$$T = B\,e^{-m_1x} + D\,e^{m_1x} + P. \tag{13.20}$$

The symbol P denotes the particular integral:

$$P = T_s + \left(\frac{m_2}{m_1}\right)^2(T_f - T_s)$$

$$= T_s + \frac{h_c}{h_c + h_r}(T_f - T_s). \tag{13.21}$$

The constants B and D of Eq. (13.20) may be found by applying the assumptions that the well temperature is T_s at $x = 0$ and that no heat flow occurs out of the end at $x = L$. That is,

$$\text{At } x = 0: \quad T = T_s.$$

$$\text{At } x = L: \quad \frac{dT}{dx} = 0.$$

Application of these conditions to Eq. (13.20) gives B and D to be

$$B = \frac{T_s - P}{e^{m_1 L} + e^{-m_1 L}} e^{m_1 L},$$

$$D = \frac{T_s - P}{e^{m_1 L} + e^{-m_1 L}} e^{-m_1 L}.$$

Substitution of these facts into Eq. (13.20) and setting $x = L$ to obtain the desired temperature T_t gives

$$T_t = \frac{(T_s - P) e^{m_1 L} e^{-m_1 L}}{e^{m_1 L} + e^{-m_1 L}} + \frac{(T_s - P) e^{-m_1 L} e^{m_1 L}}{e^{m_1 L} + e^{-m_1 L}} + P$$

$$= \frac{2(T_s - P)}{e^{m_1 L} + e^{-m_1 L}} + P$$

$$= \frac{T_s - P(1 - \cosh m_1 L)}{\cosh m_1 L}.$$

Thus,

$$T_t - T_s = (T_s - P)\frac{1 - \cosh m_1 L}{\cosh m_1 L}.$$

When P, Eq. (13.21), is introduced, this equation becomes

$$T_t - T_s = \frac{h_c}{h_c + h_r}(T_s - T_f)\frac{1 - \cosh m_1 L}{\cosh m_1 L},$$

$$\frac{T_t - T_s}{T_f - T_s} = \frac{h_c}{h_c + h_r}\frac{\cosh m_1 L - 1}{\cosh m_1 L}. \tag{13.22}$$

This equation provides a means for estimating the temperature error since T_f may be computed when T_t, T_s and the geometry of the well are given. A trial and error solution similar to that of the preceding example is usually necessary.

13.7 Heat Transfer from Radiating Fins

One of the most important problems of modern space technology is that of *thermal control*—the maintenance of a desired temperature level within a spacecraft (manned or unmanned) or satellite. Such spacecraft are subjected to heat inputs of both internal and external sources. Internal sources consist of electrically generated heat and metabolically generated heat, while the external sources consist of thermal irradiation from the sun, planetary bodies, etc. If a net gain in heat is experienced, some means of heat rejection must be provided in order that thermal control be maintained.

In outer space, the only means of rejecting heat is that of thermal radiation. For this purpose *heat rejection space radiators* have been developed. Many varieties of such systems have been developed (see Ref. 4). A typical space radiator configuration is illustrated in Fig. 13.5. Internally generated heat within the spacecraft is conveyed to the radiator by a circulating coolant (e.g., a water–glycol mixture). The coolant circulates in tubes (usually connected in a parallel circuit) which are joined by finned surfaces. Heat from the coolant conducts into the fin and is dissipated by radiation to the surroundings. The finned surface may be subjected to an external irradiation from a source such as the sun. The basic question to be answered is: Given a radiating fin of known geometry, maintained at a known base (i.e., tube temperature) and subjected to a certain external radiation, what is the rate at which heat is dissipated by the fin to the environment?

The solution for one such situation will be given here. If the fluid in adjacent radiator tubes is at the same temperature, the fin may be represented by the model suggested in Fig. 13.5—a straight fin of length L and uniform thickness w maintained, at one end, at a known temperature T_0. The other end of the fin has the condition $dT/dx = 0$ imposed by symmetry. The fin material has a

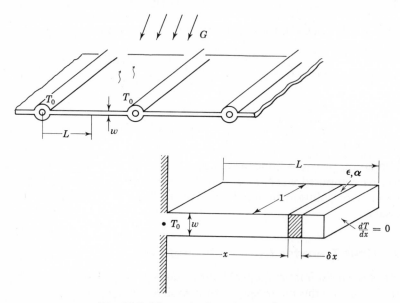

Fig. 13.5. Heat rejection space radiator.

thermal conductivity k, and the exposed surface has a total emissivity ϵ. The surface is exposed to an external irradiation G to which it exhibits a total absorptivity α. Since such radiators are often made an integral part of the spacecraft skin, only one surface of the fin is considered to be radiatively active.

An approach similar to that developed in Sec. 3.12 for a fin with a convective boundary condition may be used here. If the conduction in the fin is taken to be one-dimensional, a heat balance taken on an element dx in length yields the following relation (for a unit depth):

$$-kw\frac{dT}{dx} = -kw\frac{dT}{dx} - kw\frac{d^2T}{dx^2}\,\delta x + \cdots + \epsilon\sigma T^4\,\delta x - \alpha G\,\delta x.$$

No convective loss at the fin surface is included. For $\delta x \to 0$, the following differential equation must be satisfied at each point:

$$\frac{d^2T}{dx^2} - \frac{\epsilon\sigma}{kw}\left(T^4 - \frac{\alpha G}{\epsilon\sigma}\right) = 0.$$

The nonhomogeneous term is recognized as the equilibrium temperature, or equivalent sink temperature, defined in Example 11.2. This temperature is the equilibrium temperature the surface would achieve if isolated in the irradiation G. Thus, with

$$T_s^4 = \frac{\alpha G}{\epsilon\sigma}, \tag{13.23}$$

the differential equation for the temperature distribution in the fin is

$$\frac{d^2T}{dx^2} - \frac{\epsilon\sigma}{kw}(T^4 - T_s^4) = 0. \tag{13.24}$$

The solution of Eq. (13.24) is best discussed in terms of the following dimensionless variables:

$$\theta = \frac{T}{T_0},$$

$$\theta_s = \frac{T_s}{T_0},$$

$$\xi = \frac{x}{L}, \tag{13.25}$$

$$\lambda = \frac{\epsilon\sigma T_0^3 L^2}{kw}.$$

Introduction of these definitions into Eq. (13.24) yields

$$\frac{d^2\theta}{d\xi^2} - \lambda(\theta^4 - \theta_s^4) = 0. \tag{13.26}$$

The boundary conditions to be satisfied are

$$\text{At } \xi = 0 \, (x = 0): \quad T = T_0, \theta = 1.$$

$$\text{At } \xi = 1 \, (x = L): \quad dT/dx = d\theta/d\xi = 0. \tag{13.27}$$

The nonlinearity of Eq. (13.26) makes its solution in closed form impossible; although the first integral may readily be written

$$\frac{d\theta}{d\xi} = -\sqrt{\tfrac{2}{5}\lambda(\theta^5 - 5\theta\theta_s^4) + C}, \tag{13.28}$$

C being a constant of integration.

A complete solution of the above system, Eqs. (13.27) and (13.28), may be accomplished only by numerical means. Rather than seek the solution for the temperature distribution, the principal item of interest is the heat dissipated by the fin:

$$q = -kw\left(\frac{dT}{dx}\right)_{x=0} = -\frac{kwT_0}{L}\left(\frac{d\theta}{d\xi}\right)_{\xi=0}. \tag{13.29}$$

As in the case of the convective fins of Chapter 3, the heat dissipation is best expressed in terms of a fin effectiveness (see Sec. 3.17), which is the ratio of q to the heat that *would* be dissipated if the whole fin was maintained at T_0:

$$\kappa = \frac{q}{(\epsilon\sigma T_0^4 - \alpha G)L}$$

$$= \frac{q}{\epsilon\sigma L T_0^4(1 - \theta_s^4)}. \tag{13.30}$$

Thus, when Eqs. (13.29) and (13.30) are combined,

$$\kappa = \frac{1}{\lambda(1 - \theta_s)^4}\left(\frac{d\theta}{d\xi}\right)_{\xi=0}. \tag{13.31}$$

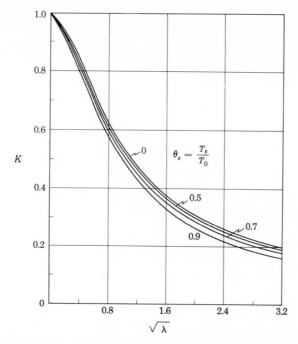

Fig. 13.6. The effectiveness of a radiating fin of uniform thickness. (From S. Lieblein, *N.A.S.A. Tech. Note D-196*, National Aeronautics and Space Administration, Washington, D.C., Nov. 1959.)

When Eq. (13.28) is solved, subject to the conditions of Eqs. (13.27), the results may be applied to Eq. (3.52) to yield the fin effectiveness in the following form:

$$\kappa = fn(\theta_s, \lambda). \tag{13.32}$$

The parameter θ_s measures the relative magnitude of the fin base temperature and the irradiation of the environment, while λ represents a combination of the thermal parameters of the fin. Lieblein, in Ref. 5, performed the numerical integration referred to above. The results are shown graphically in Fig. 13.6. By use of Fig. 13.6, along with the definitions of κ, θ_s, and λ given in Eqs. (13.23), (13.25), and (13.30), the heat rejected by the radiator may be determined.

EXAMPLE 13.4: A space radiator is constructed of fins of 3-in. half-width and thickness 0.150 in. The fin material is aluminum alloy with $k = 65$ Btu/hr-ft-°F. A surface coating on the fin has $\epsilon = 0.5$, $\alpha = 0.1$. If the fin base is maintained at 120°F, find the heat rejected per foot of length when the external irradiation is 100 Btu/hr-ft².

Solution: For the data given,

$$T_s = \left(\frac{0.1 \times 100}{0.5 \times 0.1714 \times 10^8} \right)^{1/4}$$

$$= 328°R,$$

$$\theta_s = 328/580 = 0.556,$$

$$\lambda = \frac{0.5 \times 0.1714 \times 10^{-8}(150 + 460)^4(3/12)^2}{65 \times 0.150/12}$$

$$= 9.13.$$

For $\theta_0 = 0.556$ and $\lambda = 36.5$, Fig. 13.6 yields

$$\kappa = 0.2.$$

Thus, Eq. (13.30) gives, for a 1-ft width:

$$q = 0.2 \times 0.5 \times 0.1714 \times 10^{-8} \times (3/12) \times (610)^4[1 - (0.556)^4]$$

$$= 5.35 \text{ Btu/hr-ft } (5.14 \text{ W/m}).$$

13.8   The Heat Pipe

The heat pipe is a recent (Refs. 6 and 7) development which utilizes the high latent heat of evaporation, or condensation, together with the pheno-menon of capillary pumping to transfer very high heat fluxes without the addition of external work. As a result, the device, seemingly inert, can trans-port heat fluxes many times that which can be conducted by solid metal bars of equal cross section.

The basic schematic diagram of a typical heat pipe is shown in Fig. 13.7. Structurally, the device consists of a closed vessel, or pipe, of circular cross section, a capillary wick, and a transport fluid. The wick materials are usually wire screen, woven cloth, ceramic materials, or narrow grooves machined into the pipe wall. At one end, the evaporator, heat is added at some temperature, vaporizing the liquid fluid from the wick material. The vaporized fluid flows down the central core of the pipe to the somewhat-cooler, lower-pressure end, termed the *condenser*. At the condenser end, the vapor is condensed back to a liquid with the release of the associated latent heat. The condensed liquid is "pumped" back to the evaporator section by the action of surface tension in the capillary structure of the wick material. The evaporator and condenser sections are normally separated by an adiabatic section. Because of the relatively low pressure drop in the flowing vapor, the

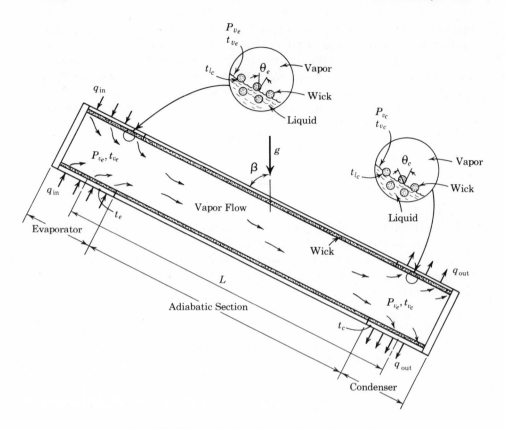

Fig. 13.7. Schematic drawing of a typical heat pipe.

difference between the evaporator temperature (where the heat is added) and the condenser temperature (where the heat is rejected) is relatively small— at least compared with the latent heat transported. These facts lead to a very high apparent conductivity between the two ends of the pipe.

Apparently, the nature of the working fluid has a significant effect on the performance of a heat pipe, as do the properties of the wick material and the relative orientation of the pipe in the gravity field. Working fluids such as water, ammonia, alcohol, molten metals, and nitrogen are used—the choice being based, in part, on the desired working temperature of the device.

In space applications in which no gravity forces act, the heat pipe may be used to transport heat over relatively large distances by lengthening the adiabatic section.

Some of the governing parameters of heat pipe performance may be obtained in the following simplified analysis. The basic transport mechanism involved is that of capillary pumping in the wick. The driving potential for this process is the capillary pressure difference which results from the surface tension at a vapor–liquid interface. Figure 13.8 shows the basic physical mechanism. By assuming that the vapor–liquid interface occurs in a circular tube of radius, r; that the liquid wets the tube wall with a contact angle θ; and that the surface tension at the interface is σ, a simple balance of forces shows that the pressure of the vapor exceeds that of the liquid by an amount

$$\Delta P = \frac{2\pi r\sigma \cos\theta}{\pi r^2}$$

$$= \frac{2\sigma \cos\theta}{r}. \tag{13.33}$$

Using the notation of Fig. 13.7, one finds that the pressure of the liquid at the condensing interface in the condenser, P_{l_c}, exceeds that at the evaporative interface in the evaporator, P_{l_e}, by an amount determined by the difference in the vapor pressure (P_{v_c} and P_{v_e}), the difference in the interfacial pressures, and the hydrostatic head imposed by gravity (if any):

$$P_{l_c} - P_{l_e} = (P_{v_c} - P_{v_e}) - (\Delta P_c - \Delta P_e) - g\rho_l L \cos\beta$$

$$= (P_{v_c} - P_{v_e}) + 2\sigma\left(\frac{\cos\theta_e}{r} - \frac{\cos\theta_c}{r}\right) - g\rho_l L \cos\beta. \tag{13.34}$$

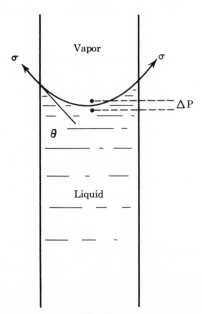

Fig. 13.8

In Eq. (13.34) the subscripts l and v denote liquid and vapor and the subscripts c and e denote condenser and evaporator. The angle β specifies the orientation of the pipe relative to the gravity vector and, for simplicity, L denotes a mean distance between the condenser and evaporator sections. It has been presumed that the wick material can be represented as being composed of circular capillaries of radius r. The symbols θ_e and θ_c represent the liquid contact angles in the evaporator and condenser. Also in Eq. (13.34) it has been presumed that the evaporator is located higher in the gravity field than the condenser—hence the negative sign on the last term.

The pressure difference given by Eq. (13.34) is the driving force to transport the liquid from the condenser back to the evaporator through the wick material. The wick, naturally, offers a resistance to this flow. Darcy's law may be used in this instance:

$$\dot{m} = (P_{l_c} - P_{l_e})\frac{\rho_l A K}{\mu_l L},$$

in which $\dot{m}$ is the mass flow rate of liquid, A is the cross-sectional area of wick material available for the liquid flow, and K is the permeability of the wick. Thus, the flow rate in the heat pipe is

$$\dot{m} = \frac{AK}{L}\frac{\sigma\rho_l}{\mu_l}\left[\frac{P_{v_c} - P_{v_e}}{\sigma} + \frac{2}{r}(\cos\theta_e - \cos\theta_c) - \frac{g\rho_l L\cos\beta}{\sigma}\right].$$

For many applications, certainly adequate for this simple analysis, one may neglect the difference between the vapor pressures at each end of the pipe ($P_{v_c} \approx P_{v_e}$). Thus, the mass flow is

$$\dot{m} = \frac{AK}{L}\frac{\sigma\rho_l}{\mu_l}\left[\frac{2}{r}(\cos\theta_e - \cos\theta_c) - \frac{g\rho_l L}{\sigma}\cos\beta\right], \tag{13.35}$$

and the heat flux between the condenser and the evaporator is

$$q = \frac{AK}{L}\frac{\sigma\rho_l\lambda}{\mu_l}\left[\frac{2}{r}(\cos\theta_e - \cos\theta_c) - \frac{g\rho_l L}{\sigma}\cos\beta\right], \tag{13.36}$$

where λ is the latent heat of vaporization.

In order to produce a positive flow rate, and thus heat flux, $(\cos\theta_e - \cos\theta_c)$ must be positive. As the power input, q, to the heat pipe increases and the rates of vaporization and condensation increase, the apparent contact angles in the evaporator and condenser must change so that $(\cos\theta_e - \cos\theta_c)$

becomes more positive. The maximum flow rate and maximum heat flux are given when $\cos \theta_e \to 1$ and $\cos \theta_c \to 0$:

$$q_{max} = \frac{2AK}{rL} \frac{\sigma \rho_l \lambda}{\mu_l} \left(1 - \frac{g \rho_l Lr}{2\sigma} \cos \beta \right). \qquad (13.37)$$

Although Eq. (13.37) may be lacking in some respects for the prediction of actual heat pipe performance, it does permit the identification of significant parameters. It is seen that q_{max} is the product of three factors. The first, $2AK/rL$, is a grouping of wick properties. The third, $g\rho_l Lr \cos \beta/2\sigma$, is a grouping of wick and liquid properties which represents the relative importance of gravitational and surface tension forces. In space, where $g = 0$, this is 0. The second term is quite significant in that it is a grouping of liquid properties called the *liquid transport factor*:

$$N_l = \frac{\rho_l \lambda \sigma}{\mu_l}.$$

The designer wishes to choose this factor to be as large as possible. It is limited on one extreme by the liquid freezing point, where $\mu_l \to \infty$, and at the other extreme by the critical point, where $\lambda \to 0$. Figure 13.9 presents values of N_l for some typical heat pipe fluids. It is apparent that the desired operating range temperature has a profound effect in the choice of the fluid to be used. Also, it may be noticed that heat pipes operating at high temperatures using, say, liquid metals, may transport heat fluxes of several orders of magnitude greater than those at lower temperatures.

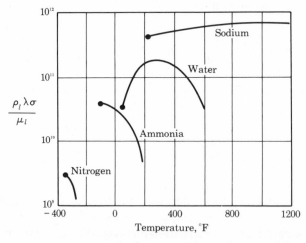

Fig. 13.9. Heat pipe, liquid transport factor for several fluids.

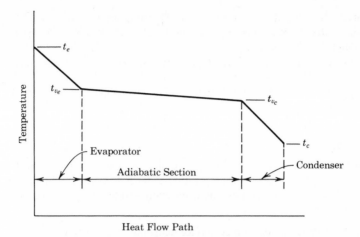

Fig. 13.10. The temperature profile of a typical heat pipe.

The maximum limitation on heat flux given by Eq. (13.37) was imposed by the maximum pumping capability of the wick material. Other considerations may override this limitation. One is associated with the fact that in practice a finite temperature difference is required between the evaporator, T_{v_e}, and the outside evaporator wall, T_e. This difference is required to drive the heat from the heat source into the heat pipe. A typical temperature distribution through a heat pipe is shown in Fig. 13.10. The over-all temperature drop is seen to exceed considerably the vapor temperature drop. For high heat loads the required difference, $t_e - t_{v_e}$, may be large enough that local superheating of the liquid will occur within the wick material. If this superheating is great enough, nucleate boiling may be initiated within the wick with a subsequent degradation of its pumping capability. Thus, the heat pipe may become "boiling-limited" rather than "pumping-limited," as discussed above.

References

1. MARKS, L. S. (ed.), *Mechanical Engineers' Handbook*, 5th ed., New York, McGraw-Hill, 1958.
2. ROBINSON, H. E., E. J. POWLITCH, and R. S. DILL, "The Thermal Insulating Value of Air Spaces," *Housing and Home Finance Agency, Housing Research Paper 32*, Washington, D.C., U.S. Government Printing Office, 1954.
3. GRÖBER, H., and S. ERK, *Die Grundgesetze der Wärmeübertragung*, Berlin, Springer, 1955.
4. MACKAY, D. B., *Design of Space Power Plants*, Englewood Cliffs, N.J., Prentice-Hall, 1963.

5. LIEBLEIN, S., "Analysis of Temperature Distribution and Radiant Heat Transfer Among a Rectangular Fin of Uniform Thickness," *N.A.S.A. Tech. Note D-196*, Washington, D.C., Nov. 1959.

6. COTTER, T. P., "Theory of Heat Pipes," *Los Alamos Sci. Lab. Rept. LA-3246-MS*, Los Alamos, N.M., 1965.

7. FELDMAN, K. T., and G. H. WHITING, "The Heat Pipe," *Mech. Eng.*, Feb. 1967, p. 30.

Problems

13.1 A bare, oxidized, wrought iron pipe (8-in. schedule 40) is placed horizontally in still air at 70°F. Its surface temperature is 200°F. Find the combined heat transfer coefficient due to radiation and convection, and find the heat loss from 100 ft of the pipe.

13.2. Repeat Prob. 13.1 for a pipe surface temperature of 870°F.

13.3. Steam at 800°F, 500 psia flows through a 6-in. nominal horizontal steel pipe at the rate of 10 ft/sec. The pipe is covered with 1 in. of 85% magnesia insulation. The emissivity of the outer insulation surface is 0.75. Find the temperature of the outer surface of the insulation and the over-all heat transfer coefficient for the pipe if the ambient air temperature is 70°F.

13.4. Is it necessary to account for radiation exchange in the determination of the over-all heat transfer coefficient of a steam condenser? Justify your answer by numerical calculations.

13.5. A 100-watt light bulb delivers approximately 10% of its energy in the visible spectrum. About 2% of the remaining infrared radiation passes unabsorbed through the glass of the bulb. Make a sound engineering estimate of the surface temperature of the glass bulb.

13.6. A surface temperature of 600°F is to be measured with a thermocouple. The ambient air temperature is 80°F. Find the error in the value indicated by the thermocouple if
(a) The surface is made of iron.
　　(1) The thermocouple is made of copper.
　　(2) The thermocouple is made of iron.
(b) The surface is made of magnesite brick.
　　(1) The thermocouple is made of copper.
　　(2) The thermocouple is made of iron.

13.7. A thermocouple is attached to a steel surface and indicates a temperature of 350°F when the ambient air is at 75°F. If the thermocouple is made of copper, what is the true surface temperature?

13.8. A thermometer well (0.25 in. O.D., 0.125 in. I.D.) made of brass extends 4 in. into a 6-in. steel pipe carrying air at 600°F, 100 psia, 15 ft/sec. The pipe wall temperature is measured to be 350°F. What temperature would be indicated by a thermocouple inserted into the well? Neglect radiation.

13.9. The thermometer well described above extends 6 in. into an 8-in. pipe carrying air at 20 psia at the rate of 1000 lb_m/hr. The pipe is bare and is located in a room at 100°F. If the thermometer indicates a temperature of 700°F, what is the temperature of the air? Neglect the effect of radiation.

13.10. A bare thermocouple ($\epsilon = 0.8$) is 0.1 in. in diameter, and it is placed in a flue duct carrying combustion gases (approximated by air) at 650°F. If the duct wall is at 250°F, what error will result in the thermocouple reading due to radiation effects? The gases are flowing at 25 ft/sec.

13.11. A thermometer is inserted normal to the flow of air (10 psia, 3000 ft/min) in a 1-ft-diameter duct. The duct wall is at 100°F and the thermometer reads 350°F. If the thermometer is 0.25 in. in diameter, what is the air temperature?

13.12. A thermocouple well (0.250 in. O.D., 0.125 in. I.D.) is made of steel and extends 6 in. into a 12-in. duct, also made of steel. The duct carries atmospheric air at 20 ft/sec, and the duct surface temperature is 200°F. If the thermocouple reads 600°F, find the air temperature if
(a) radiation is neglected;
(b) radiation is not neglected.

13.13. Verify the solution of Eq. (13.17) that is given in Eq. (13.20).

13.14. Verify that Eq. (13.28) is a first integral of Eq. (13.26).

13.15. Repeat Example 13.4 for the case of a zero irradiation environment, all other data remaining the same.

13.16. For the radiating fin described in Sec. 13.7, if the length, L, becomes very great, one might assume that $\theta_L \rightarrow \theta_s$ as well as $(d\theta/d\xi)_{x=L} = 0$. Under this circumstance, show that the effectiveness is

$$\kappa_\infty = \frac{\sqrt{\dfrac{2}{5\lambda}(1 - 5\theta_s^4 + 4\theta_s^5)}}{1 - \theta_s^4}.$$

Tables and Charts of Properties of Substances

Table A.1 Thermal Properties of Metals*

Metal	Properties at 68°F				k, Btu/hr-ft-°F									
	ρ $\dfrac{lb_m}{ft^3}$	c_p $\dfrac{Btu}{lb_m\text{-}°F}$	k $\dfrac{Btu}{hr\text{-}ft\text{-}°F}$	α $\dfrac{ft^2}{hr}$	-148°F -100°C	32°F 0°C	212°F 100°C	392°F 200°C	572°F 300°C	752°F 400°C	1112°F 600°C	1472°F 800°C	1832°F 1000°C	2192°F 1200°C
Aluminum														
Pure	169	0.214	132	3.665	134	132	132	132	132					
Al-cu (Duralumin) 94–96 Al, 3–5 Cu, trace Mg	174	0.211	95	2.580	73	92	105	112						
Al-Mg (Hydronalium) 91–95 Al, 5–9 Mg	163	0.216	65	1.860	54	63	73	82						
Al-Si (Silumin) 87 Al, 13 Si	166	0.208	95	2.773	86	94	101	107						
Al-Si (Silumin, copper bearing) 86.5 Al; 12.5 Si; 1 Cu	166	0.207	79	2.311	69	79	83	88	93					
Al-Si (Alusil) 78–80 Al; 20–22 Si	164	0.204	93	2.762	83	91	97	101	103					
Al-Mg-Si 97 Al; 1 Mg; 1 Si; 1 Mn	169	0.213	102	2.859	—	101	109	118						
Lead	710	0.031	20	0.924	21.3	20.3	19.3	18.2	17.2					
Iron														
Pure	493	0.108	42	0.785	50	42	39	36	32	28	23	21	20	21
Wrought iron (C < 0.50%)	490	0.11	34	0.634	—	34	33	30	28	26	21	19	19	19
Cast iron (C ≈ 4%)	454	0.10	30	0.666	—									
Steel (C max ≈ 1.5%)														
Carbon steel C ≈ 0.5%	489	0.111	31	0.570	—	32	30	28	26	24	20	17	17	18
1.0%	487	0.113	25	0.452	—	25	25	.24	23	21	19	17	16	17

Table A.1 (continued)

Metal	Properties at 68°F				k, Btu/hr-ft-°F									
	ρ $\dfrac{\text{lb}_m}{\text{ft}^3}$	c_p $\dfrac{\text{Btu}}{\text{lb}_m\text{-}°\text{F}}$	k $\dfrac{\text{Btu}}{\text{hr-ft-}°\text{F}}$	α $\dfrac{\text{ft}^2}{\text{hr}}$	−148°F −100°C	32°F 0°C	212°F 100°C	392°F 200°C	572°F 300°C	752°F 400°C	1112°F 600°C	1472°F 800°C	1832°F 1000°C	2192°F 1200°C
Iron (Continued)														
Carbon steel (Cont.)														
1.5%	484	0.116	21	0.376	—	21	21	21	20	19	18	16	16	17
Nickel steel Ni ≈ 0%	493	0.108	42	0.785										
10%	496	0.11	15	0.279										
20%	499	0.11	11	0.204										
30%	504	0.11	7	0.118										
40%	510	0.11	6	0.108										
50%	516	0.11	8	0.140										
60%	523	0.11	11	0.182										
70%	531	0.11	15	0.258										
80%	538	0.11	20	0.344										
90%	547	0.11	27	0.452										
100%	556	0.106	52	0.892										
Invar Ni ≈ 36%	508	0.11	6.2	0.108										
Crome steel Cr = 0%	493	0.108	42	0.785	50	42	39	36	32	28	23	21	20	21
1%	491	0.11	35	0.645	—	36	32	30	27	24	21	19	19	
2%	491	0.11	30	0.559	—	31	28	26	24	22	19	18	18	
5%	489	0.11	23	0.430	—	23	22	21	21	19	17	17	17	17
10%	486	0.11	18	0.344	—	18	18	18	17	17	16	16	17	
20%	480	0.11	13	0.258	—	13	13	13	13	14	14	15		
30%	476	0.11	11	0.204										
Cr-Ni (chrome-nickel)														
15 Cr; 10 Ni	491	0.11	11	0.204										

Table A.1 (continued)

| Metal | Properties at 68°F | | | | k, Btu/hr-ft-°F | | | | | | | | | |
	ρ $\dfrac{lb_m}{ft^3}$	c_p $\dfrac{Btu}{lb_m\text{-}°F}$	k $\dfrac{Btu}{hr\text{-}ft\text{-}°F}$	α $\dfrac{ft^2}{hr}$	-148°F -100°C	32°F 0°C	212°F 100°C	392°F 200°C	572°F 300°C	752°F 400°C	1112°F 600°C	1472°F 800°C	1832°F 1000°C	2192°F 1200°C
Cr-Ni (Continued)														
18 Cr; 8 Ni (V2A)	488	0.11	9.4	0.172	—	9.4	10	10	11	11	13	15	18	
20 Cr; 15 Ni	489	0.11	8.7	0.161										
25 Cr; 20 Ni	491	0.11	7.4	0.140										
Ni-Cr (nickel-chrome)														
80 Ni; 15 Cr	532	0.11	10	0.172										
60 Ni; 15 Cr	516	0.11	7.4	0.129										
40 Ni; 15 Cr	504	0.11	6.7	0.118										
20 Ni; 15 Cr	491	0.11	8.1	0.151	—	8.1	8.7	8.7	9.4	10	11	13		
Cr-Ni-Al: 6 Cr; 1.5 Al; 0.5 Si (Sicromal 8)	482	0.117	13	0.237										
24 Cr; 2.5 Al 0.5 Si (Sicromal 12)	479	0.118	11	0.194										
Manganese steel Ma = 0%	493	0.118	42	0.784										
1%	491	0.11	29	0.538										
2%	491	0.11	22	0.376	—	22	21	21	21	20	19			
5%	490	0.11	13	0.247										
10%	487	0.11	10	0.194										
Tungsten steel W = 0%	493	0.108	42	0.785										
1%	494	0.107	38	0.720										
2%	497	0.106	36	0.677	—	36	34	31	28	26	21			
5%	504	0.104	31	0.591										
10%	519	0.100	28	0.527										

Table A.1 (continued)

| Metal | Properties at 68°F | | | | k, Btu/hr-ft-°F | | | | | | | | | |
	ρ $\frac{lb_m}{ft^3}$	c_p $\frac{Btu}{lb_m\text{-}°F}$	k $\frac{Btu}{hr\text{-}ft\text{-}°F}$	α $\frac{ft^2}{hr}$	−148°F −100°C	32°F 0°C	212°F 100°C	392°F 200°C	572°F 300°C	752°F 400°C	1112°F 600°C	1472°F 800°C	1832°F 1000°C	2192°F 1200°C
Tungsten steel (Continued)														
20%	551	0.093	25	0.484										
Silicon steel Si = 0%	493	0.108	42	0.785										
1%	485	0.11	24	0.451										
2%	479	0.11	18	0.344										
5%	463	0.11	11	0.215										
Copper														
Pure	559	0.0915	223	4.353	235	223	219	216	—	210	204			
Aluminum bronze 95 Cu; 5 Al	541	0.098	48	0.903										
Bronze 75 Cu; 25 Sn	541	0.082	15	0.333										
Red brass 85 Cu; 9 Sn; 6 Zn	544	0.092	35	0.699	—	34	41							
Brass 70 Cu; 30 Zn	532	0.092	64	1.322	51	—	74	83	85	85				
German silver 62 Cu; 15 Ni; 22 Zn	538	0.094	14.4	0.290	11.1	—	18	23	26	28				
Constantan 60 Cu; 40 Ni	557	0.098	13.1	0.237	12	—	12.8	15						
Magnesium														
Pure	109	0.242	99	3.762	103	99	97	94	91					
Mg-Al (electrolytic) 6-8% Al; 1-2% Zn	113	0.24	38	1.397	—	30	36	43	48					
Mg-Mn 2% Mn	111	0.24	66	2.473	54	64	72	75						
Molybdenum	638	0.060	79	2.074	80	79	79							

Table A.1 (continued)

Metal	Properties at 68°F				k, Btu/hr-ft-°F									
	ρ $\dfrac{\text{lb}_m}{\text{ft}^3}$	c_p $\dfrac{\text{Btu}}{\text{lb}_m\text{-°F}}$	k $\dfrac{\text{Btu}}{\text{hr-ft-°F}}$	α $\dfrac{\text{ft}^2}{\text{hr}}$	−148°F −100°C	32°F 0°C	212°F 100°C	392°F 200°C	572°F 300°C	752°F 400°C	1112°F 600°C	1472°F 800°C	1832°F 1000°C	2192°F 1200°C
Nickel														
Pure (99.9%)	556	0.1065	52	0.882	60	54	48	42	37	34				
Impure (99.2%)	556	0.106	40	0.677	—	40	37	34	32	30	32	36	39	40
Ni-Cr: 90 Ni; 10 Cr	541	0.106	10	0.172	—	9.9	10.9	12.1	13.2	14.2	13.0			
80 Ni; 20 Cr	519	0.106	7.3	0.129	—	7.1	8.0	9.0	9.9	10.9				
Silver														
Purest	657	0.0559	242	6.601	242	241	240	238						
Pure (99.9%)	657	0.0559	235	6.418	242	237	240	216	209	208				
Tungsten	1208	0.0321	94	2.430	—	96	87	82	77	73	65	44		
Zinc, pure	446	0.0918	64.8	1.591	66	65	63	61	58	54				
Tin, pure	456	0.0541	37	1.505	43	38.1	34	33						

* From E. R. G. ECKERT, *Introduction to the Transfer of Heat and Mass*, New York, McGraw-Hill, 1950.

Table A.2 Thermal Properties of Some Nonmetals[*]

Substance	c_p $\dfrac{\text{Btu}}{\text{lb}_m\text{-}°\text{F}}$		ρ lb_m/ft^3		t $°\text{F}$	k $\text{Btu/hr-ft}^2\text{-}°\text{F}$		α ft^2/hr
Structural								
Asphalt					68	0.43	a	
Bakelite	0.38	b	79.5	b	68	0.134	b	0.0044
Bricks								
Common	0.20	d	100	d	68	0.40	a	0.02
Face			128	d	68	0.76	a	
Carborundum brick					1110	10.7	a	
					2550	6.4	a	
Chrome brick	0.20	d	188	d	392	1.34	a	0.036
					1022	1.43	a	0.038
					1652	1.15	a	0.031
Diatomaceous					400	0.14	a	
earth (fired)					1600	0.18	a	
Fire clay brick	0.23	d	128	d	932	0.60	a	0.020
(burnt 2426°F)					1472	0.62	a	0.021
					2012	0.63	a	0.021
Fire clay brick	0.23	d	145	d	932	0.74	a	0.022
(burnt 2642°F)					1472	0.79	a	0.024
					2012	0.81	a	0.024
Fire clay brick	0.23	d	165	f	392	0.58	a	0.015
(Missouri)					1112	0.85	a	0.022
					2552	1.02	a	0.027
Magnesite	0.27	d			400	2.2	a	
					1200	1.6	a	
					2200	1.1	a	
Cement, Portland			94			0.17	a	
Cement, mortar					75	0.67	a	
Concrete	0.21	b	119–144	b	68	0.47–0.81	b	0.019–0.027
Concrete, cinder					75	0.44	a	
Glass, plate	0.2	b	169	b	68	0.44	b	0.013
Glass, borosilicate			139	b	86	0.63	b	
Plaster, gypsum	0.2	d	90	d	70	0.28	a	0.016
Plaster, metal lath					70	0.27	a	
Plaster, wood lath					70	0.16	a	
Stone								
Granite	0.195	d	165	d		1.0–2.3	a	0.031–0.071
Limestone	0.217	d	155	d	210–570	0.73–0.77	a	0.022–0.023
Marble	0.193	b	156–169	b	68	1.6	b	0.054
Sandstone	0.17	b	135–144	b	68	0.94–1.2	b	0.041–0.049

Table A.2 (continued)

Substance	c_p $\dfrac{\text{Btu}}{\text{lb}_m\text{-}°\text{F}}$		ρ lb_m/ft^3		t $°\text{F}$	k Btu/hr-ft-$°$F		α ft^2/hr
Structural (Cont.)								
Wood, cross grain:								
Balsa			8.8	a	86	0.032	a	
Cypress			29	d	86	0.056	a	
Fir	0.65	d	26.0	b	75	0.063	a	0.0037
Oak	0.57	d	38–30	b	86	0.096	a	0.0049
Yellow pine	0.67	d	40	d	75	0.085	a	0.0032
White pine			27	d	86	0.065	a	
Wood, radial								
Oak	0.57	b	38–30	b	68	0.10–0.12	b	$\begin{cases} 0.0043- \\ 0.0047 \end{cases}$
Fir	0.65	b	26.0–26.3	b	68	0.08	b	0.0048
Insulating								
Asbestos			29.3	b	$\begin{cases} -328 \\ 32 \end{cases}$	0.043 0.090	b b	
Asbestos			36.0	b	$\begin{cases} 32 \\ 212 \\ 392 \\ 752 \end{cases}$	0.087 0.111 0.120 0.129	b b b b	
Asbestos			43.5	b	$\begin{cases} -328 \\ 32 \end{cases}$	0.09 0.135	b b	
Asbestos cement						1.2	a	
Asbestos cement board					68	0.43	a	
Asbestos sheet					124	0.096	a	
Asbestos felt (40 laminations per inch)					$\begin{cases} 100 \\ 300 \\ 500 \end{cases}$	0.033 0.040 0.048	a a a	
Asbestos felt (20 laminations per inch)					$\begin{cases} 100 \\ 300 \\ 500 \end{cases}$	0.045 0.055 0.065	a a a	
Asbestos, corrugated (4 plies per inch)					$\begin{cases} 100 \\ 200 \\ 300 \end{cases}$	0.05 0.058 0.069	a a a	
Balsam wool			2.2	a	90	0.023	a	
Cardboard, corrugated						0.037	a	
Celotex					90	0.028	a	
Corkboard			10	b	86	0.025	b	
Cork, expanded scrap	0.45	b	2.8–7.4	b	68	0.021	b	0.006–0.017
Cork, ground			9.4	b	86	0.025	b	

Table A.2 (continued)

Substance	$\dfrac{c_p}{\dfrac{\text{Btu}}{\text{lb}_m\text{-}°\text{F}}}$	ρ lb$_m$/ft^3		t °F	k Btu/hr-ft-°F		α ft^2/hr
Insulating							
Diatomaceous earth (powdered)		10	e	200	0.029	e	
				400	0.038	e	
				600	0.048	e	
Diatomaceous earth (powdered)		14	e	200	0.033	e	
				400	0.039	e	
				600	0.046	e	
Diatomaceous earth (powdered)		18	e	200	0.040	e	
				400	0.045	e	
				600	0.049	e	
Felt, hair		8.2	c	20	0.0237	c	
				100	0.0269	c	
				200	0.0310	c	
Felt, hair		11.4	c	20	0.0212	c	
				100	0.0254	c	
				200	0.0299	c	
Felt, hair		12.8	c	20	0.0233	c	
				100	0.0262	c	
				200	0.0295	c	
Fiber insulating board		14.8	b	70	0.028	b	
Glass wool		1.5	c	20	0.0217	c	
				100	0.0313	c	
				200	0.0435	c	
Glass wool		4.0	c	20	0.0179	c	
				100	0.0239	c	
				200	0.0317	c	
Glass wool		6.0	c	20	0.0163	c	
				100	0.0218	c	
				200	0.0288	c	
Kapok				86	0.020	a	
Magnesia, 85%		16.9	c	100	0.039	a	
				200	0.041	a	
				300	0.043	a	
				400	0.046	a	
Rock wool		4.0	c	20	0.0150	c	
				100	0.0224	c	
				200	0.0317	c	
Rock wool		8.0	c	20	0.0171	c	
				100	0.0228	c	
				200	0.0299	c	

Table A.2 (continued)

Substance	c_p $\dfrac{\text{Btu}}{\text{lb}_m\text{-}°\text{F}}$		ρ lb_m/ft^3		t °F	k Btu/hr-ft-°F		α ft²/hr
Insulating (Cont.)								
Rock wool			12.0	c	20	0.0183	c	
					100	0.0226	c	
					200	0.0281	c	
Miscellaneous								
Aerogel, silica			8.5	b	248	0.013	b	
Clay	0.21	b	91.0	b	68	0.739	b	0.039
Coal, anthracite	0.30	b	75–94	b	68	0.15	b	0.005–0.006
Coal, powdered	0.31	b	46	b	86	0.067	b	0.005
Cotton	0.31	b	5	b	68	0.034	b	0.075
Earth, coarse	0.44	b	128	b	68	0.30	b	0.0054
Ice	0.46	b	57	b	32	1.28	b	0.048
Rubber, hard			74.8	b	32	0.087	b	
Sawdust					75	0.034	a	
Silk	0.33	b	3.6	b	68	0.021	b	0.017

* Adapted from (a) A. I. Brown and S. M. Marco, *Introduction to Heat Transfer*, 3rd ed., New York, McGraw-Hill, 1958; (b) E. R. G. Eckert, *Introduction to the Transfer of Heat and Mass*, New York, McGraw-Hill, 1950; (c) R. H. Heilman, *Ind. Eng. Chem.*, Vol. 28, 1936, p. 782; (d) L. S. Marks, *Mechanical Engineers' Handbook*, 6th ed., New York, McGraw-Hill, 1958; (e) R. Calvert, *Diatomaceous Earth*, Chemical Catalog Company, Inc., 1930; (f) H. F. Norton, *J. Am. Ceram. Soc.*, Vol. 10, 1957, p. 30.

Table A.3 Specific Heats of Some Metals* (in Btu/lb$_m$-°F)

t °F	t °C	Pb	Zn	Al	Ag	Au	Cu	Ni	Fe	Co
32	0	0.0306	0.0917	0.2106	0.0557	0.0305	0.0919	0.1025	0.1051	0.1023
212	100	0.0315	0.0958	0.2225	0.0571	0.0312	0.0942	0.1132	0.1166	0.1079
392	200	0.0325	0.0999	0.2344	0.0585	0.0320	0.0965	0.1241	0.1280	0.1138
572	300	0.0335	0.1041	0.2463	0.0599	0.0327	0.0988	0.1352	0.1395	0.1192
752	400	0.0328	0.1082	0.2582	0.0612	0.0334	0.1011	0.1295	0.1508	0.1249
932	500	0.0328	0.1225	0.2702	0.0626	0.0341	0.1034	0.1310	0.1622	0.1305
1112	600	0.0328	0.1233	0.2821	0.0640	0.0349	0.1057	0.1326	0.1737	0.1362
1292	700	0.0328	0.1242	0.259	0.0654	0.0356	0.1080	0.1341	0.1853	0.1418
1472	800	0.0328	0.1250	0.259	0.0668	0.0363	0.1103	0.1356	0.1741	0.1475
1652	900	0.0328	0.1259	0.259	0.0682	0.0371	0.1126	0.1372	0.1805	0.1531
1832	1000	0.0328	0.1267	0.259	0.076	0.0378	0.1149	0.1387	0.1505	0.1588
2012	1100	—	—	—	0.076	0.0355	0.118	0.1403	0.1505	0.1644
2192	1200	—	—	—	0.076	0.0355	0.118	0.1418	0.1505	0.1701
2372	1300	—	—	—	0.076	0.0355	0.118	0.1434	0.1505	0.1757
2552	1400	—	—	—	—	—	—	0.1449	0.1505	0.1814
2732	1500	—	—	—	—	—	—	0.1455	0.1790	0.1425
2912	1600	—	—	—	—	—	—	0.1455	0.1460	0.1425
Melting points, °F		621	786	1220	1760	1945	1981	2646	2795	2696

* From W. H. McAdams, *Heat Transmission*, 3rd ed., New York, McGraw-Hill, 1954.

Table A.4   Properties of Saturated Water

t °F	$c_p{}^*$ Btu/lb$_m$-°F	$\rho\dagger$ lb$_m$/ft^3	μ^* $\dfrac{\text{lb}_m}{\text{ft-hr}}$	ν $\dfrac{\text{ft}^2}{\text{hr}}$	k^* $\dfrac{\text{Btu}}{\text{hr-ft-°F}}$	α $\dfrac{\text{ft}^2}{\text{hr}}$	β 1/°R	N_{PR}
32	1.009	62.42	4.33	0.0694	0.327	0.0052	0.03×10^{-3}	13.37
40	1.005	62.42	3.75	0.0601	0.332	0.0053	0.045	11.36
50	1.002	62.38	3.17	0.0508	0.338	0.0054	0.070	9.41
60	1.000	62.34	2.71	0.0435	0.344	0.0055	0.10	7.88
70	0.998	62.27	2.37	0.0381	0.349	0.0056	0.13	6.78
80	0.998	62.17	2.08	0.0334	0.355	0.0057	0.15	5.85
90	0.997	62.11	1.85	0.0298	0.360	0.0058	0.18	5.13
100	0.997	61.99	1.65	0.0266	0.364	0.0059	0.20	4.52
110	0.997	61.84	1.49	0.0241	0.368	0.0060	0.22	4.04
120	0.997	61.73	1.36	0.0220	0.372	0.0060	0.24	3.65
130	0.998	61.54	1.24	0.0202	0.375	0.0061	0.27	3.30
140	0.998	61.39	1.14	0.0186	0.378	0.0062	0.29	3.01
150	0.999	61.20	1.04	0.0170	0.381	0.0063	0.31	2.72
160	1.000	61.01	0.97	0.0159	0.384	0.0063	0.33	2.53
170	1.001	60.79	0.90	0.0148	0.386	0.0064	0.35	2.33
180	1.002	60.57	0.84	0.0139	0.389	0.0064	0.37	2.16
190	1.003	60.35	0.79	0.0131	0.390	0.0065	0.39	2.03
200	1.004	60.13	0.74	0.0123	0.392	0.0065	0.41	1.90
210	1.005	59.88	0.69	0.0115	0.393	0.0065	0.43	1.76
220	1.007	59.63	0.65	0.0109	0.395	0.0066	0.45	1.66
230	1.009	59.38	0.62	0.0104	0.395	0.0066	0.47	1.58
240	1.011	59.10	0.59	0.0100	0.396	0.0066	0.48	1.51
250	1.013	58.82	0.56	0.0095	0.396	0.0066	0.50	1.43
260	1.015	58.51	0.53	0.0091	0.396	0.0067	0.51	1.36
270	1.017	58.24	0.50	0.0086	0.396	0.0067	0.53	1.28
280	1.020	57.94	0.48	0.0083	0.396	0.0067	0.55	1.24
290	1.023	57.64	0.46	0.0080	0.396	0.0067	0.56	1.19
300	1.026	57.31	0.45	0.0079	0.395	0.0067	0.58	1.17
350	1.044	55.59	0.38	0.0068	0.391	0.0067	0.62	1.01
400	1.067	53.65	0.33	0.0062	0.384	0.0068	0.72	0.91
450	1.095	51.55	0.29	0.0056	0.373	0.0066	0.93	0.85
500	1.130	49.02	0.26	0.0053	0.356	0.0064	1.18	0.83
550	1.200	45.92	0.23	0.0050	0.330	0.0060	1.63	0.84
600	1.362	42.37	0.21	0.0050	0.298	0.0052	—	0.96

* From A. I. BROWN and S. M. MARCO, *Introduction to Heat Transfer*, 3rd ed., New York, McGraw-Hill, 1958, as compiled from *International Critical Tables*, *ASHVE Guide*, 1940, and SCHMIDT and SELSCHOPP, *Forsch. Geb. Ingenieurwes.*, Vol. 3, 1943, p. 227.

† From J. H. KEENAN and F. G. KEYES, *Thermodynamic Properties of Steam*, New York, Wiley, 1936.

Table A.5 Properties of Some Saturated Liquids*

$\dfrac{t}{°F}$	$\rho\,\dfrac{lb_m}{ft^3}$	$c_p\,\dfrac{Btu}{lb_m\text{-}°F}$	$\mu\,\dfrac{lb_m}{ft\text{-}hr}$	$k\,\dfrac{Btu}{hr\text{-}ft\text{-}°F}$	$\nu\,\dfrac{ft^2}{hr}$	$\alpha\,\dfrac{ft^2}{hr}$	N_{PR}	$\beta\,\dfrac{1}{°R}$
				Sulphur Dioxide				
−58	97.44	0.3247	1.831	0.140	1.88×10^{-2}	4.42×10^{-3}	4.24	
−40	95.94	0.3250	1.573	0.136	1.64	4.38	3.74	
−22	94.43	0.3252	1.360	0.133	1.44	4.33	3.31	
−4	92.93	0.3254	1.171	0.130	1.26	4.29	2.93	
14	91.37	0.3255	1.023	0.126	1.12	4.25	2.62	
32	89.80	0.3257	0.895	0.122	0.997	4.19	2.38	
50	88.18	0.3259	0.794	0.118	0.900	4.13	2.18	
68	86.55	0.3261	0.705	0.115	0.814	4.07	2.00	1.08×10^{-3}
86	84.86	0.3263	0.623	0.111	0.734	4.01	1.83	
104	82.98	0.3266	0.556	0.107	0.670	3.95	1.70	
122	81.10	0.3268	0.508	0.102	0.626	3.87	1.61	
				Methyl Chloride				
−58	65.71	0.3525	0.815	0.124	1.24×10^{-2}	5.38×10^{-3}	2.31	
−40	64.51	0.3541	0.793	0.121	1.23	5.30	2.32	
−22	63.46	0.3564	0.774	0.117	1.22	5.18	2.35	
−4	62.39	0.3593	0.749	0.113	1.20	5.04	2.38	
14	61.27	0.3629	0.723	0.108	1.18	4.87	2.43	
32	60.08	0.3673	0.703	0.103	1.17	4.70	2.49	
50	58.83	0.3726	0.677	0.099	1.15	4.52	2.55	
68	57.64	0.3788	0.651	0.094	1.13	4.31	2.63	
86	56.38	0.3860	0.631	0.089	1.12	4.10	2.72	
104	55.13	0.3942	0.606	0.083	1.10	3.86	2.83	
122	53.76	0.4034	0.570	0.077	1.06	3.57	2.97	
				Dichlorodifluoromethane				
−58	96.56	0.2090	1.159	0.039	1.20×10^{-2}	1.94×10^{-3}	6.2	
−40	94.81	0.2113	1.024	0.040	1.08	1.99	5.4	
−22	92.99	0.2139	0.910	0.040	0.979	2.04	4.8	
−4	91.18	0.2167	0.831	0.041	0.911	2.09	4.4	
14	89.24	0.2198	0.765	0.042	0.857	2.13	4.0	
32	87.24	0.2232	0.722	0.042	0.828	2.16	3.8	
50	85.17	0.2268	0.671	0.042	0.788	2.17	3.6	
68	83.04	0.2307	0.637	0.042	0.767	2.17	3.5	
86	80.85	0.2349	0.608	0.041	0.752	2.17	3.5	
104	78.48	0.2393	0.582	0.040	0.742	2.15	3.5	
122	75.91	0.2440	0.557	0.039	0.734	2.11	3.5	

Table A.5 (continued)

t °F	$\rho \dfrac{\text{lb}_m}{\text{ft}^3}$	$c_p \dfrac{\text{Btu}}{\text{lb}_m\text{-}°\text{F}}$	$\mu \dfrac{\text{lb}_m}{\text{ft-hr}}$	$k \dfrac{\text{Btu}}{\text{hr-ft-}°\text{F}}$	$\nu \dfrac{\text{ft}^2}{\text{hr}}$	$\alpha \dfrac{\text{ft}^2}{\text{hr}}$	N_{PR}	$\beta \dfrac{1}{°\text{R}}$
				Ammonia				
-58	43.93	1.066	0.742	0.316	1.69×10^{-2}	6.75×10^{-3}	2.60	
-40	43.18	1.067	0.678	0.316	1.57	6.88	2.28	
-22	42.41	1.069	0.636	0.317	1.50	6.98	2.15	
-4	41.62	1.077	0.616	0.316	1.48	7.05	2.09	
14	40.80	1.090	0.600	0.314	1.47	7.07	2.07	
32	39.96	1.107	0.579	0.312	1.45	7.05	2.05	
50	39.09	1.126	0.559	0.307	1.43	6.98	2.04	
68	38.19	1.146	0.531	0.301	1.39	6.88	2.02	1.36×10^{-3}
86	37.23	1.168	0.503	0.293	1.35	6.75	2.01	
104	36.27	1.194	0.479	0.285	1.32	6.59	2.00	
122	35.23	1.222	0.451	0.275	1.28	6.41	1.99	
				Carbon Dioxide				
-58	72.19	0.44	0.333	0.0494	4.61×10^{-3}	1.558×10^{-3}	2.96	
-40	69.78	0.45	0.319	0.0584	4.57	1.864	2.46	
-22	67.22	0.47	0.305	0.0645	4.54	2.043	2.22	
-4	64.45	0.49	0.287	0.0665	4.46	2.110	2.12	
14	61.39	0.52	0.270	0.0635	4.39	1.989	2.20	
32	57.87	0.59	0.244	0.0604	4.21	1.774	2.38	
50	53.69	0.75	0.210	0.0561	3.92	1.398	2.80	
68	48.23	1.2	0.170	0.0504	3.53	0.860	4.10	3.67×10^{-3}
86	37.32	8.7	0.116	0.0406	3.10	0.108	28.7	
				Glycerine				
32	79.66	0.540	25,650	0.163	0.322×10^3	3.81×10^{-3}	84,700	
50	79.29	0.554	9,200	0.164	0.116	3.74	31,000	
68	78.91	0.570	3,610	0.165	0.0457	3.67	12,500	0.28×10^{-3}
86	78.54	0.584	1,520	0.165	0.0194	3.60	5,380	
104	78.16	0.600	672	0.165	0.0086	3.54	2,450	
122	77.72	0.617	451	0.166	0.0058	3.46	1,630	
				Mercury				
32	850.78	0.0335	4.08	4.74	4.79×10^{-3}	166.6×10^{-3}	0.0288	
68	847.71	0.0333	3.75	5.02	4.42	178.5	0.0249	1.01×10^{-3}
122	843.14	0.0331	3.40	5.43	4.03	194.6	0.0207	
202	835.57	0.0328	3.01	6.07	3.60	221.5	0.0162	
302	828.06	0.0326	2.73	6.64	3.30	246.2	0.0134	
392	820.61	0.0325	2.55	7.13	3.11	267.7	0.0116	
482	813.16	0.0324	2.41	7.55	2.96	287.0	0.0103	

Table A.5 (continued)

t °F	$\rho \dfrac{\text{lb}_m}{\text{ft}^3}$	$c_p \dfrac{\text{Btu}}{\text{lb}_m\text{-°F}}$	$\mu \dfrac{\text{ib}_m}{\text{ft-hr}}$	$k \dfrac{\text{Btu}}{\text{hr-ft-°F}}$	$\nu \dfrac{\text{ft}^2}{\text{hr}}$	$\alpha \dfrac{\text{ft}^2}{\text{hr}}$	N_{PR}	$\beta \dfrac{1}{\text{°R}}$
				Lubricating Oil (Approx. SAE 50)				
32	56.13	0.429	9318	0.085	166	3.53×10^{-3}	47100	
68	55.45	0.449	1935	0.084	34.9	3.38	10400	0.39×10^{-3}
104	54.69	0.469	512	0.083	9.36	3.23	2870	
140	53.94	0.489	175	0.081	3.25	3.10	1050	
176	53.19	0.509	77.1	0.080	1.45	2.98	490	
212	52.44	0.530	41.3	0.079	0.788	2.86	276	
248	51.75	0.551	24.8	0.078	0.479	2.75	175	
284	51.00	0.572	15.8	0.077	0.310	2.66	116	
320	50.31	0.593	10.9	0.076	0.216	2.57	84	

* From E. R. G. ECKERT, *Introduction to the Transfer of Heat and Mass*, New York, McGraw-Hill, 1950.

Table A.6 Properties of Steam[*]

ρ in lb_m/ft^3, μ in lb_m/ft-hr, k in Btu/hr-ft-°F, c_p in Btu/lb_m-°F

Pressure-psia →	0	14.696	200	400	600	800	1000	1200	1400	1600	1800	2000	2500	3000
Temp.-°F ↓														
Saturated vapor ρ		0.03834	0.3462	0.8611	1.2990	1.7584	2.2442	2.7632	3.3201	3.9246	4.5892	5.3248	7.6511	11.655
$\mu \times 10^2$		3.015	3.856	4.195	4.432	4.630	4.810	4.982	5.153	5.329	5.518	5.725	6.401	7.756
$k \times 10^2$		1.390	2.021	2.362	2.647	2.910	3.165	3.416	3.668	3.923	4.182	4.447	5.142	5.899
c_p		0.496	0.657	0.814	0.949	1.10	1.26	1.49	1.77	2.03	2.5	3.0	4.9	7.3
N_{PR}		1.08	1.25	1.45	1.59	1.74	1.92	2.17	2.49	2.76	3.3	3.9	6.1	9.6
300°F ρ		0.03275												
$\mu \times 10^2$		3.442												
$k \times 10^2$	1.584	1.602												
c_p	0.455	0.474												
N_{PR}		1.02												
400°F ρ		0.02884	0.4235											
$\mu \times 10^2$		3.927	3.949											
$k \times 10^2$	1.845	1.856	2.045											
c_p	0.462	0.470	0.640											
N_{PR}		0.994	1.24											
500°F ρ		0.02579	0.3668	0.77815	1.2583									
$\mu \times 10^2$		4.412	4.431	4.458	4.495									
$k \times 10^2$	2.111	2.118	2.227	2.393	2.631									
c_p	0.469	0.474	0.549	0.675	0.876									
N_{PR}		0.987	1.09	1.26	1.50									

Table A.6 (continued)

↓ Temp. °F	Pressure-psia →	0	14.696	200	400	600	800	1000	1200	1400	1600	1800	2000	2500	3000
600°F	ρ		0.02333	0.3268	0.67705	1.0567	1.4751	1.9455	2.4900	3.1506					
	$\mu \times 10^2$		4.897	4.914	4.936	4.964	4.999	5.045	5.106	5.191					
	$k \times 10^2$	2.380	2.385	2.453	2.546	2.665	2.817	3.011	3.258	3.573					
	c_p	0.477	0.480	0.521	0.578	0.653	0.752	0.88	1.08	1.40					
	N_{PR}		0.986	1.04	1.12	1.22	1.33	1.47	1.69	2.03					
700°F	ρ		0.02130	0.2959	0.60577	0.93179	1.2767	1.6437	2.0370	2.4619	2.9265	3.4400	4.0177	5.9311	10.163
	$\mu \times 10^2$		5.382	5.397	5.417	5.440	5.482	5.500	5.539	5.586	5.643	5.712	5.798	6.148	7.267
	$k \times 10^2$	2.651	2.654	2.700	2.758	2.827	2.909	3.006	3.120	3.256	3.417	3.608	3.834	4.599	5.768
	c_p	0.486	0.488	0.512	0.544	0.582	0.627	0.681	0.745	0.822	0.91	1.02	1.14	1.9	6.1
	N_{PR}		0.990	1.02	1.07	1.12	1.18	1.25	1.32	1.41	1.51	1.62	1.72	2.54	7.7
800°F	ρ		0.01961	0.2708	0.55063	0.84041	1.1412	1.4539	1.7803	2.1214	2.4789	2.8555	3.2531	4.3592	5.6818
	$\mu \times 10^2$		5.867	5.881	5.899	5.919	5.941	5.968	5.998	6.032	6.072	6.116	6.167	6.331	6.567
	$k \times 10^2$	2.924	2.926	2.958	2.997	3.041	3.090	3.146	3.207	3.281	3.362	3.453	3.556	3.876	4.310
	c_p	0.494	0.496	0.512	0.533	0.555	0.580	0.608	0.639	0.674	0.71	0.76	0.81	0.97	1.17
	N_{PR}		0.995	1.02	1.05	1.08	1.12	1.15	1.20	1.24	1.28	1.3	1.4	1.6	1.8
900°F	ρ		0.01816	0.2499	0.50589	0.76846	1.0381	1.3151	1.6000	1.8936	2.1963	2.5088	2.8313	3.6900	4.6317
	$\mu \times 10^2$		6.353	6.371	6.381	6.399	6.419	6.441	6.466	6.494	6.525	6.559	6.597	6.709	6.853
	$k \times 10^2$	3.197	3.199	3.222	3.249	3.279	3.311	3.347	3.387	3.429	3.476	3.527	3.583	3.747	3.953
	c_p	0.503	0.504	0.516	0.529	0.544	0.560	0.579	0.595	0.615	0.64	0.66	0.68	0.75	0.84
	N_{PR}		1.00	1.02	1.04	1.06	1.08	1.11	1.14	1.16	1.2	1.2	1.3	1.3	1.5

Table A.6 (continued)

Pressure-psia → / ↓ Temp.-°F		0	14.696	200	400	600	800	1000	1200	1400	1600	1800	2000	2500	3000
1000°F	ρ		0.01691	0.2321	0.4686	0.70942	0.95511	1.2057	1.4613	1.7227	1.9893	2.2619	2.5413	3.2669	4.0388
	$\mu \times 10^2$	3.471	6.838	6.850	6.864	6.880	6.898	6.918	6.939	6.963	6.988	7.016	7.047	7.134	7.241
	$k \times 10^2$		3.472	3.489	3.509	3.530	3.553	3.577	3.603	3.631	3.661	3.693	3.735	3.823	3.937
	c_p	0.512	0.512	0.520	0.532	0.543	0.553	0.565	0.578	0.589	0.60	0.62	0.63	0.67	0.72
	N_{PR}		1.01	1.02	1.04	1.06	1.07	1.09	1.11	1.13	1.1	1.2	1.2	1.3	1.3
1100°F	ρ		0.01582	0.2168	0.4367	0.6600	0.88605	1.1158	1.3492	1.5860	1.8268	2.0713	2.3196	2.9595	3.6271
	$\mu \times 10^2$	3.744	7.324	7.336	7.349	7.364	7.380	7.397	7.417	7.437	7.459	7.483	7.509	7.581	7.666
	$k \times 10^2$		3.745	3.758	3.773	3.789	3.805	3.823	3.841	3.860	3.880	3.902	3.925	3.986	4.057
	c_p	0.521	0.521	0.526	0.535	0.546	0.550	0.561	0.569	0.578	0.59	0.60	0.61	0.63	0.66
	N_{PR}		1.02	1.03	1.04	1.06	1.07	1.09	1.10	1.11	1.1	1.2	1.2	1.2	1.2
1200°F	ρ		0.01487	0.2034	0.4090	0.61698	0.82727	1.0400	1.2551	1.4730	1.6931	1.9165	2.1422	2.7189	3.3135
	$\mu \times 10^2$	4.017	7.898	7.908	7.921	7.934	7.949	7.965	7.982	8.000	8.019	8.040	8.062	8.123	8.179
	$k \times 10^2$		4.018	4.028	4.040	4.052	4.064	4.077	4.090	4.104	4.119	4.134	4.150	4.192	4.239
	c_p	0.529	0.529	0.530	0.537	0.548	0.553	0.561	0.568	0.575	0.58	0.59	0.60	0.62	0.63
	N_{PR}		1.04	1.04	1.05	1.07	1.08	1.10	1.11	1.12	1.1	1.1	1.2	1.2	1.2

* ρ computed from J. H. KEENAN and F. G. KEYES, *Thermodynamic Properties of Steam*, New York, Wiley, 1936.
μ, k, c_p computed from equations recommended in *N.A.C.A. Tech. Note 3273*.

Table A.7   Properties of Dry Air at Atmospheric Pressure*

t °F	ρ lb_m/ft^3	μ $lb_m/ft\text{-}hr$	k $Btu/hr\text{-}ft\text{-}°F$	c_p $Btu/lb\text{-}°F$	ν ft^2/hr	α ft^2/hr	N_{PR}
−100	0.11028	0.03214	0.01045	0.2405	0.2914	0.3940	0.739
− 80	0.10447	0.03365	0.01099	0.2404	0.3221	0.4377	0.736
− 60	0.09924	0.03513	0.01153	0.2404	0.3540	0.4832	0.733
− 40	0.09451	0.03658	0.01207	0.2403	0.3870	0.5315	0.728
− 20	0.09021	0.03800	0.01260	0.2403	0.4212	0.5812	0.725
0	0.08629	0.03939	0.01312	0.2403	0.4565	0.6326	0.722
20	0.08269	0.04075	0.01364	0.2403	0.4928	0.6865	0.718
40	0.07938	0.04208	0.01416	0.2404	0.5301	0.7421	0.714
60	0.07633	0.04339	0.01466	0.2404	0.5685	0.7989	0.712
80	0.07350	0.04467	0.01516	0.2405	0.6078	0.8575	0.709
100	0.07087	0.04594	0.01566	0.2406	0.6482	0.9185	0.706
120	0.06843	0.04718	0.01615	0.2407	0.6895	0.9806	0.703
140	0.06614	0.04839	0.01664	0.2409	0.7316	1.0446	0.700
160	0.06401	0.04959	0.01712	0.2411	0.7747	1.1095	0.698
180	0.06201	0.05077	0.01759	0.2413	0.8187	1.1758	0.696
200	0.06013	0.05193	0.01806	0.2415	0.8636	1.2438	0.694
220	0.05836	0.05308	0.01853	0.2418	0.9095	1.3133	0.693
240	0.05669	0.05420	0.01899	0.2421	0.9561	1.3841	0.691
260	0.05512	0.05531	0.01945	0.2424	1.0034	1.4558	0.689
280	0.05363	0.05640	0.01990	0.2427	1.0517	1.5284	0.688
300	0.05221	0.05748	0.02034	0.2431	1.1009	1.6028	0.687
320	0.05087	0.05854	0.02079	0.2435	1.1508	1.6780	0.686
340	0.04960	0.05959	0.02122	0.2439	1.2014	1.7537	0.685
360	0.04839	0.06063	0.02166	0.2443	1.2529	1.8325	0.684
380	0.04724	0.06165	0.02208	0.2447	1.3050	1.9100	0.683
400	0.04614	0.06266	0.02251	0.2452	1.3580	1.9902	0.682
420	0.04509	0.06366	0.02293	0.2457	1.4118	2.0695	0.682
440	0.04409	0.06464	0.02335	0.2462	1.4660	2.1521	0.681
460	0.04313	0.06561	0.02376	0.2467	1.5212	2.2331	0.681
480	0.04221	0.06657	0.02417	0.2472	1.5771	2.3174	0.680
500	0.04133	0.06752	0.02458	0.2478	1.6337	2.4004	0.680
520	0.04049	0.06846	0.02498	0.2483	1.6908	2.4856	0.680
540	0.03968	0.06939	0.02538	0.2489	1.7487	2.5688	0.680
560	0.03890	0.07031	0.02577	0.2495	1.8075	2.6540	0.681
580	0.03815	0.07122	0.02616	0.2501	1.8668	2.7421	0.681

Table A.7 (continued)

t °F	ρ lb$_m$/ft^3	μ lb$_m$/ft-hr	k Btu/hr-ft-°F	c_p Btu/lb-°F	ν ft^2/hr	α ft^2/hr	N_{PR}
600	0.03743	0.07212	0.02655	0.2507	1.9268	2.8305	0.681
620	0.03673	0.07301	0.02694	0.2513	1.9878	2.9187	0.681
640	0.03607	0.07389	0.02732	0.2519	2.0485	3.0055	0.682
660	0.03543	0.07477	0.02770	0.2525	2.1104	3.0950	0.682
680	0.03481	0.07563	0.02807	0.2531	2.1727	3.1862	0.682
700	0.03420	0.07649	0.02844	0.2538	2.2365	3.2765	0.683
720	0.03362	0.07734	0.02881	0.2544	2.3004	3.3696	0.683
740	0.03306	0.07818	0.02918	0.2550	2.3648	3.4615	0.683
760	0.03252	0.07901	0.02954	0.2557	2.4296	3.5505	0.684
780	0.03200	0.07984	0.02990	0.2563	2.4950	3.6463	0.684
800	0.03149	0.08066	0.03026	0.2570	2.5614	3.7404	0.685
820	0.03100	0.08147	0.03062	0.2576	2.6281	3.8323	0.686
840	0.03052	0.08227	0.03097	0.2582	2.6956	3.9302	0.686
860	0.03006	0.08307	0.03132	0.2589	2.7635	4.0257	0.686
880	0.02961	0.08386	0.03167	0.2595	2.8322	4.1237	0.687
900	0.02917	0.08464	0.03201	0.2601	2.9016	4.2174	0.688
920	0.02875	0.08542	0.03235	0.2608	2.9711	4.3133	0.689
940	0.02834	0.08620	0.03269	0.2614	3.0416	4.4116	0.690
960	0.02794	0.08696	0.03303	0.2620	3.1124	4.5123	0.690
980	0.02755	0.08772	0.03337	0.2626	3.1840	4.6155	0.690
1000	0.02717	0.08847	0.03370	0.2632	3.2562	4.7133	0.691
1050	0.02627	0.09034	0.03452	0.2648	3.4389	4.9598	0.693
1100	0.02543	0.09216	0.03533	0.2663	3.6241	5.2186	0.695
1150	0.02464	0.09396	0.03613	0.2677	3.8133	5.4742	0.697
1200	0.02390	0.09572	0.03691	0.2691	4.0050	5.7403	0.698
1250	0.02320	0.09746	0.03768	0.2706	4.2009	6.0000	0.700
1300	0.02254	0.09917	0.03844	0.2719	4.3997	6.2708	0.702
1350	0.02192	0.10085	0.03919	0.2732	4.6008	6.5426	0.703
1400	0.02133	0.10250	0.03993	0.2745	4.8054	6.8140	0.705
1450	0.02077	0.10414	0.04066	0.2758	5.0140	7.0960	0.707
1500	0.02024	0.10575	0.04137	0.2771	5.2248	7.3743	0.709

* ρ: computed from ideal gas law.

μ, k, c_p: computed from recommended equations in *Handbook of Supersonic Aerodynamics*, Vol. 5, Bureau of Ordnance, Department of the Navy, Washington, D.C., 1953.

Table A.8 Properties of Air at Elevated Pressures*

k in Btu/hr-ft-°F, c_p in Btu/lb$_m$-°R, ρ in lb$_m$/ft^3

t °F		Pressure—Atmospheres						
		1.0	4.0	7.0	10.0	40.0	70.0	100.0
−100	k	0.01053			0.01069			0.01234
	c_p	0.2405			0.2499			0.4170
	ρ	0.1104			1.1282			13.594
32	k	0.01359			0.01361			
	c_p	0.2403			0.2446			
	ρ	0.08069			0.8110			
80	k	0.01459			0.01461			0.01495
	c_p	0.2405			0.2439			0.2776
	ρ	0.07348			0.7366			7.395
260	k	0.01816			0.01817			0.01830
	c_p	0.2424			0.2441			0.2599
	ρ	0.05507			0.5498			5.350
440	k	0.02150	0.02150	0.02150	0.02150	0.02152		
	c_p	0.2462	0.2465	0.2469	0.2476	0.2506		
	ρ	0.04406	0.1761	0.3078	0.4392	1.7377		
620	k	0.02473	0.02473	0.02473	0.02473	0.02476		
	c_p	0.2513	0.2515	0.2517	1.2520	0.2542		
	ρ	0.03671	0.1467	0.2564	0.3659	1.4463		
980	k	0.03098	0.03098	0.03098	0.03098	0.03098	0.03100	
	c_p	0.2626	0.2627	0.2629	0.2630	0.2642	0.2653	
	ρ	0.02753	0.1100	0.1923	0.2744	1.0856	1.8788	
1340	k	0.03963	0.03693	0.03693	0.03693	0.03693	0.03693	
	c_p	0.2730	0.2731	0.2732	0.2733	0.2739	0.2746	
	ρ	0.02203	0.08803	0.1539	0.2197	0.8700	1.5075	
1700	k	0.04231	0.04231	0.04231	0.04231	0.04234	0.04234	0.04233
	c_p	0.2819	0.2820	0.2820	0.2820	0.2825	0.2830	0.2833
	ρ	0.01836	0.07337	0.1283	0.1831	0.7260	1.2597	1.7840
2060	k	0.04771	0.04771	0.04771	0.04771	0.04772	0.04772	0.04772
	c_p	0.2902	0.2902	0.2902	0.2903	0.2906	0.2910	0.2912
	ρ	0.01574	0.06290	0.1100	0.1570	0.6232	1.0822	1.5343
2420	k	0.05326	0.05326	0.05326	0.05326	0.05324	0.05323	0.05324
	c_p	0.2986	0.2985	0.2985	0.2985	0.2987	0.2989	0.2992
	ρ	0.01377	0.05504	0.09625	0.1374	0.5459	0.9487	1.3461
3140	k	0.06550	0.06509	0.06500	0.06497	0.06487	0.06485	0.06482
	c_p	0.3198	0.3173	0.3167	0.3165	0.3158	0.3159	0.3159
	ρ	0.01101	0.04404	0.07702	0.1099	0.4374	0.7612	1.0815
4940	k	0.1544	0.1244	0.1166	0.1126	0.1020	0.09922	0.09774
	c_p	0.6835	0.5217	0.4796	0.4613	0.4052	0.3895	0.3843
	ρ	0.00717	0.02900	0.05087	0.07275	0.2912	0.5082	0.7236

* Adapted from *Handbook of Supersonic Aerodynamics*, Vol. 5, Bureau of Ordnance, Department of the Navy, Washington, D.C., 1953.

Table A.9   Thermal Conductivities* of Some Gases at Low Pressures†

Gas	t °F	k Btu/hr-ft-°F	Gas	t °F	k Btu/hr-ft-°F
Acetone (1)	32	0.0057	Carbon monoxide	−328	0.0037
	115	0.0074	(3)	−148	0.0088
	212	0.0099		32	0.0134
	363	0.0147		212	0.0176
Acetylene (2)	−103	0.0068	Carbon		
	32	0.0108	tetrachloride	115	0.0041
	122	0.0140	(1)	212	0.0052
	212	0.0172		363	0.0065
Ammonia (3)	−58	0.0097	Dichlorodifluoro-		
	32	0.0126	methane	32	0.0048
	212	0.0192	(F-12)	122	0.0064
	392	0.0280		212	0.0080
	572	0.0385		302	0.0097
	752	0.0509			
			Ethane (2)	−94	0.0066
Argon (3)	−148	0.0063		−29	0.0086
	32	0.0095		32	0.0106
	212	0.0123		212	0.0175
	392	0.0148			
	572	0.0171	Ethyl alcohol	68	0.0089
			(1)	212	0.0124
Butane (n—) (4)	32	0.0078			
	212	0.0135	Ether (1)	32	0.0077
				115	0.0099
Butane (iso-) (4)	32	0.0080		212	0.0131
	212	0.0139		363	0.0189
				413	0.0209
Benzene (1)	32	0.0052			
	115	0.0073	Ethylene (2)	−96	0.0064
	212	0.0103		32	0.0101
	363	0.0152		122	0.0131
	413	0.0176		212	0.0161
Carbon dioxide	−58	0.0064	Helium (3)	−328	0.0338
(3)	32	0.0084		−148	0.0612
	212	0.0128		32	0.0818
	392	0.0177		212	0.0988
	572	0.0229			

Table A.9 (continued)

Gas	t °F	k Btu/hr-ft-°F	Gas	t °F	k Btu/hr-ft-°F
Hydrogen (3)	−328	0.0293	Methylene		
	−148	0.0652	chloride (1)	32	0.0039
	32	0.0966		115	0.0049
	212	0.1240		212	0.0063
	392	0.1484		413	0.0095
	572	0.1705			
			Nitrogen (3)	−328	0.0040
Methane (3)	−328	0.0045		−148	0.0091
	−148	0.0109		32	0.0139
	32	0.0176		212	0.0181
	212	0.0255		392	0.0220
	392	0.0358		572	0.0255
	572	0.0490		752	0.0287
Methyl chloride	32	0.0053	Oxygen (3)	−328	0.0038
(1)	115	0.0072		−148	0.0091
	212	0.0094		32	0.0142
	363	0.0130		122	0.0166
	413	0.0148		212	0.0188
			Sulphur dioxide	32	0.0050
			(5)	212	0.0069

* Adapted from W. H. MCADAMS, *Heat Transmission*, 3rd ed., New York, McGraw-Hill, 1954, as compiled from (1) MOSER, Dissertation, Berlin, 1913; (2) A. EUCKEN, *Physik Zeitschr.*, Vol. 12, 1911, p. 1101; (3) F. G. KEYES, *Tech. Rept. 37*, Project Squid, 1952; (4) W. B. MANN and B. G. DICKENS, *Proc. Royal Soc.*, A134, 1931, p. 77; (5) B. G. DICKENS, *Proc. Royal Soc.*, A143, 1934, p. 517.

† These values may be interpreted as applying at a pressure of 1 atm except when the gaseous phase cannot exist at the temperature quoted—in which cases the values should be interpreted as applying at the corresponding saturation pressure.

Table A.10   Specific Heat Equations, Critical Constants, van der Waals' Constants, and Sutherland's Constants for Various Gases*

Gas	Molecular Weight	Specific Heat Equation c_p in Btu/lb$_m$-°R, T in °R	Critical Constants (1) p_c atm	Critical Constants (1) T_c °R	van der Waals' Constants (1) a $\dfrac{\text{atm}-\text{ft}^6}{(\text{lb-mole})^2}$	van der Waals' Constants (1) b $\dfrac{\text{ft}^3}{(\text{lb-mole})}$	Sutherland Constants (1) $C_1 \times 10^8$ Slugs $\overline{\text{ft-sec-}(\text{°R})^{1/2}}$	Sutherland Constants (1) C_2 °R
Air	29	(2) $c_p = 0.219 + 0.342 \times 10^{-4}T - 0.293 \times 10^{-8}T^2$ (500 − 2700°R)	37.2	238.4	343.5	0.585	2.270	198.7
Oxygen	32	(3) $c_p = 0.36 - 5.375T^{-1/2} + 47.8T^{-1}$ (540 − 5000°R)	49.7	277.8	349.5	0.510	2.57	198
Nitrogen	28.02	(3) $c_p = 0.338 - 123.8T^{-1} + 4.14 \times 10^4 T^{-2}$ (540 − 9000°R)	33.5	226.9	346	0.618	2.16	184
Carbon dioxide	44	(3) $c_p = 0.368 - 148.4T^{-1} + 3.2 \times 10^4 T^{-2}$ (540 − 6300°R)	73.0	547.7	925	0.686	2.42	420
Carbon monoxide	28	(3) $c_p = 0.338 - 117.5T^{-1} + 3.82 \times 10^4 T^{-2}$ (540 − 9000°R)	35.0	241.5	381	0.639	2.18	196
Hydrogen	2.02	(4) $c_p = 2.857 + 2.867 \times 10^{-4}T + 9.92T^{-1/2}$ (540 − 4000°R)	12.8	59.9	62.8	0.426	1.01	127
Methane	16.03	(5) $c_p = 0.211 + 6.25 \times 10^{-4}T - 8.28 \times 10^{-8}T^2$ (540 − 2700°R)	45.8	343.2	581.2	0.6855	1.53	279

Table A.10 (continued)

Gas	Molecular Weight	Specific Heat Equation c_p in Btu/lb$_m$-°R, T in °R	Critical Constants (1)		van der Waals' Constants (1)		Sutherland Constants (1)	
			p_c atm	T_c °R	a $\dfrac{\text{atm}-\text{ft}^6}{(\text{lb-mole})^2}$	b $\dfrac{\text{ft}^3}{(\text{lb-mole})}$	$C_1 \times 10^8$ Slugs ft-sec-(°R)$^{1/2}$	C_2 °R
Ethyl alcohol	46.0	(7) $c_p = 0.0978 + 0.459 \times 10^{-3}T$ $(680 - 1120°\text{R})$	63.1	929.3	3077	1.344	1.82	720
Benzene	78.0	(7) $c_p = 0.0833 + 0.371 \times 10^{-3}T$ $(520 - 1120°\text{R})$	47.7	1011	4820	1.935	1.61	726
Freon-12	120.92	(8) $c_p = 0.0775 + 0.128 \times 10^{-3}T$ $(460 - 600°\text{R})$	39.6	692.4	2726	1.595	2.31	570
Ammonia	17.03	(6) $c_p = 0.363 + 2.57 \times 10^{-4}T - 1.319 \times 10^{-8}T^2$ $(540 - 1800°\text{R})$	111.5	730.0	1076	0.598	2.40	849
Helium	4.00	$c_p = 1.24$ (all temperatures)	2.26	946.8	8.57	0.372	2.36	176
Mercury vapor	200.61	$c_p = 0.0248$ $(540 - 3600°\text{R})$	> 200	> 3281.7	5100	1.070	9.81	1792

* Adapted from (1) N. A. HALL, *Thermodynamics of High Velocity Flow*, Englewood Cliffs, N.J., Prentice-Hall, 1951; (2) H. M. SPENCER and J. L. JUSTICE, *J. Am. Chem. Soc.* Vol. 56, 1934, p. 2311; (3) R. L. SWEIGERT and M. W. BEARDSLEY, *Ga. Inst. Tech., Eng. Expt. Sta. Bull.* 2, 1938; (4) J. CHIPMAN and M. G. FONTANA, *J. Am. Chem. Soc.*, Vol. 57, 1935, p. 48; (5) H. M. SPENCER, *J. Am. Chem. Soc.*, Vol. 67, 1945, p. 1859; (6) H. M. SPENCER and G. N. FLANNAGAN, *J. Am. Chem. Soc.*, Vol. 64, 1942, p. 2511; (7) G. S. PARKS and H. M. HUFFMAN, *Am. Chem. Soc. Mon. 60*, 1932; (8) R. M. BUFFINGTON and J. FLEISCHER, *Ind. Eng. Chem.*, Vol. 23, 1931, p. 1290.

Table A.11 Properties of Some Liquid Metals*

Metal	t °F	ρ lb_m/ft^3	μ $lb_m/ft\text{-}hr$	k Btu/hr-ft-°F	c_p Btu/lb_m-°R	ν ft²/hr	α ft²/hr	N_{PR}
	600	625	3.88	9.5	0.0345	0.0062	0.44	0.014
	800	616	3.26	9.0	0.0357	0.0053	0.41	0.013
Bismuth	1000	608	2.68	9.0	0.0369	0.0044	0.40	0.011
	1200	600	2.22	9.0	0.0381	0.0037	0.39	0.0094
	1400	591	1.95	9.0	0.0393	0.0033	0.39	0.0084
	700	658	5.86	9.3	0.038	0.0089	0.37	0.024
	850	652	4.82	9.0	0.037	0.0074	0.37	0.020
Lead	1000	646	4.07	8.9	0.037	0.0063	0.37	0.017
	1150	639	3.77	8.7	0.037	0.0059	0.37	0.016
	1300	633		8.6				
	50	847	3.90	4.7	0.033	0.0046	0.17	0.027
	200	834	2.92	6.0	0.033	0.0035	0.22	0.016
Mercury	300	826	2.48	6.7	0.033	0.0030	0.25	0.012
	400	817	2.45	7.2	0.032	0.0030	0.27	0.011
	600	802	2.08	8.1	0.032	0.0026	0.31	0.0084
	300	50.4	0.907	26.0	0.19	0.018	2.7	0.0066
	500	48.7	0.584	24.7	0.19	0.012	2.7	0.0043
Potassium	800	46.3	0.440	22.8	0.18	0.0095	2.7	0.0035
	1100	43.8	0.364	20.6	0.18	0.0083	2.6	0.0032
	1300	42.1	0.328	19.1	0.18	0.0078	2.5	0.0031
	200	58.0	1.68	49.8	0.33	0.029	2.6	0.011
	400	56.3	1.07	46.4	0.32	0.019	2.6	0.0072
Sodium	700	53.7	0.70	41.8	0.31	0.013	2.5	0.0050
	1000	51.2	0.49	37.8	0.30	0.0096	2.4	0.0040
	1300	48.6	0.44	34.5	0.30	0.0091	2.4	0.0038
	850	431	7.76	33.7	0.119	0.018	0.66	0.027
Zinc	1000	428	6.42	33.2	0.116	0.015	0.67	0.022
	1200	422	5.06	32.8	0.113	0.012	0.69	0.017
	1500	408	4.08	32.6	0.107	0.010	0.74	0.014
	300	657		5.23	0.035		0.227	
55.5% Bi	550	646	4.26	6.20	0.035	0.0066	0.274	0.024
44.5% Pb	700	639	3.71	6.85	0.035	0.0058	0.306	0.019
(Eutectic)	1100	620						
	1200	614						
	200	55.4	1.39	14.8	0.270	0.025	0.994	0.026
44% K	400	53.8	0.86	15.3	0.261	0.016	1.09	0.015
56% Na	700	51.3	0.56	15.9	0.252	0.011	1.23	0.0090
	1000	48.8	0.43	16.4	0.248	0.0088	1.35	0.0065
	1300	46.2	0.39	16.7	0.249	0.0084	1.45	0.0058

* Adapted from M. JAKOB, *Heat Transfer*, Vol. II, New York, Wiley, 1957, p. 564. Original data is contained in *Liquid Metals Handbook*, 2nd ed. (rev.), *NAVEXOS P-733 (rev.)*, Atomic Energy Commission, Department of the Navy, Washington, D.C., Jan. 1954.

Table A.12 Emissivities of Various Surfaces*

Surface Description (Reference)	Emissivity, ϵ/T, °F
METALS AND METAL PLATING	
Aluminum	
Foil, bright side, as received (1)	0.04/700
75S-T alloy, weathered (1)	0.16/1500
1100-0, commercially pure (1)	0.05/200; 0.05/400; 0.05/600; 0.05/800
1100-0, commercially pure, oxidized at 600°F (1)	0.04/200; 0.04/400; 0.05/600; 0.05/800
24S-T81, with chromic acid anodize (1)	0.17/300; 0.17/335; 0.17/365; 0.17/400
24S-T81, with H_2SO_4 anodize (1)	0.85/300; 0.82/335; 0.80/365; 0.78/800
Beryllium (1)	0.16/300; 0.21/700; 0.26/900; 0.30/1100
Beryllium, anodized (1)	0.90/300; 0.88/700; 0.85/900; 0.82/1100
Brass	
Highly polished (2)	0.030/530
Polished (8)	0.10/100; 0.10/600
Rolled plate (3)	0.06/72
Chromium	
Polished (6)	0.08/100; 0.36/2000
0.1 mil thick plate on 0.5 mil nickel on 321 stainless steel (1)	0.12/200; 0.13/340; 0.14/480; 0.15/620; 0.15/750
Gold, evaporated on fiber glass (1)	0.05/200; 0.05/300; 0.75/500
Gold, coated on stainless steel (1)	0.09/200; 0.09/340; 0.11/490; 0.15/600; 0.14/750
Iron	
Cast iron, polished (5)	0.21/392
Cast iron, oxidized (5)	0.64/390; 0.78/1100
Iron plate, completely rusted (3)	0.69/67
Wrought iron, polished (7)	0.28/100; 0.28/480
Wrought iron, oxidized (7)	0.94/70; 0.94/680
Platinum on polished steel (1)	0.13/200; 0.15/340; 0.14/480; 0.15/620; 0.15/750

Table A.12 (continued)

Surface Description (Reference)	Emissivity, $\epsilon/T, °F$
Silver	
Pure, polished (2)	0.020/440; 0.032/1160
5 mil silver plate on 0.5 mil nickel on 321 stainless steel (1)	0.06/200; 0.06/340; 0.06/480; 0.07/620; 0.08/750
Silver plate on 321 stainless steel (1)	0.11/200; 0.11/400; 0.11/600; 0.13/800
Steel	
Polished plate (4)	0.066/212
Rough plate (8)	0.94/100; 0.97/700
Stainless steel, type 301 (1)	0.14/200; 0.15/400; 0.16/600; 0.18/800
Stainless steel, type 321, oxidized at 500°F (1)	0.27/200; 0.27/400; 0.28/600; 0.32/800
Stainless steel, type 321 with black oxide (1)	0.66/200; 0.66/400; 0.69/600; 0.76/800
Inconel X, oxidized at 1925°F (1)	0.61/200; 0.79/340; 0.81/480; 0.86/620; 0.81/750
Tungsten filament (9)	0.39/6000
PAINTS AND LACQUERS	
Aluminized silicone resin paint, on 321 stainless steel, baked at 600°F (1)	0.20/200; 0.20/400; 0.21/600; 0.22/800
Black lacquer on iron (3)	0.88/76
Black lacquer, flat black (8)	0.96/100; 0.98/200
Black, heat resistant, on 321 stainless (1)	0.81/200; 0.76/400; 0.76/600; 0.80/800
Black, high heat, Dixon 208, on stainless (1)	0.93/300; 0.92/500; 0.93/700; 0.93/900; 0.93/1100
International orange on 2024-T4A1 (1)	0.72/200; 0.68/400; 0.53/600
White enamel, fused on iron (3)	0.90/66
White acrylic resin paint on A1 (1)	0.90/200; 0.87/400

Table A.12 (continued)

Surface Description (Reference)	Emissivity, $\epsilon/T,°F$
MISCELLANEOUS	
Asbestos board (3)	0.96/74
Asbestos paper (8)	0.93/100; 0.94/700
Brick	
Red, rough (3)	0.93/70
Building (10)	0.45/1832
Fire clay (10)	0.75/1832
Magnesite refractory (10)	0.38/1832
Lampblack, thick coat on iron (3)	0.97/68
Graphite, pressed (11)	0.98/480; 0.98/950
Plaster, rough lime (7)	0.91/50; 0.91/190
Glass (3)	0.94/72
Concrete tile (10)	0.63/1832
Roofing paper (3)	0.91/69

* When more than one value is given, linear interpolation is permissible.
Compiled from data given in (1) D. K. EDWARDS, K. E. NELSON, R. D. RODDICK, and G. T. GIER, "Basic Studies on the Use and Control of Solar Energy," *Univ. Calif. Dept. Eng. Rept. 60–93*, Los Angeles, 1960; and from data given in W. H. McADAMS, *Heat Transmission*, 3rd ed., New York, McGraw-Hill, 1954, as compiled by H. C. HOTTEL, from (2) H. SCHMIDT and E. FURTHMANN, *Mitt. Kaiser-Wilhelm-Inst. Eisenforsch., Abhandl.*, Vol. 109, 1928, p. 225; (3) E. SCHMIDT, *Gesundh-Ing.*, Beiheft 20, Reihe 1, 1927; (4) B. T. BARNES, FORSYTHE, and ADAMS, *J. Opt. Soc. Am.*, Vol. 37, 1947, p. 804; (5) C. F. RANDOLPH and M. J. OVERHOLTZER, *Phys. Rev.*, Vol. 2, 1913, p. 144; (6) E. O. HULBERT, *Astrophys. J.*, Vol. 42, 1915, p. 205; (7) F. WAMSLER, *Zeitschr. V.D.I.*, Vol. 55, 1911, p. 599; *Mitt Forsch.*, Vol. 98, 1911, p. 1; (8) HEILMANN, *Trans. ASME*, Vol. 32, 1944, p. 239; (9) C. ZWIKKER, *Arch. Neerland Sci.* Vol. 9, 1925, p. 207; (10) M. W. THRING, *Sciences of Flames and Furnaces*, London, Chapman & Hall, 1952; (11) M. PIRANI, *J. Sci. Instr.*, Vol. 16, 1939.

Table A.13   Solar Absorptivity of Various Surfaces*

Surface Description	Solar Absorptivity α_s	Infrared Emissivity				
		450°R	555°R		1000°R	
		ϵ	ϵ	α_s/ϵ	ϵ	α_s/ϵ
Aluminum foil, coated with 10μ silicon	0.522		0.12	4.35	0.12	4.35
Silicon solar cell, 1 mm thick	0.938		0.316	2.97	0.497	1.89
Chromium plate 0.1 mil thick on 0.5 mil nickel on 321 stainless steel	0.778		0.150	5.18	0.182	4.27
Stainless steel, type 410	0.764		0.130	5.88	0.180	4.24
Titanium 75A, heated 850°F 300 hr	0.798		0.211	3.78	0.294	2.72
Titanium C-110, heated 800°F 100 hr	0.524		0.162	3.24	0.202	2.59
Titanium, anodized	0.515	0.866		0.59	0.835	0.62
Ebanol C on copper	0.908		0.11	8.25		
Ebanol S on steel	0.848		0.10	8.48		
Aluminum, 6061-T-6, 1 mil anodize	0.923	0.841		1.10	0.847	1.09
Inconel X, oxidized	0.898	0.711		1.26	0.809	1.11
Stainless steel 301, with Armco black oxide	0.891	0.746		1.19	0.756	1.18
Graphite, on sodium silicate on polished aluminum	0.960	0.908		1.06	0.930	1.03
Glass, 3 mils on silicon solar cell	0.925	0.843		1.10	0.877	1.05
Titanox, 2 mils on black paint	0.154	0.885		0.17	0.905	0.17
Flat black epoxy paint on aluminum	0.951	0.888		1.07	0.924	1.03
White epoxy paint on aluminum	0.248	0.882		0.28	0.912	0.27

* Compiled from data given in D. K. EDWARDS, K. E. NELSON, R. D. RODDICK, and J. T. GIER, "Basic Studies on the Use and Control of Solar Energy," *Univ. Calif. Dept. Eng. Rept. 60-93*, Los Angeles, 1960.

Table A.14   Planck Radiation Functions*

λT	$\dfrac{W_{b\lambda}}{\sigma T^5} \times 10^5$	$\dfrac{W_{b(0-\lambda)}}{\sigma T^4}$	λT	$\dfrac{W_{b\lambda}}{\sigma T^5} \times 10^5$	$\dfrac{W_{b(0-\lambda)}}{\sigma T^4}$	λT	$\dfrac{W_{b\lambda}}{\sigma T^5} \times 10^5$	$\dfrac{W_{b(0-\lambda)}}{\sigma T^4}$
1000	0.0000394	0	7200	10.089	0.4809	13400	2.714	0.8317
1200	0.001184	0	7400	9.723	0.5007	13600	2.605	0.8370
1400	0.01194	0	7600	9.357	0.5199	13800	2.502	0.8421
1600	0.0618	0.0001	7800	8.997	0.5381	14000	2.416	0.8470
1800	0.2070	0.0003	8000	8.642	0.5558	14200	2.309	0.8517
2000	0.5151	0.0009	8200	8.293	0.5727	14400	2.219	0.8563
2200	1.0384	0.0025	8400	7.954	0.5890	14600	2.134	0.8606
2400	1.791	0.0053	8600	7.624	0.6045	14800	2.052	0.8648
2600	2.753	0.0098	8800	7.304	0.6195	15000	1.972	0.8688
2800	3.872	0.0164	9000	6.995	0.6337	16000	1.633	0.8868
3000	5.081	0.0254	9200	6.697	0.6474	17000	1.360	0.9017
3200	6.312	0.0368	9400	6.411	0.6606	18000	1.140	0.9142
3400	7.506	0.0506	9600	6.136	0.6731	19000	0.962	0.9247
3600	8.613	0.0667	9800	5.872	0.6851	20000	0.817	0.9335
3800	9.601	0.0850	10000	5.619	0.6966	21000	0.702	0.9411
4000	10.450	0.1051	10200	5.378	0.7076	22000	0.599	0.9475
4200	11.151	0.1267	10400	5.146	0.7181	23000	0.516	0.9531
4400	11.704	0.1496	10600	4.925	0.7282	24000	0.448	0.9589
4600	12.114	0.1734	10800	4.714	0.7378	25000	0.390	0.9621
4800	12.392	0.1979	11000	4.512	0.7474	26000	0.341	0.9657
5000	12.556	0.2229	11200	4.320	0.7559	27000	0.300	0.9689
5200	12.607	0.2481	11400	4.137	0.7643	28000	0.265	0.9718
5400	12.571	0.2733	11600	3.962	0.7724	29000	0.234	0.9742
5600	12.458	0.2983	11800	3.795	0.7802	30000	0.208	0.9765
5800	12.282	0.3230	12000	3.637	0.7876	40000	0.0741	0.9881
6000	12.053	0.3474	12200	3.485	0.7947	50000	0.0326	0.9941
6200	11.783	0.3712	12400	3.341	0.8015	60000	0.0165	0.9963
6400	11.480	0.3945	12600	3.203	0.8081	70000	0.0092	0.9981
6600	11.152	0.4171	12800	3.071	0.8144	80000	0.0055	0.9987
6800	10.808	0.4391	13000	2.947	0.8204	90000	0.0035	0.9990
7000	10.451	0.4604	13200	2.827	0.8262	100000	0.0023	0.9992
							0	1.0000

* From R. V. DUNKLE, *Trans. ASME*, Vol. 76, No. 549, 1954.

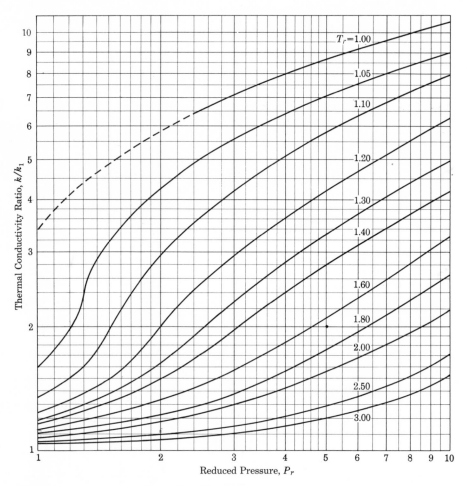

Fig. A.1. A generalized correlation chart of the thermal conductivity of gases at high pressures: k = thermal conductivity at high pressure, k_1 = thermal conductivity at atmospheric pressure and same temperature. (Reprinted from E. W. Comings and M. F. Nathan, *Ind. Eng. Chem.* Vol. 39, 1947, pp. 964–970. Copyright 1947 by the American Chemical Society and reprinted by permission of the copyright owner.)

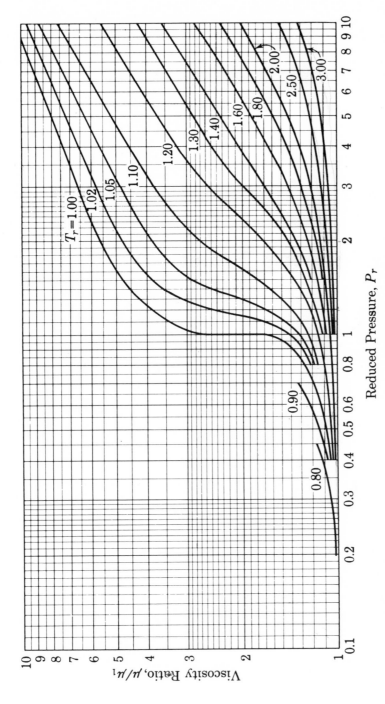

Fig. A.2. A generalized correlation chart of the dynamic viscosity of gases at high pressures: μ = viscosity at high pressure, μ_1 = viscosity atmospheric pressure and same temperature. (From E. W. Comings, B. J. Mayland, and R. S. Egly, *Univ. Illinois Eng. Expt. Stat. Bull. 354,* Urbana, Ill., 1944.)

Miscellaneous Tables

Table B.1 Conversion Factors for Units of Viscosity

Multiply by Entry ↓ to Obtain→	Centipoise	$\dfrac{g_f\text{-sec}}{cm^2}$	$\dfrac{lb_f\text{-sec}}{in^2}$	$\dfrac{lb_f\text{-hr}}{ft^2}$	$\dfrac{lb_m}{ft\text{-hr}}$	$\dfrac{kg}{m\text{-sec}}$
Centipoise	1	1.0197×10^{-5}	1.4504×10^{-7}	5.8016×10^{-9}	2.4191	1.0000×10^{-3}
$\dfrac{g_f\text{-sec}}{cm^2}$	9.8067×10^{4}	1	1.4224×10^{-2}	5.6895×10^{-4}	2.3723×10^{5}	9.8067×10
$\dfrac{lb_f\text{-sec}}{in^2}$	6.8947×10^{6}	7.0305×10^{1}	1	4.0000×10^{-2}	1.6679×10^{7}	6.8947×10^{3}
$\dfrac{lb_f\text{-hr}}{ft^2}$	1.7237×10^{8}	1.7577×10^{3}	2.5001×10^{1}	1	4.1698×10^{8}	1.7237×10^{5}
$\dfrac{lb_m}{ft\text{-hr}}$	4.1338×10^{-1}	4.2152×10^{-6}	5.9957×10^{-8}	2.3983×10^{-9}	1	4.1338×10^{-4}
$\dfrac{kg}{m\text{-sec}}$	1.0000×10^{3}	1.0197×10^{-2}	1.4504×10^{-4}	5.8016×10^{-6}	2.4191×10^{3}	1

Table B.2   Dimensions of Standard Tube Sizes (in inches)

O.D.	Gage	I.D.	O.D.	Gage	I.D.
5/8	18	0.527	3/4	18	0.652
	16	0.495		16	0.620
	14	0.459		14	0.584
1	18	0.902	1¼	18	1.152
	16	0.870		16	1.120
	14	0.834		14	1.084
	12	0.782		12	1.032
	10	0.732		10	0.982
1½	13	1.310	2	13	1.810
	12	1.282		12	1.782
	11	1.260		11	1.760
	10	1.232		10	1.732
	9	1.204		9	1.704
	8	1.170		8	1.670

Table B.3   Dimensions of Commercial Pipe (in inches)

Nominal Size	Approximate Diameters			
	Schedule 40		Schedule 80	
	O.D.	I.D.	O.D.	I.D.
½	0.840	0.622	0.840	0.546
1	1.315	1.049	1.315	0.957
2	2.375	2.067	2.375	1.939
3	3.500	3.068	3.500	2.900
4	4.500	4.026	4.500	3.826
5	5.563	5.047	5.563	4.813
6	6.625	6.065	6.625	5.761
8	8.625	7.981	8.625	7.625
10	10.750	10.020	10.750	9.564
12	12.750	11.938	12.750	11.376

A Short Summary of Bessel Functions with Brief Tables

An equation of some importance which is encountered quite frequently in certain phases of applied mathematics (Chapters 3 and 4 of this book) is

$$\frac{d^2y}{dx^2} + \frac{1}{x}\frac{dy}{dx} + \left(1 - \frac{n^2}{x^2}\right)y = 0, \tag{C.1}$$

in which n is a constant. This is called *Bessel's equation of order n.* The various forms of the solution to this equation are termed *Bessel functions.*

Much of the difficulty that is often experienced by many students in gaining an understanding of Bessel functions may be avoided if one approaches their definitions as simply the solutions to a particular differential equation. The same approach may be used to define other more familiar functions. For example, the method of Frobenius for expressing the solutions of a differential equation as infinite series shows that the two solutions to

$$\frac{d^2y}{dx^2} + y = 0$$

are

$$y_1 = 1 - \frac{x^2}{2!} + \frac{x^4}{4!} - \frac{x^6}{6!} + \cdots,$$

$$y_2 = x - \frac{x^3}{3!} + \frac{x^5}{5!} - \cdots.$$

For convenience one could define the symbols $\cos x = y_1$ and $\sin x = y_2$ as abbreviations for the above series. Without any reference to the sides of a right triangle, one could then derive all the familiar trigonometric properties and identities from these definitions.

In a similar fashion, the Bessel functions are simply defined as the solutions to Bessel's equation, Eq. (C.1). This equation is a linear, second-order, ordinary, homogeneous differential equation and as such has two linearly independent solutions. The general solution is then obtained by adding these two solutions, each solution being multiplied by an arbitrary constant. The method of Frobenius gives one such solution to be

$$J_n(x) = \sum_{p=0}^{\infty} (-1)^p \frac{1}{\Gamma(n + p + 1)p!}\left(\frac{x}{2}\right)^{n+2p}, \qquad (C.2)$$

where $\Gamma(n + p + 1)$ represents Euler's gamma function. This series is represented by the symbol $J_n(x)$ for convenience and is called the "Bessel function of the first kind, order n." The second solution is found to be

$$J_{-n}(x) = \sum_{p=0}^{\infty} (-1)^p \frac{1}{\Gamma(-n + p + 1)p!}\left(\frac{x}{2}\right)^{-n+2p}. \qquad (C.3)$$

Thus, a general solution to Eq. (C.1) is

$$y = AJ_n(x) + BJ_{-n}(x).$$

However, this solution is not always the general one, for if the constant n is an *integer*, an expansion of the two series given in Eqs. (C.2) and (C.3) will show [since $\Gamma(m) = (m - 1)!$ for $m =$ an integer]

$$J_{-n}(x) = (-1)^n J_n(x), \qquad \text{for } n \text{ an integer.}$$

So, when n is an integer (usually the case in physical problems involving Bessel functions), the two solutions given above are *not* linearly independent. The result is a single solution to Bessel's equation, but since it is a second-order equation, two independent solutions are required. Thus, in the case of $n =$ an integer (including 0), a second solution must be sought.

The second solution is

$$Y_n(x) = \frac{2}{\pi}\left[\ln\left(\frac{x}{2}\right) + \gamma\right]J_n(x) - \frac{1}{2}\sum_{p=0}^{n-1}\frac{(n - p - 1)!}{p!}\left(\frac{x}{2}\right)^{2p-n}$$

$$+ \frac{1}{2}\sum_{p=0}^{\infty} (-1)^{p+1}\frac{(x/2)^{n+2p}}{(n + p)!p!}[\varphi(n + p) + \varphi(p)]\Bigg]. \qquad (C.4)$$

In this expression,

$$\varphi(k) = 1 + \frac{1}{2} + \frac{1}{3} + \cdots + \frac{1}{k} = \sum_{r=1}^{k} \frac{1}{r},$$

$$\gamma = \text{Euler's constant} = \lim_{m \to \infty} [\varphi(m) - \ln m] = 0.5772157.$$

Also, note that

$$Y_{-n}(x) = (-1)^n Y_n(x).$$

For $n = 0$ the second solution is

$$Y_0(x) = \frac{2}{\pi} \left[\ln \left(\frac{x}{2} + \gamma \right) J_0(x) + \sum_{p=0}^{\infty} (-1)^{p+1} \frac{(x/2)^{2p}}{p!\,p!} \varphi(p) \right]. \qquad \text{(C.5)}$$

Thus, the general solution to Eq. (C.1) is

$$y = AJ_n(x) + BY_n(x), \qquad \text{for } n = 0 \text{ or an integer.}$$

Consider now the following variation of Bessel's equation:

$$\frac{d^2 y}{dx^2} + \frac{1}{x} \frac{dy}{dx} + \left(c^2 - \frac{n^2}{x^2} \right) y = 0. \qquad \text{(C.6)}$$

This may be altered by use of the transformation

$$z = cx,$$

giving

$$\frac{d^2 y}{dz^2} + \frac{1}{z} \frac{dy}{dz} + \left(1 - \frac{n^2}{z^2} \right) y = 0.$$

This latter form is the standard form of Bessel's equation as stated in Eq. (C.1). Thus, the solution to Eq. (C.6) is

$$y = AJ_n(cx) + BJ_{-n}(cx), \qquad n \neq \text{integer},$$

$$y = AJ_n(cx) + BY_n(cx), \qquad n = \text{integer}.$$

Consider the equation

$$\frac{d^2y}{dx^2} + \frac{1}{x}\frac{dy}{dx} + \left(-1 - \frac{n^2}{x^2}\right)y = 0. \tag{C.7}$$

This equation may be transformed by $z = ix \, (i = \sqrt{-1})$ into the standard form of Bessel's equation:

$$\frac{d^2y}{dz^2} + \frac{1}{z}\frac{dy}{dz} + \left(1 - \frac{n^2}{z^2}\right)y = 0.$$

So the solution to Eq. (C.7) may be written

$$y = AJ_n(ix) + BJ_{-n}(ix), \qquad n \neq \text{integer},$$
$$y = AJ_n(ix) + BY_n(ix), \qquad n = \text{integer}. \tag{C.8}$$

It is conventional to define new functions of real argument rather than leave the solution in the forms given above. These definitions are:

"Modified Bessel function of the first kind": $I_n(x) = i^{-n}J_n(ix)$. (C.9)

"Modified Bessel function of the second kind":

$$K_n(x) = \frac{\pi}{2}i^{n+1}[i^n I_n(x) + iY_n(ix)]. \tag{C.10}$$

Substitution of Eqs. (C.9) and (C.10) into the solution of Eq. (C.8) gives the following way of expressing the solution of Eq. (C.7) (using the case of $n =$ integer as an example):

$$y = Ai^n I_n(x) + B\frac{1}{i}\left[\frac{2}{\pi}\frac{1}{i^{n+1}}K_n(x) - i^n I_n(x)\right]$$

$$= CI_n(x) + DK_n(x),$$

in which C and D are other arbitrary constants involving A, B, and i. For $n \neq$ an integer,

$$y = CI_n(x) + DI_{-n}(x).$$

Summarizing: For

$$\frac{d^2y}{dx^2} + \frac{1}{x}\frac{dy}{dx} + \left(c^2 - \frac{n^2}{x^2}\right)y = 0,$$

the solution is

$$y = AJ_n(cx) + BJ_{-n}(cx), \qquad n \neq \text{integer},$$

$$y = AJ_n(cx) + BY_n(cx), \qquad n = \text{integer}.$$

For

$$\frac{d^2y}{dx^2} + \frac{1}{x}\frac{dy}{dx} + \left(-c^2 - \frac{n^2}{x^2}\right)y = 0,$$

the solution is

$$y = CI_n(cx) + DI_{-n}(cx), \qquad n \neq \text{integer},$$

$$y = CI_n(cx) + DK_n(cx), \qquad n = \text{integer}.$$

Also:

$$\left.\begin{aligned}
J_{-n}(x) &= (-1)^n J_n(x) \\
Y_{-n}(x) &= (-1)^n Y_n(x) \\
K_{-n}(x) &= K_n(x) \\
I_{-n}(x) &= I_n(x)
\end{aligned}\right\} n = \text{integer}.$$

Values of J_n, Y_n, I_n, and K_n for $n = 0$ and $n = 1$ are given in Tables C.1 and C.2. These orders, 0 and 1, are those most frequently encountered in physical problems which are described mathematically by Bessel's equation.

From the definition of J_n in Eq. (C.2) one may write

$$u^n J_n(u) = \sum_{p=0}^{\infty} (-1)^p \frac{1}{\Gamma(p + n + 1)p!}\left(\frac{1}{2}\right)^{n+2p} u^{2n+2p}.$$

If u is a function of x and if the above function is differentiated with respect to x,

$$\frac{d}{dx}[u^n J_n(u)] = \sum_{p=0}^{\infty}(-1)^p\frac{1}{\Gamma(n+p+1)p!}\left(\frac{1}{2}\right)^{n+2p}2(n+p)u^{2n+2p-1}\frac{du}{dx}$$

$$= u^n\sum_{p=0}^{\infty}(-1)^p\frac{n+p}{\Gamma(n+p+1)p!}\left(\frac{u}{2}\right)^{n+2p-1}\frac{du}{dx}$$

$$= u^n\sum_{p=0}^{\infty}(-1)^p\frac{1}{\Gamma(n+p)p!}\left(\frac{u}{2}\right)^{n+2p-1}\frac{du}{dx}$$

$$= u^n J_{n-1}(u)\frac{du}{dx}.$$

After expanding the left side of this equation, the following formula for the derivative of the Bessel function is obtained:

$$\frac{d}{dx}J_n(u) = \left[J_{n-1}(u) - \frac{n}{u}J_n(u)\right]\frac{du}{dx}. \qquad (C.11)$$

By similar reasoning, the following analogous equations for the other Bessel functions may be obtained:

$$\frac{d}{dx}Y_n(u) = \left[Y_{n-1}(u) - \frac{n}{u}Y_n(u)\right]\frac{du}{dx}, \qquad (C.12)$$

$$\frac{d}{dx}I_n(u) = \left[I_{n-1}(u) - \frac{n}{u}I_n(u)\right]\frac{du}{dx}, \qquad (C.13)$$

$$\frac{d}{dx}K_n(u) = \left[-K_{n-1}(u) - \frac{n}{u}K_n(u)\right]\frac{du}{dx}. \qquad (C.14)$$

The following special cases are of particular use:

$$\frac{d}{dx}[J_0(ax)] = aJ_{-1}(ax) = -aJ_1(ax),$$

$$\frac{d}{dx}[Y_0(ax)] = aY_{-1}(ax) = -aY_1(ax),$$

$$\frac{d}{dx}[I_0(ax)] = aI_{-1}(ax) = aI_1(ax), \qquad (C.15)$$

$$\frac{d}{dx}[K_0(ax)] = -aK_{-1}(ax) = -aK_1(ax).$$

Two more general differential equations which lead to solutions involving Bessel functions are given below. The solutions are also given and may be verified by substitution.

$$\frac{d^2y}{dx^2} + \frac{1 - 2a}{x}\frac{dy}{dx} + \left[(bcx^{c-1})^2 + \frac{a^2 - n^2c^2}{x^2}\right]y = 0,$$

$$y = x^a[AJ_n(bx^c) + B_{Y_n}^{J}{}_{-n}(bx^c)],$$

$$\frac{d^2y}{dx^2} + \left[(1 - 2a)\frac{1}{x} + 2\alpha\right]\frac{dy}{dx} + \left[(bcx^{c-1})^2\right.$$

$$\left. + \frac{\alpha(2a - 1)}{x} + \frac{a^2 - n^2c^2}{x^2} + \alpha^2\right]y = 0,$$

$$y = x^a\, e^{\alpha x}[AJ_n(bx^c) + B_{Y_n}^{J}{}_{-n}(bx^c)].$$

In the above solutions, if the arguments of the Bessel functions are imaginary, the modified functions (I_n, K_n) should replace those shown.

Table C.1 Selected Values of the Bessel Functions of the First and Second Kinds, Orders Zero and One

x	$J_0(x)$	$J_1(x)$	$Y_0(x)$	$Y_1(x)$
0.0	1.00000	0.00000	$-\infty$	$-\infty$
0.2	+.99002	0.09950	−1.0811	−3.3238
0.4	+.96039	+.19603	−.60602	−1.7809
0.6	+.91200	+.28670	−.30851	−1.2604
0.8	+.84629	+.36884	−.08680	−.97814
1.0	+.76520	+.44005	+.08825	−.78121
1.2	+.67113	+.49830	+.22808	−.62113
1.4	+.56686	+.54195	+.33790	−.47915
1.6	+.45540	+.56990	+.42043	−.34758
1.8	+.33999	+.58152	+.47743	−.22366
2.0	+.22389	+.57672	+.51038	−.10703
2.2	+.11036	+.55596	+.52078	+.00149
2.4	+.00251	+.52019	+.51042	+.10049
2.6	−.09680	+.47082	+.48133	+.18836
2.8	−.18503	+.40971	+.43591	+.26355
3.0	−.26005	+.33906	+.37685	+.32467
3.2	−.32019	+.26134	+.30705	+.37071
3.4	−.36430	+.17923	+.22962	+.40101
3.6	−.39177	+.09547	+.14771	+.41539
3.8	−.40256	+.01282	+.06450	+.41411
4.0	−.39715	−.06604	−.01694	+.39792
4.2	−.37656	−.13864	−.09375	+.36801
4.4	−.34226	−.20278	−.16333	+.32597
4.6	−.29614	−.25655	−.22345	+.27374
4.8	−.24042	−.29850	−.27230	+.21357
5.0	−.17760	−.32760	−.30851	+.14786
5.2	−.11029	−.34322	−.33125	+.07919
5.4	−.04121	−.34534	−.34017	+.01013
5.6	+.02697	−.33433	−.33544	−.05681
5.8	+.09170	−.31103	−.31775	−.11923
6.0	+.15065	−.27668	−.28819	−.17501
6.2	+.20175	−.23292	−.24830	−.22228
6.4	+.24331	−.18164	−.19995	−.25955
6.6	+.27404	−.12498	−.14523	−.28575
6.8	+.29310	−.06252	−.08643	−.30019
7.0	+.30007	−.00468	−.02595	−.30267
7.2	+.29507	+.05432	+.03385	−.29342
7.4	+.27859	+.10963	+.09068	−.27315
7.6	+.25160	+.15921	+.14243	−.24280
7.8	+.25541	+.20136	+.18722	−.20389
8.0	+.17165	+.23464	+.22352	−.15806
8.2	+.12222	+.25800	+.25011	−.10724
8.4	+.06916	+.27079	+.26622	−.05348
8.6	+.01462	+.27275	+.27146	−.00108
8.8	−.03923	+.26407	+.26587	+.05436
9.0	−.09033	+.24531	+.24994	+.10431
9.2	−.13675	+.21471	+.22449	+.14911
9.4	−.17677	+.18163	+.19074	+.18714
9.6	−.20898	+.13952	+.15018	+.21706
9.8	−.23227	+.09284	+.10453	+.23789
10.0	−.24594	+.04347	+.05567	+.24902

Table C.2 Selected Values of the Modified Bessel Functions of the First and Second Kinds, Orders Zero and One

x	$I_0(x)$	$I_1(x)$	$(2/\pi)K_0(x)$	$(2/\pi)K_1(x)$
0.0	1.000	0.0000	$+\infty$	$+\infty$
0.2	1.0100	0.1005	1.1158	3.0405
0.4	1.0404	0.2040	0.70953	1.3906
0.6	1.0920	0.3137	0.49498	0.82941
0.8	1.1665	0.4329	0.35991	0.54862
1.0	1.2661	0.5652	0.26803	0.38318
1.2	1.3937	0.7147	0.20276	0.27667
1.4	1.5534	0.8861	0.15512	0.20425
1.6	1.7500	1.0848	0.11966	0.15319
1.8	1.9896	1.3172	0.92903×10^{-1}	0.11626
2.0	2.2796	1.5906	0.72507	0.89041×10^{-1}
2.2	2.6291	1.9141	0.56830	0.68689
2.4	3.0493	2.2981	0.44702	0.53301
2.6	3.5533	2.7554	0.35268	0.41561
2.8	4.1573	3.3011	0.27896	0.32539
3.0	4.8808	3.9534	0.22116	0.25564
3.2	5.7472	4.7343	0.17568	0.20144
3.4	6.7848	5.6701	0.13979	0.15915
3.6	8.0277	6.7028	0.11141	0.12602
3.8	9.5169	8.1404	0.8891×10^{-2}	0.9999×10^{-2}
4.0	11.3019	9.7595	0.7105	0.7947
4.2	13.4425	11.7056	0.5684	0.6327
4.4	16.0104	14.0462	0.4551	0.5044
4.6	19.0926	16.8626	0.3648	0.4027
4.8	22.7937	20.2528	0.2927	0.3218
5.0	27.2399	24.3356	0.2350	0.2575
5.2	32.5836	29.2543	0.1888	0.2062
5.4	39.0088	35.1821	0.1518	0.1653
5.6	46.7376	42.3283	0.1221	0.1326
5.8	56.0381	50.9462	0.9832×10^{-3}	0.1064
6.0	67.2344	61.3419	0.7920	0.8556×10^{-3}
6.2	80.7179	73.8859	0.6382	0.6879
6.4	96.9616	89.0261	0.5146	0.5534
6.6	116.537	107.305	0.4151	0.4455
6.8	140.136	129.378	0.3350	0.3588
7.0	168.593	156.039	0.2704	0.2891
7.2	202.921	188.250	0.2184	0.2331
7.4	244.341	227.175	0.1764	0.1880
7.6	294.332	274.222	0.1426	0.1517
7.8	354.685	331.099	0.1153	0.1424
8.0	427.564	399.873	0.9325×10^{-4}	0.9891×10^{-4}
8.2	515.593	483.048	0.7543	0.7991
8.4	621.944	583.657	0.6104	0.6458
8.6	750.461	705.377	0.4941	0.5220
8.8	905.797	852.663	0.4000	0.4221
9.0	1093.59	1030.91	0.3239	0.3415
9.2	1320.66	1246.68	0.2624	0.2763
9.4	1595.28	1507.88	0.2126	0.2236
9.6	1927.48	1824.14	0.1722	0.1810
9.8	2329.39	2207.13	0.1396	0.1465
10.0	2815.72	2670.99	0.1131	0.1187

A Short Summary of the Orthogonal Functions Used in Chapter 4

Given a countably infinite set of functions: $g_1(x), g_2(x), g_3(x), \ldots, g_n(x), \ldots,$ $g_m(x), \ldots,$ the functions are termed orthogonal in the interval $a \le x \le b$ if

$$\int_a^b g_m(x)g_n(x)\, dx = 0 \qquad \text{for } m \ne n. \qquad (D.1)$$

A set of orthogonal functions has particular value in the possible representation of an arbitrary function as an infinite series of the orthogonal set in the specified interval. If $f(x)$ denotes the arbitrary function, consider the possibility of expressing it as a linear combination of the orthogonal functions:

$$f(x) = C_1 g_1(x) + C_2 g_2(x) + \cdots + C_n g_n(x) + \cdots + C_m g_m(x) + \cdots$$

$$= \sum_{n=1}^{\infty} C_n g_n(x). \qquad (D.2)$$

The C's are constants to be determined. If the series of Eq. (D.2) is convergent and integrable after multiplication by one of the functions, say $g_n(x)$, then

$$\int_a^b f(x)g_n(x)\,dx = C_1 \int_a^b g_1(x)g_n(x)\,dx + C_2 \int_a^b g_2(x)g_n(x)\,dx + \cdots$$

$$+ C_n \int_a^b g_n^2(x)\,dx + \cdots + C_m \int_a^b g_m(x)g_n(x)\,dx + \cdots.$$

The orthogonality definition given in Eq. (D.1) makes all the integrals on the right side of the equation above vanish except for the one term when $m = n$. Thus,

$$\int_a^b f(x)g_n(x)\,dx = 0 + 0 + \cdots + C_n \int_a^b g_n^2(x)\,dx + 0 + \cdots,$$

and the constant C_n may be calculated:

$$C_n = \frac{\displaystyle\int_a^b f(x)g_n(x)\,dx}{\displaystyle\int_a^b g_n^2(x)\,dx}. \tag{D.3}$$

Thus, when the function $f(x)$ is given, Eq. (D.3) enables one to calculate the constants, C_n, to be used in the series representation of $f(x)$. These constants are expressed in terms of the given set of orthogonal functions, $g_n(x)$.

A set of functions $[g_1(x), g_2(x), \ldots]$ may form an orthogonal set in the interval $a \le x \le b$ with respect to a weighting factor, $p(x)$, if

$$\int_a^b p(x)g_n(x)g_m(x)\,dx = 0 \qquad \text{for } m \ne n. \tag{D.4}$$

As before, if an arbitrary function, $f(x)$ can be represented as an infinite series of the functions,

$$f(x) = C_1 g_1(x) + C_2 g_2(x) + \cdots + C_n g_n(x) + \cdots + C_m g_m(x) + \cdots$$

$$= \sum_{n=1}^{\infty} C_n g_n(x),$$

the constants are given by

$$C_n = \frac{\displaystyle\int_a^b p(x)f(x)g_n(x)\,dx}{\displaystyle\int_a^b p(x)g_n^2(x)\,dx}. \tag{D.5}$$

Some of the orthogonal sets of functions used in Chapter 4 will now be discussed as examples.

The Sine and Cosine Functions

Consider the following set of functions in the interval $0 \leq x \leq L$:

$$\sin \frac{\pi x}{L}, \sin \frac{2\pi x}{L}, \sin \frac{3\pi x}{L}, \dots, \sin \frac{n\pi x}{L}, \dots.$$

This may also be expressed as

$$\sin \lambda_1 x, \sin \lambda_2 x, \sin \lambda_3 x, \dots, \sin \lambda_n x, \dots, \qquad \lambda_n = \frac{n\pi}{L}; n = 1, 2, 3, \dots. \quad \text{(D.6)}$$

Now, in the interval,

$$\int_a^b \sin \lambda_n x \sin \lambda_m x \, dx = \left[-\frac{\sin (\lambda_n + \lambda_m)x}{2(\lambda_m + \lambda_n)} + \frac{\sin (\lambda_n - \lambda_m)x}{2(\lambda_n - \lambda_m)} \right]$$

$$= 0 \qquad \text{for } \lambda_m \neq \lambda_n, \qquad\qquad \text{(D.7)}$$

since

$$\lambda_n = \frac{n\pi}{L}, \qquad \lambda_m = \frac{m\pi}{L}.$$

Thus, the set of functions in Eq. (D.6) is an orthogonal set. Also, for $m = n$:

$$\int_0^L \sin^2 \lambda_n x \, dx = \left. \frac{1}{2\lambda_n} (\lambda_n x - \sin \lambda_n x \cos \lambda_n x) \right|_0^L$$

$$= \frac{L}{2}. \qquad\qquad \text{(D.8)}$$

Thus, an arbitrary function, $f(x)$, may, if the series converges, be represented as a series of the functions of Eq. (D.6):

$$f(x) = C_1 \sin \lambda_1 x + C_2 \sin \lambda_2 x + \cdots,$$

or

$$f(x) = \sum_{n=1}^{\infty} C_n \sin \lambda_n x. \qquad\qquad \text{(D.9)}$$

The C_n's will be, from Eqs. (D.3) and (D.8),

$$C_n = \frac{2}{L} \int_0^L f(x) \sin \lambda_n x \, dx,$$

$$\lambda_n = \frac{n\pi}{L}, \qquad n = 1, 2, 3. \tag{D.10}$$

Thus, the function $f(x)$ is representable by the series

$$f(x) = \frac{2}{L} \sum_{n=1}^{\infty} \sin \lambda_n x \int_0^L f(x) \sin \lambda_n x \, dx. \tag{D.11}$$

In a similar fashion, one can show that the set of functions

$$\{\cos \lambda_n x\}, \lambda_n = \frac{n\pi}{L}, \qquad n = 0, 1, 2, 3, \ldots, \tag{D.12}$$

is an orthogonal set in $0 \le x \le L$. Also, then, an arbitrary function $f(x)$, may be represented as a convergent series of these functions:

$$f(x) = \frac{A_0}{2} + \sum_{n=1}^{\infty} A_n \cos \lambda_n x, \tag{D.13}$$

if

$$A_n = \frac{2}{L} \int_0^L f(x) \cos \lambda_n x \, dx \tag{D.14}$$

with

$$\lambda_n = \frac{n\pi}{L}, \qquad n = 0, 1, 2, 3, \ldots.$$

Thus,

$$f(x) = \frac{1}{L} \int_0^L f(x) \, dx + \frac{2}{L} \sum_{n=1}^{\infty} \cos \lambda_n x \int_0^L f(x) \cos \lambda_n x \, dx. \tag{D.15}$$

In many instances in heat conduction problems it may be necessary to express a function as an infinite series of the sines or cosines, such as in Eqs. (D.9) or (D.13), but in which the λ_n's are defined by relations other than that specified by Eqs. (D.6) and (D.12). These characteristic equations defining the λ_n's arise out of the application of the boundary conditions of the particular problem under consideration. One such case is discussed in Chapter 4, wherein one wishes to represent a function as a sine series in the interval $0 \le x \le L$ as in Eq. (D.9), when the λ_n's are defined as the roots of the equation:

$$(\lambda_n L) \tan (\lambda_n L) - B = 0, \qquad n = 1, 2, 3, \ldots; B = \text{constant.} \qquad \text{(D.16)}$$

Now, since Eq. (D.16) may be written

$$(\lambda_n L) \sin \lambda_n L = B \cos \lambda_n L,$$

some algebra will show that the integral expressed in Eq. (D.7) will again vanish. Also, Eq. (D.8) gives, instead of $L/2$, that

$$\int_0^L \sin^2 \lambda_n x \, dx = \frac{L}{2} - \frac{\sin \lambda_n L \cos \lambda_n L}{2\lambda_n}.$$

Thus, if λ_n is a root of

$$(\lambda_n L) \tan \lambda_n L - B = 0,$$

then an arbitrary function, $f(x)$, may be expressed as a sine series

$$f(x) = \sum_{n=1}^{\infty} C_n \sin \lambda_n x, \qquad \text{(D.17)}$$

where

$$C_n = \frac{\displaystyle\int_0^L f(x) \sin \lambda_n x \, dx}{\dfrac{L}{2} - \dfrac{\sin \lambda_n L \cos \lambda_n L}{2\lambda_n}}. \qquad \text{(D.18)}$$

Thus, in terms of these constants, $f(x)$ may be represented by

$$f(x) = 2 \sum_{n=1}^{\infty} \sin \lambda_n x \frac{\lambda_n \displaystyle\int_0^L f(x) \sin \lambda_n x \, dx}{\lambda_n L - \sin \lambda_n L \cos \lambda_n L}.$$

Similarly when Eq. (D.16) holds, an expansion in terms of cosines may be made:

$$f(x) = \sum_{n=1}^{\infty} A_n \cos \lambda_n x, \tag{D.19}$$

where

$$A_n = \frac{\displaystyle\int_0^L f(x) \cos \lambda_n x \, dx}{\displaystyle\frac{L}{2} + \frac{\sin \lambda_n L \cos \lambda_n L}{2\lambda_n}}. \tag{D.20}$$

The corresponding representation of $f(x)$ is, then,

$$f(x) = 2 \sum_{n=1}^{\infty} \cos \lambda_n L \frac{\lambda_n \displaystyle\int_0^L f(x) \cos \lambda_n x \, dx}{\lambda_n L + \sin \lambda_n L \cos \lambda_n L}.$$

The Bessel Functions

In heat conduction problems in cylindrical coordinate systems, the solutions are often expressed in terms of the Bessel functions (Appendix C). To express an arbitrary function as an infinite series of such functions it will be necessary to show their orthogonality. The Bessel functions are orthogonal with respect to the weighting factor: $p(x) = x$. For example, considering J_0, it will be shown that an arbitrary function may be expressed, in an interval, as a linear combination of the set $J_0(\lambda_1 x), J_0(\lambda_2 x), J_0(\lambda_3 x), \ldots, J_0(\lambda_n x), \ldots,$ where the parameters denoted by λ_n are defined, in some way, by the boundary conditions of the problem.

In other words, it will be shown that a function $f(x)$ may be represented in the following way, provided that the λ_n's are properly defined:

$$f(x) = C_1 J_0(\lambda_1 x) + C_2 J_0(\lambda_2 x) + \cdots$$

$$= \sum_{n=1}^{\infty} C_n J_0(\lambda_n x). \tag{D.21}$$

In order to be able to do this, Eq. (D.4) shows that [for $p(x) = x$] the following condition must be satisfied:

$$\int_a^b x J_0(\lambda_n x) J_0(\lambda_m x) \, dx = 0, \qquad m \neq n. \tag{D.22}$$

Then Eq. (D.5) shows that the constants C_n are

$$C_n = \frac{\int_a^b xf(x)J_0(\lambda_n x)\,dx}{\int_a^b xJ_0^2(\lambda_n x)\,dx}. \tag{D.23}$$

In order to prove the orthogonality condition of Eq. (D.22) and to evaluate the constants given by Eq. (D.23), one needs expressions for

$$\int_a^b xJ_0(\lambda_n x)J_0(\lambda_m x)\,dx \qquad \text{and} \qquad \int_a^b xJ_0^2(\lambda_n x)\,dx.$$

These two integrals may readily be evaluated by repeated "integration by parts," utilizing the following formulas resulting from Eq. (C.11):

$$\frac{d}{dx}[J_0(\lambda_n x)] = -\lambda_n J_1(\lambda_n x), \qquad \frac{d}{dx}[xJ_1(\lambda_n x)] = \lambda_n xJ_0(\lambda_n x),$$

$$\int J_1(\lambda_n x)\,dx = -\frac{1}{\lambda_n}J_0(\lambda_n x), \qquad \int xJ_0(\lambda_n x)\,dx = \frac{x}{\lambda_n}J_1(\lambda_n x).$$

The results are

$$\int xJ_0(\lambda_n x)J_0(\lambda_m x)\,dx = \frac{x}{\lambda_n^2 - \lambda_m^2}[\lambda_n J_0(\lambda_m x)J_1(\lambda_n x) - \lambda_m J_0(\lambda_n x)J_1(\lambda_m x)], \tag{D.24}$$

$$\int xJ_0^2(\lambda_n x)\,dx = \frac{x^2}{2}[J_0^2(\lambda_n x) + J_1^2(\lambda_n x)]. \tag{D.25}$$

As a particular example, consider the set of functions $J_0(\lambda_1 x),\ J_0(\lambda_2 x),\ \ldots,$ $J_0(\lambda_n x),\ \ldots,$ in which the λ_n's are defined in the following way for the interval $0 \le x \le R$. Let the λ_n's be the roots of the equation

$$J_0(\lambda_n R) = 0. \tag{D.26}$$

Examination of the tables of $J_0(x)$ given in Appendix C shows that J_0 has a succession of zeros that differ by an interval approaching π as $x \to \infty$. Hence, there are a countably infinite set of the λ_n's defined in Eq. (D.26). For the interval $0 \leq x \leq R$, Eq. (D.24) reduces to

$$\int_0^R x J_0(\lambda_n x) J_0(\lambda_m x)\, dx = \frac{R}{\lambda_n^2 - \lambda_m^2}[\lambda_n J_0(\lambda_m R) J_1(\lambda_n R) - \lambda_m J_0(\lambda_n R) J_1(\lambda_m R)].$$

By virtue of Eq. (D.26), $J_0(\lambda_n R) = J_0(\lambda_m R) = 0$, so

$$\int_0^R x J_0(\lambda_n x) J_0(\lambda_m x)\, dx = 0.$$

Thus, the functions $J_0(\lambda_1 x)$, $J_0(\lambda_2 x), \ldots$, are orthogonal in the interval $0 \leq x \leq R$ if λ_n is a root of Eq. (D.26). To use Eq. (D.23) to obtain the constants of the linear series expansion, Eq. (D.25) must be evaluated for the particular definition of λ_n. Thus,

$$\int_0^R x J_0^2(\lambda_n x)\, dx = \frac{R^2}{2}[J_0^2(\lambda_n R) + J_1^2(\lambda_n R)]$$

$$= \frac{R^2}{2}[0 + J_1^2(\lambda_n R)]$$

$$= \frac{R^2}{2} J_1^2(\lambda_n R).$$

Summarizing, an arbitrary function may be expressed, in the interval $0 \leq x \leq R$, as a series of J_0's:

$$f(x) = \sum_{n=1}^{\infty} C_n J_0(\lambda_n x).$$

The constants, C_n, will be given by Eq. (D.23):

$$C_n = \frac{\displaystyle\int_0^R x f(x) J_0(\lambda_n x)\, dx}{\dfrac{R^2}{2} J_1^2(\lambda_n R)} \tag{D.27}$$

if the λ_n's are the roots of

$$J_0(\lambda_n R) = 0.$$

In a similar fashion it may be shown that the same expression,

$$f(x) = \sum_{n=1}^{\infty} C_n J_0(\lambda_n x),$$

$$(D.28)$$

may be written in the interval $0 \le x \le R$ if the λ_n's are the roots of

$$J_1(\lambda_n R) = 0.$$

$$(D.29)$$

In this case, the C_n's are given by

$$C_n = \frac{\displaystyle\int_0^R xf(x)J_0(\lambda_n x)\,dx}{\dfrac{R^2}{2}J_0^2(\lambda_n R)}.$$

$$(D.30)$$

As a final example, Sec. 4.7 considers the possibility of expressing an arbitrary function, $f(x)$, as a series expansion in $J_0(\lambda_n x)$, when λ_n is defined as the nth root of the transcendental equation

$$\lambda_n R \frac{J_1(\lambda_n R)}{J_0(\lambda_n R)} - B = 0.$$

$$(D.31)$$

In the latter equation B is a constant. That the functions with λ_n thus defined are orthogonal in the interval $0 \le x \le R$ can be seen by substitution of Eq. (D.31) into Eq. (D.24):

$$\int_0^R x J_0(\lambda_n x)J_0(\lambda_m x)\,dx$$

$$= \frac{R}{\lambda_m^2 - \lambda_n^2}\left[\lambda_n\left(\lambda_m R \frac{J_1(\lambda_m R)}{B}\right)J_1(\lambda_n R) - \lambda_m \lambda_n R\left(\frac{J_1(\lambda_n R)}{B}\right)J_1(\lambda_m R)\right]$$

$$= 0.$$

The fact that this equation equals zero results from the definition of the λ_n's (and λ_m's) given in Eq. (D.31). Equation (D.25) yields, then,

$$\int_0^R x J_0^2(\lambda_n x)\,dx = \frac{R^2}{2}[J_0^2(\lambda_n R) + J_1^2(\lambda_n R)],$$

so Eq. (D.23) gives the C_n's to be

$$C_n = \frac{\dfrac{2}{R^2} \displaystyle\int_0^R xf(x)J_0(\lambda_n x)\,dx}{J_0^2(\lambda_n R) + J_1^2(\lambda_n R)}. \tag{D.32}$$

Determination of the Shape Factor

The shape factor, F_{1-2}, was introduced in Sec. 11.5 as the fraction of diffuse radiation leaving one surface, A_1, which directly strikes a second surface, A_2. This definition led to the following expression for the shape factor, Eq. (11.48),

$$F_{1-2} = \frac{1}{A_1} \int_{A_2} \int_{A_1} \frac{\cos \theta_1 \cos \theta_2 \, dA_1 \, dA_2}{\pi r^2}. \tag{E.1}$$

The geometrical variables $\theta_1, \theta_2, r, A_1$, and A_2 are illustrated in Fig. 11.14. The shape factor is seen to be a purely geometrical quantity—dependent only upon the size, shape, and relative orientation of the two surfaces.

The discussions of Sec. 11.5, and succeeding sections, emphasized, over and over, the importance of this shape factor in the calculation of radiant exchange. Whether dealing with individual surfaces, enclosures, etc., the shape factor entered into almost every consideration. Thus, it is important to have at hand numerical values of the shape factor for geometric configurations of practical significance. Some of the fundamental configurations will be considered here. References 1, 2, 3, and 4 may be consulted for more extensive tabulations of the shape factor and for a great variety of cases. In addition to consideration of basic configurations, application may be made of the

630

reciprocal and additive properties of the shape factor, discussed in Sec. 11.5, to calculate F for other, more complex, configurations.

The Shape Factor for Finite Parallel, Opposed Rectangles

Of very great application in engineering is the configuration of two equal, parallel, directly opposed rectangles. Figure E.1 shows two rectangles $W \times L$ in size and spaced a distance D apart. By arbitrarily denoting the upper area by A_1 and the lower by A_2, the two coordinate systems (x_1, y_1) and (x_2, y_2) may be chosen as shown. Selecting area elements dA_1 and dA_2 on each surface, one finds that their centers have coordinates (x_1, y_1) and (x_2, y_2), respectively. The distance r between dA_1 and dA_2 is then given by

$$r^2 = D^2 + (x_1 - x_2)^2 + (y_1 - y_2)^2.$$

The cosines of the two angles θ_1 and θ_2 are identical and equal to $\cos \theta_1 = \cos \theta_2 = D/r$. Equation (E.1) for the shape factor F_{1-2} (or F_{2-1} in this case of equal areas) gives

$$F_{1-2} = \frac{1}{WL} \int_0^W \int_0^L \int_0^W \int_0^L \frac{D^2 \, dx_1 \, dy_1 \, dx_2 \, dy_2}{\pi[D^2 + (x_1 - x_2)^2 + (y_1 - y_2)^2]^2}.$$

The result of the integration of the above equation may be written in terms of the dimensionless ratios W/D and L/D. The result is given below with the following symbols introduced for simplicity:

$$R_1 = \frac{L}{D},$$

$$R_2 = \frac{W}{D},$$

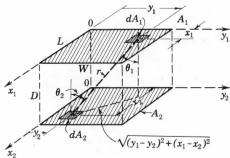

Fig. E.1

$$F_{1-2} = \frac{1}{\pi}\left[\frac{1}{R_1 R_2}\ln\frac{(1 + R_1^2)(1 + R_2^2)}{(1 + R_1^2 + R_2^2)} - \frac{2}{R_1}\tan^{-1}R_2 - \frac{2}{R_2}\tan^{-1}R_1\right.$$

$$\left. + 2\sqrt{1 + \frac{1}{R_1^2}}\tan^{-1}\frac{R_2}{\sqrt{1 + R_1^2}} + 2\sqrt{1 + \frac{1}{R_2^2}}\tan^{-1}\frac{R_1}{\sqrt{1 + R_2^2}}\right].$$

$$\text{(E.2)}$$

For rapid determination of the shape factor expressed by this equation, Fig. 11.15 presents F_{1-2} as a function of $R_1 = L/D$ and $R_2 = W/D$.

The Shape Factor for Perpendicular Rectangles Having a Common Edge

Another configuration of particular engineering importance is that of two perpendicular rectangles with a common edge. Such a configuration exists between the walls of a room (or furnace) and the floor or ceiling.

The approach to the evaluation of the shape factor is identical to that used in the preceding section. Figure E.2 depicts a rectangle, call it A_1, of dimensions $D \times L$ located normal to A_2 with dimensions $D \times W$. The dimension D is, then, the length of the common edge. Figure E.2 shows the coordinate system selected, and the various quantities needed to evaluate F_{1-2} are indicated. Without discussion, the shape factor is expressed by

$$r^2 = y_1^2 + y_2^2 + (x_1 - x_2)^2,$$

$$\cos\theta_1 = \frac{y_2}{r}, \qquad \cos\theta_2 = \frac{y_1}{r}.$$

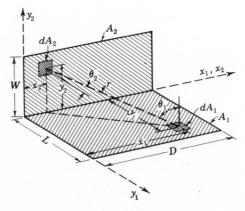

Fig. E.2

In terms of the two dimensionless parameters,

$$R_1 = \frac{L}{D},$$

$$R_2 = \frac{W}{D},$$

integration gives

$$F_{1-2} = \frac{1}{\pi R_1} \left\{ R_1 \tan^{-1} \frac{1}{R_1} + R_2 \tan^{-1} \frac{1}{R_2} - \sqrt{R_1 + R_2^2} \tan^{-1} \frac{1}{\sqrt{R_1^2 + R_2^2}} \right.$$

$$+ \frac{1}{4} \ln \left[\frac{(1 + R_1^2)(1 + R_2^2)}{(1 + R_1^2 + R_2^2)} \left(\frac{R_2^2(1 + R_1^2 + R_2^2)}{(1 + R_2^2)(R_1^2 + R_2^2)} \right)^{R_2^2} \right.$$

$$\left. \left. \times \left(\frac{R_1^2(1 + R_1^2 + R_2^2)}{(1 + R_1^2)(R_1^2 + R_2^2)} \right)^{R_1^2} \right] \right\}. \tag{E.3}$$

The results of this expression are presented graphically in Fig. 11.16. As should be intuitively apparent, Fig. 11.16 shows that F_{1-2} must always be less than 0.5 for the configuration under consideration.

EXAMPLE: A square black plane, 10 ft × 10 ft, is parallel and directly opposed to another square plane of the same size located 5 ft away. Compare the amount of radiation exchanged in this case to that obtained for a square plane, 10 ft × 10 ft, located normal to another rectangle, 10 ft × 5 ft, with its 10 ft side a common edge with the square.

Solution: For $R_1 = R_2 = 2$, Fig. 11.15 gives F_{1-2} in the parallel case to be $F_{1-2} = 0.415$. In the perpendicular case, for $R_1 = 1$, $R_2 = 0.5$, Fig. 11.16 gives $F_{1-2} = 0.15$. These calculations show that a plane radiates more of its energy directly overhead than to the sides.

Other Configurations Derivable from Perpendicular Rectangles with a Common Edge

As examples of the application of the additive property of shape factors, this section will consider a few cases of practical value for which the shape factor can be deduced from the results obtained in Eq. (E.3) and Fig. 11.16. The same principles of analysis may be applied to other basic configurations.

Case 1. Let it be desired to find the shape factor F_{1-2} for the configuration shown in Fig. E.3(a) with one rectangle displaced from the common intersection line. Considering the area A_2 plus the fictitious area A_3 to compose a single area, call it $A_{(2,3)}$, then one obtains

$$A_1 F_{1-(2,3)} = A_1 F_{1-2} + A_1 F_{1-3}. \qquad (E.4)$$

Both $F_{1-(2,3)}$ and F_{1-3} may be found from the results of Eq. (E.3) or Fig. 11.16. To obtain F_{2-1} for this configuration one could use the reciprocal property

$$F_{2-1} = F_{1-2} \frac{A_1}{A_2},$$

or one could do the following:

$$A_{(2,3)} F_{(2,3)-1} = A_2 F_{2-1} + A_3 F_{3-1},$$
$$F_{2-1} = \frac{A_{(2,3)} F_{(2,3)-1} - A_3 F_{3-1}}{A_2}. \qquad (E.5)$$

Case 2. By similar reasoning the shape factor F_{1-2} for the situation depicted in Fig. E.3(b) is

$$F_{1-2} = \frac{A_{(1,3)} F_{(1,3)-(2,4)} + A_3 F_{3-4} - A_3 F_{3-(2,4)} - A_{(1,3)} F_{(1,3)-4}}{A_1}. \qquad (E.6)$$

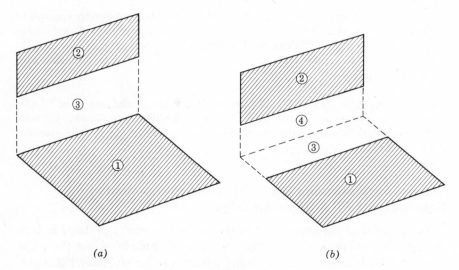

(a)   (b)

Fig. E.3

All the F's in this equation are of the type given in Fig. 11.16. The areas A_3 and A_4 are, of course, fictitious surfaces.

Case 3. The determination of the shape factor F_{1-2} for the configuration shown in Fig. E.4(a) is a little more involved than the above two cases. If one defines the fictitious areas A_3 and A_4 as shown, the additive principle gives

$$F_{1-2} = \frac{1}{A_1}[A_{(1,3)}F_{(1,3)-(2-4)} - A_1F_{1-4} - A_3F_{3-2} - A_3F_{3-4}]. \quad (E.7)$$

One notes now that the shape factor F_{3-4} required in this expression is for a configuration of the same form as that for F_{1-2}—which is what is sought. This dilemma may be solved by developing another reciprocal relation of considerable importance.

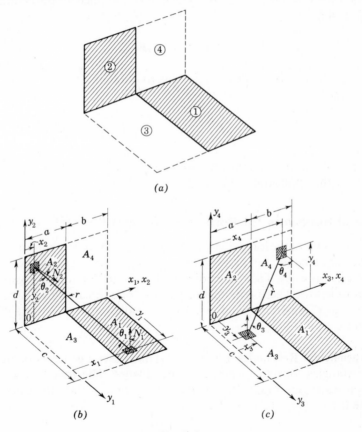

(a)

(b)   (c)

Fig. E.4

The integral representation of the two shape factors F_{1-2} and F_{3-4} must be formed. Figure E.4(b) and (c) shows the dimensions and coordinates used for this purpose. The integrands of these expressions will be the same as for the basic configuration discussed in the preceding section, only the limits of integration will be different. The resulting expressions for F_{1-2} and F_{3-4} are

$$A_1 F_{1-2} = \frac{1}{\pi} \int_0^d \int_0^a \int_0^c \int_a^{a+b} \frac{y_1 y_2 \, dx_1 \, dy_1 \, dx_2 \, dy_2}{[y_1^2 + y_2^2 + (x_1 - x_2)^2]^2},$$

$$A_3 F_{3-4} = \frac{1}{\pi} \int_0^d \int_a^{a+b} \int_0^c \int_0^a \frac{y_3 y_4 \, dx_3 \, dy_3 \, dx_4 \, dy_4}{[y_3^2 + y_4^2 + (x_3 - x_4)^2]^2}.$$

These two integrals are of identical form except for the order of integration. Since the nature of the integrand permits the interchange of the order of the integration, the following reciprocal formula is obtained for the configuration of Fig. E.4(a):

$$A_1 F_{1-2} = A_3 F_{3-4}.$$

This reciprocal relation then enables one to obtain F_{1-2} since Eq. (E.7) now reduces to

$$F_{1-2} = \frac{1}{2A_1} [A_{(1,3)} F_{(1,3)-(2-4)} - A_1 F_{1-4} - A_3 F_{3-2}]. \tag{E.8}$$

This involves only shape factors obtainable from Fig. 11.16. Example 11.5 illustrates the application of this relation.

General Relations for Perpendicular and Parallel Rectangles

The methods of the foregoing sections may be applied to obtain expressions for the shape factor for perpendicular and parallel rectangles oriented in rather general ways. The results obtained by Hamilton and Morgan (Ref. 1) are given below. These relations will enable one to calculate shape factors between any portions (windows, doors, walls, etc.) of a parallelepiped structure.

Perpendicular Rectangles. Figure E.5(a) illustrates a general orientation of two rectangles located in perpendicular planes. The following reciprocal relations exist [see Fig. E.5(a) for the explanation of the subscripts used to denote the various areas involved]:

$$A_1 F_{1-3'} = A_3 F_{3-1'} = A_{3'} F_{3'-1} = A_{1'} F_{1'-3}. \tag{E.9}$$

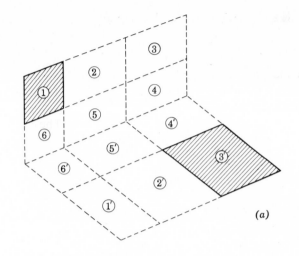

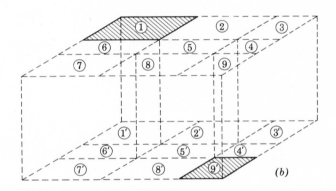

Fig. E.5

The shape factor, $F_{1-3'}$, is given by

$$A_1 F_{1-3'} = K_{1-3'} = \tfrac{1}{2}[K_{(1,2,3,4,5,6)^2} - K_{(2,3,4,5)^2} - K_{(1,2,5,6)^2} + K_{(4,5,6)^2}$$

$$- K_{(4,5,6)-(1',2',3',4',5',6')} - K_{(1,2,3,4,5,6)-(4',5',6')}$$

$$+ K_{(1,2,5,6)-(5'6')} + K_{(2,3,4,5)-(4',5')} + K_{(5,6)-(1',2',5',6')}$$

$$+ K_{(4,5)-(2',3',4',5')} + K_{(2,5)^2} - K_{(2,5)-5'} - K_{(5,6)^2}$$

$$- K_{(4,5)^2} - K_{5-(2',5')} + K_{5^2}]. \tag{E.10}$$

In Eq. (E.10) K_{m-n} is used to symbolize $K_{m-n} = A_m F_{m-n}$, and $K_{(m)^2} = A_m F_{m-m'}$.

Parallel Rectangles. Similar relations for opposed parallel rectangles are given below, using the notation depicted in Fig. E.5(b). The following reciprocal relations exist:

$$A_1 F_{1-9'} = A_3 F_{3-7'} = A_9 F_{9-1'} = A_7 F_{7-3'}. \tag{E.11}$$

The shape factor $F_{1-9'}$ is given by

$$A_1 F_{1-9'} = K_{1-9'} = \tfrac{1}{4}\{K_{(1,2,3,4,5,6,7,8,9)^2} - K_{(1,2,5,6,7,8)^2} - K_{(2,3,4,5,8,9)^2}$$

$$- K_{(1,2,3,4,5,6)^2} + K_{(1,2,5,6)^2} + K_{(2,3,4,5)^2} + K_{(4,5,8,9)^2}$$

$$- K_{(4,5)^2} - K_{(5,8)^2} - K_{(5,6)^2} - K_{(4,5,6,7,8,9)^2}$$

$$+ K_{(5,6,7,8)^2} + K_{(4,5,6)^2} + K_{(2,5,8)^2} - K_{(2,5)^2}\}. \tag{E.12}$$

Other Configurations of Interest

The number of configurations which may be considered is nearly limitless. However, three other situations of practical value will be presented here. No derivations are given; only the analytical results are presented, with corresponding graphical displays given in the main body of the text.

Directly Opposed Disks. For two disks, radii r_1 and r_2, placed L apart in parallel planes and having collinear centers

$$F_{1-2} = \tfrac{1}{2}[x - \sqrt{x^2 - 4(R_2/R_1)^2}],$$

$$x = 1 + (1 + R_2^2)/R_1^2, \tag{E.13}$$

$$R_1 = \frac{r_1}{L}, \qquad R_2 = \frac{r_2}{L}.$$

Figure 11.17 presents this shape factor in graphical form.

Concentric Cylinders. For two concentric cylinders of equal length L, the shape factor F_{2-1} for direct radiation from the inner surface of the outer cylinder, radius r_2, to the outer surface of the inner cylinder, radius r_1, Ref. 3 gives

$$F_{2-1} = R_1 \left[1 - \frac{1}{\pi} \cos^{-1}\left(\frac{\chi_1}{\chi_2}\right) \right] + \frac{\sqrt{(\chi_1 + 2)^2 - 4R_1^2}}{2\pi R_2} \cos^{-1}\left(R_1 \frac{\chi_1}{\chi_2}\right)$$

$$+ \frac{\chi_1}{2\pi R_2} \sin^{-1} R_1 - \frac{\chi_2}{4R_2}.$$

$$\chi_1 = R_2^2 + R_1^2 - 1, \qquad \chi_2 = R_2^2 - R_1^2 + 1,$$

$$R_1 = \frac{r_1}{r_2}, \qquad R_2 = \frac{L}{r_2}.$$

(E.14)

The geometry of the configuration and a plot of F_{2-1} are given in Fig. 11.18.

Small Spheres or Planes Irradiated by a Large Sphere. Present-day applications in space technology require knowledge of the amounts of heat acquired by satellites and spacecraft from planetary bodies. In most instances the size of the receiving body may be taken to be quite small compared to that of the planetary body. Figure 11.19 illustrates the geometrical parameters involved in consideration of the radiant exchange between a large planetary body and a small sphere or plane element. The small receiving surface is denoted by dA_1 and the large emitting one by A_2. In the case of the plane element (which might be part of a larger space structure), the direction of the surface normal with respect to the line drawn to the planetary center, λ, must be known. Because the expressions for the shape factor F_{dA_1-2} cannot be written in simple closed forms, only the results of numerical integrations are given in Fig. 11.19 (Ref. 5) for the two cases.

One must note that the shape factors given are valid only if the large sphere has a uniform radiosity. In the event that only part of the sphere is radiating (as might be the case for reflected solar radiation) one must take this fact into account. Results of such calculations are given in Refs. 3 and 5.

References

1. HAMILTON, D. C., and W. R. MORGAN, "Radiant Interchange Configuration Factors," *N.A.C.A. Tech. Note 2836*, Dec. 1952.
2. KREITH, F., *Radiation Heat Transfer for Spacecraft and Solar Power Plant Design*, Scranton, Pa., International Textbook, 1962.
3. STEVENSON, J. A., and J. C. GRAFTON, "Radiation Heat Transfer Analysis for Space Vehicles," Flight Accessories Laboratory, Aeronautical Systems Division, Wright-Patterson Air Force Base, Ohio, *Report ASD 61-119, Part I*, Dec. 1961.

4. PLAMONDON, J. A., "Numerical Determination of Radiation Configuration Factors for Some Common Geometrical Situations," *Tech. Report 32–127*, Jet Propulsion Laboratory, Calif. Inst. Technology, 1961.

5. CLARK, L. G., and E. C. ANDERSON, "Geometric Shape Factors for Planetary-Thermal and Planetary-Reflected Radiation Incident upon Spinning and Non-Spinning Spacecraft," *N.A.S.A. Tech. Note D-2835*, May 1965.

SI-Unit Conversion Table for Heat Transfer Calculations

The Eleventh General Conference on Weights and Measures in 1960 defined and officially sanctioned an international system of units (the Système International d'Unités) to be designated as SI in all languages. This system, based on the meter, kilogram, second, and ampere, was accepted by the representatives of the 36 participating countries, including the United States. Although the United States has not officially adopted SI units by congressional action, their use in this country has grown to such an extent that it is necessary that engineers be proficient in using them and be able to convert between them and the traditional English engineering system.

The conversion factors given in this appendix are limited to those physical quantities occurring commonly in engineering heat-transfer calculations (i.e., no conversions are given for electrical quantities, etc.). Likewise, conversions are given only for English units encountered in this text (i.e., gallons, tons, etc., are omitted).

SI units are based on the fundamental definitions:

Quantity	Unit	Unit Abbreviation
Mass	kilogram	kg
Length	meter	m
Time	second	s
Temperature	degree Kelvin	°K

The following *derived* units are then defined:

Quantity	Unit	Unit Abbreviation
Force	$1 \text{ newton} = 1\dfrac{\text{kg-m}}{\text{s}^2}$	N
Energy	$1 \text{ joule} = 1 \text{ N-m}$	J
Power	$1 \text{ watt} = 1\dfrac{\text{J}}{\text{s}}$	W
Temperature	degree Celsius = degree Kelvin -273.15	

The equivalency between the above quantities and their English-unit counterparts may be obtained from various sources (e.g., Ref. 1) to yield the following conversion factors:

Acceleration:	1 ft/sec^2	$= 3.0480 \times 10^{-1} \text{ m/s}^2$
	1 ft/hr^2	$= 2.3519 \times 10^{-8} \text{ m/s}^2$
Area:	1 ft^2	$= 9.2903 \times 10^{-2} \text{ m}^2$
	1 in.^2	$= 6.4516 \times 10^{-4} \text{ m}^2$
Conductance, thermal:	$1 \text{ Btu/hr-ft}^2\text{-}°\text{F}$	$= 5.6784 \text{ W/m}^2\text{-}°\text{C}$
Conductivity, thermal:	$1 \text{ Btu/hr-ft-}°\text{F}$	$= 1.7308 \text{ W/m-}°\text{C}$
Density:	$1 \text{ lb}_m/\text{ft}^3$	$= 1.6018 \times 10 \text{ kg/m}^3$
Diffusivity, thermal:	$1 \text{ ft}^2/\text{sec}$	$= 9.2903 \times 10^{-2} \text{ m}^2/\text{s}$
	$1 \text{ ft}^2/\text{hr}$	$= 2.5806 \times 10^{-5} \text{ m}^2/\text{s}$
Energy:	1 Btu	$= 1.0551 \times 10^3 \text{ J}$
	1 kw-hr	$= 3.6000 \times 10^6 \text{ J}$
	1 ft-lb_f	$= 1.3558 \text{ J}$
	1 hp-hr	$= 2.6845 \times 10^6 \text{ J}$
Force:	1 lb_f	$= 4.4482 \text{ N}$
Heat:	1 Btu	$= 1.0551 \times 10^3 \text{ J}$
Heat flow rate:	1 Btu/sec	$= 1.0551 \times 10^3 \text{ W}$
	1 Btu/hr	$= 2.9308 \times 10^{-1} \text{ W}$
Heat flux:		
(unit area):	1 Btu/hr-ft^2	$= 3.1546 \text{ W/m}^2$
(unit length):	1 Btu/hr-ft	$= 9.6152 \times 10^{-1} \text{ W/m}$
Heat generation rate:		
(unit mass):	1 Btu/hr-lb_m	$= 6.4612 \times 10^{-1} \text{ W/kg}$
(unit volume):	1 Btu/hr-ft^3	$= 1.0350 \times 10 \text{ W/m}^3$

Heat transfer coefficient:	1 Btu/hr-ft^2-°F	= 5.6784 W/m^2-°C
Latent heat:	1 Btu/lb$_m$	= 2.3260 × 10^3 J/kg
Length:	1 ft	= 3.0480 × 10^{-1} m
	1 micron	= 1.0000 × 10^{-6} m
	1 in.	= 2.5400 × 10^{-2} m
	1 mile	= 1.6093 × 10^3 m
Mass:	1 lb$_m$	= 4.5359 × 10^{-1} kg
Mass flow rate:	1 lb$_m$/sec	= 4.5359 × 10^{-1} kg/s
	1 lb$_m$/hr	= 1.2600 × 10^{-4} kg/s
Mass flux:	1 lb$_m$/sec-ft^2	= 4.8824 kg/s-m^2
	1 lb$_m$/hr-ft^2	= 1.3562 × 10^{-3} kg/s-m^2
	1 lb$_m$/sec-in.^{2}	= 7.0362 × 10^2 kg/s-m^2
	1 lb$_m$/hr-in.^{2}	= 1.9545 × 10^{-1} kg/s-m^2
Momentum, linear:	1 lb$_m$-ft/sec	= 1.3825 × 10^{-1} kg-m/s
	1 lb$_m$-ft/hr	= 3.8404 × 10^{-5} kg-m/s
Power:	1 Btu/sec	= 1.0551 × 10^3 W
	1 ft-lb$_f$/sec	= 1.3558 W
	1 Btu/hr	= 2.9308 × 10^{-1} W
	1 hp	= 7.4570 × 10^2 W
Pressure:	1 lb$_f$/ft^2	= 4.7880 × 10 N/m^2
	1 lb$_f$/in.^{2}	= 6.8948 × 10^3 N/m^2
	1 standard atmosphere	= 1.0133 × 10^5 N/m^2
	1 in. water	= 2.4909 × 10^2 N/m^2
	1 ft water	= 2.9891 × 10^3 N/m^2
	1 in. mercury	= 3.3866 × 10^3 N/m^2
Resistance, thermal:		
(total):	1 hr-°F/Btu	= 1.8956°C/W
(unit):	1 hr-ft^2-°F/Btu	= 1.7611 × 10^{-1} m^2-°C/W
Specific energy:	1 Btu/lb$_m$	= 2.3260 × 10^3 J/kg
	1 ft-lb$_f$/lb$_m$	= 2.9891 J/kg
Specific heat:	1 Btu/lb$_m$-°F	= 4.1868 × 10^3 J/kg-°C
Specific volume:	1 ft^3/lb$_m$	= 6.2428 × 10^{-2} m^3/kg
Surface tension:	1 lb$_f$/in.	= 1.7513 N/m
Temperature:	°R	°K = $\frac{5}{9}$ × °R
	°F	°C = $\frac{5}{9}$ × (°F − 32)

Temperature difference:	$1°F(°R)$	$= \frac{5}{9}°C(°K)$
Time:	1 hr	$= 3.6000 \times 10^3$ s
	1 min	$= 6.0000 \times 10$ s
Velocity:	1 ft/sec	$= 3.0480 \times 10^{-1}$ m/s
	1 ft/hr	$= 8.4667 \times 10^{-5}$ m/s
	1 mph	$= 4.4704 \times 10^{-1}$ m/s
Viscosity, dynamic: (see also Table B.1)	1 poise (g/cm-sec)	$= 1.0000 \times 10^{-1}$ kg/ms (N-s/m^2)
	1 lb$_m$/ft-sec	$= 1.4882$ kg/m-s
	1 lb$_m$/ft-hr	$= 4.1338 \times 10^{-4}$ kg/m-s
	1 lb$_f$-sec/in.^{2}	$= 6.8947 \times 10^3$ kg/m-s
	1 lb$_f$-hr/ft^2	$= 1.7237 \times 10^5$ kg/m-s
Viscosity, kinematic:	1 stoke (cm^2/sec)	$= 1.0000 \times 10^{-4}$ m^2/s
	1 ft^2/sec	$= 9.2903 \times 10^{-2}$ m^2/s
	1 ft^2/hr	$= 2.5806 \times 10^{-5}$ m^2/s
Volume:	1 ft^3	$= 2.8317 \times 10^{-2}$ m^3
	1 in.^{3}	$= 1.6387 \times 10^{-5}$ m^3
Volume flow rate:	1 ft^3/sec	$= 2.8317 \times 10^{-2}$ m^3/s
	1 ft^3/min	$= 4.7195 \times 10^{-4}$ m^3/s
	1 ft^3/hr	$= 7.8658 \times 10^{-6}$ m^3/s

Reference

1. MECHTLY, E. A., "The International System of Units, Physical Constants and Conversion Factors," *N.A.S.A. SP-7012*, National Aeronautics and Space Administration, Washington, D.C., 1964.

Index